The Boundary Element Methods in Engineering

The Boundary Element Methods in Engineering

P. K. Banerjee
State University of New York at Buffalo

McGRAW-HILL BOOK COMPANY

London · New York · St Louis · San Francisco · Auckland · Bogotá · Caracas
Lisbon · Madrid · Mexico · Milan · Montreal · New Delhi · Panama · Paris
San Juan · São Paulo · Singapore · Sydney · Tokyo · Toronto

Published by
McGRAW-HILL Book Company Europe
Shoppenhangers Road, Maidenhead, Berkshire, SL6 2QL, England
Telephone 0628 23432
Fax 0628 770224

British Library Cataloguing-in-Publication Data
Banerjee, P.K.
 Boundary Element Methods in Engineering. – 2 Rev. ed
 I. Title
 620.001

 ISBN 0-07-707769-5

Library of Congress Cataloging-in-Publication Data
Banerjee, P.K. (Prasanta Kumar),
 The boundary element methods in engineering/
P.K. Banerjee. –2nd ed.
  p. cm.
 Rev. ed. of: Boundary element methods engineering science/
P.K. Banerjee and R. Butterfield. c1981.
 Includes bibliographical references and index.
 ISBN 0-07-707769-5
 1. Boundary element methods. 2. Engineering mathematics.
I. Banerjee, P.K. (Prasanta Kumar), Boundary element methods
in engineering science. II. Title.
TA347.B69B36 1993
620′.001′51535–dc20   93-13475 CIP

Second edition of Boundary Element Methods in Engineering
Science, P.K. Banerjee and R. Butterfield, McGraw-Hill, 1981

1234 CUP 97654

Typeset by Alden Multimedia
and printed and bound in Great Britain at the University Press, Cambridge

This book is dedicated to all the women in my life

My wife Barbara
My daughters Nina, Sara and Lisa

CONTENTS

PREFACE

Computational methods have made significant contributions in all fields of engineering mechanics. Of these the finite element and the finite difference methods have become widely known and gained widespread acceptance among the engineering analysis community. One of the major difficulties in using these very powerful analyses is the formidable demand on the data preparation effort and thus the total computing costs.

The last two decades have seen the emergence of an equally versatile and powerful method of computational engineering mechanics, namely the boundary element method. The mathematical background of the boundary element method has been known for nearly one hundred years. Indeed, some of the boundary integral formulations for the elastic, elastodynamic wave propagation and potential flow equations have existed in the literature for at least fifty years. With the emergence of digital computers the method had began to gain popularity as 'the panel method', 'the boundary integral equation method' and 'the integral equation method' during the sixties. The name was changed to 'the boundary element method' (BEM) by Banerjee and Butterfield in 1975, so as to make it more popular in the engineering analysis community. It was at that time the first four chapters of the previous edition of this book were used as course notes in a graduate course at the Southampton University, UK. It was indeed gratifying to note that as a result several established finite element workers at that institution were inspired to adapt this method in their research. After publication of the first edition many have made substantial contributions to the development of the method and a number of textbooks and monographs have been written and several thousand archival journal articles have been published, indicating a very high level of activity.

The second edition of this book essentially incorporates the massive development of the BEM technology that has occurred during the last decade and is organized in such a manner that the first five chapters (to be studied sequentially) together with Chapter 17 dealing with computer programming could be used in an introductory course (as has been done at the State University of New York at Buffalo over the last decade) while the remaining chapters, which have the obvious flavour of a monograph, could be utilized in an advanced instructional course or research. The book is also written specifically for engineers who like to learn only enough of the mathematics involved to enable them to examine practical engineering problems.

In writing this second edition the author had to rely very heavily on the ideas of former colleagues (Professor R. Butterfield, Drs G.R. Tomlin, J.O. Watson, D.N. Cathie, T.G. Davies and G.W. Mustoe) as well as current ones (Drs R.B. Wilson,

S.T. Raveendra, S. Ahmad, G.F. Dargush, D.P. Henry, H.C. Wang, A.S.M. Israil, A. Deb, K. Honkala, M. Chopra and Y. Shi). It has been the author's privilege to have known them and worked with them. The author is particularly indebted to the Boundary Element Software Technology Corporation of Getsville, New York, for the use of their general purpose boundary element system, GP-BEST, to create several examples in this book. The author is especially grateful to Ms Carmella Gosden for her skilful typing of the manuscript and to the editors, David Crowther and Ros Comer, and to the copy-editor for their continuous encouragement.

Finally the author would like to acknowledge the financial supports of the National Science Foundation, National Aeronautics and Space Administration, United Technologies Corporation, General Motors Corporation, Ford Motor Company of USA and Mercedes Benz of Germany, who have been generous in providing funding for research on BEM. This thoroughly revised edition would not have been possible without their support which opened up many areas of BEM applications.

P.K. Banerjee
Department of Civil Engineering
State University of New York at Buffalo
Buffalo, New York

ONE

AN INTRODUCTION TO BOUNDARY ELEMENT METHODS

1.1 BACKGROUND

When an engineer or scientist constructs a quantitative mathematical model of almost any kind of system he or she usually starts by establishing the behaviour of an infinitesimally small, differential element of it based on assumed relationships between the major variables involved. This leads to a description of the system in the form of a set of differential equations. Once the basic model has been constructed and the properties of the particular differential equation understood, subsequent efforts are then directed towards obtaining a solution of the equations within a particular region, which is often a very complicated shape and composed of zones of different materials each with complex properties. Various conditions will have been specified on the boundaries of the region, and these may be either constant or variable with time. It is not at all surprising, therefore, that the solution of such differential equations has been a major concern of analysts for over two centuries.

The irregular boundaries of the majority of practical problems preclude any analytical solution of the governing equations and numerical methods become the only feasible means of obtaining adequately precise and detailed results.

The numerical methods most widely used at present tackle the differential equations directly in the form in which they were derived, without any further mathematical manipulation, in one of two ways: either by approximating the differential operators in the equations by simpler, localized algebraic ones valid at a series of nodes within the region or by representing the region itself by non-

1

infinitesimal (i.e. finite) elements of material which are assembled to provide an approximation to the real system.

The finite difference method[1] is the progenitor of the former approach and was, until about twenty-five years ago when it began to be superseded by methods of the second kind, much the most commonly used technique. Finite difference methods have attractions in that they can, in principle, be applied to any system of differential equations, but, unfortunately, incorporation of the problem boundary conditions is very often an inconvenient computerized operation. The precision of the numerical solution obtained is entirely dependent on the fineness of the mesh used to define the nodal points and consequently large systems of simultaneous algebraic equations are always generated as part of the solution procedure.

By far the most popular approach at present is the alternative one of reverting to physical subdivision of the body into elements of finite size. Each element reproduces approximately the behaviour of the small region of the body which it represents, but complete continuity between the elements is only enforced in an overall minimum energy sense. The finite element method[2] epitomizes this approach and, in recent years, has reached such a stage of development that many would doubt that any equivalent, let alone superior, technique might ever appear. The range and power of finite element methods, together with the relative ease with which realistic boundary conditions can be incorporated, does indeed present a formidable challenge to any other contending system. Its weakest aspects are that it is conceptually a whole body discretization scheme which inevitably leads to very large numbers of finite elements, especially in three-dimensional problems with distant boundaries, within each of which the solution variables do not all vary continuously.[2]

1.2 AN ALTERNATIVE APPROACH

An obvious alternative approach to the above methods would be to attempt to integrate the differential equations analytically in some way before either proceeding to any discretization scheme or introducing any approximations. We are, of course, attempting to integrate the differential equations to find a solution whatever method we use, but the essence of boundary element methods is the transformation of the differential equations into equivalent sets of integral ones as the first step in their solution. Intuitively one would expect from such an operation (if it were successful) a set of equations which would involve only values of the variable at the extremes of the range of integration (i.e. on the boundaries of the region). This, in turn, would imply that any discretization scheme needed subsequently would only involve subdivisions of the bounding surface of a body. This is exactly what happens, with the consequence that any homogeneous region requires only surface, rather than whole body, discretization (hence the name proposed by Banerjee and Butterfield, the boundary element method[3,4]) and therefore a homogeneous region becomes merely one large, sophisticated 'element' in the finite element sense. The solution variables will then vary continuously throughout the region and all approximations of geometry, etc., will only occur on its outer boundaries.

Another intuitive expectation would be that the development of the boundary element methods and their solution might well involve more mathematical complexity than the other methods mentioned above. Fortunately this is only partially true, although it does account for the fact that these methods have been developed mainly by mathematicians in the past. The published literature,[5-7] although extensive, tends to have an obvious mathematical bias without the final compensatory 'carrot' that at the end of it all a comprehensive technique will emerge which the analyst can use in a general way. However, the situation has improved, from the utilitarian point of view, over the last two decades, and boundary element methods (BEMs) of analysis developed essentially from integral equation ideas are now available in a generally applicable form without recourse to proofs of existence and uniqueness for each individual solution. As a result the method is now gaining very considerable popularity and being incorporated into high-speed digital computer algorithms immediately useful to the practising analyst.[4, 8-21]

1.3 A SIMPLE EXAMPLE

In order to introduce the ideas underlying the use of BEMs, and the character of the unit solutions of the governing differential equations that arise, in the simplest possible way the following section presents, in some detail, the solutions to a simple one-dimensional problem. We would emphasize that we are not in any way recommending the BEM as a preferred way of solving such problems. Indeed, for one-dimensional systems generally, BEMs are not efficient problem-solving tools at all. Nevertheless, the endpoint reached via the one-dimensional examples is a series of systematic steps, a solution algorithm, which illustrates all the essential features of procedures that can be used almost unchanged to solve vastly more complicated two- and three-dimensional problems.

1.3.1 Potential Flow in One Dimension

Figure 1.1 represents a one-dimensional, homogeneous field of length L and unit cross-section. The boundaries of the system are merely the two endpoints P, Q, with only one 'boundary element' at each point. These are maintained at zero potential, $p(Q) = p(P) = 0$. A point source, intensity ψ, is applied at any position B, a 'load point', and located by the coordinate ξ. A general point P', a 'field point', within the body is located by coordinate x.

We could be considering here either the flow of electricity or heat along a uniform conductor or non-viscous incompressible flow along a uniform conduit. In all these cases the potential $p(x)$, representing voltage, temperature or total head respectively, will be governed by the Laplace equation at all points within PQ

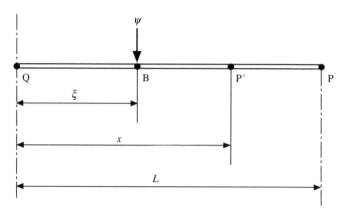

Figure 1.1 Definition of the problem.

other than B. Thus in one dimension

$$\frac{d^2p}{dx^2} = 0 \tag{1.1}$$

If the conductivity of the medium is k then the current intensity, heat flux or fluid flow velocity $v(x)$ will be given by

$$v = -k\frac{dp}{dx} \tag{1.2}$$

Again, Eqs (1.1) and (1.2) could equally well describe the deflection $p(x)$ and the slope $v(x)$ of a tightly stretched horizontal weightless string under a high tension k and a small vertical load $\psi =$ unity, as shown in Fig. 1.2. It is a very elementary exercise to solve Eqs (1.1) and (1.2) under the given boundary conditions at the ends of P, Q to obtain

$$p(x) = \frac{(L-\xi)}{L}\frac{x}{k} \qquad v(x) = -\frac{(L-\xi)}{L} \qquad \text{for} \quad 0 \le x < \xi \tag{1.3a}$$

and $$p(x) = \frac{(L-x)}{L}\frac{\xi}{k} \qquad v(x) = \frac{\xi}{L} \qquad \text{for} \quad \xi < x \le L \tag{1.3b}$$

At $x = \xi, p(x)$ is uniquely defined by either Eqs (1.3a) or (1.3b), but there is a step change in $v(x)$ equal to unity, as P$'$ moves from one side to the other of B (i.e. as x increases from $(\xi - \varepsilon)$ to $(\xi + \varepsilon)$ with $\varepsilon \to 0$). Such a step change in one of the dependent variables as the 'load point' and 'field point' coincide is a key feature of all BEMs.

The governing differential equation for the problem can be written in more general form as

$$\frac{d^2p(x)}{dx^2} + \psi(x) = 0 \tag{1.4}$$

where now ψ is a specified distributed source intensity along x and with $k =$ unity, the velocity $v(x)$ is simply $-dp(x)/dx$. In order to investigate the possibility of

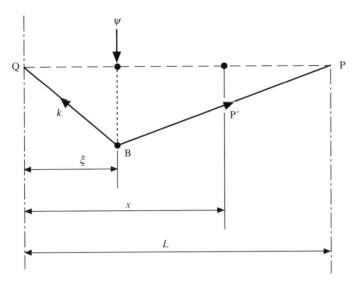

Figure 1.2 Fundamental solution due to source.

actually integrating Eq. (1.4) over the range $0 \leq x \leq L$, we shall introduce a function $G(x, \xi)$ which is yet undefined except that it is sufficiently continuous to be differentiable as often as required. If we multiply both sides of Eq. (1.4) by G and begin to integrate by parts we obtain[4]

$$\int_0^L \frac{d^2 p(x)}{dx^2} G \, dx + \int_0^L \psi(x) G \, dx = 0$$

Therefore

$$\left[G \frac{dp(x)}{dx} \right]_0^L - \int_0^L \left[\frac{dG}{dx} \frac{dp(x)}{dx} \right] dx + \int_0^L \psi(x) G \, dx = 0$$

and

$$\left[G \frac{dp(x)}{dx} - p(x) \frac{dG}{dx} \right]_0^L + \int_0^L p(x) \frac{d^2 G}{dx^2} dx + \int_0^L \psi(x) G \, dx = 0 \qquad (1.5)$$

We now specify G to be a solution of

$$\frac{d^2 G(x, \xi)}{dx^2} + \delta(x, \xi) = 0 \qquad (1.6)$$

where $\delta(x, \xi)$ is the Dirac 'delta function' (or impulse function) which is mathematically equivalent to the effect of a unit concentrated source applied at the point ξ. The key property of the delta function is that it is zero at all x except in the neighbourhood of $x = \xi$, where it becomes infinitely large in such a way that[22, 23]

$$\int_{-\infty}^{\infty} \delta(x, \xi) \, dx = \int_0^L \delta(x, \xi) \, dx = 1(\xi)$$

The delta function is therefore an operator with a sifting property, a 'needle'

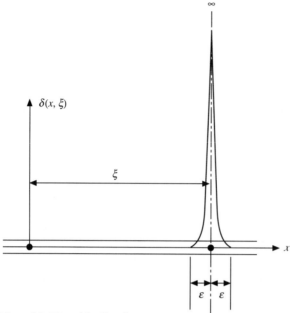

Figure 1.3 Dirac delta function.

(Fig. 1.3), which 'points' to a specific value, $p(\xi)$ say, of any function $p(x)$ which it operates on, as in

$$\int_0^L p(x)\delta(x,\xi)\,dx = p(\xi)$$

If we substitute Eq. (1.6) into Eq. (1.5) we get

$$\left[G\frac{dp(x)}{dx} - p(x)\frac{dG}{dx}\right]_0^L + \int_0^L \psi(x)G\,dx = \int_0^L p(x)\delta(x,\xi)\,dx$$

which simplifies, using the above property of the δ operator, to

$$\left[G\frac{dp(x)}{dx} - p(x)\frac{dG}{dx}\right]_0^L + \int_0^L \psi(x)G\,dx = p(\xi) \tag{1.7}$$

Our function G, which is the solution of Eq. (1.6), is clearly precisely that previously quoted in Eq. (1.3) with $\psi = 1$ (i.e. the fundamental solution of the governing differential equation); thus

$$G(x,\xi) = 0.5(l - |r|) \qquad\qquad k = 1 \tag{1.8a}$$

and
$$-\frac{dG(x,\xi)}{dx} = F(x,\xi) = 0.5(\text{sgn } r) \qquad k = 1 \tag{1.8b}$$

where r is the distance $(x - \xi)$ and at $r = \pm l$ the function $G = 0$, with l being an arbitrary distance.

If we choose the arbitrary length $l = L$ and recall that here the velocity $v(x) = -dp(x)/dx$, we can write Eq. (1.7) as

$$p(\xi) = - [G(x,\xi)v(x) - F(x,\xi)p(x)]_0^L + \int_0^L \psi(x)G(x,\xi)\,dx \qquad (1.9)$$

or $\quad p(\xi) = - [G(L,\xi)v(L) - G(0,\xi)v(0)] + [F(L,\xi)p(L) - F(0,\xi)p(0)]$

$$+ \int_0^L \psi(x)G(x,\xi)\,dx \qquad (1.10)$$

Equation (1.10) is seen to provide the potential $p(\xi)$ at any interior field point (ξ) consistent with a set of boundary potentials $(p(0), p(L))$, velocities $(v(0), v(L))$ and a specified internal distribution of sources $\psi(x)$. If there is merely a single point source $\psi(x_0)$ at a point x_0 within the region then the integral on the right-hand side of Eq. (1-8) becomes simply $\psi(x_0)G(x_0,\xi)$.

By taking the first derivative of $p(\xi)$ with respect to this variable ξ we get

$$-\frac{dp(\xi)}{d\xi} = v(\xi) = \left[\frac{dG}{d\xi}(L,\xi)v(L) - \frac{dG}{d\xi}(0,\xi)v(0)\right]$$

$$- \left[p(L)\frac{dF}{d\xi}(L,\xi) - p(0)\frac{dF(0,\xi)}{d\xi}\right] - \int_0^L \psi(x)\frac{dG}{d\xi}(x,\xi)\,dx \qquad (1.11)$$

Equation (1.11) thus provides the velocity $v(\xi)$ at any field point (ξ) consistent with the boundary potentials and velocities and known internal source distribution.

Now by taking the field point ξ to the boundary points P and Q such that $\xi = L - \varepsilon$ for P and $\xi = 0 + \varepsilon$ for Q we can use Eq. (1.10) to write

$$\begin{Bmatrix} p(L-\varepsilon) \\ p(0+\varepsilon) \end{Bmatrix} = - \begin{bmatrix} G(L, L-\varepsilon) & -G(0, L-\varepsilon) \\ G(L, 0+\varepsilon) & -G(0, 0+\varepsilon) \end{bmatrix} \begin{Bmatrix} v(L) \\ v(0) \end{Bmatrix}$$

$$+ \begin{bmatrix} F(L, L-\varepsilon) & -F(0, L-\varepsilon) \\ F(L, 0+\varepsilon) & -F(0, 0+\varepsilon) \end{bmatrix} \begin{Bmatrix} p(L) \\ p(0) \end{Bmatrix} + \begin{Bmatrix} \int_0^L \psi(x)G(x, L-\varepsilon)\,dx \\ \int_0^L \psi(x)G(x, 0+\varepsilon)\,dx \end{Bmatrix} \qquad (1.12)$$

It is of the utmost importance to appreciate the fact that we must approach the boundary points from inside our region PQ. Since G is unambiguously defined within the region we really do not have to pay much attention to it. The function F, on the other hand, is multi-valued whenever its arguments x and ξ coincide and we must take care in evaluating this limit. Throughout our BEM analyses we shall find that the function F will require some closer attention.

Substituting values for G and F from Eqs (1.8a,b) we obtain in the limit, as $\varepsilon \to 0$,

$$\begin{Bmatrix} p(L) \\ p(0) \end{Bmatrix} = - \begin{bmatrix} 0.5L & 0 \\ 0 & -0.5L \end{bmatrix} \begin{Bmatrix} v(L) \\ v(0) \end{Bmatrix}$$

$$+ \begin{bmatrix} 0.5 & 0.5 \\ 0.5 & 0.5 \end{bmatrix} \begin{Bmatrix} p(L) \\ p(0) \end{Bmatrix} + \begin{Bmatrix} \int_0^L \psi(x)G(x, L)\,dx \\ \int_0^L \psi(x)G(x, 0)\,dx \end{Bmatrix}$$

in which, by absorbing the left-hand side into the appropriate terms on the right-hand side, we get

$$
-\begin{bmatrix} 0.5L & 0 \\ 0 & -0.5L \end{bmatrix} \begin{Bmatrix} v(L) \\ v(0) \end{Bmatrix} - \begin{bmatrix} 0.5 & -0.5 \\ -0.5 & 0.5 \end{bmatrix} \begin{Bmatrix} p(L) \\ p(0) \end{Bmatrix}
$$
$$
+ \begin{Bmatrix} \int_0^L \psi(x)G(x,L)\,dx \\ \int_0^L \psi(x)G(x,0)\,dx \end{Bmatrix} = 0 \qquad (1.13)
$$

Note that in Eq. (1.13) if we consider a problem in which the potentials $p(L) = p(0) =$ unity, then the velocities $v(L) = v(0) = 0$ (since there can not be any flow). Our equation (1.13) does obviously satisfy this condition and as a result each row of the second matrix sums to zero.

Equation (1.13) can be used to calculate the initially unknown boundary value data from the known boundary values and a prescribed value of $\psi(x)$ for our one-dimensional field. For example,

1. With $v(L), v(0)$ and $\psi(x)$ specified, solution of Eq. (1.13) will yield the unknown value at $p(L)$ with respect to the specified datum $p(0) = 0$.
2. With $p(L), p(0)$ and $\psi(x)$ specified, the solution will provide the unknowns $v(L)$ and $v(0)$.
3. With $v(L), p(0), \psi(x)$ or $v(0), p(L), \psi(x)$ specified, the solution process will provide respectively the values of the set $v(0), p(L)$ or $v(L), p(0)$.

If we return to our initial problem, shown in Fig. 1.1 [that is $p(L) = p(0) = 0$ and a point source of intensity ψ specified at a distance ξ_1 from the left-hand end], then Eq. (1.13) becomes

$$
-\begin{bmatrix} 0.5L & 0 \\ 0 & -0.5L \end{bmatrix} \begin{Bmatrix} v(L) \\ v(0) \end{Bmatrix} = -\begin{Bmatrix} G(\xi_1, L) \\ G(\xi_1, 0) \end{Bmatrix} = -0.5 \begin{Bmatrix} \psi(\xi_1) \\ \psi(L - \xi_1) \end{Bmatrix} \qquad (1.14)
$$

or
$$
v(L) = \psi\xi_1 \qquad v(0) = -\psi(L - \xi_1)
$$

The potentials and velocities at selected interior points can then be obtained by substituting the values of $p(L), p(0), v(L), v(0)$ in Eqs (1.10) and (1.11) respectively.

On the other hand, if we wish to solve the mixed boundary value problem, that is $p(0) = p^*, v(L) = v^*$, and a point source ψ applied at a distance ξ_1 from the left-hand end, we can write Eq. (1.13) as

$$
\begin{bmatrix} 0.5L & 0 \\ 0 & -0.5L \end{bmatrix} \begin{Bmatrix} v^* \\ v(0) \end{Bmatrix} + \begin{bmatrix} 0.5 & -0.5 \\ -0.5 & 0.5 \end{bmatrix} \begin{Bmatrix} p(L) \\ p^* \end{Bmatrix} = 0.5 \begin{Bmatrix} \psi\xi_1 \\ \psi(L - \xi_1) \end{Bmatrix} \qquad (1.15)
$$

the solution of which provides

$$
p(L) = \psi\xi_1 - Lv^* + p^* \qquad v(0) = v^* - \psi
$$

Once again Eqs (1.10) and (1.11) can be used to obtain the potentials and velocities at any internal point (ξ) by simple substitution of these boundary values.

It may be helpful to summarize the few very simple steps involved in the solution process described above, since they inevitably become rather submerged in the

explanations that accompany them. There are really six steps in the entire procedure as outlined above:

1. Statement of the fundamental solution for a unit source [Eqs (1.3a,b)].
2. Integration by parts by multiplying the governing equations with the fundamental to derive the basic integral identity [Eq. (1.9)].
3. Deriving the system matrices by taking the field point ξ on to the boundary [Eq. (1.12)].
4. Formation of the final system of coefficient matrices by substituting coordinates etc. [Eq. (1.13)].
5. Solution of the boundary equations (1.13) for the prescribed boundary values.
6. Backsubstitution of the boundary values into Eqs (1.10) and (1.11) to obtain the interior quantities.

We shall see that the above six steps can be used for vastly more complex two- and
three-dimensional problems. Thus, although the complexity of problems will
change enormously that of the underlying algorithm to solve them will not.

1.4 THE HISTORICAL DEVELOPMENT OF BOUNDARY ELEMENT METHODS

Whereas the major properties of differential equations were well established by the
nineteenth century the first rigorous investigation of the classical kinds of integral
equation was published by Fredholm as late as 1905. Since then they have been
studied intensively, particularly in connection with field theory, and there are
many texts dealing with these developments.[5-8]

A major contribution to the formal understanding of integral equations generally has been made more recently by Mikhlin[7] who discusses such equations
with both scalar and vector (multi-dimensional) integrands and in particular
those with singularities and discontinuities within the range of integration. All
of this is presented within a rigorous mathematical framework, most of which is
rather unfamiliar to the majority of applied scientists. Despite the great advances
that have been made in the classification and analysis of the properties of integral
equations, none of the major authors appears to have considered the possibility
that a general numerical algorithm for solving a wide range of practical problems
might be based on them. The impetus for such a development has been provided by
the high-speed digital computer, and one result has been the emergence of the
boundary element method.[4, 9-21]

One of the earliest boundary element works for the complete solution for the
non-linear Navier–Stokes equation in fluid dynamics can be found in the remarkable 1927 monograph by Oseen,[24] who treated the highly non-linear incompressible Navier–Stokes equation by identifying a linearized operator which contained
much of the physics of the high-speed flow, often called Oseen flow. The remaining
non-linear terms were then treated as pseudo-body forces. Unfortunately this
occurred much before the advent of digital computers. With the advent of

modern digital computers Massonnet[25] in 1956, Hess[26] in 1962 showed how indirect BEMs could be used respectively in solutions of elasticity and potential flow. The first direct BEMs formulations for acoustics, elasticity and elastodynamics were presented by Banaugh and Goldsmith[27, 28] in 1963. It is interesting to note that the accuracies of their numerical implementation were not matched by others until the work of Lachat and Watson[29] was published in 1976. Meanwhile, somewhat independently, Jaswon, Symm and Ponter[30-32] examined the application of the direct BEM formulations to problems of potential flow and shaft torsion.

In principle these methods can be applied to any problem for which the governing differential equation is either linear or can be split into a linear and a nonlinear part. In problems involving elliptic differential equations the solutions are direct, whereas for parabolic and hyperbolic systems of equations marching processes in time have to be introduced. Thus a very wide range of physical problems is encompassed: e.g. those of elastostics,[32-34] steady state and transient potential flow,[31, 32, 35-39] elastodynamics,[40-49] elastoplasticity,[50-54] finite inelastic deformation,[55-57] Navier–Stokes flow,[58,59] etc., can all be solved by the BEM. They can also be used in conjunction with other numerical techniques, such as finite element or finite difference methods, in a hybrid formulation. Such composite solutions extend the range of application almost indefinitely since BEMs have very distinct advantages for problems of large physical dimensions whereas finite element methods are an attractive means of incorporating finite sized bodies into such systems or fine detail in regions with rapidly varying properties. A more comprehensive comparison of these attributes will now be provided as a conclusion to this introductory chapter.

1.5 COMPARISON OF THE ATTRIBUTES OF FINITE ELEMENT METHODS AND BOUNDARY ELEMENT METHODS

1.5.1 Applicability

All boundary element methods utilize the fundamental solutions in infinite space and are therefore only applicable to either completely linear systems or those for which the differential equations can be approximated as a summation of linear and non-linear operators. This latter category therefore extends their compass to a great many problems of interest in engineering mechanics. There appears to be very few problems solvable by finite element methods that cannot be solved reasonably efficiently by BEMs. There are problems in which either the properties of almost every individual material element are different or those in which the general geometry of the problem is such that one or more spatial dimension is disproportionately small in relation to others but not sufficiently so to genuinely reduce its effective dimensionality (e.g. moderately thick plates and shells, narrow thin strips, etc.), which are difficult by BEMs. Even in this more difficult latter case a careful numerical implementation can overcome some of these limitations.

1.5.2 Problem Dimensionality

BEMs reduce the dimensionality of the basic process by one; i.e. for two-dimensional problems the analysis generates a one-dimensional boundary integral equation and for three-dimensional problems only two-dimensional surface integral equations arise (see the self-explanatory Figs 1.4 and 1.5 which illustrate this point).

Each distinct bounded zone in a BEM analysis has to be treated as homogeneous and therefore, for problems in which the inhomogeneity is so great that very large numbers of small homogeneous zones are needed to model it adequately, the BEM zonal boundary scheme degenerates in efficiency into one of essentially whole body subdivision. If, in a homogeneous region problem, either distributed body forces are to be included or the governing differential equations are non-linear (as in the cases of elastoplasticity and Navier–Stokes flow, for example), then the boundary integrals have to be augmented by a volume integral involving arbitrary subdivisions of the interior of the body. However, in these cases the internal subdivisions do not result in any increase in the order of the final system of algebraic equations to be solved. The reader should distinguish carefully between the latter situation, in which the interior subdivisions arise from known distributions of body forces (volume integrals of continuously distributed conservative body forces can very often be transformed into equivalent boundary integrals using particular integrals or the Gauss theorem) or pseudo-incremental body forces in plasticity in otherwise homogeneous zones, and the former, which reflects the fundamental initial inhomogeneity of the problem.

Thus for the great majority of practical cases the simple boundary discretization necessarily leads to a very much smaller system of simultaneous equations than any scheme of whole body discretization (see Fig. 1.5). On the other hand, the system matrices generated by BEMs are fully populated for a homogeneous

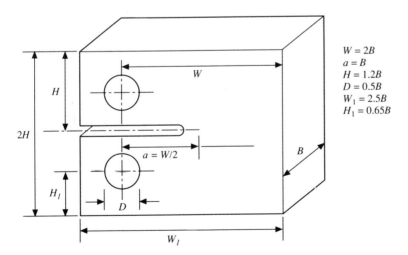

$W = 2B$
$a = B$
$H = 1.2B$
$D = 0.5B$
$W_1 = 2.5B$
$H_1 = 0.65B$

Figure 1.4 A fracture test specimen CT.15.

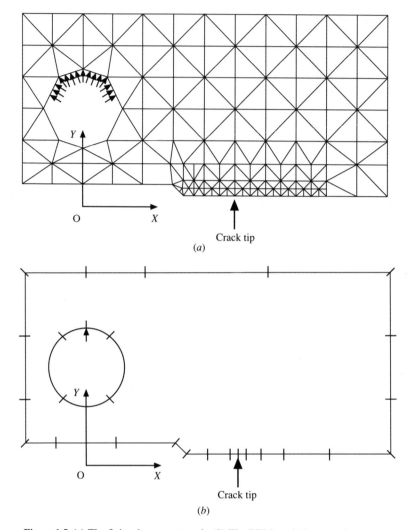

Figure 1.5 (*a*) The finite element network. (*b*) The BEM analysis network.

region and block banded when more than one region is involved, whereas the much larger matrices that arise in finite element solutions are relatively sparsely populated.

The evaluation of each component of the matrices in a BEM solution does, however, involve much more arithmetic calculation than its finite element counterpart, which offsets some of the computer time saved by the much reduced matrix reduction requirements. Nevertheless, this also means that, as bigger and bigger problems[60] are tackled, the overall computer costs increase very much less dramatically with problem size with BEMs than for finite element schemes. From studies undertaken by various workers it can be concluded that comparable solution times between finite element and boundary element methods on three-dimensional problems solved with similar precision generally show a time advantage in favour of the latter. This difference could be very much greater in certain classes of problems which are particularly amenable to BEMs, for example:

1. Systems with some boundaries at infinity. Since the BEM solution procedure automatically satisfies admissible boundary conditions at infinity, no subdivisions of these boundaries arise, whereas with the finite element method infinite boundaries have to be approximated by an appreciable number of distant elements.
2. Those involving semi-infinite regions with portions of the free surface 'unloaded'. Again by choosing the appropriate singular solution to use with BEMs the 'unloaded' areas, which are usually the greater part of the free surface, do not need to be discretized at all.[61]

1.5.3 Continuous Interior Modelling

BEMs involve modelling of only the boundary geometry of the system. Once the necessary boundary information has been derived, values of the solution variables can then be calculated at any subsequently selected interior points. Furthermore, the solution is fully continuous throughout the interior of the body. Both of these features appear to be unique to BEMs among the possible alternatives. As a result of the latter facility the analyst can obtain values of variables at any specific interior point he or she may care to choose subsequent to the main analysis and with very high resolution—near, for example, stress concentrations in elastic or elastoplastic bodies.

1.5.4 Accuracy and Error Distribution

The boundary integral equation itself is a statement of the exact solution to the problem posed. Errors due to discretization and numerical approximations arise only on, and adjacent to, the boundaries due to our inability to carry out the required integrations in closed form. If the numerical integration procedure is made sufficiently sophisticated (by using, for example, curved boundary elements and continuously varying distributions of function over the boundary), then the errors so introduced can be very small indeed. Numerical integration is, of course, always a much more stable and precise process than numerical differentiation and the BEM does not require any differentiation of numerical quantities whatsoever.

It should by now be quite clear that, in the absence of body forces, the analyst need only specify the boundary geometry data of a region (in addition to the boundary conditions, material properties, etc., common to all solution methods). The effort devoted to data preparation is therefore substantially less than that required by any method involving internal geometrical modelling (Fig. 1.5).

Thus for the great majority of practical problems the BEM offers very substantial advantages which can only enhance its use in industry.[60]

1.6 CONCLUDING REMARKS

In this chapter we have described the historical development of BEMs as a practical problem-solving tool and discussed their usefulness in comparison to other currently popular alternatives. From these comparisons we conclude that BEMs have great potential advantages over other methods, some of which have now been realized. It is our hope that the subsequent chapters of this book will help to accelerate this process both by illustrating their power through the solution of a wide range of practical problems and by emphasizing the very simple physical ideas, already familiar to most engineers and applied scientists, on which they are based.

REFERENCES

1. Southwell, R.V. (1946) *Relaxation Methods in Theoretical Physics*, Oxford University Press.
2. Zienkiewicz, O.C. (1971) *The Finite Element Method in Engineering Science*, McGraw-Hill, London.
3. Banerjee, P.K. and Butterfield, R. (1976) Boundary element methods in geomechanics, Chapter 16 in *Finite Elements in Geomechanics*, Ed. G. Gudehus, John Wiley and Sons, London.
4. Banerjee, P.K. and Butterfield, R. (1981) *Boundary Element Methods in Engineering Sciences*, McGraw-Hill, London; Russian Edition (1984), Mir Publishers, Moscow; Chinese Edition (1988), National Defense Press, Beijing.
5. Kellog, O.D. (1929) *Foundations of Potential Theory*, Springer-Verlag, Berlin; also published by Dover, New York, in 1953.
6. Kupradze, V.D. (1963) *Potential Methods in the Theory of Elasticity*, translated from Russian by Israel Program for Scientific Translation, Jerusalem.
7. Mikhlin, S.G. (1965) *Multidimensional Singular Integrals and Integral Equations*, Pergamon Press, Oxford.
8. Jaswon, M.A. and Symm, G.T. (1977) *Integral Equation Methods in Potential Theory and Elastostatics*, Academic Press, London.
9. Banerjee, P.K. and Butterfield, R. (Editors) (1979) *Developments in Boundary Element Methods*, Volume I, Elsevier Applied Science, London.
10. Banerjee P.K. and Mukherjee, S. (Editors) (1984) *Developments in Boundary Element Methods*, Volume 3, Elsevier Applied Science, London.
11. Mukherjee, S. (1982) *Boundary Element Methods in Creep and Fracture*, Elsevier Applied Science, London.
12. Crouch, S. and Starfields, A.M. (1983) *Boundary Element Methods in Solid Mechanics*, Allen and Unwin, London.
13. Liggett, J.A. and Liu, P.L.F. (1983) *Boundary Integral Equation Method for Porous Media Flow*, Allen and Unwin, London.
14. Brebbia, C, Telles, J.C.F. and Wrobel, L.C. (1984) *Boundary Element Technique*, Springer-Verlag, Berlin.
15. Brebbia, C. (Editor) (1984) *Topics in Boundary Element Research*, Volume 2, *Time Dependent and Vibration Problems*, Springer-Verlag, Berlin.
16. Hartmann, F. (1989) *Introduction to Boundary Element Method*, Springer-Verlag, Berlin.

17. Banerjee, P.K. and Wilson, R.B. (Editors) (1989) *Industrial Applications of Boundary Element Method*, Elsevier Applied Science, London.
18. Banerjee, P.K. and Morino, L. (Editors) (1990) *Boundary Element Methods in Nonlinear Fluid Dynamics*, Elsevier Applied Science, London.
19. Banerjee, P.K. and Kobayashi, S. (Editors) (1992) *Advanced Dynamic Analysis by BEM*, Elsevier Applied Science, London.
20. Becker, A.A. (1992) *The Boundary Element Method in Engineering*, McGraw-Hill, London.
21. Cruse, T.A. (1988) *Boundary Element Analysis in Computational Fracture Mechanics*, Kluwer Academic Press, Nederland.
22. Wylie (Jr), C.R. (1960) *Advanced Engineering Mathematics*, 3rd edn., McGraw-Hill, New York.
23. Malvern, L.E. (1969) *Introduction to the Mechanics of Continuous Medium*, Prentice-Hall, Englewood, New Jersey.
24. Oseen, C.W. (1927) *Neure Methoden und Ergebnisse in der Hydrodynamik*, Akad Verlagegellschaft, Leipzig, Germany.
25. Massonnet, Ch. (1956) Solution génerale du problème aux tensions de élasticité tridemensonnelle, Proceedings of 9th International Congress in *Applied Mechanics*, Brussels, pp. 168–180.
26. Hess, J.L. (1962) Calculation of potential flow about bodies of revolution having axes perpendicular to the free stream direction, *Journal of Aerospace Sciences*, Vol. 29, pp. 726–742.
27. Banaugh, R.P. and Goldsmith, W. (1963) Defractions of steady acoustic waves by surfaces of arbitrary shape, *Journal of Acoustical Society of America*, Vol. 42, No. 2, pp. 391–397.
28. Banaugh, R.P. and Goldsmith, W. (1963) Defractions of steady elastic waves by surfaces of arbitrary shape, *Journal of Applied Mechanics*, Vol. 30, pp. 589-597.
29. Lachat, J.C. and Watson, J.O. (1976) Effective numerical treatment of boundary integral equations: a formulation for three-dimensional elasto-statics, *International Journal of Numerical Methods in Engineering*, Vol. 10, pp. 991–1005.
30. Jaswon, M.A. (1963) Integral equation methods in potential theory, I, *Proceedings of Royal Society, London*, Vol. 275(A), 23-32.
31. Symm, G.T. (1963) Integral equation in potential theory, II, *Proceedings of Royal Society, London*, Vol. 275(A), pp. 33–46.
32. Jaswon, M.A. and Ponter, A.R. (1963) An integral equation for the torsion problem, *Proceedings of Royal Society, London*, Vol. 275(A), pp. 237–246.
33. Rizzo, F.J. and Shippy, D.J. (1977) An advanced boundary integral equation method for three-dimensional thermoelasticity, *International Journal of Numerical Methods in Engineering*, Vol. 11, pp. 1753–1790.
34. Wilson, R.B. and Cruse, T.A. (1978) Efficient implementation of anisotropic three-dimensional boundary integral equation stress analysis, *International Journal of Numerical Methods in Engineering*, Vol. 12, pp. 1383–1397.
35. Chang, Y.P., Kang, C.S. and Chen, S.J. (1973) The use of fundamental Green function for solution of problems of heat conduction in anisotropic media, *International Journal of Heat and Mass Transfer*, Vol. 16, pp. 1905–1918.
36. Dargush, G. and Banerjee, P.K. (1991) Application of boundary element method to transient heat conduction, *International Journal of Numerical Methods in Engineering*, Vol. 31, pp. 1231–1247.
37. Dargush, G. and Banerjee, P.K. (1989) A time domain boundary element method in poroelasticity, *International Journal of Numerical Methods in Engineering*, Vol. 28, pp. 2423–2449.
38. Wrobel, L.C. and Brebbia, C. (1981) A formulation of the boundary element method for axisymmetric heat conduction, *International Journal of Heat and Mass Transfer*, Vol. 24, pp. 843–850.
39. Banerjee, P.K., Butterfield, R. and Tomlin, G.R. (1981) Boundary element method for two-dimensional problems of ground water flow, *International Journal of Numerical Analytical Methods in Geomechanics*, Vol. 5, pp. 843–850.
40. Ahmad, S. and Banerjee, P.K. (1988) Time domain transient elasto-dynamic analysis of 3-D solids by BEM, *International Journal of Numerical Methods in Engineering*, Vol. 26, pp. 1709–1728.
41. Wang, H.C. and Banerjee, P.K. (1990) Axisymmetric transient elastodynamic analysis by Boundary Element Method, *International Journal of Solids and Structures*, Vol. 26, No. 4, pp. 401–415.

42. Ahmad, S. and Banerjee, P.K. (1988) Multi-domain BEM for two-dimensional problems of elasto-dynamics, *International Journal of Numerical Methods in Engineering*, Vol. 26, pp. 891–911.

43. Israil, A.S.M. and Banerjee, P.K. (1991) Advanced time-domain formulation of BEM for two-dimensional transient elastodynamics, *International Journal of Numerical Methods in Engineering*, Vol. 29, pp. 1421–1440.

44. Wilson, R.B., Miller, N. and Banerjee, P.K. (1990) Free vibration analysis of three-dimensional solids by BEM, *International Journal of Numerical Methods in Engineering*, Vol. 29, pp. 1737–1757.

45. Wang, H.C. and Banerjee, P.K. (1988) Axisymmetric free vibration analysis by BEM, *Journal of Applied Mechanics*, Vol. 55, pp. 437–442.

46. Ahmad, S. and Banerjee, P.K. (1986) Free vibration analysis by BEM using particular integrals, *Journal of Engineering Mechanics, ASCE*, Vol. 112, pp. 682–695.

47. Wang, H.C. and Banerjee, P.K. (1990) Free vibration analysis of axisymmetric solids by BEM, *International Journal of Numerical Methods in Engineering*, Vol. 29, pp. 985–1001.

48. Wang, H.C. and Banerjee, P.K. (1990) Generalized axisymmetric elastodynamic analysis by BEM, *International Journal of Numerical Methods in Engineering*, Vol. 30, pp. 115–131.

49. Banerjee, P.K., Ahmad, S. and Chen, K. (1988) Advanced application of BEM to wave barriers in multi-layered three-dimensional soil media, *Earthquake Engineering and Structural Dynamics*, Vol. 16, pp. 1040–1060.

50. Banerjee, P.K. and Raveendra, S.T. (1986) Advanced developments of BEM of two and three-dimensional problems of elastoplasticity, *International Journal of Numerical Methods in Engineering*, Vol. 23, pp. 985–1002.

51. Banerjee, P.K. and Raveendra, S.T. (1987) A new boundary element formulation for two-dimensional elastoplastic analysis, *Journal of Engineering Mechanics, ASCE*, Vol. 113, pp. 671–688.

52. Henry, D.P. and Banerjee, P.K. (1988) A variable stiffness type boundary element formulation for axisymmetric elastoplastic media, *International Journal of Numerical Methods in Engineering*, Vol. 26, pp. 1005–1027.

53. Banerjee, P.K., Henry, D.P. and Raveendra, S.T. (1989) Advanced inelastic analysis of solids by BEM, *International Journal of Mechanical Science*, Vol. 31, pp. 309–322.

54. Henry, D.P. and Banerjee, P.K. (1988) A new boundary element formulation for two- and three-dimensional elastoplasticity using particular integrals, *International Journal of Numerical Methods in Engineering*, Vol. 26, pp. 2079–2096.

55. Chandra, S. and Mukherjee, S. (1987) A boundary element analysis of metal extrusion processes, *ASME Journal of Applied Mechanics*, Vol. 54, pp. 335–340.

56. Chandra, S. and Saigal, S. (1991) A boundary element analysis of the axisymmetric extrusion process, *International Journal of Nonlinear Mechanics*, Vol. 26, pp. 1–13.

57. Okada, H., Rajiyah, H. and Atluri, S.N. (1990) A full tangent stiffness field boundary element formulation for geometric and material nonlinear problems of solid mechanics, *International Journal of Numerical Methods in Engineering*, Vol. 29, pp. 15–35.

58. Dargush, G. and Banerjee, P.K. (1991) A boundary element method for steady incompressible thermo-viscous flow, *International Journal of Numerical Methods in Engineering*, Vol. 31, pp. 1605–1626.

59. Dargush, G. and Banerjee, P.K. (1991) A time dependent incompressible viscous BEM for moderate Reynolds numbers, *International Journal of Numerical Methods in Engineering*, Vol. 31, pp. 1627–1648.

60. Butenschön, H.J., Möhrmann, W. and Bauer, W. (1989) Advanced stress analysis by a commercial BEM code, Chapter 8 in *Industrial Application of BEM*, Eds P.K. Banerjee and R.B. Wilson, Elsevier Applied Science, London.

61. Butterfield, R. and Banerjee, P.K. (1971) The problem of pile cap-pile group interaction, *Geotechnique*, Vol. 21, June, pp. 135–142.

TWO

SOME BASIC MATHEMATICAL PRINCIPLES

2.1 INTRODUCTION

The reader unfamiliar with indicial notation, the summation convention and elementary tensor transformation rules will find that the book progresses from a minimal use of these ideas in the initial chapters to a progressively more elaborate notation adorned with a multiplicity of suffixes in the later ones. This is virtually unavoidable if we are to handle efficiently symbols with very many components which combine, one with the other, according to very precisely defined rules.

In this chapter we set out the basic features of indicial notation and summation convention which, in combination with some basic integral identities, allows us to deal with arrays of quantities in a manner ideally suited to the development of BEM formulations. Since the whole BEM concept rests upon geometrical descriptions of boundaries, internal cells and functions distributed over them, in the later part of the chapter we introduce to our reader some simple concepts of geometric transformation, interpolation, so that one becomes used to the idea of integrating over transformed space using numerical quadrature.

2.2 INDICIAL NOTATION

The key concept here is that all entities which can be best defined by a number of components are to be labelled by a suffix (or superfix) which indicates the number and form of the components. Thus, for example, coordinate components would be labelled x_i, which in three dimensions means the set of components (x_1, x_2, x_3). The range of $i = 1, 2, 3$ here, whereas in two dimensions the range would be $i = 1, 2,$

representing (x_1, x_2). Similarly, components of a vector $\mathbf{u}$ would be shown simply as u_i, which represents the whole set of them (u_1, u_2, u_3). Note that:

1. Suffixes, i, j, k, etc., can have any specified range.
2. We now no longer use different labels for components of the same set of quantities [such as (x, y, z) or (u, v, w), etc.].

More complicated quantities might be usefully characterized by multiple suffixes, for example, $\sigma_{ij}, T_{ijk}, C_{ijkl}$, etc. The first of these (σ_{ij}) could represent compactly all the nine stress components at a point by permuting all the combinations of the $i, j = 1, 2, 3$ suffixes ($\sigma_{11}, \sigma_{12}, \ldots, \sigma_{23}, \sigma_{33}$); the known symmetry of $\sigma_{ij} = \sigma_{ji}$ means that only six of the components are independent. [In two dimensions, of course, the same symbol $\sigma_{ij}\,(i, j = 1, 2)$ represents just $\sigma_{11}, \sigma_{12}, \sigma_{21}, \sigma_{22}$.] These ideas are already familiar to most applied scientists through matrix algebra, but they can be extended indefinitely to provide a convenient way of handling entities such as T_{ijk} (with 3^3 components in three dimensions) and the elastic compliance C_{ijkl} with $3^4 = 81$ components, particularly when used in conjunction with the following summation convention.

2.2.1 The Summation Convention for Indices

We shall assume here $i, j, k, \ldots = 1, 2, 3$ unless otherwise stated. Consider first the implication of (outer) products of our indexed symbols (for example $u_i v_j$ or $\sigma_{ij} n_k$ or $\sigma_{ij} \varepsilon_{kl}$); clearly there are nine combinations of u_i and v_j components and we shall have

$$u_i v_j = v_j u_i = w_{ij} \qquad \text{(say)} \qquad (2.1)$$

in which the ordering of the symbols (u, v) is of no importance. Similarly $\sigma_{ij} n_k = s_{ijk}$ and $\sigma_{ij} \varepsilon_{kl} = E_{ijkl}$, etc. However, we frequently encounter other (inner) products in which some suffixes are repeated, such as, for example, the scalar product $\mathbf{u}$ and $\mathbf{v}$ where

$$\phi = \mathbf{u} \cdot \mathbf{v} = u_1 v_1 + u_2 v_2 + u_3 v_3, \qquad \text{i.e.} \sum_{i=1}^{3} u_i v_i = \phi \qquad \text{(say)} \qquad (2.2)$$

or a strain energy (U) term (with ε_{ij} strain components) such as

$$U = (\sigma_{11}\varepsilon_{11} + \sigma_{12}\varepsilon_{12} + \sigma_{13}\varepsilon_{13} + \sigma_{21}\varepsilon_{21} + \cdots + \sigma_{33}\varepsilon_{33})$$

$$= \sum_{i=1}^{3} \sum_{j=1}^{3} \sigma_{ij}\varepsilon_{ij} \qquad (2.3)$$

or, indeed, our usual matrix product for an $n \times n$ system of algebraic equation, which can be written as

$$\sum_{j=1}^{n} A_{ij} x_j = b_i \qquad (2.4)$$

If, in all of these examples, we stipulate that repeated suffixes imply summation on them over their full range of values we can immediately write $\phi = u_i v_i$, $U = \sigma_{ij}\varepsilon_{ij}$ and $A_{ij}x_j = b_i$ without ambiguity, providing we ensure that no suffix appears more than twice in any expression. The summed indices are known as dummy indices and the symbols used for them are interchangeable, thus:

$$A_{ij}u_j + B_{ik}u_k = A_{il}u_l + B_{il}u_l \tag{2.5}$$

Equations (2.4) and (2.5) are important in that they show how the free indices must 'match' on each side of correctly written equations. More elaborate expressions obey the same rules; e.g. the general linear stress–strain relationship in elasticity then becomes, say,

$$\sigma_{ij} = C_{ijkl}\varepsilon_{kl} \tag{2.6}$$

and Laplace's equation governing potential flow (heat transfer, electrostatics, etc.) is

$$\frac{\partial^2\phi}{\partial x_1^2} + \frac{\partial^2\phi}{\partial x_2^2} + \frac{\partial^2\phi}{\partial x_3^2} = \frac{\partial^2\phi}{\partial x_i \partial x_i} = \phi_{,ii} \tag{2.7}$$

where the utmost abbreviation is often achieved by using a comma to indicate partial differentiation, as above, or $\partial\phi/\partial x_i = \phi_{,i}$, $\partial u_i/\partial x_j = u_{i,j}$, etc. Note that whereas $A_{ij}B_{kl} \equiv B_{kl}A_{ij}$, etc., and the sequence of writing A, B is not important, in general $A_{ij} \neq A_{ji}$ unless A happens to be symmetrical in i and j.

Two symbols play a major role in manipulating our indexed quantities:

1. The Kronecker delta (or substitution tensor)

$$\delta_{ij} = \begin{cases} 1 & \text{when } i = j \\ 0 & \text{when } i \neq j \end{cases} \tag{2.8}$$

has, by definition, the property that

$$\left. \begin{array}{l} u_i\delta_{ij} = u_j, \quad u_i v_j\delta_{ik} = u_k v_j, \text{etc., and also} \\ u_i v_j\delta_{ij} = u_i v_i, \quad \sigma_{ij}\varepsilon_{kl}\delta_{ik}\delta_{lj} = \sigma_{ij}\varepsilon_{ij} \end{array} \right\} \tag{2.9}$$

The latter expressions are 'contracted' by multiplication with δ_{ij} (i.e. reduced in rank by two by each multiplication). Scalars (ϕ) are called rank 0; u_i 'vectors', rank 1; etc.; C_{ijkl} rank 4. Thus the final stress–strain product in (2.9) is reduced from rank 4 to a scalar quantity (energy) U.

2. The permutation symbol

$$e_{ijk} = \begin{cases} 0 & \text{if any two suffixes are equal} \\ 1 & \text{for cyclic suffix order, } 12312\ldots \\ -1 & \text{for anticyclic suffix order, } 321321\ldots \end{cases}$$

is rather less convenient to deal with than δ_{ij} and arises in evaluating determinants: namely,

$$\det \| J_{ij} \| = e_{rst}J_{1r} J_{2s} J_{3t} \tag{2.10}$$

or components of vector cross-products, $\mathbf{W} = \mathbf{U} \times \mathbf{V}$ say, as

$$W_i = e_{ijk} U_j V_k \tag{2.11}$$

Whereas $\delta_{ii} = 3$ is an obvious result it is a worthwhile exercise to check that $e_{ijk} e_{jki} = 6$.

Sometimes summation over indices is applied unless 'suppressed', for example, by writing $A_{(ii)}$ which then means any of $A_{11}, A_{22}, \ldots$, not their sum.

2.2.2 Cartesian Tensors and Transformation Rules

If we confine ourselves for the moment to orthogonal Cartesian coordinate systems and a set of axes y_i which are translated (by h_i) and rotated with respect to x_i, then, if the direction cosine array of the rotation is λ_{ij} [with $\lambda_{12} = \cos(y_1, x_2)$, etc.], we have the transformation rule

$$y_i = \lambda_{ij}(x_j - h_j) \tag{2.12}$$

or

$$dy_i = \lambda_{ij}\, dx_j = \frac{\partial y_i}{\partial x_j} dx_j = \frac{\partial x_j}{\partial y_i} dx_j \tag{2.13}$$

since, in our special orthogonal Cartesian coordinates, $\lambda_{ij} = \partial y_i / \partial x_j = \partial x_j / \partial y_i$. Note that the length (ds) of a line element in y_i can be written as

$$ds^2 = dy_i dy_i = \left(\frac{\partial y_i}{\partial x_j} dx_j\right)\left(\frac{\partial x_k}{\partial y_i} dx_k\right) = \delta_{jk} dx_j dx_k = dx_j dx_j$$

as expected.

We now define Cartesian tensors as entities that transform according to the following rules (dashed symbols refer to the tensor components in y; undashed in x):

1. Zero rank (scalars):

$$\phi'(y) = \phi(x) \tag{2.14a}$$

2. First rank (true vectors, forces, displacements, gradients of scalars, etc.):

$$V'_i = \lambda_{ij} V_j = \frac{\partial y_i}{\partial x_j} V_j = \frac{\partial x_j}{\partial y_i} V_j \tag{2.14b}$$

3. Second rank (stress, strain, conductivity, etc.):

$$\sigma'_{ij} = \lambda_{ik}\lambda_{jl}\sigma_{kl}, \text{etc.} \tag{2.14c}$$

4. Fourth rank (elastic compliance, etc.):

$$C'_{ijkl} = \lambda_{im}\lambda_{jn}\lambda_{kp}\lambda_{lq} C_{mnpq}, \text{etc.} \tag{2.14d}$$

Again, as an exercise, it is useful to demonstrate the invariance of strain energy under transformation:

$$U = \sigma'_{ij}\varepsilon'_{ij} = \lambda_{ik}\lambda_{jl}\sigma_{kl}\lambda_{im}\lambda_{jn}\varepsilon_{mn} = \delta_{km}\delta_{ln}\sigma_{kl}\varepsilon_{mn}$$
$$= \sigma_{mn}\varepsilon_{mn} = \sigma_{ij}\varepsilon_{ij}$$

2.2.3 An Illustrative Example

As an illustrative example[1] let us consider the derivation of the strain kernel function defined in Chapter 4. The displacement field $u_i(x)$ due to a unit force vector $e_k(\xi)$ is given by

$$u_i(x) = c_1 \left[c_2 \ln r \delta_{ik} - \frac{y_i y_k}{r^2} \right] e_k(\xi) \tag{2.15}$$

where

$$y_i = (x - \xi)_i \qquad r^2 = y_i y_i$$

We can obtain $\partial u_i(x)/\partial x_j$ from Eq. (2.15) as

$$\frac{\partial u_i}{\partial x_j} = c_1 \left[c_2 \delta_{ik} \frac{\partial}{\partial x_j}(\ln r) - \frac{\partial}{\partial x_j} \frac{y_i y_k}{r^2} \right] e_k(\xi)$$
$$= c_1 \left[c_2 \delta_{ik} \frac{\partial}{\partial r}(\ln r) \frac{\partial r}{\partial x_j} - y_i y_k \frac{\partial}{\partial r}\left(\frac{1}{r^2}\right) \frac{\partial r}{\partial x_j} - \frac{y_i}{r^2}\frac{\partial y_k}{\partial x_j} - \frac{y_k}{r^2}\frac{\partial y_i}{\partial x_j} \right] e_k(\xi) \tag{2.16}$$

Noting that here $\partial y_k/\partial x_j = \delta_{jk}$ and $\partial y_i/\partial x_j = \delta_{ij}$ we can rewrite Eq. (2.16) as

$$\frac{\partial u_i}{\partial x_j} = c_1 \left[c_2 \delta_{ik}\left(\frac{1}{r}\right)\frac{y_j}{r} - y_i y_k \frac{-2 y_j}{r^3}\frac{}{r} - \frac{y_i}{r^2}\delta_{jk} - \frac{y_k}{r^2}\delta_{ij} \right] e_k(\xi)$$
$$= \frac{c_1}{r} e_k(\xi)\left(c_2 \delta_{ik}\frac{y_j}{r} + \frac{2 y_i y_j y_k}{r^3} - \frac{y_i}{r}\delta_{jk} - \frac{y_k}{r}\delta_{ij} \right) \tag{2.17}$$

Interchanging indices i and j we get

$$\frac{\partial u_j}{\partial x_i} = \frac{c_1}{r} e_k(\xi)\left(c_2 \delta_{jk}\frac{y_i}{r} + \frac{2 y_i y_j y_k}{r^3} - \frac{y_j}{r}\delta_{ik} - \frac{y_k}{r}\delta_{ij} \right) \tag{2.18}$$

Then using Eqs (2.17) and (2.18) we arrive at the strain field as

$$\varepsilon_{ij} = \frac{1}{2}\left(\frac{\partial u_i}{\partial x_j} + \frac{\partial u_j}{\partial x_i} \right)$$
$$= \frac{c_1}{r} e_k(\xi)\left[0.5(c_2 - 1)\left(\delta_{ik}\frac{y_j}{r} + \delta_{jk}\frac{y_i}{r} \right) - \delta_{ij}\frac{y_k}{r} + \frac{2 y_i y_j y_k}{r^3} \right] \tag{2.19}$$

Results of this kind are very cumbersome to derive without using indicial notation.

2.3 INTEGRAL IDENTITIES

2.3.1 The General Form of Gauss' Theorem

Consider the function F to be a typical scalar component of a general tensor field $F_{jkl...}$ of any rank, all components of which are defined, continuous and differentiable as required, within the region V and on its surface S. Then, from Fig. 2.1, considering a volume element with its axis along x_1, we have

$$\int_V \frac{\partial F}{\partial x_1} dV = \int_S (F^* - F^{**}) dx_2 dx_3 \qquad (2.20)$$

where, throughout, single-starred quantities refer to their values at the 'right-hand' end of the volume element and double-starred ones to those at the 'left-hand' end. Thus, for outward unit normals **n** and surface area elements ds, we have

$$dx_2 dx_3 = n_1^* ds^* = -n_1^{**} ds^{**}$$

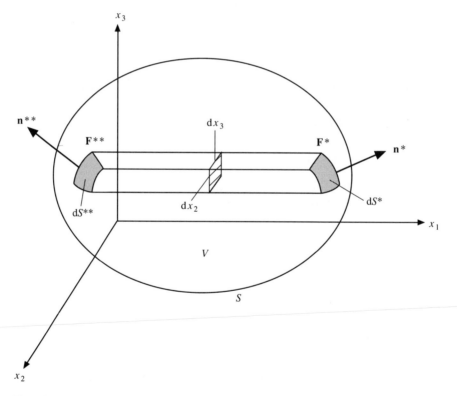

Figure 2.1 Elemental volume core parallel to the x_1 axis with vector field (**F**) components on surface area elements (dS).

and, therefore,

$$\int_V \frac{\partial F}{\partial x_1} \, \mathrm{d}V = \int_S (F^* n_1^* \, \mathrm{d}s^* + F^{**} n_1^{**} \, \mathrm{d}s^{**}) = \int_S F n_1 \, \mathrm{d}S$$

whence

$$\int_V \frac{\partial F}{\partial x_i} \, \mathrm{d}V = \int_S F n_i \, \mathrm{d}S \tag{2.20a}$$

and similarly for all other components of $F_{jkl...}$ to give, finally,

$$\int_V \frac{\partial F_{jkl...}}{\partial x_i} \, \mathrm{d}V = \int_S F_{ijkl...} n_i \, \mathrm{d}S \tag{2.20b}$$

a general form of Gauss' theorem, which is a fundamental theorem in engineering mechanics.

Many other classical integral theorems and statements of conservation laws follow directly from Eq. (2.20b) which is, of course, applicable in any number of dimensions:

1. For a scalar function F:

$$\int_V F_{,i} \, \mathrm{d}V = \int_S F n_i \, \mathrm{d}S \qquad \text{or} \qquad \int_V \mathrm{grad} \, F \, \mathrm{d}V = \int_S F \mathbf{n} \, \mathrm{d}S \tag{2.21}$$

2. For a vector function $\mathbf{F}$:
   (a) The divergence theorem,

$$\int_V F_{i,i} \, \mathrm{d}V = \int_S F_i n_i \, \mathrm{d}S \quad \text{or} \quad \int_V \mathrm{div} \, \mathbf{F} \, \mathrm{d}V = \int_S \mathbf{F} \cdot \mathbf{n} \, \mathrm{d}S \tag{2.22}$$

Thus if, say, $F_i = v_i = -kp_{,i}$ and $q = v_i n_i$ and $F_{i,i} = -kp_{,ii} = \psi$, which are the potential flow equations of Chapter 3, then we can write

$$-k \int_V p_{,ii} \, \mathrm{d}V = \int_V \psi \, \mathrm{d}V = \int_S v_i n_i \, \mathrm{d}S = \int_S q \, \mathrm{d}S$$

where q is the normal flux on the boundary. Then

$$\int_V \psi \, \mathrm{d}V = \int_S q \, \mathrm{d}S$$

expresses the conservation equation in potential flow.
   (b) Stokes' theorem,

$$\int_V \varepsilon_{ijk} F_{k,j} \, \mathrm{d}V = \int_S \varepsilon_{ijk} F_k n_j \, \mathrm{d}S \qquad \int_V \mathrm{curl} \, \mathbf{F} \, \mathrm{d}V = \int_S \mathbf{n} \times \mathbf{F} \, \mathrm{d}S \tag{2.23}$$

This theorem is extremely useful in evaluating surface integrals in BEMs. For example, Epton[2] discusses a slightly different form of this theorem to reduce certain surface integrals to a more manageable form.

3. For second-rank tensor components F_{ij}:

$$\int_V F_{ij,j} \, \mathrm{d}V = \int_S F_{ij} n_j \, \mathrm{d}S$$

if, say, $F_{ij} = \sigma_{ij}$ (i.e. stress components) and $\sigma_{ij,j} + \psi_i = 0$ and $t_i = \sigma_{ij}n_j$ (the stress equilibrium equations), then

$$\int_V \psi_i \, \mathrm{d}V + \int_S t_i \, \mathrm{d}S = 0 \qquad (2.24)$$

which can be considered as either a stress-flux conservation equation or a statement of overall equilibrium.

2.3.2 Green's Identities

These follow directly from Eq. (2.22). If a vector field **F** is related to two scalar fields (ϕ, χ), say, via

$$\mathbf{F} = \phi \, \mathrm{grad} \, \chi \qquad \text{or} \qquad F_i = \phi\chi_{,i}$$

then

$$F_{i,i} = \phi_{,i}\chi_{,i} + \phi\chi_{,ii} \qquad F_i n_i = \phi\chi_{,i} n_i$$

Substitution into Eq. (2.22) yields Green's first identity:

$$\int_V (\phi\chi_{,ii} + \phi_{,i}\chi_{,i}) \, \mathrm{d}V = \int_S (\phi\chi_{,i} n_i) \, \mathrm{d}S$$

or $\qquad \displaystyle\int_V (\phi\nabla^2\chi + \mathrm{grad} \, \phi \cdot \mathrm{grad} \, \chi) \, \mathrm{d}V = \int_S \left(\phi \frac{\partial\chi}{\partial n}\right) \mathrm{d}S$ $\qquad (2.25)$

Interchanging ϕ and χ and subtracting the result from Eqs (2.25) eliminates the $(\mathrm{grad} \, \phi \cdot \mathrm{grad} \, \chi)$ terms to produce Green's second identity:

$$\int_V (\phi\chi_{,ii} - \chi\phi_{,ii}) \, \mathrm{d}V = \int_S (\phi\chi_{,i} - \chi\phi_{,i}) n_i \, \mathrm{d}S$$

or $\qquad \displaystyle\int_V (\phi\nabla^2\chi - \chi\nabla^2\phi) \, \mathrm{d}V = \int_S \left(\phi \frac{\partial\chi}{\partial n} - \chi \frac{\partial\phi}{\partial n}\right) \mathrm{d}S$ $\qquad (2.26)$

The idea of expressing F_i as a product of two functions, used to develop Green's identities, can also be used with products of higher-rank quantities. For example, writing $F_j = u_i\sigma_{ij}^*$ and $F_j^* = u_i^*\sigma_{ij}$ where (u_i, σ_{ij}) and (u_i^*, σ_{ij}^*) are two corresponding differentiable displacement and stress fields in an elastic body then, using Eqs (2.22) again exactly as for the development of Eqs (2.25), we obtain

$$\int_V (u_{i,j}\sigma_{ij}^* + u_i\sigma_{ij,j}^*) \, \mathrm{d}V = \int_S (u_i\sigma_{ij}^* n_j) \, \mathrm{d}S = \int_S (u_i t_i^*) \, \mathrm{d}S$$

and $\qquad \displaystyle\int_V (u_{i,j}^*\sigma_{ij} + u_i^*\sigma_{ij,j}) \, \mathrm{d}V = \int_S (u_i^*\sigma_{ij} n_j) \, \mathrm{d}S = \int_S (u_i^* t_i) \, \mathrm{d}S$ $\qquad (2.27a)$

where we have used the relation $\sigma_{ij} n_j = t_i$. If we subtract these two equations the first terms on the left-hand side cancel since

$$u_{i,j}\sigma_{ij}^* = \tfrac{1}{2} C_{ijkl} (u_{i,j} u_{k,l}^* + u_{i,j} u_{l,k}^*) \tag{2.27b}$$

and

$$u_{i,j}^* \sigma_{ij} = \tfrac{1}{2} C_{ijkl} (u_{i,j}^* u_{k,l} + u_{i,j}^* u_{l,k}) \tag{2.27c}$$

are identical, because of the symmetries of C_{ijkl} in the various suffixes. Therefore we deduce that

$$\int_V (u_i \sigma_{ij,j}^* - u_i^* \sigma_{ij,j}) \, dV = \int_S (u_i t_i^* - u_i^* t_i) \, dS$$

For equilibrium, $\sigma_{ij,j} + \psi_i = 0$, etc. Therefore this equation is the Betti–Maxwell identity

$$\int_V (u_i^* \psi_i - u_i \psi_i^*) \, dV = \int_S (u_i t_i^* - u_i^* t_i) \, dS \tag{2.28}$$

which states that: if two distinct elastic equilibrium states, $A(u_i, t_i$ and $\psi_i)$ and $B(u_i^*, t_i^*$ and $\psi_i^*)$ exist in a body V bounded by the surface S, then the work done by the forces of A on the displacements of state B is equal to the work done by the forces of state B on the displacements of state A. This is the well-known reciprocal work theorem.

2.3.3 Integration by Parts

In Sec. 2.3.1 we have seen that Gauss' theorem could be applied to any sufficiently smooth function F which can be scalar, a vector, i.e. a tensor of any rank. Therefore, if we consider the product of two such functions A and B we can write

$$\int_V \frac{\partial}{\partial x_i} (AB) \, dV = \int_S AB n_i \, dS$$

or

$$\int_V A \frac{\partial B}{\partial x_i} \, dV + \int_V B \frac{\partial A}{\partial x_i} \, dV = \int_S AB n_i \, dS$$

Therefore

$$\int_V A \frac{\partial B}{\partial x_i} \, dV = \int_S AB n_i \, dS - \int_V B \frac{\partial A}{\partial x_i} \, dV \tag{2.29}$$

Equation (2.29) is a generalization of the integration by parts used in Chapter 1 to derive BEM formulation, and can be used to derive integral identities which form the basis of many finite element and all direct BEM formulations.

Let us consider the potential flow equation in three-dimensions:

$$\frac{\partial^2 p}{\partial x_i \partial x_i} + \psi = 0 \tag{2.30}$$

We now multiply Eq. (2.30) by a function p^* and integrate over the volume, i.e.

$$\int_V \frac{\partial^2 p}{\partial x_i \partial x_i} p^* \, \mathrm{d}V + \int_V \psi p^* \, \mathrm{d}V = 0$$

We apply Eq. (2.29) to evaluate the first volume integral by parts to obtain

$$\int_S \frac{\partial p}{\partial x_i} p^* n_i \, \mathrm{d}V - \int_V \frac{\partial p}{\partial x_i} \frac{\partial p^*}{\partial x_i} \, \mathrm{d}V + \int_V \psi p^* \, \mathrm{d}V = 0$$

(2.31)

or

$$\int_S \frac{\partial p}{\partial n} p^* \, \mathrm{d}S + \int_V \psi p^* \, \mathrm{d}V = \int_V \frac{\partial p}{\partial x_i} \frac{\partial p^*}{\partial x_i} \, \mathrm{d}V$$

which can be used together with a locally based approximation for both p and p^* to derive a finite element formulation. If we integrate by parts the right-hand side once more we obtain

$$\int_S \frac{\partial p}{\partial n} p^* \, \mathrm{d}S + \int_V \psi p^* \, \mathrm{d}V = \int_S p \frac{\partial p^*}{\partial x_i} n_i \, \mathrm{d}S - \int_V p \frac{\partial^2 p^*}{\partial x_i \partial x_i} \, \mathrm{d}V$$

(2.32)

the identity that would be used in a BEM formulation for potential flow problems in Chapter 3.

We now consider the equilibrium equation in the theory of elasticity:

$$\frac{\partial \sigma_{ij}}{\partial x_j} + \psi_i = 0$$

Once again multiplying both sides with a sufficiently smooth function u_i^* and integrating over the volume, we get

$$\int_V \frac{\partial \sigma_{ij}}{\partial x_j} u_i^* \, \mathrm{d}V + \int_V \psi_i u_i^* \, \mathrm{d}V = 0$$

To this we can now apply Eq. (2.29) to evaluate the volume integral by parts to obtain

$$\int_S \sigma_{ij} u_i^* n_j \, \mathrm{d}S - \int_V \sigma_{ij} \frac{\partial u_i^*}{\partial x_j} \, \mathrm{d}V + \int_V \psi_i u_i^* \, \mathrm{d}V = 0$$

Utilizing the symmetry of σ_{ij} and the relation $\sigma_{ij} n_j = t_i$ we can express the above as

$$\int_S t_i u_i^* \, \mathrm{d}S + \int_V \psi_i u_i^* \, \mathrm{d}V = \int_V \sigma_{ij} \varepsilon_{ij}^* \, \mathrm{d}V$$

(2.33)

where

$$\varepsilon_{ij}^* = \frac{1}{2}\left(\frac{\partial u_i^*}{\partial x_j} + \frac{\partial u_j^*}{\partial x_i}\right)$$

Equation (2.33) is now the well-known virtual work equation which forms the basis of the FEM formulation in elasticity.

The right-hand side of Eq. (2.33) can be integrated by parts once again by using the constitutive and strain-displacement relations together with the assumption that our * state stresses satisfy

$$\frac{\partial \sigma_{ij}^*}{\partial x_j} + \psi_i^* = 0 \qquad \text{and} \qquad \sigma_{ij}^* n_j = t_i^*$$

to produce

$$\int_S t_i u_i^* \, dS + \int_V \psi_i u_i^* \, dV = \int_S t_i^* u_i \, dS + \int_V \psi_i^* u_i \, dV \qquad (2.34)$$

the well-known reciprocal work theorem derived earlier [Eq. (2.28)] using a slightly different approach.

It is very important to note that in deriving Eq. (2.33) we only had to make assumptions of equilibrium and sufficient smoothness of u_i^* in order to differentiate once. However, in Eq. (2.34), which we shall later use in developing BEM formulations, we had to propose that both states satisfy equilibrium and the constitutive relations.

The above process of integration by parts could be applied to all differential equations, linear or non-linear to develop the equivalent integral identities for use in BEMs. Thus, if our general differential equation were

$$L(u) + \psi = 0 \qquad (2.35)$$

and we were to investigate the integration by parts of the product of Eqs (2.35) with some other, adequately differentiable and continuous, function (v) over V, we could, step by step, transpose the differential operations from u to v while simultaneously generating integrations of functions of u and v over S. The general form of equation obtained will always be

$$\int_V v L(u) \, dV = \int_V u L^*(v) \, dV + \int_S [M^*(v)N(u) - M(u)N^*(v)] \, dS \qquad (2.36)$$

where the operator L^* is called the adjoint of L and (M, N, M^*, N^*) are also differential operators which arise from the integration by parts procedure. In all cases considered above $L = L^*$, $M = M^*$, $N = N^*$ and therefore the operator is self-adjoint.

In some BEM literature the above procedure is described as a 'weighted residual' method. This is unfortunate because the weighted residual method is much more than the simple 'integration by parts'.

2.4 INTERPOLATION FUNCTIONS AND GEOMETRIC TRANSFORMATIONS

In order to make practical uses of boundary element formulations it is necessary to represent surfaces and, sometimes in a non-linear problem, the volume by an assemblage of surface elements and volume cells. The technology associated with such processes is well established in the finite element method and is discussed in the following sections.

2.4.1 Geometrical Transformations

If we consider, initially, points located by x_i in a global Cartesian space (X) and by ζ_i in local curvilinear coordinates associated with a space (Z) of the same dimensionality, the x_i will be expressible as a function of the ζ_i, $x_i = f_i(\zeta_1, \zeta_2, \zeta_3)$ and conversely the $\zeta_i = g_i(x_1, x_2, x_3)$. Such transformation equations are usually written in the abbreviated form $x_i \equiv x_i(\zeta)$ and $\zeta_i \equiv \zeta_i(x)$, when differential components of line elements in X and Z will be interrelated by

$$\mathrm{d}x_i = \frac{\partial f_i}{\partial \zeta_j}\,\mathrm{d}\zeta_j = \frac{\partial x_i}{\partial \zeta_j}\,\mathrm{d}\zeta_j \qquad (2.37)$$

or
$$\mathrm{d}\mathbf{x} = \mathbf{J}\,\mathrm{d}\zeta$$

The matrix $\mathbf{J}$ is known as the Jacobian of the transformation

$$\mathbf{J} = \left[\frac{\partial x_i}{\partial \zeta_j}\right]$$

The inverse operation, if it exists, will be defined by

$$\mathrm{d}\zeta = \mathbf{J}^{-1}\mathrm{d}\mathbf{x}$$

with $\qquad\qquad\qquad\qquad\qquad\qquad\qquad\qquad\qquad\qquad\qquad (2.38)$

$$\mathrm{d}\zeta_i = \frac{\partial \zeta_i}{\partial x_j}\,\mathrm{d}x_j$$

where

$$\mathbf{J}^{-1} = \left[\frac{\partial \zeta_i}{\partial x_j}\right]$$

The determinant of $\mathbf{J}, J = \|\mathbf{J}\|$, is known as the Jacobian (determinant) of the transformation. One could use, equally validly, either transformation (2.37) or (2.38) to define a Jacobian and when reading other texts (e.g. Fung[3]) it is important to note their starting point.

In order that any transformation $x \to \zeta$ is reversible (has an inverse) and that (x_i, ζ_i) points are in one-to-one correspondence in some region of interest (i.e. that any set of numbers x_i defines a unique set of numbers ζ_i and vice versa) it is sufficient that:[3]

1. The f_i are single-valued, continuous functions of x_i with continuous first partial derivatives.
2. J must be finite (that is $1/J \neq 0$) at any point.

These conditions alone ensure that the transformation is admissible. Since we also require that right-handed coordinate systems shall remain right handed when transformed, our transformation must also be proper, which requires that J be positive everywhere (e.g. for simple transformations between the orthogonal Cartesian system $J = +1$). There are a few basic transformation operations which we shall use subsequently; these are given below, with primed symbols referring to functions in Z and unprimed ones in X.

Scalar fields   A scalar field $\psi(x)$ in X transforms identically to $\psi'(\zeta)$ in Z at corresponding (x_i, ζ_i) points. Thus,

$$\psi(x_1, x_2, x_3) = \psi'(\zeta_1, \zeta_2, \zeta_3)$$

$$(2.39a)$$

or
$$\psi(x) = \psi'(\zeta) \quad \text{or} \quad \psi = \psi'$$

Vector fields   The gradient of a scalar field $(\partial \psi / \partial x_i) = \mathbf{v}$, say, in X transforms by the differentiation chain rule as

$$\frac{\partial \psi}{\partial x_i} = \frac{\partial \psi}{\partial \zeta_j} \frac{\partial \zeta_j}{\partial x_i}$$

that is

$$\mathbf{v} = \mathbf{J}^{-1} \mathbf{v}' \qquad \mathbf{v}' = \mathbf{J} \mathbf{v} \qquad (2.39b)$$

Such transformation rules can be extended systematically to higher-rank tensor quantities, a very lucid account of which will be found in Fung[3] [see also Eq. (2.14)]. The Jacobian matrix of a transformation is seen to be fundamental to both geometrical mappings and the manner in which tensor components change between the x_i and ζ_i coordinate systems.

2.4.2 Transformation of Differential Volume, Area, and Line Elements Internal Cells

Figure 2.2 shows a number of differential length, area and volume elements mapped between X and Z. Since they all take simpler geometrical forms in Z we shall find it more convenient to evaluate the quantities $dV(x)$ [and $dA(x)$, $dS(x)$, etc.] appearing in the basic BEM statements in terms of their mappings in Z, which can all be identical 'unit' elements irrespective of their size in X. Although the surface elements (in three dimensions) and boundary line segments (in two dimensions) are the major features of the BEM it is simpler to deal with them after the transformation of volume (in three dimensions) cells. Consider a differential volume element $(d\zeta_1 d\zeta_2 d\zeta_3)$ in Z (Fig. 2.2a) which transforms to the curvilinear elemental cell of volume $dV(x)$ in X. Then with $\mathbf{e}$ as base vectors in Z, elementary vector algebra[4] provides the volume of the elemental cell $dV(x)$ in X as

$$dV(x) = (\mathbf{e}_2 \times \mathbf{e}_3) \cdot \mathbf{e}_1 (d\zeta_1 d\zeta_2 d\zeta_3) \qquad (2.40)$$

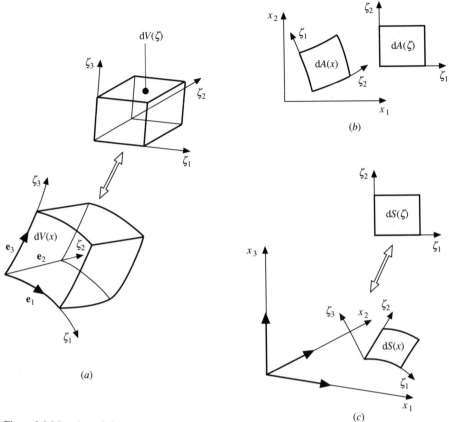

Figure 2.2 Mapping of elements in X to Z.

or

$$dV(x) = \begin{Vmatrix} \dfrac{\partial x_1}{\partial \zeta_1} & \dfrac{\partial x_1}{\partial \zeta_2} & \dfrac{\partial x_1}{\partial \zeta_3} \\[2ex] \dfrac{\partial x_2}{\partial \zeta_1} & \dfrac{\partial x_2}{\partial \zeta_2} & \dfrac{\partial x_2}{\partial \zeta_3} \\[2ex] \dfrac{\partial x_3}{\partial \zeta_1} & \dfrac{\partial x_3}{\partial \zeta_2} & \dfrac{\partial x_3}{\partial \zeta_3} \end{Vmatrix} (d\zeta_1 d\zeta_2 d\zeta_3) = J_V(d\zeta_1 d\zeta_2 d\zeta_3)$$

where the V suffix has been added to the Jacobian determinant J to indicate its relevance to volume transformation. If we are considering a flat, two-dimensional elemental cell of area $(d\zeta_1 d\zeta_2)$ in Z (Fig. 2.2b), the corresponding transformation rule is easily shown to be

$$dA(x) = \begin{Vmatrix} \dfrac{\partial x_i}{\partial \zeta_i} \end{Vmatrix} (d\zeta_1 d\zeta_2) = J_A(d\zeta_1 d\zeta_2) \qquad i,j = 1,2 \tag{2.42}$$

i.e. only the range of the suffixes has changed between Eqs (2.41) and (2.42).

Boundary patches A boundary patch (in three dimensions) of elemental area $dS(x)$ must be carefully distinguished from a two-dimensional cell $dA(x)$, for $dS(x)$ is not flat in X although it is flat and bounded by (ζ_1, ζ_3) in Z (Fig. 2.2c). Again we use the well-known vector algebra to obtain

$$dS(x) = |\mathbf{e}_1 \times \mathbf{e}_2|(d\zeta_1 d\zeta_2)$$

or
$$dS(x) = \sqrt{(d_1)^2 + (d_2)^2 + (d_3)^2}(d\zeta_1 d\zeta_2) = J_S(d\zeta_1 d\zeta_2) \tag{2.43}$$

where

$$d_1 = \left(\frac{\partial x_2}{\partial \zeta_1} \frac{\partial x_3}{\partial \zeta_2} - \frac{\partial x_3}{\partial \zeta_1} \frac{\partial x_2}{\partial \zeta_2} \right)$$

and d_2, d_3 are the other lower minors of the basic Jacobian matrix $[\mathbf{J}]$ (i.e. those that involve ζ_1, ζ_2 only).

Equation (2.43) is clearly very different from Eq. (2.42) and only reduces to it when all derivatives related to x_3 are zero and $dS(x) = d_3(d\zeta_1 d\zeta_2) = J_A(d\zeta_1 d\zeta_2)$ again.

Line segments The transformation of a general length element, $dl(x)$, from X to Z follows immediately

$$dl(x) = J_l\, dl\,(\eta) \tag{2.44}$$

By analogy with the development of Eq. (2.43) we need to evaluate line integrals involving $dl(x)$ which will be expressed in terms of ζ. Therefore, since $dl(\zeta)$ will be directed along ζ, say,

$$J_l = \left[\left(\frac{\partial x_1}{\partial \zeta} \right)^2 + \left(\frac{\partial x_2}{\partial \zeta} \right)^2 \right]^{1/2} \tag{2.45}$$

2.4.3 Linear Interpolation Functions

Interpolation functions enable us to generate useful curvilinear shape functions in a systematic way. Again the simplest introduction is by considering linear internals cells, this time parallelograms (two dimensional) and parallelipipeds (three dimensional).

Consider a simple parallelogram cell (Fig. 2.3a). We could seek a set of functions $f(\Phi_\alpha)$, where Φ_α are the values of some function ϕ at each of the nodes $\alpha = 1, 2, 3, 4$ in X, which map ϕ over the square of η through

$$\phi = f_0 + f_1\eta_1 + f_2\eta_2 + f_3\eta_1\eta_2 \tag{2.46}$$

Function ϕ will thereby vary linearly along each $\eta_1 = \pm 1$ boundary and we need to determine the four (f) functions to satisfy $\phi = \Phi_\alpha$ at each node α. The necessary four equations can be obtained by writting Eq. (2.46) for the four nodal points.

This interpolation function can be used to produce shape functions for so-called 'linear' rectangular elements. Higher-order expressions provide a means of

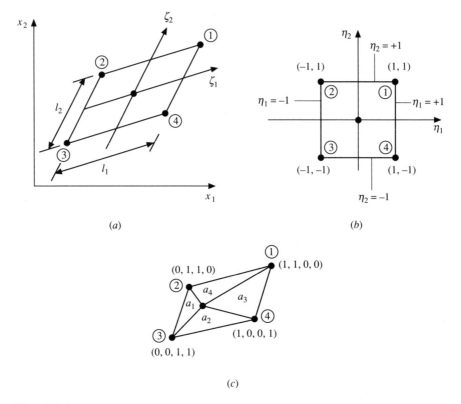

Figure 2.3 A parallelogram cell.

generating most of the commoner shape functions used in finite element analysis. Having found the $f = f(\Phi_\alpha)$, Eq. (2.46) can be rewritten as

$$\phi = \phi(\eta) = \Phi_\alpha N_\alpha \qquad (2.47)$$

where each of the shape functions $N_\alpha(\eta)$ is now a function of η_i. The result we obtain from these operations is

$$\left.\begin{array}{l} 4N_1 = (1 + \eta_1)(1 + \eta_2) \\ 4N_2 = (1 - \eta_1)(1 + \eta_2) \\ 4N_3 = (1 - \eta_1)(1 - \eta_2) \\ 4N_4 = (1 + \eta_1)(1 - \eta_2) \end{array}\right\} \quad \text{or} \quad 4N_\alpha = (1 + s_{\alpha 1}\eta_1)(1 + s_{\alpha 2}\eta_2) \qquad (2.48)$$

where the second, composite expression is a most useful and convenient way of writing the whole set of N_α functions; $s_{\alpha 1}$ (always ± 1) takes the sign of the η_1 coordinate at node α, etc. (Fig. 2.3b). No sum over α is implied in any equations involving $s_{\alpha i}$. We note two important features of Eqs (2.48) which are common to all such shape functions:

1. Each function N_α has unit value at node α and is zero at all other nodes (i.e. the effects of all the Φ_α are uncoupled one from the other).

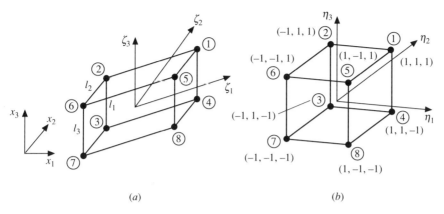

Figure 2.4 A parallelepiped cell.

2. The sum of the $N_\alpha = 1$ at any point. This ensures that the interpolation function contains a 'complete' polynomial in η_i which is a necessary condition for such a shape function transformation to be admissible.

We now see our way forward to more sophisticated curvilinear elements in X for we can, in principle, use more complicated complete polynomials in Eq. (2.46), match the number of terms (α) with nodal values ϕ_α and arrive at a shape function equation corresponding to Eqs (2.48) in any number of spatial dimensions.

The equations corresponding to Eqs (2.46) and (2.48) for the three-dimensional parallelepiped transformation shown in Fig. 2.4a,b and the associated 'linear' shape functions can be derived easily from the following interpolation function which contains the requisite eight terms:

$$\phi = f_0 + f_1\eta_1 + f_2\eta_2 + f_3\eta_3 + f_4\eta_1\eta_2 + f_5\eta_2\eta_3 + f_6\eta_3\eta_1 + f_7\eta_1\eta_2\eta_3 \qquad (2.49)$$

The shape functions then follow as

$$8N_\alpha = (1 + s_{\alpha 1}\eta_1)(1 + s_{\alpha 2}\eta_2)(1 + s_{\alpha 3}\eta_3) \qquad (2.50)$$

where the $s_{\alpha i}$ signs agree with the η_i coordinate component signs in the nodal numbering scheme adopted (Fig. 2.4b) [for example $(-1, +1, +1)$ for node 2, etc.]. Once again, $N_\alpha = 1$ at node α is zero at other nodes and $\sum_{\alpha=1}^{8} N_\alpha = 1$ always.

An alternative mapping or interpolation using area and volume coordinates in two- and three-dimensional cells respectively is shown in Fig. 2.5a,b. The first diagram defines what are usually called 'area' coordinates for a triangle a_α ($\alpha = 1, 2, 3$) where $a_1 = A_1/A$ with A the total triangle area and A_1 the area of the subtriangle opposite node 1. We necessarily have $a_1 + a_2 + a_3 = 1$ and any point x_i can be defined unambiguously in terms of a_α. By requiring

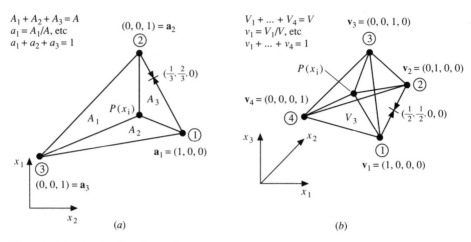

Figure 2.5 'Area' and 'volume' coordinates.

$\mathbf{x}_1$ (the coordinates of node 1) to transform to $(1, 0, 0)^T$ in terms of a_α we obtain a linearly varying x_i field given by

$$x_i = \left\{ \begin{array}{c} x_1 \\ x_2 \end{array} \right\} = [\mathbf{x_1}\ \mathbf{x_2}\ \mathbf{x_3}] \left\{ \begin{array}{c} a_1 \\ a_2 \\ a_3 \end{array} \right\} = x_{i\alpha} a_\alpha \qquad (2.51)$$

Therefore, a_α and η_α are identical.

If we define 'volume' coordinates on the tetrahedron (Fig. 2.5b) v_α ($\alpha = 1, 2, 3, 4$) with $v_1 = V_1/V$ by analogy with the triangle, the outer terms in Eq. (2.51) still equate with v_α replacing a_α and $i = 1, 2, 3$. Therefore for the 'linear' tetrahedron $v_\alpha = \eta_\alpha$.

This idea could equally well be applied to a line element defined by $l_1 + l_2 = 1$ with

$$x = [x_1 x_2] \left\{ \begin{array}{c} l_1 \\ l_2 \end{array} \right\}$$

or extended to specify a rectangle by four triangular area coordinates (see, for example, Fig. 2.3c) with $a_1 + a_3 = 1$, $a_2 + a_4 = 1$ and $a_1 = 2A_1/A$, etc., for the parallelogram shown.

2.4.4 Curvilinear Transformations and Shape Functions

Line elements We shall take as our paradigm the transformation of a vector of coordinates x_i ($i = 1, 2$) specified by its components $= X_{i\alpha}$ at a number of geometric nodes (α) along each line element sufficient to describe the order of variation required by a complete polynomial interpolation function. For any line element, all this means is that we must have two nodes per element for a linear variation of each x_i component, three for quadratic, four for

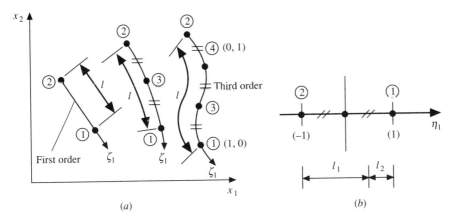

Figure 2.6 Line elements.

cubic, five for quartic, etc. Thus for ζ_i along the element (Fig. 2.6a) and quadratic variation, a suitable interpolation function would be $\phi = f_0 + f_1 \zeta_1 + f_2 \zeta_1^2$, augmented by $f_3 \zeta_1^3$ for third order, etc. Following the same pattern of calculation that led to Eq. (2.47) we arrive at the following sets of shape functions N_α (that is $x_i = X_{i\alpha} N_\alpha$). These are given in terms of normalized coordinates $\eta_1 = \zeta_1/l$, for nodes numbered as shown in Fig. 2.6 with a central origin in η.

Linear variation $(\alpha = 1, 2)$:

$$\left. \begin{array}{l} N_1 = \frac{1}{2}(1 + \eta_1) \\ N_2 = \frac{1}{2}(1 - \eta_1) \end{array} \right\} N_\alpha = \frac{1}{2}(1 + s_{\alpha 1} \eta_1) \qquad (2.52)$$

Quadratic (second-order) variation $(\alpha = 1, 2, 3)$:

$$\left. \begin{array}{l} N_1 = \frac{1}{2}\eta_1(1 + \eta_1) \\ N_2 = -\frac{1}{2}\eta_1(1 - \eta_1) \\ N_3 = (1 + \eta_1)(1 - \eta_1) = 1 - \eta_1^2 \end{array} \right\} = \frac{1}{2}s_{\alpha 1}\eta_1(1 + s_{\alpha 1}\eta_1) \qquad (2.53)$$

Cubic (third-order) variation $(\alpha = 1, 2, 3, 4)$:

$$\left. \begin{array}{l} N_1 = \frac{1}{16}(1 + \eta_1)(1 + 3\eta_1)(3\eta_1 - 1) \\ N_2 = \frac{1}{16}(1 - \eta_1)(1 + 3\eta_1)(3\eta_1 - 1) \end{array} \right\} = \frac{1}{16}(9\eta_1^2 - 1)(1 + s_{\alpha 1}\eta_1)$$

$$\left. \begin{array}{l} N_3 = \frac{9}{16}(1 - \eta_1)(1 - \eta_1)(1 + 3\eta_1) \\ N_4 = \frac{9}{16}(1 + \eta_1)(1 - \eta_1)(1 - 3\eta_1) \end{array} \right\} = \frac{9}{16}(1 - \eta_1)(1 + 3s_{\alpha 1}\eta_1)$$

$$(2.54)$$

Plane triangular cells Since these are all minor extensions of the foregoing argument the results will be simply listed in terms of area coordinates ($a_1 = \eta_1$, $a_2 = \eta_2$ and $a_3 = 1 - \eta_1 - \eta_2$).

Linear variation ($\alpha = 1, 2, 3$):

$$N_\alpha = a_\alpha \tag{2.55}$$

Quadratic variation The complete polynomial interpolation function now has six terms and we therefore require the six nodes shown in Fig. 2.7a:

$$\phi = f_0 + f_1\eta_1 + f_2\eta_2 + f_3\eta_1\eta_2 + f_4\eta_1^2 + f_5\eta_2^2$$

which leads to (for $\alpha = 1, 2, \ldots, 6$)

$$\mathbf{N}^{\mathrm{T}} = \{a_1(2a_1 - 1), a_2(2a_2 - 1), a_3(2a_3 - 1), 4a_1a_2, 4a_2a_3, 4a_3a_1\} \tag{2.56}$$

Cubic variation The complete cubic polynomial has ten terms and therefore the four nodes per triangle side (to allow cubic variation) have to be augmented by a tenth (centroidal) node (Fig. 2.7a). The shape functions are

$$
\begin{aligned}
N_1 &= \tfrac{1}{2}a_1(3a_1 - 1)(3a_1 - 2) & N_2, N_3 \text{ similarly} \\
\left. \begin{aligned} N_4 &= \tfrac{9}{2}a_1a_2(3a_1 - 1) \\ N_5 &= \tfrac{9}{2}a_1a_2(3a_2 - 1) \end{aligned} \right\} & N_6, N_7, N_8, N_9 \text{ similarly} \\
N_{10} &= 27a_1a_2a_3
\end{aligned}
\tag{2.57}
$$

All the above are taken from finite element sources[5,6] where a recurrence relationship is deduced for 'triangular' elements relating the N_α^{n+1} [for the $(n + 1)$th-order element] to N_α^n (for the nth-order element) which enables shape functions to be developed for 'triangles' of any order.

Plane parallelipiped cells Our second-order variation nodal pattern (Fig. 2.8a) on page 38 requires three nodes per side (totalling eight) and therefore the interpolation function has to expand to, say,

$$\phi = f_0 + f_1\eta_1 + f_2\eta_2 + f_3\eta_1\eta_2 + f_4\eta_1^2 + f_5\eta_2^2 + f_6\eta_1^2\eta_2 + f_7\eta_2^2\eta_1$$

whence shape functions, calculated as before for equally divided rectangle sides, are as follows.

Linear variation ($\alpha = 1, 2, 3, 4$):

$$
\left. \begin{aligned}
N_1 &= \tfrac{1}{4}(1 + \eta_1)(1 + \eta_2) \\
N_2 &= \tfrac{1}{4}(1 - \eta_1)(1 + \eta_2) \\
N_3 &= \tfrac{1}{4}(1 - \eta_1)(1 - \eta_2) \\
N_4 &= \tfrac{1}{4}(1 + \eta_1)(1 - \eta_2)
\end{aligned} \right\} \qquad N_\alpha = \tfrac{1}{4}(1 + s_{\alpha 1}\eta_1)(1 + s_{\alpha 2}\eta_2) \tag{2.58}
$$

It should be noted that a general quadrilateral (Fig. 2.8c) can also be transformed into a square in η by Eqs (2.58) and therefore we are not confined to parallelogram cells.

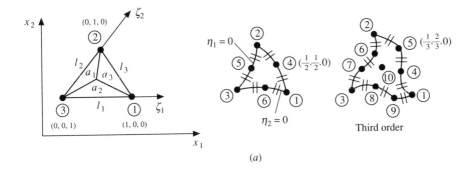

(a)

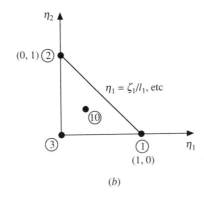

(b)

Figure 2.7 Planar cells.

Quadratic variation $(\alpha = 1, 2, \ldots, 8)$:

$$\left. \begin{aligned} N_1 &= \tfrac{1}{4}(1 + \eta_1)(1 + \eta_2)(\eta_1 + \eta_2 - 1) \\ N_2 &= \tfrac{1}{4}(1 - \eta_1)(1 + \eta_2)(-\eta_1 + \eta_2 - 1) \\ N_\alpha &= \tfrac{1}{4}(1 + s_{\alpha 1}\eta_1)(1 + s_{\alpha 2}\eta_2)(s_{\alpha 1}\eta_1 + s_{\alpha 2}\eta_2 - 1) \end{aligned} \right\} \quad N_3, N_4 \text{ similarly}$$

(2.59)

$$\left. \begin{aligned} N_5 &= \tfrac{1}{2}(1 - \eta_1)(1 + \eta_1)(1 + \eta_2) = \tfrac{1}{2}(1 - \eta_1^2)(1 + s_{\alpha 2}\eta_2) \\ N_6 &= \tfrac{1}{2}(1 - \eta_1)(1 + \eta_2)(1 - \eta_2) = \tfrac{1}{2}(1 - \eta_2^2)(1 + s_{\alpha 1}\eta_1) \end{aligned} \right\} \quad N_7, N_8 \text{ similarly}$$

Cubic variation $(\alpha = 1, 2, 3, \ldots, 12)$:

$$\left. \begin{aligned} N_1 &= \tfrac{1}{32}(1 + \eta_1)(1 + \eta_2)[9(\eta_1^2 + \eta_2^2) - 10] \\ N_2 &= \tfrac{1}{32}(1 - \eta_1)(1 + \eta_2)[9(\eta_1^2 + \eta_2^2) - 10] \\ N_\alpha &= \tfrac{1}{32}(1 + s_{\alpha 2}\eta_1)(1 + s_{\alpha 2}\eta_2)[9(\eta_1^2 + \eta_2^2) - 10] \end{aligned} \right\} \quad N_3, N_4 \text{ similarly} \quad (2.60)$$

$$N_5 = \tfrac{9}{32}(1 + \eta_1)(1 - \eta_1)(1 + \eta_2)(1 + 3\eta_1)$$
$$N_6 = \tfrac{9}{32}(1 + \eta_1)(1 - \eta_1)(1 + \eta_2)(1 - 3\eta_1)$$

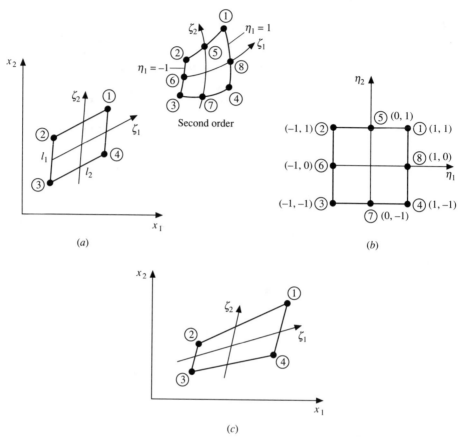

Figure 2.8 Quadrilateral cells.

Here

$$N_\alpha = \tfrac{9}{32}(1 - \eta_1^2)(1 + s_{\alpha 2}\eta_2)(1 + s_{\alpha 1}\eta_1)$$

and $N_7, \ldots, N_{12}$ follow a corresponding pattern.

Three-dimensional cells

(a) Tetrahedral cells. These are a straightforward extension of the curvilinear triangle in the same way that the straight-sided tetrahedral transformation followed on from the triangle. Thus, in volume coordinates $(v_1 + v_2 + v_3 + v_4 = 1)$ with equal divisions of the 'tetrahedral' edges we have (Fig. 2.9) the following variations.

Linear variation $(\alpha = 1, 2, 3, 4)$:

$$N_\alpha = v_\alpha \tag{2.61}$$

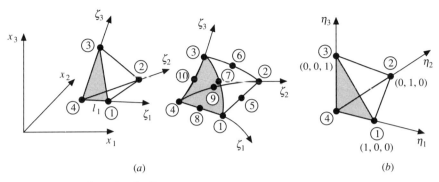

Figure 2.9 Three-dimensional cells

Quadratic variation $(\alpha = 1, 2, \ldots, 10)$:

$$\left.\begin{array}{l} N_1 = v_1(2v_1 - 1) \\ N_5 = 4v_1v_2 \\ N_6 = 4v_2v_3 \end{array}\right\} \qquad \begin{array}{l} N_2, N_3, N_4 \text{ similarly} \\[6pt] N_7, \ldots, N_{10} \text{ similarly} \end{array} \qquad (2.62)$$

Cubic variation $(\alpha = 1, 2, \ldots, 20)$:

$$N_1 = \tfrac{1}{2}v_1(3v_1 - 1)(3v_1 - 2) \quad N_2, N_3, N_4 \text{ similarly}$$

$$\left.\begin{array}{l} N_5 = \tfrac{9}{2}v_1v_2(3v_1 - 1) \\[4pt] N_6 = \tfrac{9}{2}v_1v_2(3v_2 - 1) \end{array}\right\} \qquad N_7, N_8, N_9, N_{10} \text{ and all other mid-side nodes similarly}$$

$$N_{11} = 27v_1v_2v_3 \qquad\qquad \text{and the other three 'centroidal' nodes similarly}$$

(b) Hexahedral cells. These are an extension of the 'quadrilateral' cells with nodes identified as shown in Fig. 2.10.

Linear case $(\alpha = 1, 2, \ldots, 8)$:

$$N_\alpha = \tfrac{1}{8}(1 + s_{\alpha 1}\eta_1)(1 + s_{\alpha 2}\eta_2)(1 + s_{\alpha 3}\eta_3) \qquad (2.63)$$

Quadratic case $(\alpha = 1, 2, 3, \ldots, 20)$:

Nodes $1, 2, \ldots, 8$:

$$N_\alpha = \tfrac{1}{8}(1 + s_{\alpha 1}\eta_1)(1 + s_{\alpha 2}\eta_2)(1 + s_{\alpha 3}\eta_3)(s_{\alpha 1}\eta_1 + s_{\alpha 2}\eta_2 + s_{\alpha 3}\eta_3 - 2)$$

Nodes 9, 10, 11, 12:

$$N_\alpha = \tfrac{1}{4}(1 - \eta_1^2)(1 + s_{\alpha 2}\eta_2)(1 + s_{\alpha 3}\eta_3) \qquad (2.64)$$

Nodes 13, 14, 15, 16:

$$N_\alpha = \tfrac{1}{4}(1 - \eta_2^2)(1 + s_{\alpha 1}\eta_1)(1 + s_{\alpha 3}\eta_3)$$

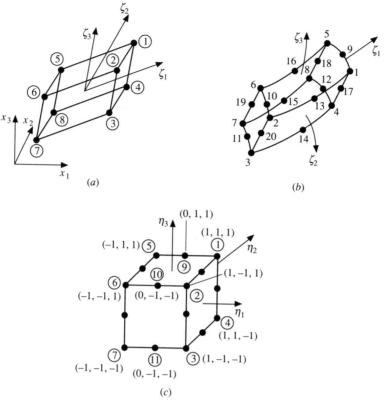

Figure 2.10 Hexahedral cells.

Nodes 17, 18, 19, 20:

$$N_\alpha = \tfrac{1}{4}(1 - \eta_3^2)(1 + s_{\alpha 1}\eta_1)(1 + s_{\alpha 2}\eta_2)$$

All the foregoing shape functions have been listed here for convenient reference although one might note:

1. By substituting $s_{\alpha 3}\eta_3 = \eta_3 = 1$ in the hexahedral shape functions the 'quadrilateral' values are recovered and replacement of v_α by a_α in the tetrahedral set produces those for the 'triangles'.
2. Other forms of contractions from higher- to lower-dimensional elements are possible. Watson,[7] for example, points out that by collapsing nodes 2, 3, 6 for the second-order quadrilateral to a single node, quadratic triangular shape functions will be generated. The reference axes will, of course, remain the ζ_i set of Fig. 2.8a, which will not be aligned along the 'triangle' sides. Similar operations on hexahedral cells (Fig. 2.10) could be used to generate 'triangular' prisms, although it would probably be more convenient to combine triangular shape functions in the (ζ_1, ζ_3) plane with, say, functions from a hexahedral cell

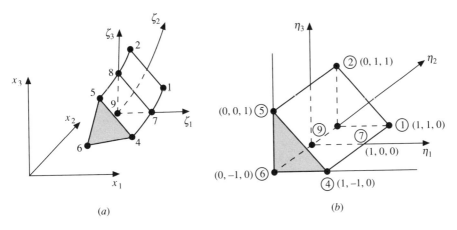

Figure 2.11 Mapping a triangular prism.

along (ζ_2). For example (Fig. 2.11), the prism of linear triangle cross-section and quadratic variation along its length will have for

Nodes 1, 2, ... , 6:

$$N_\alpha = \tfrac{1}{2} a_\alpha (1 + s_{\alpha 2} \eta_2) s_{\alpha 2} \eta_2 \tag{2.65}$$

Nodes 7, 8, 9:

$$N_\alpha = \tfrac{1}{2} a_\alpha (1 - \eta_2^2)$$

2.4.5 Curvilinear Boundary Elements

Most of the above material relating to shape functions has been taken from finite element sources[5,6] in which the emphasis is on interior (cell) rather than surface discretization schemes. We have already discussed curvilinear boundaries in two-dimensional problems (Fig. 2.6) and all that remains to be explained is the three-dimensional counterpart of curved planar boundary patches (elements), as shown in Fig. 2.12.

We could develop the shape functions for the doubly curved quadratic triangle of Fig. 2.12a by contracting the second-order tetrahedron (Fig. 2.11) . Nodes 3, 10, 4 will all have identical coordinates and coalesce into, say, node 3′. The shape function N_3' for this node will then be the sum of $N_3 + N_{10} + N_4$ with $\eta_3 = v_3 = 0$. Recollecting that $v_1 = \eta_1, v_2 = \eta_2, v_3 = \eta_3$ but $v_4 = 1 - \eta_1 - \eta_2 - \eta_3$, and using N_α from Eqs (2.62), we obtain, with $\eta_3 = 0$,

$$N_3' = (1 - \eta_1 - \eta_2)(1 - 2\eta_1 - 2\eta_2)$$

N_1, N_2 and $N_5 = N_4'$ are unaffected, but $N_5' = N_6 + N_9$ and $N_6' = N_7 + N_8$. If we evaluate these and note that on the curvilinear surface element $\eta_\alpha = a_\alpha \ (\alpha = 1, 2, 3)$ with $a_1 + a_2 + a_3 = 1$, the shape functions we require (Fig. 2.12a) are found to be

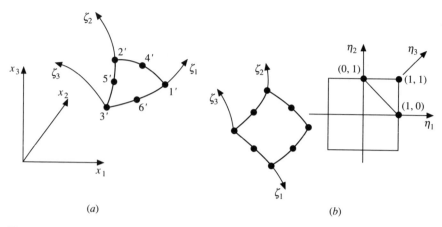

Figure 2.12 Curvilinear boundary elements.

identical to those for a second-order triangle in two dimensions. Exactly in a similar way the curvilinear quadrilateral elements can be handled and the shape functions we require can be shown to be those listed in Eqs (2.58) to (2.60). On reflection, this must be so, and to deal with such curvilinear surface elements we need to notice only that the Cartesian coordinates x_i of an arbitrary point on the element are defined in terms of the nodal Cartesian coordinates x_i^k ($k = 1, 2, \ldots, n$, for example) and shape functions $N^k(\eta)$:

$$x_i(\eta) = N^k(\eta) x_i^k \tag{2.66}$$

where the shape functions $N^k(\eta)$ can be linear (flat elements), quadratic, cubic, etc., defined in Eqs (2.52) to (2.54), (2.55) to (2.57) and (2.58) to (2.60).

By differentiating Eq. (2.66) with respect to η (the local axis directed along a line boundary element) we obtain a vector defining the tangent to the element at the point η:

$$s_i(\eta) = \frac{dx_i}{d\eta} = \frac{dN^k(\eta)}{d\eta} x_i^k \tag{2.67}$$

The Jacobian $J(\eta)$ is equal to $\sqrt{(dx_i/d\eta)(dx_i/d\eta)}$ and the normals are given by either $(s_2, -s_1)$ or $(-s_2, s_1)$ depending on which side of the element the body lies.

For a surface boundary element in a three-dimensional analysis we can similarly construct a local axes system η_j ($j = 1, 2$), and differentiating Eq. (2.66) with respect to η_j vectors, tangent to the coordinate lines, are obtained

$$s_{ij}(\eta) = \frac{\partial x_i}{\partial \eta_j} = \frac{\partial N^k(\eta)}{\partial \eta_j} x_i^k \tag{2.68}$$

A vector normal to the element at η can be obtained by taking the cross-product of the two vectors defined by Eq. (2.68), and the Jacobian is the modulus of this cross-product.

2.5 NUMERICAL INTEGRATION

2.5.1 Introduction

The purpose of this section is simply to present a short summary of nodal positions and their weights for Gaussian quadrature formulae in the transformed space, which will be useful in evaluating various integrals over elements and cells in BEMs. The fundamental improvement which Gauss–Legendre methods of numerical integration achieve over the commoner methods (Simpson's trapezoidal rules, etc.) is that equivalently precise results can be obtained while using essentially only half the number of ordinates. This is a consequence of adopting as parameters in the formulae not only the weights applied to each ordinate but also the location (nodes) of the ordinates with the range of integration (Fig. 2.13).

Therefore, if we wish to approximate the following integral (over a one-dimensional domain) by the summation shown, as we modify the number of ordinates used not only will their weights (A_i) change but also the nodal coordinates (x_i) will have specific optimal values:

$$I_1 = \int_{-1}^{1} f(x)\, dx \simeq \sum A_i f(x_i) \qquad (2.69)$$

As an example consider three nodes ($i = 3$); they will be located at points (x_i)($0, \pm \sqrt{15}/5$) within the range of integration with weights (A_i)($8/9, 5/9, 5/9$). In equations like (2.69) summation over the range (i) will be implied. Such Gaussian formulae will integrate polynomial expressions exactly up to the order ($2i - 1$) (that is $i = 3$ is exact for a quintic polynomial) and the errors are therefore of order $d^{2i}f/dx^{2i}$.

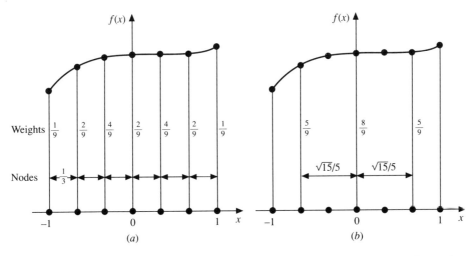

Figure 2.13 (*a*) Equally spaced ordinates with Simpson's rule weights. (*b*) Gaussian scheme, three nodes, exact for fifth-order polynomial.

Discussion of the underlying mathematics will be found in Lanczos[8] and Stroud and Secrest[9]. All the tables in this section have been taken from Refs 8 to 10.

2.5.2 Numerical Integration Formulae

Repeated application of Eq. (2.69) allows us to use the basic weights and nodes given in Table 2.1 for both two- and three-dimensional integration via, for rectangles,

$$I_2 = \int_{-1}^{1} \int_{-1}^{1} f(x_1, x_2) \, dx_1 dx_2 \simeq \sum_i \sum_j A_i A_j f(x_i, x_j) \qquad (2.70)$$

and, for hexahedral cells,

$$I_3 = \int_{-1}^{1} \int_{-1}^{1} \int_{-1}^{1} f(x_1, x_2, x_3) \, dx_1 dx_2 dx_3 \simeq \sum_i \sum_j \sum_k A_i A_j A_k f(x_i, x_j, x_k) \quad (2.71)$$

By combining a triangular pattern with linear ones it is clearly possible to derive integration schemes for triangular-prismatic elements much as related parametric representations were combined in Sec. 2.4.

Finally, Table 2.2 provides weights and nodal coordinates for a form particularly useful in planar BEM problems involving the singular function log r:

$$I_4 = \int_0^1 \ln\left(\frac{1}{x}\right) f(x) \, dx \simeq \sum A_i f(x_i) \qquad (2.72)$$

Since the fundamental solutions in two-dimensional problems all contain logarithmic terms, Eq. (2.72) is useful for integrating products involving them when the load and field points fall within the same element. Note that the range of integration is $(0, 1)$ and therefore the boundary element may have to be divided through the singular point into two sub-elements in order to use Eq. (2.72).

2.6 SOME EXAMPLES OF GEOMETRIC TRANSFORMATION AND INTEGRATION

In order to illustrate essential elements of transformation and integration we consider below two simple exercises involving the determination of lengths and areas.

2.6.1 Length of an Arc

Figure 2.14 shows the arc defined by three nodes in the physical space which is mapped into the intrinsic coordinate space in Fig. 2.14b. The coordinates in the real space are then given by

$$x_i = [N_1 \ N_2 \ N_3] \begin{Bmatrix} x_i^1 \\ x_i^2 \\ x_i^3 \end{Bmatrix} \qquad i = 1, 2$$

Table 2.1 Table of weights and nodal coordinates for $\int_{-1}^{1} f(x)\,\mathrm{d}x \simeq A_i f(x_i)$

x_i		A_i
i = 2		
0.577 350 269 1	(1)	0.100 000 000 0
i = 3		
0.774 596 669 2		0.555 555 555 5
0.000 000 000 0		0.888 888 888 8
i = 4		
0.861 136 311 5		0.347 854 845 1
0.339 981 043 5		0.652 145 154 8
i = 5		
0.906 179 845 9		0.236 926 885 0
0.538 469 310 1		0.478 628 670 4
0.000 000 000 0		0.568 888 888 8
i = 6		
0.932 469 514 2		0.171 324 492 3
0.661 209 386 4		0.360 761 573 0
0.238 619 186 0		0.467 913 934 5
i = 7		
0.949 107 912 3		0.129 484 966 1
0.741 531 118 5		0.279 705 391 4
0.405 845 151 3		0.381 830 050 5
0.000 000 000 0		0.417 959 183 6
i = 8		
0.960 289 856 5		0.101 228 536 2
0.796 666 477 4		0.313 706 645 8
0.525 532 409 9		0.313 706 645 8
0.183 434 642 6		0.362 683 783 3
i = 9		
0.968 160 239 5	(−1)	0.812 743 883 6
0.836 031 107 3		0.180 648 160 6
0.613 371 432 7		0.260 610 696 4
0.324 253 423 4		0.312 347 077 0
0.000 000 000 0		0.330 239 355 0
i = 10		
0.973 906 528 5	(−1)	0.666 713 443 0
0.865 063 366 6		0.149 451 349 1
0.679 409 568 2		0.219 086 362 5
0.433 395 394 1		0.269 266 719 3
0.148 874 338 9		0.295 524 224 7
i = 11		
0.978 228 658 1	(−1)	0.556 685 671 1
0.887 062 599 7		0.125 580 369 4
0.730 152 005 5		0.186 290 210 9
0.519 096 129 2		0.233 193 764 5
0.269 543 155 9		0.262 804 544 5
0.000 000 000 0		0.272 925 086 7

Table 2.2 Table of weights and nodal coordinates for $\int_0^1 \ln(1/x) f(x)\,dx \simeq A_i f(x_i)$

	x_i		A_i
	i = 2		
	0.112 008 806 1		0.718 539 319 0
	0.602 276 908 1		0.281 460 680 9
	i = 3		
(−1)	0.638 907 930 8		0.513 404 552 2
	0.368 997 063 7		0.391 980 041 2
	0.766 880 303 9	(−1)	0.946 154 065 6
	i = 4		
(−1)	0.414 484 801 9		0.383 464 068 1
	0.245 274 914 3		0.386 875 317 7
	0.556 165 453 5		0.190 435 126 9
	0.848 982 394 5	(−1)	0.392 254 871 2
	i = 5		
(−1)	0.291 344 721 5		0.297 893 471 7
	0.173 977 213 3		0.349 776 226 5
	0.411 702 520 5		0.234 488 290 0
	0.677 314 174 5	(−1)	0.989 304 595 1
	0.894 771 361 0	(−1)	0.189 115 521 4
	i = 6		
(−1)	0.216 340 058 4		0.238 763 662 5
	0.129 583 391 1		0.308 286 573 2
	0.314 020 449 9		0.245 317 426 5
	0.538 657 217 3		0.142 008 756 5
	0.756 915 337 3	(−1)	0.554 546 223 2
	0.922 668 851 3	(−1)	0.101 689 586 9
	i = 7		
(−1)	0.167 193 554 0		0.196 169 389 4
	0.100 185 677 9		0.270 302 644 2
	0.246 294 246 2		0.239 681 873 0
	0.433 463 493 2		0.165 775 774 8
	0.632 350 988 0	(−1)	0.889 432 271 3
	0.811 118 626 7	(−1)	0.331 943 043 5
	0.940 848 166 7	(−2)	0.593 278 701 5
	i = 8		
(−1)	0.133 202 441 6		0.164 416 604 7
(−1)	0.797 504 290 1		0.237 525 610 0
	0.197 871 029 3		0.226 841 984 4
	0.354 153 994 3		0.175 754 079 0
	0.529 458 575 2		0.112 924 030 2
	0.701 814 529 9	(−1)	0.578 722 107 1
	0.849 379 320 4	(−1)	0.209 790 737 4
	0.953 326 450 0	(−2)	0.368 640 710 4

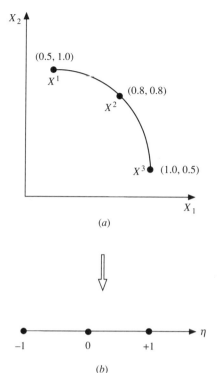

(a)

(b)

Figure 2.14 An arc (a) in physical space and (b) in the mapped space.

where

$$N_1 = \tfrac{1}{2}\eta(\eta - 1)$$
$$N_2 = (1 - \eta)(1 + \eta)$$
$$N_3 = \tfrac{1}{2}\eta(\eta + 1)$$

and

$$\frac{\mathrm{d}N_1}{\mathrm{d}\eta} = \eta - \frac{1}{2}$$
$$\frac{\mathrm{d}N_2}{\mathrm{d}\eta} = -2\eta$$
$$\frac{\mathrm{d}N_3}{\mathrm{d}\eta} = \eta + \frac{1}{2}$$

The Jacobian is then

$$|J| = \sqrt{\left(\frac{\mathrm{d}x_1}{\mathrm{d}\eta}\right)^2 + \left(\frac{\mathrm{d}x_2}{\mathrm{d}\eta}\right)^2}$$

where

$$\frac{\mathrm{d}x_1}{\mathrm{d}\eta} = \left[\frac{\mathrm{d}N_1}{\mathrm{d}\eta} \ \frac{\mathrm{d}N_2}{\mathrm{d}\eta} \ \frac{\mathrm{d}N_3}{\mathrm{d}\eta}\right] \begin{Bmatrix} x_1^1 \\ x_1^2 \\ x_1^3 \end{Bmatrix} = -0.1\eta + 0.25$$

and

$$\frac{\mathrm{d}x_2}{\mathrm{d}\eta} = \left[\frac{\mathrm{d}N_1}{\mathrm{d}\eta} \ \frac{\mathrm{d}N_2}{\mathrm{d}\eta} \ \frac{\mathrm{d}N_3}{\mathrm{d}\eta}\right] \begin{Bmatrix} x_2^1 \\ x_2^2 \\ x_2^3 \end{Bmatrix} = -0.1\eta - 0.25$$

The explicit expression for the Jacobian is then

$$|J| = \sqrt{(-0.1\eta + 0.25)^2 + (-0.1\eta - 0.25)^2}$$

$$\ell = \int_{-1}^{1} |J| d\eta = \int_{-1}^{1} \sqrt{0.02} \, (\eta^2 + 2.5^2)^{1/2} d\eta$$

which can be evaluated by integration formula to give $\ell = 0.7254$.

2.6.2 Area of a Curvilinear Triangle

The curvilinear area in the physical space (x_i) as shown in Fig. 2.15a can be transformed to the transformed space η_i as shown in Fig. 2.15b. The shape functions associated with this geometric transformation are:

$$N_1 = \eta_1(2\eta_1 - 1) \qquad N_2 = \eta_2(2\eta_2 - 1) \qquad N_3 = \eta_3(2\eta_3 - 1)$$
$$N_4 = 4\eta_1\eta_2 \qquad\qquad N_5 = 4\eta_2\eta_3 \qquad\qquad N_6 = 4\eta_3\eta_1$$

where $\eta_3 = 1 - \eta_1 - \eta_2$.

In the transformed space the elemental area is given by

$$dA = |J| d\eta_1 d\eta_2$$

where

$$|J| = \begin{vmatrix} \dfrac{\partial x_1}{\partial \eta_1} & \dfrac{\partial x_1}{\partial \eta_2} \\[2mm] \dfrac{\partial x_2}{\partial \eta_1} & \dfrac{\partial x_2}{\partial \eta_2} \end{vmatrix} = \dfrac{\partial x_1}{\partial \eta_1}\dfrac{\partial x_2}{\partial \eta_2} - \dfrac{\partial x_1}{\partial \eta_2}\dfrac{\partial x_2}{\partial \eta_1}$$

with

$$\frac{\partial \eta_3}{\partial \eta_1} = \frac{\partial \eta_3}{\partial \eta_2} = -1$$

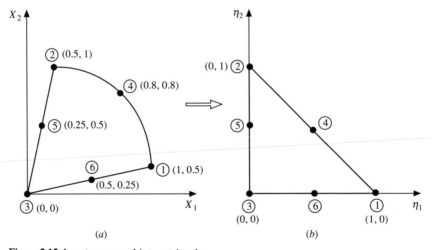

(a)　　　　　　　　　　　　　　(b)

Figure 2.15 A sector mapped into a triangle.

By utilizing the relationship between the spaces (x_i, η_i),

$$x_i = \sum_{j=1}^{6} N_j x_i^j \qquad i = 1, 2$$

we can calculate the necessary derivatives:

$$\frac{\partial x_1}{\partial \eta_1} = 0.2\eta_2 + 1$$

$$\frac{\partial x_2}{\partial \eta_1} = 0.2\eta_2 + 0.5$$

$$\frac{\partial x_1}{\partial \eta_2} = 0.2\eta_1 + 0.5$$

$$\frac{\partial x_2}{\partial \eta_2} = 0.2\eta_1 + 1$$

Consequently,

$$|J| = \frac{\partial x_1}{\partial \eta_1}\frac{\partial x_2}{\partial \eta_2} - \frac{\partial x_1}{\partial \eta_2}\frac{\partial x_2}{\partial \eta_1} = 0.1\eta_1 + 0.1\eta_2 + 0.75$$

The area A can therefore be determined by numerically integrating

$$A = \int_0^1 (0.1\eta_1 + 0.1\eta_2 + 0.75)\,\mathrm{d}\eta_1\mathrm{d}\eta_2 = 0.4083$$

2.7 CONCLUDING REMARKS

The two exercises shown above should allow readers to see that the operation over the transformed space is really quite simple.

It is very important to absorb the simple principles involved in indicial notations, integration by parts, geometric transformation and integration over the transformed space for line elements, surface patches and volume cells because all of these would be required in subsequent chapters in developing BEM solutions for practical problems.

REFERENCES

1. Banerjee, P.K. and Butterfield, R. (1981) *Boundary Element Methods in Engineering Science*, McGraw-Hill, London.
2. Epton, M.A. (1992) *Integration by parts for boundary element methods*, AIAA Journal, Vol. 30, No. 2, pp. 496–504.
3. Fung, Y.C. (1965) *Foundations of Solid Mechanics*, Prentice-Hall, Englewood Cliffs, New Jersey.
4. McCrea, W.H. (1953) *Analytical Geometry of Three-dimensions*, Oliver and Boyd, Edinburgh and London, and *Interscience*, New York.

5. Zienkiewicz, O.C. (1971) *The Finite Element Method in Engineering Science*, McGraw-Hill, London.
6. Norrie, H.N. and Vires, G. de (1973) *The Finite Element Method*, Academic Press, London.
7. Watson, J. (1979) Advanced implementation of the boundary element method for two and three-dimensional elasto-statics, Chapter III in *Developments in Boundary Element Methods*, Eds P.K. Banerjee and R. Butterfield, Elsevier Applied Science Publishers, London.
8. Lanczos, C. (1961) *Linear Differential Operators*, Van Nostrand, New York and London.
9. Stroud, A.H. and Secrest, D. (1966) *Gaussian Quadrature Formulas*, Prentice-Hall, Englewood Cliffs, New Jersey.
10. Hammer, P.C., Marlowe, O.P. and Stroud, A.H. (1956) Numerical integration over simplexes and cones', *Mathematical Tables Aids Compendium*, Vol. 10, pp. 130–137.

THREE

TWO- AND THREE-DIMENSIONAL PROBLEMS OF STEADY STATE POTENTIAL FLOW

3.1 INTRODUCTION

The main objective of the book is to guide the reader through classes of important problems which can be solved efficiently by BEMs and which become progressively more complicated, either by virtue of their dimensionality, their governing equations or purely by the use of higher-order numerical discretization schemes.

This chapter is the first of these steps and in it linear two- and three-dimensional steady state potential flow are analysed. The governing partial differential equations for such problems are elliptical (Laplace's equation or Poisson's equation) and relate to the simplest mathematical models of hydraulic flow and the flow of electricity, heat, etc. Each of these differential equations has to be satisfied by a potential function p (electrical or hydraulic potential or temperature), the spatial gradient of which is linearly related by a conductivity or permeability parameter to a flux or flow (electrical current density, fluid flow velocity or heat flux respectively).

The idea of formulating the solution to problems in potential theory in terms of integral equations appears in early texts,[1] but it is only very much more recently that this approach has been translated into a general numerical solution procedure.[2–17]

3.2 GOVERNING EQUATIONS

The general problem which we are to solve can be posed as follows (Fig. 3.1). A homogeneous region (V) with isotropic permeability (k) is bounded by a surface

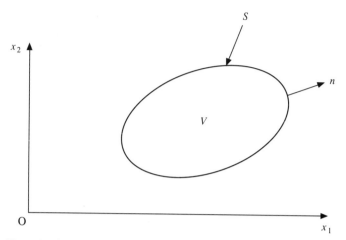

Figure 3.1 A volume V enclosed by a surface S.

(S), over one portion of which the boundary potential (p) is specified and the normal velocity (flux) component is q over the remainder of S. Sources, or sinks, of known ψ intensity per unit area or volume may also exist within V.

We are to determine the flow velocity and the potential existing at any specified point within V or on S. Referred to a rectangular Cartesian coordinate system ($x_i, i = 1, 2$ for two dimensions and $i = 1, 2, 3$ in three dimensions) the continuity equation for the flow at all points, other than sources or sinks, is Laplace's equation which has to be satisfied by the potential $p(x)$ at any point (x_i):

$$\frac{\partial^2 p(x)}{\partial x_1^2} + \frac{\partial^2 p(x)}{\partial x_2^2} = \frac{\partial^2 p(x)}{\partial x_i \, \partial x_i} = 0 \tag{3.1}$$

The corresponding flow 'velocity' vector components $v_i(x)$ are given variously by Darcy's law, Ohm's law, Fourier's law, etc., as

$$v_i(x) = -k \frac{\partial p(x)}{\partial x_i} \tag{3.2}$$

If all the boundary potentials were given around S (the Dirichlet problem) we would have, say,

$$p(x_0) = g(x_0) \qquad x_0 \text{ on } S \tag{3.3a}$$

or if all the velocity components normal to the boundary were specified (the Neumann problem) we would have, say,

$$q(x_0) = h(x_0) \qquad x_0 \text{ on } S \tag{3.3b}$$

where

$$q(x_0) = v_1 n_1 + v_2 n_2 = v_i(x_0) n_i(x_0)$$

with $n_i(x_0)$ the components of the outward normal unit vector at (x_0). The more general, mixed boundary value problem would have either p or q specified on portions of S or an inter-relationship of the form $q = -(p - p_0)/R$, where p_0 is the ambient potential and R is a resistance would be provided.

Before considering our fundamental solution of Eq. (3.1) we should note that all the following analysis is directly applicable to homogeneous anisotropic regions for which the quantity k will be a second-rank tensor quantity k_{ij}. The generalization of the flux equation becomes

$$v_i(x) = -k_{ij}\frac{\partial p(x)}{\partial x_j}$$

which leads to the continuity equation, via $\partial v_i(x)/\partial x_i = 0$,

$$k_{ij}\frac{\partial^2 p(x)}{\partial x_i \partial x_j} = 0$$

for spatially constant k_{ij}.

However, if we ensure that the axes of x_i are directed along the principal axes of k_{ij} (that is k_{ij} is diagonalized with principal values in two dimensions of k_1 and k_2, say) this equation simplifies to

$$k_1\frac{\partial p(x)}{\partial x_1^2} + k_2\frac{\partial p(x)}{\partial x_2^2} = 0$$

whence by substituting scaled coordinates (η_i) such that $[\eta_1 = x_1, \eta_2 = x_2(k_1/k_2)]$ Eq. (3.1) is recovered as $\partial^2 p(\eta)/\partial \eta_i \partial \eta_i = 0$ and the problem has been transformed into an isotropic one with $k = \sqrt{k_1 k_2}$. Similarly, in the three-dimensional case if the principal values of k_{ij} are $k, k\alpha, k\beta$ and the axes are scaled by $(1, 1/\alpha^{0.5}, 1/\beta^{0.5})$ the scaled equivalent isotropic permeability then becomes equal to $k(\alpha\beta)^{0.5}$. If this prior operation is assumed then the following analyses cover problems involving both isotropic and anisotropic regions.

3.3 SINGULAR SOLUTIONS

The fundamental solution of Eq. (3.1), which is a basic component of all the subsequent analyses, relates the potential $p(x)$ generated at a 'field point' (x_i) by a unit source $e(\xi)$, say, applied at a load point (ξ_i). Although the origin for the coordinates (ξ_i) is identical to that for (x_i) it is absolutely essential that each be reserved for a specific purpose. Thus the classical singular solution can be written as[1]

$$p(x) = G(x, \xi)e(\xi) \tag{3.4}$$

where

$$G(x, \xi) = -\frac{1}{2\pi k}\ln r \quad \text{and} \quad G(x, \xi) = \frac{1}{4\pi k}\frac{1}{r}$$

for two- and three-dimensional problems respectively and r is the distance between

x and ξ. Thus in two dimensions, for example,

$$r^2 = (x_1 - \xi_1)^2 + (x_2 - \xi_2)^2 = (x - \xi)_i(x - \xi)_i = y_i y_i$$

if $y_i = (x - \xi)_i$. $G(x, \xi)$ is a 'two-point' function involving the coordinates (x, ξ) of two points and since we shall sometimes need to differentiate, or integrate, G with respect to x and sometimes with respect to ξ, a distinction must be preserved between the two arguments.

By differentiating Eq. (3.4) with respect to x_i we obtain the flux (velocity) vector components $v_i(x)$:

$$v_i(x) = -k \frac{\partial p(x)}{\partial x_i} = \frac{1}{2\pi} \frac{y_i}{r^2} e(\xi) \qquad \left(= \frac{1}{4\pi} \frac{y_i}{r^3} e(\xi) \text{ for three dimensions} \right) \quad (3.5)$$

If $n_i(x)$ are the components of a unit vector at x_i defining the outward normal direction to an element through (x_i) the velocity $q(x)$ along n is then

$$q(x) = v_i n_i = \frac{y_i n_i}{2\pi r^2} e(\xi) = \frac{n_1 y_1 + n_2 y_2}{2\pi r^2} e(\xi) \quad (3.6a)$$

or $\qquad\qquad q(x) = F(x, \xi)e(\xi) \qquad$ say

where

$$F(x, \xi) = \frac{(x - \xi)_i n_i(x)}{2\pi r^2} = \frac{y_i n_i(x)}{2\pi r^2} \qquad \left(= \frac{y_i n_i}{4\pi r^3} \text{ in three dimensions} \right) \quad (3.6b)$$

It is worth noting that whereas $G(x, \xi)$ is symmetrical in its arguments (x, ξ), $F(x, \xi)$ is antisymmetrical and changes sign when x and ξ are interchanged. Also as $x_i \to \xi_i$ (i.e. as the field and load points coalesce) both the singular expressions, Eqs (3.4) and (3.6), become infinite. The first expression, involving $\log r$ (or $1/r$ in three dimensions), is only 'weakly singular' when integrated along a line element (or over an area in three dimensions) and it will be found that the singularity is suppressed when such a function is integrated across it in the normal way. However, the second expression, $F(x, \xi)$, involving a singularity of order $1/r$ (or $1/r^2$ in three dimensions) is 'strongly singular' when integrated along a line (or over an area in three dimensions) through the singular point and cannot be integrated successfully in the same way. Such singular functions play a major role in all BEMs and, as will be seen later, ensure the diagonal dominance of the resulting matrix equations.

3.4 DIRECT BEM FOR A HOMOGENEOUS REGION

Once more we start from the governing differential equation equivalent to Eq. (1.1) in two and three dimensions:

$$k \frac{\partial^2 p(x)}{\partial x_i \, \partial x_i} = -\psi(x) \quad (3.7)$$

where $\psi(x)$ is now a specified distribution of source strengths over the domain (V) of our problem.

The following analysis is a generalization of the one developed in Sec. 1.3 where now $G(x, \xi)$ is a two-point function specified to be the solution of the equation

$$k\frac{\partial^2 G(x, \xi)}{\partial x_i\, \partial x_i} + \delta(x, \xi) = 0 \qquad (3.8)$$

Here $\delta(x, \xi)$ is the Dirac delta function which is zero unless all corresponding components of x_i and ξ_i are identical. When $x_i \equiv \xi_i$, $\delta(x, \xi)$ has a sifting property such that, for example,

$$\int_V p(x)\delta(x, \xi)\, dV(x) = p(\xi) \qquad \text{with } \xi \text{ within } V \qquad (3.9)$$

i.e. when the left-hand-side product in this equation is integrated over V the value of $p(\xi)$ at the specific point $x_i = \xi_i$ is sifted out as the only non-zero resultant. In Eq. (3.8) the function G has to be the ubiquitous singular solution for the potential generated at x_i by a unit point source applied at the point ξ_i in an infinite body given by Eq. (3.4).

We now multiply both sides of Eq. (3.7) by G and integrate by parts twice over the region V (cf. Secs 1.3 and 2.3). This will again generate an equation expressing $p(\xi)$ in terms of the boundary information on S and derivatives of G. Thus for example:

$$\int_V \left(G\frac{\partial^2 p}{\partial x_i\, \partial x_i}\right) dV = \int_S \left(G\frac{\partial p}{\partial x_i} - p\frac{\partial G}{\partial x_i}\right) n_i\ dS + \int_V \left(p\frac{\partial^2 G}{\partial x_i\, \partial x_i}\right) dV \qquad (3.10)$$

where from Eqs (3.8) and (3.9) the last term reduces to $-p(\xi)/k$.

Thus by using Eqs (3.7) and (3.8) in the above we obtain, when written out fully,

$$p(\xi) = k\int_S \left[G(x, \xi)\frac{\partial p(x)}{\partial x_i} - p(x)\frac{\partial G(x, \xi)}{\partial x_i}\right] n_i(x)\ dS(x) + \int_V G(x, \xi)\psi(x)\, dV(x) \qquad (3.11)$$

If, as before, we define

$$F(x, \xi) = -k\left[\frac{\partial G(x, \xi)}{\partial x_i}\right] n_i(x)$$

and recall that the flux in the n_i direction is

$$q(x) = -k\left[\frac{\partial p(x)}{\partial x_i}\right] n_i(x)$$

then Eq. (3.11) can be written more concisely as

$$p(\xi) = \int_S \left[p(x)F(x, \xi) - q(x)G(x, \xi)\right] dS(x) + \int_V \psi(x)G(x, \xi)\, dV(x) \qquad (3.12)$$

an equation that enables us to calculate the potential at any point ξ from a knowledge of both the potential and the flux at all points around the boundary S and the specified internal source distribution $\psi(x)$.

If we now imagine the point ξ_i to approach the boundary S from inside V, then Eq. (3.12) becomes the limiting value, as $\xi \rightarrow \xi_0$, of

$$p(\xi_0) = \int_S p(x)F(x, \xi_0)\,dS(x) - \int_S q(x)G(x, \xi_0)\,dS(x) + \int_V \psi(x)G(x, \xi_0)\,dV(x)$$

$$(3.13)$$

All the integrals have singularities whenever x and ξ_0 coincide, although it will be found subsequently that those involving G, with a weak singularity, can be evaluated without difficulty either analytically or numerically. However, the F boundary integral is improper and has a strong singularity of the order $1/r$ in two dimensions and $1/r^2$ in three dimensions. Therefore this has to be evaluated as

$$\oint_S p(x)F(x, \xi_0)\,dS(x) = \tfrac{1}{2}p(\xi_0) + \int_S p(x)F(x, \xi_0)\,dS(x) \qquad (3.14)$$

for a smooth surface point ξ_0.

The following simple evaluation of the 'free term' $\tfrac{1}{2}p(\xi_0)$ in Eq. (3.14) may help the reader to understand why it is necessary to interpret the integral involving F in this way. Figure 3.2a shows a smooth portion of the boundary S of a two-dimensional region near ξ_0 which has been extended by a small semi-circular region S_ϵ, of radius $\epsilon \rightarrow 0$, to envelope ξ_0, which is actually on the boundary. We are to calculate the singular portion of $\oint_S p(\xi_0)F(x, \xi_0)\,dS(x)$ which arises when $\xi \rightarrow \xi_0$ on S from inside V. If ξ_0 is the origin of a local coordinate system then,

$$F(x, \xi_0) = \frac{(x - \xi_0)_i}{2\pi r^2}n_i(x)$$

$$= \frac{x_i n_i}{2\pi r^2} = \frac{x_i(x_i/r)}{2\pi r^2} = \frac{1}{2\pi\epsilon}$$

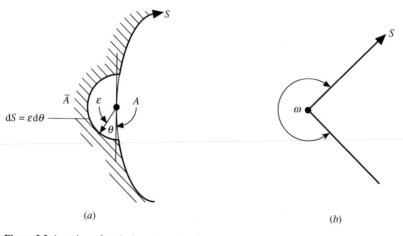

$$dS = \epsilon\,d\theta$$

(a)   (b)

Figure 3.2 A region of exclusion around a sharp corner.

and $dS(x) = \varepsilon\,d\theta$, whence

$$\int_{S_\varepsilon} p(\xi_0)F(x,\xi_0)\,dS(x) = p(\xi_0)\int_0^\pi \frac{1}{2\pi\varepsilon}\varepsilon\,d\theta = \frac{1}{2}p(\xi_0) \tag{3.15}$$

Equation (3.14) therefore accounts for this 'free term' separately. It is worth noting here (Fig. 3.2b) that if S does not have a unique tangent at ξ_0 then the limits of integration for the free term will not be 0 to π and the coefficient $p(\xi_0)$ will not be $\frac{1}{2}$. This situation can arise at corners or junctions of linearly approximated boundary elements; e.g. from Fig. 3.2b the limits of integration are now 0 and ω and the 'free term' is $(\omega/2\pi)p(\xi_0)$. It is perhaps appropriate to mention also that, for potential flow problems in three dimensions, the 'free term' can be evaluated in an almost identical fashion by considering a hemi-spherical region, of radius $\varepsilon \to 0$, enclosing ξ_0. We then have $G(x,\xi) = 1/(4\pi k r)$ and therefore $F(x,\xi) = y_i n_i(x)/4\pi r^3$ so that, as $\xi \to \xi_0$,

$$F(x,\xi_0) = \frac{x_i n_i}{4\pi\varepsilon^3} = \frac{1}{4\pi\varepsilon^2} \qquad \text{and now} \qquad dS = 2\pi\varepsilon^2 \cos\theta\,d\theta$$

whence

$$\int_{\Delta S} p(\xi_0)F(x,\xi_0)\,dS(x) = p(\xi_0)\int_0^{\pi/2} \frac{\cos\theta}{2}\,d\theta = \frac{1}{2}p(\xi_0) \tag{3.15a}$$

once more. For the equivalent three-dimensional case corresponding to Fig. 3.2b the 'free term' equals $(\omega/4\pi)p(\xi_0)$ where ω is now the solid angle at the point ξ_0.

Substitution of Eq. (3.14) into Eq. (3.13) produces the boundary integral equation in the form we require:

$$cp(\xi_0) = \int_S [p(x)F(x,\xi_0) - q(x)G(x,\xi_0)]\,dS(x) + \int_V \psi(x)G(x,\xi_0)\,dV(x) \tag{3.16}$$

where $c = \frac{1}{2}$ for a smooth surface point x_0.

In principle this identity enables us to use the specified boundary values and internal sources to calculate the remaining unspecified boundary data. Once all the boundary information is known (i.e. both p and q around S) it can be used in Eq. (3.12) to calculate $p(\xi)$ at any point inside V. We can also calculate $q(\xi)$ by recalling that $q(\xi) = -k[\partial p(\xi)/\partial\xi_i]n_i(\xi)$ and differentiating the ξ dependent terms in Eq. (3.12) under the integral sign to produce

$$q(\xi) = \int_S [p(x)H(x,\xi) - q(x)F(x,\xi)]\,dS(x) + \int_V \psi(x)F(x,\xi)\,dV(x) \tag{3.17}$$

which now involves the differentiation of $F(x,\xi)$ at ξ_i and a new expression

$$H(x,\xi) = -k\frac{\partial F(x,\xi)}{\partial\xi_j}n_j(\zeta) = k^2\frac{\partial^2 G(x,\xi)}{\partial x_i\,\partial\xi_j}n_i(x)n_j(\xi)$$

whence for the two-dimensional case

$$H(x,\xi) = \frac{k}{2\pi r^2}\left(\frac{2y_i y_j}{r^2} - \delta_{ij}\right)n_i(x)n_j(\xi) \tag{3.18}$$

with $y_i = (x - \xi)_i$ and δ_{ij} the Kronecker delta symbol. Equation (3.18) is obtained by differentiating $F(x, \xi) = y_j n_j(x)/2\pi r^2$ as follows:

$$H(x, \xi) = \frac{-k}{2\pi} n_i(x) n_j(\xi) \frac{\partial(y_i/r^2)}{\partial \xi_j} = \frac{-k}{2\pi} n_i(x) n_j(\xi) \left(\frac{1}{r^2} \frac{\partial y_i}{\partial \xi_j} - \frac{2y_i}{r^3} \frac{\partial r}{\partial \xi_j} \right)$$

but $\partial y_i / \partial \xi_j = -\delta_{ij}$ and since $r^2 = y_k y_k$, $\partial r / \partial y_k = y_k/r$. Therefore,

$$\frac{\partial r}{\partial \xi_j} = \frac{\partial r}{\partial y_k} \frac{\partial y_k}{\partial \xi_j} = -\frac{y_k}{r} \delta_{jk} = -\frac{y_j}{r}$$

and

$$H(x, \xi) = \frac{k}{2\pi r^2} n_i(x) n_j(\xi) \left(\frac{2 y_i y_j}{r^2} - \delta_{ij} \right)$$

Results of this kind are very tedious to derive without using indicial notation since the above contains eight distinct product terms when written out fully (two of which will be zero). The reader is therefore urged to master the few simple rules involved shown in Chapter 2 at this stage, in preparation for the rather more extensive use of the notation that arises in elasticity, etc.

3.4.1 Discretization of the Surface Integral and Formation of the System Matrices

The key problem is the solution of Eq. (3.16) using the specified boundary conditions to calculate the remaining, initially unknown, boundary values of p and q.

We shall consider first the simplest flat boundary element discretization scheme (Fig. 3.3) with constant distributions of variables over the elements and ignore for the time being the contribution of ψ over volume. If we discretize the boundary S into N such elements with n being a typical element, then Eq. (3.16) can be written for the mth element on the boundary as

$$\tfrac{1}{2} p(\xi_0^m) = \sum_{n=1}^{N} p^n \int_{\Delta S} F(x^n, \xi_0^m) \, dS(x) - \sum_{n=1}^{N} q^n \int_{\Delta S} G(x^n, \xi_0^m) \, dS(x) \qquad (3.19)$$

The matrix equivalent for Eq. (3.19) is then

$$\tfrac{1}{2} p^m = \left(\int_{\Delta S} F^{nm} \, dS \right) p^n - \left(\int_{\Delta S} G^{nm} \right) q^n \qquad (3.20)$$

where p^m and q^n are $N \times 1$ columns and the bracketed terms compatible row vectors.

The integral terms in the row vectors may be evaluated either numerically, in all cases, or analytically for the simpler F and G functions, as shown later in this chapter. After carrying out the summations the matrix equation can then be written

$$\tfrac{1}{2} p^m = (\bar{F}^{nm}) p^n - (G^{nm}) q^n$$

or

$$(F^{nm}) p^n - (G^{nm}) q^n = 0 \qquad (3.21)$$

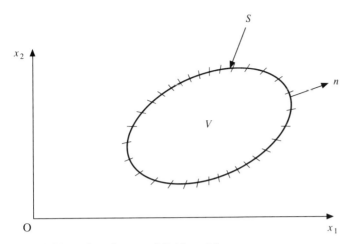

Figure 3.3 Boundary element subdivision of S.

if the $\frac{1}{2}p^m$ term is absorbed into the $m = n$ position of the F matrix. By allowing m to range over $1, \ldots, N$ we obtain the complete set of all equations

$$\mathbf{F}^s\mathbf{p} - \mathbf{G}^s\mathbf{q} = 0 \tag{3.22}$$

Again the superfix s denotes a matrix derived from surface integrals of F and G over surface elements, and $\mathbf{p}$ and $\mathbf{q}$ are vectors of the boundary potentials and fluxes.

The boundary value problems will have either the potentials specified at some boundary points (hence p is known) and the fluxes specified on the remaining ones (in which case q is known). The unspecified boundary information, $\mathbf{p}$ or $\mathbf{q}$ respectively, can be calculated directly from Eq. (3.22). Equation (3.22) should preferably be scaled by $\alpha \simeq k$, to ensure that the terms in the F and G matrices are of comparable magnitude: namely

$$\mathbf{F}^s\mathbf{p} - \alpha\mathbf{G}^s(\mathbf{q}/\alpha) = 0 \tag{3.22a}$$

Such a scaling can of course also be achieved automatically by working throughout with a suitable set of dimensionless variables.

It is then a very simple operation to rearrange Eq. (3.22a) so that the known p and q values form one vector $\mathbf{Y}$ of size $(N \times 1)$ and the unknown p and q values another $N \times 1$ vector $\mathbf{X}$, whence Eq. (3.22a) can be rewritten as

$$\mathbf{AX} = \mathbf{BY} \tag{3.23}$$

and solved for $\mathbf{X}$, with the result that all p and q components are now known around S.

As mentioned previously, if the boundary does not have a unique tangent at any point where $F(x, \xi_0)$ has to be evaluated (this is required in all higher-order

BEM analyses), then the coefficient of the free term in Eq. (3.16), etc., will not be $\frac{1}{2}$ but an unknown quantity, c say. Equation (3.21) would then be

$$cp^m = \bar{F}^{nm}p^n - (G^{nm})q^n \tag{3.24}$$

and if the cp^m term is absorbed into the leading diagonal terms of $\bar{F}^{nm}$ we arrive at Eq. (3.22) once more, from which the value of the diagonal term (including c) can be deduced directly by considering the effect of a uniform unit potential, applied around the whole boundary S. Clearly this produces zero boundary flux ($q^n = 0$) and Eq. (3.22) then reduces to $F^s = 0$, which implies that the sum of the components in each individual row of F^s has to be zero [see, for example, Eq. (1.10b)]. Therefore the diagonal term is found directly from this requirement, which involves only the summation of the off-diagonal terms of $\bar{F}^{nm}$ with a change in sign and therefore, if round-off errors are not significant, eliminates the need to consider these complex integrals. In fact it is possible to view this process differently using the exact integral representation given by Eq. (3.12) or Eq. (3.16) as opposed to the final discretized form of Eq. (3.24). Thus if we consider the application of uniform potential p with no source distribution present in the body (that is, $\psi = 0$) we shall have $q = 0$. Applying this condition to Eq. (3.12), for example, leads to

$$p(\xi) = \int_S F(x, \xi)p(x)\,\mathrm{d}S(x)$$

By taking the field point ξ to a surface point ξ_0 we can express this in the usual manner (i.e. considering a small exclusion S_ϵ around ξ_0 with limit $S_\epsilon \to 0$):

$$p(\xi_0) = \int_{S_\epsilon} F(x, \xi)p(\xi_0)\,\mathrm{d}S(x) + \int_{S-S_\epsilon} F(x, \xi)p(x)\,\mathrm{d}S(x)$$

or

$$p(\xi_0)\left[1 - \int_{S_\epsilon} F(x, \xi)\,\mathrm{d}S(x)\right] = \int_{S-S_\epsilon} F(x, \xi)p(x)\,\mathrm{d}S(x)$$

where the terms within the square bracket equal c. Thus,

$$cp(\xi_0) = \int_S F(x, \xi)p(x)\,\mathrm{d}S(x)$$

or

$$c = \int_S F(x, \xi)\,\mathrm{d}S(x)$$

since $p(x) = p(\xi_0) =$ unity.

Clearly we can now substitute c back into the original boundary integral representation (3.16) to produce

$$p(\xi_0)\int_S F(x, \xi)\,\mathrm{d}S(x) = \int_S [F(x, \xi)p(x) - G(x, \xi)q(x)]\,\mathrm{d}S(x) + \int_V \psi(x)G(x, \xi)\,\mathrm{d}V(x)$$

or

$$\int_S \{F(x, \xi)[p(x) - p(\xi_0)] - G(x, \xi)q(x)\}\,\mathrm{d}S(x) + \int_V \psi(x)G(x, \xi)\,\mathrm{d}V(x) = 0 \tag{3.25}$$

which looks like a 'new' integral formulation in which all of the integrals involved are no worse than weakly singular. In reality, however, this is not a new formulation but an application of the 'equipotential principle', i.e. the operations described below Eq. (3.24), to the intact integral representation. Thus, if Eq. (3.25) is discretized by the same numerical process which led to Eq. (3.24) we shall find that the coefficients of the final matrices would be identical.

3.4.2 Calculation of Internal Potentials and Velocities

Once all the components of p and q (on the boundary) are known then backsubstitution into Eq. (3.12) or Eq. (3.17) will yield the potential $p(\xi)$ or the n direction flux $q(\xi)$ at any subsequently selected point ξ_i within V. These equations are again most usefully expressed in a discretized form. Following the discretization procedure explained previously, two equations then become, for, say, a selected internal point at ξ^r,

$$p(\xi^r) = \sum_{n=1}^{N} p^n \int_{\Delta S} F(x^n, \xi^r) \, dS(x) - \sum_{n=1}^{N} q^n \int_{\Delta S} G(x^n, \xi^r) \, dS(x)$$

that is

$$p^r = (F^{nr})p^n - (G^{nr})q^n \tag{3.26}$$

where all the components of the F and G matrices can be evaluated by standard integration since ξ^r never coincides with x^q. When treated similarly Eq. (3.17) becomes

$$q^r = (H^{rn})p^n - (F^{rn})q^n \tag{3.27}$$

A further point of interest arises from Eq. (3.22) which can be written as (note, however, that we only can do this here because $\mathbf{G}^s$ is a square matrix, which is not often the case)

$$\mathbf{q} = [\mathbf{G}^s]^{-1}\mathbf{F}^s\mathbf{p} = \mathbf{K}\mathbf{p} \tag{3.28}$$

Equation (3.28) then relates boundary fluxes and potentials in a way closely similar to a finite element formulation, although we now have just a 'super-element' representing the whole of one homogeneous region of any shape. We shall show, later, how to assemble such zonal super-elements together to solve problems involving 'piecewise' homogeneous bodies by the BEM.

However, before doing this we shall establish, below, a complete set of all the intermediate integrals of the type $\int_{\Delta S} G^{mn} \, dS$, etc., which have to be evaluated to utilize Eqs (3.19), (3.26) and (3.27). These are analytical solutions for planar elements with functions uniformly distributed over them.

3.4.3 Integrals over Boundary Elements

Even when the boundary S of any problem has been discretized, ready for a matrix approximation to the governing integral equations, the fundamental solution G

and its derivatives F, H have still to be integrated along individual boundary elements. The following intermediate integrals of this kind will be required:

$$\int_{\Delta S} G^{nm}\,dS \qquad \int_{\Delta S} F^{nm}\,dS \qquad \text{and} \qquad \int_{\Delta S} H^{nr}\,dS$$

For the simpler constitutive equations and simple discretization schemes, particularly those incorporating uniformly distributed potentials and fluxes, etc., over straight line boundary elements in two dimensions and flat triangular or rectangular surface elements in three dimensions all the above integrals can be evaluated analytically. They can, of course, also be approximated to any degree of accuracy by numerical integration methods, as can integrals involving more complex functions distributed over curved elements. We shall discuss numerical quadrature, higher-order elements, etc., in detail later in this chapter.

Meanwhile, the F, G, H functions which occur in potential flow are particularly amenable to analytical integration when distributed over line and triangular or rectangular elements. This analytical treatment also helps to develop an understanding of the problems associated with singular functions.

Two-dimensional Cases   In all cases we shall set up a local coordinate system (Fig. 3.4a), usually with our field point ξ^m at the origin. Whenever a unit vector $n_i(x)$ is introduced we assume that it will have been transformed from the global to the local coordinate system.

1. $\int_{\Delta S} G(x^n, \xi^m)\,dS(x)$, where

$$G(x, \xi) = -\frac{1}{2\pi k}\ln\left|\frac{r}{r_0}\right|$$

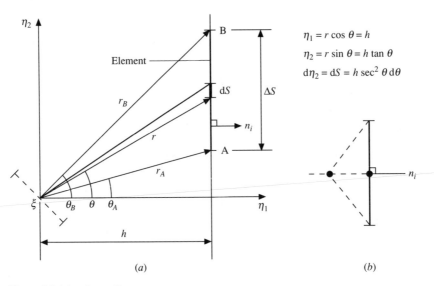

Figure 3.4 A local coordinate system over a line element.

with r_0 an arbitrary constant which can be taken as unity in the present case. In local coordinates, we therefore have

$$\int_{\Delta S} G(x^n, \xi^m)\, dS = \int_{\Delta S} G(\eta, 0)\, d\eta_2 = \int_{\theta_A}^{\theta_B} -\frac{h}{2\pi k} \ln \left| \frac{h \sec \theta}{r_0} \right| \sec^2\theta\, d\theta$$

$$= -\frac{h}{2\pi k} \left[\tan\theta \left(\ln \left| \frac{h \sec\theta}{r_0} \right| - 1 \right) + \theta \right]_{\theta_A}^{\theta_B}$$

Thus

$$\int_{\Delta S} G(x^n, \xi^m)\, dS = -\frac{1}{2\pi k} \left[r \sin\theta \left(\ln \left| \frac{r}{r_0} \right| - 1 \right) + \theta h \right]_{\theta_A, r_A}^{\theta_B, r_B} \tag{3.29}$$

Equation (3.29) provides the value of the potential generated at any field point ξ^m by a uniformly distributed source of unit intensity along AB. The field point ξ^m could be the midpoint of, say, the nth boundary element as required in Eq. (3.19). The summations over all boundary elements always involve the situation when $n = m$ (that is ξ^m approaches the midpoint of the nth boundary element) and such diagonal components of the $[G]$ matrices, etc., must be evaluated with care because all the kernel functions are singular when $x_i = \xi_i$.

By approaching ξ^m from inside V (Fig. 3.4b) we have $h \to 0$, $\theta_B = -\theta_A \to \pi/2$, $r_B = r_A \to \Delta S/2 = b$ (say), when Eq. (3.29) becomes

$$\int_{\Delta S} G(x, \xi)\, dS = -\frac{1}{2\pi k} \left[2b \left(\ln \frac{b}{r_0} - 1 \right) \right] = -\frac{b}{\pi k} \left(\ln \frac{b}{r_0} - 1 \right) \tag{3.30}$$

2. $\int_{\Delta S} F(x^n, \xi^m)\, dS(x)$. Now $F(x, \xi) = [(x - \xi)_i/2\pi r^2]n_i(x)$ and we have $n_1(\eta) = 1$, $n_2(\eta) = 0$. Therefore,

$$\int_{\Delta S} F(x, \xi)\, dS = \int_{\Delta S} F(\eta, 0)\, d\eta_2 = \frac{1}{2\pi} \int_{\theta_A}^{\theta_B} \frac{h(h \sec^2\theta)}{h^2 \sec^2\theta}\, d\theta = \left[\frac{\theta}{2\pi} \right]_{\theta_A}^{\theta_B} \tag{3.31}$$

and when ξ^m approaches the midpoint of the element ΔS,

$$\int_{\Delta S} F(x^n, \xi^m)\, dS(x) = -\tfrac{1}{2}$$

as x^n and ξ^m coalesce. In this particular case, this is the only non-zero contribution arising from the strongly singular integral and forms the leading diagonal of the F^{nm} matrix.

3. $\int_{\Delta S} H(x^n, \xi^r)\, dS(x)$. Here we have used ξ^r for the field point to emphasize that this integral only arises in the calculations by direct BEMs, of fluxes at points ξ^m not on the boundary S. From Eq. (3.18),

$$H(x, \xi) = -\frac{k}{2\pi} n_i(\xi) \left[\frac{n_j(x)}{r^2} \left(\delta_{ij} - \frac{2y_i y_j}{r^2} \right) \right]$$

where $y_i = (x_i - \xi_i)$. In our local coordinate system, $n_1(\eta) = 1$, $n_2(\eta) = 0$, this equation becomes

$$H(\eta, 0) = -\frac{k}{2\pi} \left[n_1(\xi) \left(\frac{1}{r^2}\right) \left(1 - \frac{h^2}{r^2}\right) - n_2(\xi) \left(\frac{1}{r^2}\right) \left(\frac{h\eta_2}{r^2}\right) \right]$$

Therefore, from Fig. 3.4a and recollecting that $d\eta_2/r^2 = d\theta/h$, we have

$$\int_{\Delta S} H(x, \xi) \, dS = -\int_{\Delta S} H(\eta, 0) \, d\eta_2$$

$$= -\int_{\theta_A}^{\theta_B} \frac{k}{2\pi h} \left[n_1(1 - \cos^2 \theta) - n_2(\sin \theta \cos \theta) \right] d\theta$$

$$= -\frac{k}{8\pi h} \left[n_1(2\theta - \sin 2\theta) + n_2 \cos 2\theta \right]_{\theta_A}^{\theta_B} \qquad (3.32)$$

Equation (3.32) presents no problems for any field point within V which is what we require. However, it is of interest to note that the strong singularity in H of order $1/r^2$ does mean that for near surface points the evaluation of this integral presents a formidable numerical difficulty in the direct BEM (DBEM). Equations (3.29) to (3.32) provide analytical expressions for all the element line integrals required to compile all the matrices G^{nm}, F^{nm} and H^{rm}.

The difficulty of evaluating the flux at the near surface points mentioned above can, however, be easily overcome if we apply the 'equipotential technique' developed earlier (that is $p = $ unity and therefore $q = 0$) to Eq. (3.27) which then reduces to

$$H^{rn} = 0$$

It is then clearly possible to evaluate the most difficult coefficient of the H matrix as a sum of the coefficients of all other components of the row r with a change in sign.

Three-dimensional integrals　In three-dimensional cases the simplest surface elements are usually flat triangles or rectangles covering the entire surface. We now have to integrate our G and F functions over these elements.

First we note the following property of r in three-dimensions:

$$\frac{\partial^2 r}{\partial x_i \, \partial x_i} = \frac{2}{r}$$

Therefore

$$\Delta G = \int_{\Delta S} G(x, \xi_0) \, dS = \frac{1}{4\pi k} \int_{\Delta S} \frac{1}{r} \, dS = \frac{1}{8\pi k} \int_{\Delta S} \frac{\partial^2 r}{\partial x_i \, \partial x_i} \, dS$$

$$= \frac{1}{8\pi k} \int_C \frac{\partial r}{\partial x_i} e_i \, dC$$

where C is the contour surrounding the elemental surface ΔS and e_i the global components of the outward normal to C.

Similarly for the flux kernel since the normal does not vary over our element we have

$$\Delta F = \int_{\Delta S} F(x, \xi)\, dS = -\frac{1}{4\pi} \int_{\Delta S} \frac{\partial}{\partial x_i}\left(\frac{1}{r}\right) n_i\, dS$$

$$= -\frac{n_i}{4\pi} \int_{\Delta S} \frac{\partial}{\partial x_i}\left(\frac{1}{r}\right) dS = -\frac{n_i}{4\pi} \int_C \frac{1}{r} e_i\, dC$$

By constructing a local polar coordinate system in the plane of the element as shown in Fig. 3.5 the line integrals over the side L_m between the nodes $m, m+1$ can be evaluated as[18]

$$\Delta G = \frac{1}{4\pi k}\left[G_{m, m+1} + |z|\alpha\right] \qquad (3.33a)$$

$$\Delta F = \frac{1}{4\pi}\left[F_{m, m+1} + \text{sign}(z)\alpha\right] \qquad (3.33b)$$

where $\alpha = 0$ if the projected point does not lie on L_m and $\alpha = 2\pi$, if it does fall on

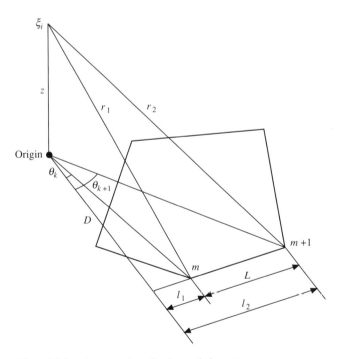

Figure 3.5 Local parameters of polygonal element.

the line L_m and

$$G_{m,m+1} = D \log \frac{r_1 + r_2 + L}{r_1 + r_2 - L} - |z| \tan^{-1}(|z|y)$$

$$F_{m,m+1} = \tan^{-1}(zy)$$

$$y = \frac{D(r_1 \ell_2 - r_2 \ell_1)}{D^2 r_1 r_2 + z^2 \ell_1 \ell_2}$$

The total result can be then obtained by successively considering each side and adding the results together.

3.4.4 Zoned Inhomogeneous Bodies

So far we have dealt with problems involving one homogeneous zone of either isotropic or anisotropic material. In most practical situations the regions concerned contain a number of contiguous zones of materials each having different but homogeneous properties (i.e. they are zoned or piecewise homogeneous bodies).

Whereas the matrices which arise in the basic equation (3.22) are fully populated we shall find that zoned bodies lead to block-banded matrix systems with one block for each zone and overlaps between blocks when zones have a common interface. In general there will be a number of zones V^m, $m = 1, 2, \ldots$, each enclosed by its surface S^m. Where two regions (see Fig. 3.6) have a common interface, e.g. regions 1, 2, we have to ensure that corresponding points on the S^1 and S^2 boundaries are at the same potential and that there is continuity of flow across corresponding interface elements. Thus on the interface

$$p_{12} - p_{21} = 0 \quad \text{and} \quad q_{12} + q_{21} = 0 \tag{3.34}$$

where the number of components in each of these interface potential and flux vectors is equal to the chosen number of interface boundary elements, say R.

The simplest possible way to assemble the final system matrix for our two-zone problem would be to rewrite the system equations for each zone as

$$\{q\} = [G]^{-1}[F]\{p\} \tag{3.35}$$

This is of course possible only if the matrix G is square. For the simple uniformly distributed quantities q and p over each element the matrix can be inverted.

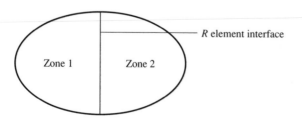

R element interface

Zone 1   Zone 2

Figure 3.6 A two-zone problem.

We can separate the peripheral fluxes and potentials for zone 1, say, from the interface values, partition the above equation and write

$$\left\{ \begin{matrix} q_1 \\ q_{12} \end{matrix} \right\} = \left[\begin{matrix} A_1 & A_{12} \\ B_{12} & C_{12} \end{matrix} \right] \left\{ \begin{matrix} p_1 \\ p_{12} \end{matrix} \right\} \tag{3.36}$$

and, similarly for zone 2,

$$\left\{ \begin{matrix} q_2 \\ q_{21} \end{matrix} \right\} = \left[\begin{matrix} A_2 & A_{21} \\ B_{21} & C_{21} \end{matrix} \right] \left\{ \begin{matrix} p_2 \\ p_{21} \end{matrix} \right\} \tag{3.37}$$

By applying the interface conditions (3.34) we can assemble Eqs (3.36) and (3.37) into a single system

$$\left\{ \begin{matrix} q_1 \\ 0 \\ q_2 \end{matrix} \right\} = \left[\begin{matrix} A_1 & A_{12} & 0 \\ B_{12} & C_{12} + C_{21} & B_{21} \\ 0 & A_{21} & A_2 \end{matrix} \right] \left\{ \begin{matrix} p_1 \\ p_{12} \\ p_2 \end{matrix} \right\} \tag{3.38}$$

of $(N_1 + N_2 - R)$ equations.

Note that the above assembly process is essentially identical to that used in a finite element system. Equation (3.38) can now be solved for any combination of specified boundary conditions involving half of the boundary quantities p_1, p_2, q_1 and q_2. Having obtained the unknown boundary information, the solutions for interior quantities can be achieved by treating each zone as entirely independent entities.

By proceeding in the same manner, the four-zone problem such as that shown in Fig. 3.7 leads to the system equations shown in Eq. 3.39.

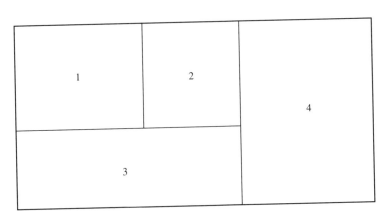

Figure 3.7 A general four-zone problem.

The assembly process is one of the simplest possible ones to use in the boundary element method. However, we shall see later that for higher-order elements a slightly different approach is needed.

$$
\begin{Bmatrix} q_1 \\ 0 \\ 0 \\ q_2 \\ 0 \\ 0 \\ q_3 \\ 0 \\ q_4 \end{Bmatrix} =
\begin{bmatrix}
A_1 & A_{12} & & A_{13} & & & & & \\
B_{12} & C_{12}+C_{21} & & B_{21} & & & & & \\
B_{13} & & C_{13}+C_{31} & & & B_{31} & & & \\
& & & A_2 & A_{23} & A_{24} & & & \\
& & & B_{23} & C_{23}+C_{32} & & B_{32} & & \\
& & & B_{24} & & C_{24}+C_{42} & & B_{42} & \\
& & & & & & A_3 & & \\
& & & & & B_{34} & C_{34}+C_{43} & B_{43} & \\
& & & & & & & & A_4
\end{bmatrix}
\begin{Bmatrix} p_1 \\ p_{12} \\ p_{13} \\ p_2 \\ p_{23} \\ p_{24} \\ p_3 \\ p_{34} \\ p_4 \end{Bmatrix}
$$

$$(3.39)$$

3.5 HIGHER-ORDER ELEMENTS

Although the simple elements discussed earlier allow one to develop insight into the boundary element solution procedures, the real advantage and accuracy of boundary element methods become available to a user only when higher-order elements are considered. In order to introduce our reader to this process we note that without introducing any approximations we can write Eq. (3.16) with $\psi = 0$ as

$$
cp(\xi) = -\sum_{n=1}^{N} \int_{\Delta S_n} [G(x,\xi)q(x) - F(x,\xi)p(x)]\,dS(x)
\tag{3.40}
$$

in which c is a constant whose value depends on the position of the field point ξ and the integral over S has been replaced by piecewise summation over the surface. In our treatment of the simplest boundary element procedures earlier we were able to assume q and p remaining constant over elements so that they could be taken outside the integral sign. This enabled us to reduce the problem of having to evaluate simple integrals of functions G and F.

However, if we assume that functions $q(x)$ and $p(x)$ vary in a certain specified manner over a typical boundary element ($n = 1, \ldots, N$) with each element having K nodes ($k = 1, \ldots, K$), we can express the variations over each element as

$$
q(x) = N^k(x)q^k
\tag{3.41a}
$$

$$
p(x) = N^k(x)p^k
\tag{3.41b}
$$

where $N^k(x)$ are shape functions discussed in Chapter 2, with q^k and p^k being the nodal values of q and p respectively. We can then write Eq. (3.40) for a typical field point ξ^m as

$$
cp(\xi^m) = -\sum_{n} \left\{ \left[\int_{\Delta S_n} G(x,\xi^m)N^k(x)\,dS \right] q_n^k - \left[\int_{\Delta S_n} F(x,\xi^m)N^k(x)\,dS \right] p_n^k \right\}
$$

$$(3.42)$$

The problem is now reduced to the evaluation of the integrals within the brackets. It is immediately evident that now, over each element, K integrals involving the function G and the same number involving the function F need to be evaluated.

The potential p can be considered to have a unique value at a node and there-fore continuous across the element boundaries. The flux q, however, can have a discontinuity at a common node between adjacent boundary elements, since this is an element-related quantity. In view of this it is only possible, after integra-tion, to sum the coefficients of common nodal values of p^k in Eq. (3.42) to express it for a field point ξ as

$$cp(\xi) = -G^{nk}q^{nk} + F^l p^l \qquad (3.43)$$

By successively taking ξ to all M boundary nodes we can express the above as

$$(cp)^m = -G^{mnk}q^{nk} + \bar{F}^{ml}p^l \qquad (3.44)$$

where
p^m = $M \times 1$ vector of M nodal values of p
q^{nk} = $(N \times K) \times 1$ vector of nodal values of q of all N elements
G^{mnk} = $M \times (N \times K)$ matrix of coefficients
$\bar{F}^{ml}$ = $M \times M$ matrix of coefficients

We can now absorb the coefficients of the p^m term on the left-hand side of Eq. (3.44) with the leading diagonals of the matrix $\bar{F}^{mm}$ to obtain

$$G^{mnk}q^{nk} - F^{mm}p^m = 0 \qquad (3.45)$$

in which we can once more use the uniform potential solution to determine the leading diagonals of F.

Equation (3.45) can now be scaled such that coefficients of G and F matrices are of the same order of magnitude by introducing a factor α such that

$$(\alpha G)^{mnk}(q/\alpha)^{nk} - F^{mm}p^m = 0 \qquad (3.46)$$

The final system equations can now be formed from the above by noting the fol-lowing:

1. For each node only one unknown quantity is admissible (either q or p). All other quantities for that node are either prescribed or must somehow be elimi-nated.
2. For a node shared by multiple elements, the fluxes, if prescribed, can be multi-plied with the appropriate columns of the G matrix and be transferred to the right-hand side. If the fluxes at these nodal locations are not known and the node is located on a smooth surface they must be made equal to provide a con-tinuous distribution.
3. For an edge or a corner node, if the multi-valued fluxes are not prescribed then one must invoke an auxiliary relationship of the form $q = -(1/R)(p - p_0)$, where R is a suitable resistance, p the unknown potential at the node and p_0 the ambient potential which can be assigned to the prescribed value of the potential at the node. This will leave p as the final unknown value. Alterna-tively these multi-valued fluxes could be expressed in terms of the extrapolated values of fluxes at the neighbouring nodes and thus removed from the system equations.

The final system equation can then be expressed once more as

$$\mathbf{Ax} = \mathbf{b} \tag{3.47}$$

where the matrix $\mathbf{A}$ now is a non-symmetric and fully populated square matrix.

3.5.1 Evaluation of the Integrals

In this higher-order formulation the integrals involved are now of the form:

$$\int_{\Delta S_n} G(x, \xi^m) N^k(x) \, \mathrm{d}S(x)$$

$$\int_{\Delta S_n} F(x, \xi^m) N^k(x) \, \mathrm{d}S(x)$$

over the element ΔS_n having $k = 1, \ldots, K$ nodes. For flat elements, using the local coordinate systems shown in Figs 3.4 and 3.5 it is possible to evaluate all the required integrals analytically for two- and three-dimensional cases with both linear and quadratic variations of q and p over the elements. Since these involve lengthy algebraic expressions, we leave these as exercises for our readers. Whenever practical, analytical integrations must be preferred.

For curved elements, particularly for any general curvilinear element, numerical integration is necessary. Since this is a costly process, it is essential to optimize this effort and yet make sure that all the integrals are evaluated to at least four to five correct digits. In general an effective integration scheme must recognize the behaviour of the kernel, shape function and Jacobian products[19] when the integrals are transformed into expressions like

$$I^m = \int_{\Delta S(\eta)} G(x(\eta), \xi^m) N^k(\eta) J(\eta) \, \mathrm{d}S(\eta) \tag{3.48}$$

where our element over which integrations have to be performed have been transformed (see Chapter 2) into an intrinsic coordinate space η_i. Generally three classes of integrals are involved:

1. The field point ξ^m some distance away from the element ΔS_n when the integrals are called 'well behaved'. This class represents the largest number of integrals involved and hence the major cost.
2. The field point ξ^m on the element ΔS_n, which has two quite different cases, that is ξ^m coincides with the node k and ξ^m not coinciding with node k.
3. The field point ξ^m is only a small distance away from the element ΔS_n.

The process of numerical integration of 'well-behaved' integrals essentially involve the following steps:

1. The element in the physical space is first transformed into the intrinsic coordinate space.

2. An optimal order of numerical integration P is selected which automatically defines the locations and the associated weights at integration points in the intrinsic coordinate space.
3. Values of the shape functions N^k and Jacobian at the integration points in the intrinsic coordinate space are then determined.
4. The locations x_i of the integration points on the physical space and the values of the normals at the same locations are calculated so that P values of the kernel functions G and F can be obtained.
5. Typically, the integral such as the one in Eq. (3.48) is then approximated as

$$I^m = \sum_{p=1}^{P} [G(x^p, \xi^m) N^k(\eta^p) J(\eta^p)] \omega^p \qquad (3.49)$$

where x^p, η^p and ω^p are physical coordinates, intrinsic coordinates and weights respectively of the integration points ($p = 1, \ldots, P$) on the element ΔS_n.

The above process is equally applicable to both two- and three-dimensional cases provided the appropriate dimensionality of x_i and η_i are taken into account (see Chapter 2). However, the order P must be chosen such that these integrals are calculated accurately with a minimum of computing efforts. This can be achieved by reducing P as the distance (R_n) between ξ^m and the element ΔS_n in the relation to a characteristic dimension L_n of the element increases. Thus typically for potential flow problems for $R_n/L_n = 2, 4, 6$ or greater, the order P can be chosen as 4, 3 and 2 respectively. Experience, however, indicates that it is not desirable to reduce the order below 2 regardless of how large R_n/L_n is.

While the above technique is satisfactory for all integrals of class 1 it is not possible to apply this to those belonging to class 3, even if a higher value of P is used. It is necessary to subdivide the element, beginning from a point on the element that is nearest to the field point ξ, as shown for a three-dimensional problem in Fig. 3.8. As an alternative, which works satisfactorily only for two-dimensional problems, one can apply an additional coordinate transformation:

$$\int_0^1 f(\eta) \, d\eta = \int_0^1 f(\eta(\gamma)) J \, d\gamma \qquad (3.50)$$

where

$$J = d\eta/d\gamma$$

One can choose the transformation such that more integration points are clustered towards the point ξ^m. Mustoe[20] discussed such a transformation of pth order and Telles[21] described an implementation for $p = 3$ (cubic transformation).

When the field point ξ^m falls on ΔS_n in a two-dimensional problem the integral involving G can be determined accurately by a special log-weighted (ω_ℓ) quadrature, i.e.

$$\int_0^1 \ln\left(\frac{1}{\eta}\right) f(\eta) \, d\eta \approx \sum_{p=1}^{P} f(\eta^p) \omega_\ell^p \qquad (3.51)$$

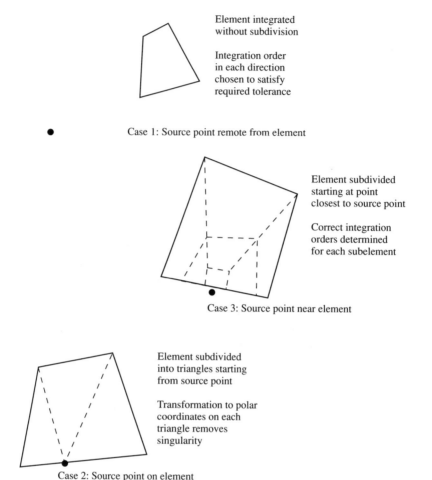

Element integrated
without subdivision

Integration order
in each direction
chosen to satisfy
required tolerance

Case 1: Source point remote from element

Element subdivided
starting at point
closest to source point

Correct integration
orders determined
for each subelement

Case 3: Source point near element

Element subdivided
into triangles starting
from source point

Transformation to polar
coordinates on each
triangle removes
singularity

Case 2: Source point on element

Figure 3.8 Self-adaptive integration system.

It must be noted that such a formula cannot be applied to non-singular or near-singular integrals. Moreover, if other non-singular functions have been added to G to satisfy some boundary conditions (e.g. in a half-space problem a mirror image of G can be used to satisfy some free surface boundary conditions) these must be separated out and treated accordingly. In addition, the element must be subdivided through the point ξ_m to apply the above formula.

For two-dimensional problems involving functions G and F, if the field point does not coincide with the node k the integrals are then non-singular and hence may be subdivided through ξ and a normal integration over each segment is possible. If ξ^m coincides with the node k, the integral involving F together with the jump term is determined indirectly using the 'equipotential condition'.

For a similar three-dimensional situation when the field point ξ falls on the element ΔS_n, the element can be divided into several triangular elements each with a common vertex at ξ_m. Each triangular sub-element is then mapped into a square. This transformation process, which is identical to a $r-\theta$ polar coordinate transformation through the field point ξ^m, yields a Jacobian which approaches to zero at ξ^m and therefore eliminates the singularity in the integral involving $G(x, \xi^m)$. For convenience the integrals involving $F(x, \xi^m)$ is handled in the same manner, except for one term which is obtained indirectly using the 'equipotential condition'.

3.5.2 General Remarks

To summarize the above numerical integration process, we note that it relies very heavily on the behaviour of the kernel-shape function–Jacobian products. If it is bounded it can be integrated accurately by suitable sub-segmentation to match the variations over the elements as the point ξ^m is approached. In addition, sub-segmentation also reduces errors in integration resulting from bad element aspect ratios. If it is singular then a suitable transformation of Jacobian is used to eliminate the singularity. Strong singularities are handled indirectly by imposing a physical solution on the discretized system, which also provides the stability to the discretized system since there is consistency of errors for all integrated coefficients.

3.6 ADVANCED MULTI-REGION ASSEMBLY

Although we were able to introduce a very simple assembly process in Sec. 3.4.4, which one can use to develop multi-region analyses, the use of higher-order elements makes the G matrix rectangular. Therefore, two new approaches are introduced in this section. With reference to our two-region problem shown in Fig. 3.6, we express Eq. (3.46) for each region as

$$\begin{bmatrix} G_{11}^1 & G_{12}^1 \\ G_{21}^1 & G_{22}^1 \end{bmatrix} \begin{Bmatrix} q_1 \\ q_{12} \end{Bmatrix} = \begin{bmatrix} F_{11}^1 & F_{12}^1 \\ F_{21}^1 & F_{22}^1 \end{bmatrix} \begin{Bmatrix} p_1 \\ p_{12} \end{Bmatrix} \tag{3.52a}$$

and for region 1 and for region 2 as

$$\begin{bmatrix} G_{11}^2 & G_{12}^2 \\ G_{21}^2 & G_{22}^2 \end{bmatrix} \begin{Bmatrix} q_2 \\ q_{21} \end{Bmatrix} = \begin{bmatrix} F_{11}^2 & F_{12}^2 \\ F_{21}^2 & F_{22}^2 \end{bmatrix} \begin{Bmatrix} p_2 \\ p_{21} \end{Bmatrix} \tag{3.52b}$$

By invoking the interface relations equation (3.34) we can assemble these equations into a single system as

$$\begin{bmatrix} G_{11}^1 & G_{12}^1 & 0 \\ G_{21}^1 & G_{22}^1 & 0 \\ 0 & -G_{12}^2 & G_{11}^2 \\ 0 & -G_{22}^2 & G_{21}^2 \end{bmatrix} \begin{Bmatrix} q_1 \\ q_{12} \\ q_2 \end{Bmatrix} = \begin{bmatrix} F_{11}^1 & F_{12}^1 & 0 \\ F_{21}^1 & F_{22}^1 & 0 \\ 0 & F_{12}^2 & F_{11}^2 \\ 0 & F_{22}^2 & F_{21}^2 \end{bmatrix} \begin{Bmatrix} p_1 \\ p_{12} \\ p_2 \end{Bmatrix} \tag{3.53}$$

Note that now the coefficients are not added up. Equation (3.53) can be used to form the final system equations (3.46) and (3.47), where all the discussions therein are also relevant. However, it should be noted that although the final system matrix will still be non-symmetric and blockbanded a very large amount of storage is now necessary because at the interfaces both fluxes q_{12} and potentials p_{12} have to be retained as unknowns in the final system equations. Moreover, for the four-zone problem shown in Fig. 3.7, this assembly process is extremely cumbersome.

An improved assembly process[22] could be developed if we observe that in a multi-zone model the source points can be collected into two distinct sets. The first set includes all source points residing on a region-to-region inter-face. These nodes will be labelled with the subscript a, denoting analysis-set. Let N_a^m represent the number of such nodes. The remaining source points are each associated with a single region, and consequently may be eliminated at the zonal level. The latter nodes are considered part of the omitted set and identified by the subscript o. Typically, the o-set, which con-tains N_o^m nodes, is much larger than the a-set. Introducing this notation into Eq. (3.52a) produces

$$\begin{bmatrix} G_{aa}^m & G_{ao}^m \\ G_{oa}^m & G_{oo}^m \end{bmatrix} \begin{Bmatrix} q_a^m \\ q_o^m \end{Bmatrix} = \begin{bmatrix} F_{aa}^m & F_{ao}^m \\ F_{oa}^m & F_{oo}^m \end{bmatrix} \begin{Bmatrix} p_a^m \\ p_o^m \end{Bmatrix} \tag{3.54}$$

Next, boundary conditions can be applied at the zonal level. All of the quan-tities $\{q_a^m\}$ and $\{p_a^m\}$ are unknown and can be combined to form the vector $\{x_a^m\}$. However, the boundary conditions must specify all but N_o^m of the quan-tities $\{q_o^m\}$ and $\{p_o^m\}$. The unknown components form $\{x_o^m\}$. Then, Eq. (3.54) becomes

$$\begin{bmatrix} A_{aa}^m & A_{ao}^m \\ A_{oa}^m & A_{oo}^m \end{bmatrix} \begin{Bmatrix} x_a^m \\ x_o^m \end{Bmatrix} = \begin{Bmatrix} b_a^m \\ b_o^m \end{Bmatrix} \tag{3.55}$$

for each region. The matrix A^m is not square. It is of size $(N_a^m + N_o^m) \times (2N_a^m + N_o^m)$. On the other hand, A_{oo}^m is a square matrix of order N_o^m. By applying Gaussian elimination, Eq. (3.55) can be reduced to

$$[\bar{A}_{aa}^m]\{x_a^m\} = \{\bar{b}_a^m\} \tag{3.56}$$

where symbolically

$$[\bar{A}_{aa}^m] = [A_{aa}^m] - [A_{ao}^m][A_{oo}^m]^{-1}[A_{oa}^m] \tag{3.57a}$$

$$[\bar{b}_a^m] = \{b_a^m\} - [A_{ao}^m][A_{oo}^m]^{-1}\{b_o^m\} \tag{3.57b}$$

Once this is accomplished for each zone, then $[\bar{A}_{aa}^m]$ and $[\bar{b}_a^m]$ for the regions are assembled together. After the interface conditions are imposed the residual equa-tions can be written as

$$[\bar{A}_{aa}]\{x_a\} = \{\bar{b}_a\} \tag{3.58}$$

However, by this time, since many equations have been eliminated, the resulting system matrices are much smaller. In fact the order of the square $[\bar{A}_{aa}]$ matrix is only N_a, where for an M region problem

$$N_a = \sum_{m=1}^{M} N_a^m$$

After Eq. (3.58) is solved for $\{x_a\}$, the remaining unknowns in the individual regions, $\{x_o^m\}$, can be easily determined via backsubstitution. Specifically,

$$\{x_o^m\} = [A_{oo}^m]^{-1}\{\bar{b}_o^m\} \tag{3.59}$$

where

$$\{\bar{b}_o^m\} = \{b_o^m\} - [A_{oa}^m]\{x_a^m\} \tag{3.60}$$

With this procedure, often called a 'sub structured solution', none of the zero blocks present in Eq. (3.53) ever enter into the formulation. This greatly reduces the disk storage requirements, particularly for large problems, and also eliminates unnecessary operations. In addition, by reordering operations during integration and assembly, further savings can be realized.

If all interfaces are assumed smooth, both the potential and the flux at each interface node are unique. In defining the size of A^m in Eq. (3.55) a similar assumption was made. More generally, there are N_a^m components in $\{p_a^m\}$; however $\{q_a^m\}$ contains a component for each node of each interface element. Only for the case in which a unique interfacial normal exists at each interface node will the number of components in $\{q_a^m\}$, say Q_a^m, reduce to N_a^m. Thus, in general, $Q_a^m \geq N_a^m$ and, of course, Q_a^m interface relationships are needed to eliminate the unknowns $\{q_a^m\}$. This can be accomplished by writing a resistance relationship at each node of each interface element. For example, for a connection between regions 1 and 2, the equations

$$q^1 = -\frac{1}{R}(p^2 - p^1) \tag{3.61a}$$

$$q^2 = -\frac{1}{R}(p^1 - p^2) \tag{3.61b}$$

can be utilized to eliminate q^1 and q^2 in terms of the nodal potentials. In the case of perfect contact, the resistance R is set to a suitably small value to ensure that $p^1 \cong p^2$ while avoiding ill-conditioning. This element-based approach permits the interconnection of any number of sub-structures at a node.

3.7 INDIRECT BEM FORMULATIONS

For convenience we repeat the basic Eq. (3.12) from Sec. 3.4, which is the DBEM statement for the potential $p(x)$ at any point x_i interior to V, but we now interchange the x and ξ symbols: namely

$$\int_V p(\xi)\delta(\xi, x)\,dV(\xi) = p(x)$$

$$= \int_S [p(\xi)F(\xi, x) - q(\xi)G(\xi, x)]\,dS(\xi) + \int_V \psi(\xi)G(\xi, x)\,dV(\xi)$$

$$(3.62)$$

Consider the region $\bar{V}$, bounded by S but exterior to V, within which there are no sources and assume $\bar{p}(x)$ to be a solution to the Laplace equation $\partial^2\bar{p}/\partial x_i\,\partial x_i = 0$ within $\bar{V}$. An exact repetition of the analysis leading up to Eq. (3.13), carrying out integrations over S and V but still with the field point at x_i within V, develops an equivalent equation

$$0 = \int_S [-\bar{p}(\xi)F(\xi, x) - \bar{q}(\xi)G(\xi, x)]\,dS(\xi)$$

where, in comparison to Eq. (3.12):

1. The last term is zero, since $\bar{\psi}(x) = 0$.
2. The sign of the term involving $F(\xi, x)$ is reversed since the sense of the outward normal to $\bar{V}$ is opposite to that of V.
3. The left-hand side term is zero, since x_i is now exterior to $\bar{V}$ [that is $\int_V \bar{p}(\xi)\delta(\xi, x)\,dV(\xi) = 0$].

If $\bar{p}(x)$ is specified to be that solution, in $\bar{V}$, which establishes on S exactly the same boundary potentials as those in our initial interior region problem [that is, $\bar{p}(\xi) = p(\xi), \xi \in S$], then substitution of $p(\xi)$ for $\bar{p}(\xi)$ in our second equation and adding it to Eq. (3.62) produces

$$p(x) = -\int_S [q(\xi) + \bar{q}(\xi)]G(\xi, x)\,dS(\xi) + \int_V \psi(\xi)G(\xi, x)\,dV(\xi)$$

Therefore,

$$p(x) = \int_S \phi(\xi)G(\xi, x)\,dS(\xi) + \int_V \psi(\xi)G(x, \xi)\,dV(\xi) \qquad (3.63)$$

where

$$\phi(\xi) = -[q(\xi) + \bar{q}(\xi)]$$

which is a new integral statement with an arbitrary function ϕ distributed over the boundary S. Since this function ϕ does not usually have any physical relevance to the problem, such statements are often referred to as an indirect boundary integral statement leading to an IBEM (indirect BEM) formulation. The interested reader may care to refer to Lamb[23] where the above argument is developed almost identically, although in recent years many appeared to have independently discovered it.

Such IBEM statements can also be developed using simple physical reasoning such as superposition of the fundamental solution G in terms of the arbitrary density function ϕ as shown in Banerjee and Butterfield.[11] It is interesting to note that if we adopt $\bar{q}$ as that solution in $\bar{V}$ that establishes, on $S, \bar{q} = q$ then we can obtain a second indirect formulation

$$p(x) = \int_S F(x,\xi)\mu(\xi)\,\mathrm{d}S + \int_V \psi(\xi)G(x,\xi)\,\mathrm{d}V(\xi) \qquad (3.64)$$

which is often referred to a higher-order source formulation involving a new fictitious function $\mu(\xi)$ where $\mu(\xi) = p(\xi) - \bar{p}(\xi)$. Unfortunately the subsequent numerical treatment of Eq. (3.64) presents considerable numerical difficulties; therefore this alternative indirect formulation has not been very popular.

For the purposes of developing our IBEM solutions (here, we restrict our attention specifically to two-dimensional problems) we need to express Eq. (3.63) by using ξ and z as integration points on S and V respectively:

$$p(x) = \int_S G(x,\xi)\phi(\xi)\,\mathrm{d}s(\xi) + \int_V G(x,z)\psi(z)\,\mathrm{d}V(z) + C \qquad (3.65)$$

where the constant C arises from the fact that G, in a two-dimensional problem, does not vanish at a large distance but can only be referred to its value (datum) at $r = r_0$. This requires that to guarantee the uniqueness of our solution we must ensure that C is determined to satisfy the auxiliary relation

$$\int_S \phi(x)\,\mathrm{d}S + \int_V \psi(z)\,\mathrm{d}V = 0 \qquad (3.66)$$

We can now differentiate Eq. (3.65) with respect to x_i to determine the flux q from

$$q(x) = \int_S F(x,\xi)\phi(\xi)\,\mathrm{d}S(\xi) + \int_V F(x,z)\psi(z)\,\mathrm{d}V(z) \qquad (3.67)$$

In principle the only remaining formal step to arrive at a 'solution' to the problem is to bring the point (x_i) on to the boundary S (that is $x \to x_0$), whence Eqs (3.65) and (3.67) become

$$p(x_0) = \int_S G(x_0,\xi)\phi(\xi)\,\mathrm{d}S(\xi) + \int_V G(x,z)\psi(z)\,\mathrm{d}V(z) + C \qquad (3.68)$$

$$q(x_0) = \oint_S F(x_0,\xi)\phi(\xi)\,\mathrm{d}S(\xi) + \int_V F(x_0,z)\psi(z)\,\mathrm{d}V(z) \qquad (3.69)$$

where $\oint$ represents an improper integral due to the singularity of F as $\xi \to x_0$. In a 'well-posed' problem, one of either $p(x_0)$ or $q(x_0)$ will be known at every point of the boundary and Eqs (3.68) and (3.69) are two simultaneous integral equations which can be solved for the only unknown $\phi(\xi)$.

On the basis of our experience with the direct formulation outlined earlier, we would anticipate that all the integrals (or numerical summations) required can be evaluated in the normal sense with the exception of the first line integral in Eq. (3.69) which has a strong, $1/r$, singularity when $\xi = x_0$ and will lead to a jump term as before. The value of the jump term is, in fact, obvious in some cases on physical grounds. For example, in our problem when $\xi = x_0$ at a smooth boundary point the source $\phi(\xi)$ bifurcates half into the interior region (V) and half into the exterior region $(\bar{V})$. Its contribution to $q(x)$ as x approaches x_0 from inside V is therefore clearly $-\frac{1}{2}\phi(x_0)$, provided that x_0 is not located on any corner of S (that is x_0 on S must possess a unique tangent direction). The value of the 'free term' on corners was discussed earlier. We can now rewrite Eq. (3.69) with x_0 approaching S as

$$q(x_0) = -\tfrac{1}{2}\phi(x_0) + \int_S F(x_0, \xi)\phi(\xi)\,dS(\xi) + \int_V F(x_0, z)\psi(z)\,dV(z) \qquad (3.70)$$

while Eq. (3.68) remains unchanged.

3.7.1 Discretization of the Surface

The discretization scheme used here utilizes flat boundary elements, characterized by their midpoints, along any one of which, say the nth element, the fictitious source $\phi(\xi^n)$ is uniformly distributed. For simplicity we shall also consider the internal source term $\psi = 0$. We can then write discrete approximations to Eqs (3.68) and (3.70) for $p(x_0^m)$ and $q(x_0^m)$, the potential and normal velocity components on the mth boundary element, as

$$p(x_0^m) = \sum_{n=1}^{N} \phi(\xi^n) \int_{\Delta S} G(x_0^m, \xi^n)\,dS(\xi^n) + C \qquad (3.71)$$

$$q(x_0^m) = -\tfrac{1}{2}\phi(x_0^m) + \sum_{n=1}^{N} \phi(\xi^n) \int_{\Delta S} F(x_0^m, \xi^n)\,dS(\xi^n) \qquad (3.72)$$

3.7.2 Formation of the System Matrices

We have deliberately allowed the notation to become quite elaborate so that the roles of the various coordinates, the location of field and load points, etc., are defined unambiguously. Once this has been achieved it becomes very much more convenient to rewrite Eqs (3.71) and (3.72) in matrix notation as

$$p^m = \left(\int_{\Delta S} G^{mn}\,dS \right) \phi^n + C \qquad (3.73a)$$

$$q^m = -\tfrac{1}{2}\phi^m + \left(\int_{\Delta S} F^{mn}\,dS \right) \phi^n \qquad (3.73b)$$

where ϕ^n is an $N \times 1$ column vector and the terms in brackets are compatible row vectors. Each term in these row vectors is the result of integrating the G and F kernels function over ΔS, etc. The detailed evaluation of these intermediate integrals will be discussed later; for the moment we note that the final form of Eqs (3.73a,b) will always be equivalent to

$$p^m = (G^{mn})\phi^n + C \tag{3.74a}$$

$$q^m = (F^{mn})\phi^n \tag{3.74b}$$

where the $-\frac{1}{2}\phi^m$ term in Eq. (3.73b) has now been added into the $(m = n)$ term of F^{mn}.

If precisely similar operations are carried out for all elements $n (n = 1, 2, \ldots, N)$ the total sets of all such equations for $\mathbf{p}$ and $\mathbf{q}$ can be assembled to yield simply

$$\mathbf{p} = \mathbf{G}\phi + \mathbf{I}C \tag{3.75a}$$

$$\mathbf{q} = \mathbf{F}\phi \tag{3.75b}$$

where clearly $\mathbf{p}, \mathbf{q}$ and ϕ are $N \times 1$ vectors of the surface quantities and $\mathbf{I}$ is an $N \times 1$ unit column vector.

Before assembling Eqs (3.75a,b) into a global set we need to return to our discussion of C. As mentioned previously, C has arisen as an arbitrary parameter related to our freedom of choice of datum from which to measure the potentials. If we choose C such that the algebraic sum of all sources ϕ applied over S is zero, then we shall have overcome the intrinsic problem that the $\ln(r)$ term in the G kernel does not go to zero as $r \to \infty$.

We therefore need an auxiliary Eq. (3.66) to ensure this: namely

$$\sum_{n=1}^{N} \int_{\Delta S} \phi(\xi^n)\,\mathrm{d}S(\xi^n) = \sum_{n=1}^{N} (\phi^n \Delta S^n) = 0$$

or

$$(\mathbf{b}_n)\phi = 0 \tag{3.76}$$

where $(\mathbf{b}_n)$ is an $N \times 1$ row vector whose components are merely element lengths.

When we assemble a global set of equations from Eqs (3.75a,b) to solve a specific problem we shall not require every component equation. In general, boundary potentials will only be specified on some, $\mathbf{p}^s$ say, of the total set of $\mathbf{p}$ boundary potentials; similarly, the normal velocity will be specified on $\mathbf{q}^s$ of the total $\mathbf{q}$ set. However, the total number of components in $\mathbf{p}^s$ and $\mathbf{q}^s$ together will always be N. We therefore select from Eqs (3.74a,b) just the N equations for which boundary information has been provided and assemble them, together with Eq. (3.76), into a global set as

$$\begin{Bmatrix} \alpha\mathbf{p} \\ \mathbf{q} \\ 0 \end{Bmatrix} = \begin{bmatrix} \mathbf{I}_0 & \alpha\mathbf{G} \\ \mathbf{0} & \mathbf{F} \\ \mathbf{0} & \mathbf{b}_n \end{bmatrix} \begin{Bmatrix} \alpha C \\ \phi \end{Bmatrix} \tag{3.77}$$

where $I_0, 0$ are unit and zero column vectors with numbers of components corresponding to p and q respectively, and α is a scaling factor which multiplies the components shown such that the coefficients are of the same order of magnitude. Equation (3.77) can now be solved to obtain boundary source density ϕ and the constant C, backsubstitution of which in Eqs (3.75a,b) recovers all boundary values. Multi-zone problems can also be solved by IBEM[11] by forming system matrices for each zone independently and forming the final system matrix by satisfying interface conditions.

3.7.3 Integration of Kernel Functions

Note that as a result of the change of arguments x and ξ the integration now is carried with respect to ξ and x is now the field point where the normal vector is calculated. In the DBEM formulation the normal n_i was at the integration point!

Since the function $G(x, \xi)$ is symmetric with respect to its arguments x and ξ the results of integration outlined in Sec. 3.4.3 can still be used. Those involving the $F(x, \xi)$ kernel, however, cannot be used here.

For the two-dimensional case we note that

$$F(x, \xi) = \frac{(x - \xi)_i n_i(x)}{2\pi r^2}$$

and therefore in our local coordinate system, defined in Fig. 3.4,

$$\int_{\Delta S} F(x, \xi)\, dS \rightarrow \int_{\Delta S} F(0, \eta)\, d\eta_2 = -\frac{1}{2\pi} \int_{\theta_A}^{\theta_B} n_i(x) n_i \frac{h \sec^2 \theta}{h^2 \sec^2 \theta}\, d\theta$$

$$= -\frac{1}{2\pi} \int_{\theta_A}^{\theta_B} (n_1 + n_2 \tan \theta)\, d\theta$$

$$= -\frac{1}{2\pi} \left[n_1 \theta - n_2 \ln \left| \frac{h}{r} \right| \right]_{\theta_A, r_A}^{\theta_B, r_B} \qquad (3.78)$$

Again, if we let x^m approach x_0^n from inside V, $r_A = r_B$, $n_2 \rightarrow 0$, $n_1 \rightarrow 1$, $\theta_B = -\theta_A \rightarrow \pi/2$ and

$$\int_{\Delta A} F(x_0^q, \xi^q)\, dS \Rightarrow -\frac{1}{2\pi} (n_i)(\theta_B - \theta_A) = -\frac{1}{2} \qquad (3.79)$$

Notice, once again that the jump term is the only contribution on the diagonal of the F matrix.

3.7.4 Three-dimensional Problems

Three-dimensional problems with the IBEM do not have the added complexity of Eq. (3.66) since $G(x, \xi)$ decays to zero at infinity. Therefore, if errors are not present the integral effects of the distributed sources ϕ and known sources ψ automatically vanish at a large distance from S. The system equations equivalent to Eq. (3.77) then simply become

$$\left\{ \begin{matrix} \alpha\mathbf{p} \\ \mathbf{q} \end{matrix} \right\} = \left[\begin{matrix} \alpha\mathbf{G} \\ \mathbf{F} \end{matrix} \right] \{\phi\} \tag{3.80}$$

which can be solved for ϕ.

Once again, since the function $G(x, \xi)$ is symmetric with respect to its arguments all the results of analytical integration of G over surface elements in Eq. (3.33a) can be used. It is, however, necessary to repeat the integration of the F kernel, i.e.

$$\Delta F = \int_{\Delta S} F(x, \xi)\, dS(\xi) = -\frac{1}{4\pi} n_i(x) \int_{\Delta S} \frac{\partial}{\partial x_i} \left(\frac{1}{r} \right) dS(\xi)$$

$$= \frac{1}{4\pi} n_i(x) \int_{\Delta S} \frac{\partial}{\partial \xi_i} \left(\frac{1}{r} \right) dS(\xi)$$

$$= \frac{n_i(x)}{4\pi} \int_C \frac{1}{r} e_i\, dC \tag{3.81}$$

which now needs to be evaluated over the boundary of ΔS, i.e. over C.

3.8 SOURCE DISTRIBUTION WITHIN VOLUME

So far we have deliberately avoided the treatment of the source term which, of course, is known and its effects at a source point ξ^m (for example) could be determined by evaluating the volume integral

$$\int G(x, \xi^m)\psi(x)\, dV(x)$$

to form a vector that could be added to the right-hand side of the final system equations (3.47). Since this process is expensive an alternative method is desirable.

If we apply the well-known theory of particular integral to the inhomogeneous differential Eq. (3.7), we have

$$\frac{\partial^2 p^c}{\partial x_i\, \partial x_i} = 0 \tag{3.82}$$

and

$$\frac{\partial^2 p^o}{\partial x_i\, \partial x_i} + \psi = 0 \tag{3.83}$$

where the total solution $p = p^c + p^o$, and p^c, the complementary solution, satisfies the homogeneous differential Eq. (3.82) while the particular integral, p^o, satisfies the complete differential equation [inhomogeneous equation (3.83)].

The particular integral is any solution, without any due regard to the geometry or the boundary conditions of the problem, that can be made to satisfy the differential Eq. (3.83). The particular solutions are never unique (but the total solution p is unique) and can be obtained by inspection, by trial and error or by using the method of undetermined coefficients.

Thus, if a particular solution p is found then the BEM could be applied to Eq. (3.82) to produce the final system matrix equations

$$\mathbf{Gq^c} - \mathbf{Fp^c} = \mathbf{0} \tag{3.84}$$

which could be modified by the particular solutions $\mathbf{p}^o$ and $\mathbf{q}^o$ to the required form

$$\mathbf{Gq} - \mathbf{Fp} = \mathbf{Gq^o} - \mathbf{Fp^o} \tag{3.85}$$

where the relations $\mathbf{q^c} = \mathbf{q} - \mathbf{q}^o$ and $\mathbf{p^c} = \mathbf{p} - \mathbf{p}^o$ have been used.

Clearly the right-hand side involves the known quantities that could be taken as an additional right-hand side vector $\mathbf{b}^o$ and the left-hand side could be manipulated into a set involving the known and the unknown quantities $\{\mathbf{x}\}$ to produce

$$[\mathbf{A}]\{\mathbf{x}\} = \{\mathbf{b}\} + \{\mathbf{b}^o\} \tag{3.86}$$

3.8.1 Particular Integrals for Simple Sources

If the distribution of sources ψ in the interior could be approximated by simple polynomials over the whole region, the particular solutions could be easily determined. Thus if ψ is approximated as

$$\psi(x) = \sum_m K_m(x)\phi_m \tag{3.87}$$

where

$$K_m(x) = \begin{cases} [1, x_1, x_2, x_1^2, x_2^2, x_2 x_2] & \text{in two-dimensions} \\ [x, x_1, x_2, x_3, x_1^2, x_2^2, x_3^2, x_1 x_2, x_1 x_3, x_2 x_3] & \text{in three dimensions} \end{cases}$$

and ϕ_m = the coefficients of the polynomial

we can easily determine the particular solution of Eq. (3.83) as

$$p^o(x) = \sum_m P_m(x)\phi_m \tag{3.88}$$

where

$$P_1 = A_1 x_i x_i$$

$$P_m = A_2 x_i x_i x_j \qquad \text{for } m = 2, 3, 4 \text{ and } j = m - 1$$

$$P_m = A_3 x_j^4 \qquad \text{for } m = 5, 6, 7 \text{ and } j = m - 4$$

$$P_m = A_4 x_i x_i x_j x_k \qquad \text{for } m = 8(j = 1, k = 2), m = 9(j = 1, k = 3)$$

$$\text{and } m = 10(j = 2, k = 3)$$

$$A_1 = -1/(2d)$$

$$A_2 = -1/[2(d+2)]$$

$$A_3 = -1/2$$

$$A_4 = -1/[2(d+4)]$$

$d = $ the dimensionality of the problem

Using Eq. (3.88) and the relation

$$q^o(x) = -k \frac{\partial p^o(x)}{\partial x_i} n_i(x)$$

we can determine the particular solution for the flux $q^o(x)$.

If a very complex distribution of ψ is prescribed it is often preferable to subdivide the region into several zones such that within each zone the distribution could be reasonably approximated by Eq. (3.87). However, a very localized source distribution must be exactly represented by the volume integral

$$p^o(\xi^m) = \int_D G(x, \xi^m)\psi(x)\,\mathrm{d}D \tag{3.89}$$

where D represents this local volume within V. Obviously Eq. (3.89) is also a particular integral because it satisfies Eq. (3.83) at ξ.

3.9 RELATED PROBLEMS

3.9.1 Free Surface Flow

Many practical groundwater flow problems involve free surface boundary conditions (S_f in Fig. 3.9a). In such problems the free surface is a boundary along which the pressure head is zero, relative to atmospheric pressure, and the potential at a point on S_f is therefore simply its height (h) above an arbitrary datum (Fig. 3.9a). This figure shows a typical free surface for an earth dam which has two distinct sections with rather different boundary conditions on them. Along the whole of $S_f(1, 2, 3)$ we require $p(h) = h$ and, additionally, since $S_f(1, 2)$, the phreatic surface, is a streamline we have $q(h) = 0$ along $(1, 2)$. Section $(2, 3)$ is a seepage surface along which water emerges from the face of the dam and here we have h specified unambiguously with $q(h) \neq 0$ along $(2, 3)$. The precise location of $S_f(1, 2)$ is not known initially and has to be determined as part of the solution by an iterative

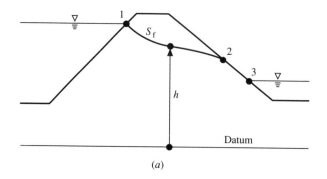

(a)

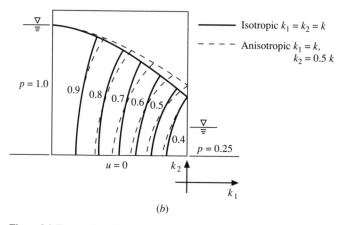

(b)

Figure 3.9 Free surface flow.

process. Niwa, Kobayashi and Fukui[9] were the first to tackle such a problem using the DBEM in conjunction with the following simple iterative scheme.

The location of $S_f(1, 2, 3)$ is guessed initially and $p(h) = h$ assumed along it. This problem is then solved by the direct BEM which generates automatically the values $q(h)$ along S_f (with, of course, $q(h) \neq 0$ in general). These values of $q(h)$ are used as the specified boundary conditions along $S_f(1, 2)$ and the problem solved again with no other changes. The output is now $p(h) = h'$, say, along S_f which generates an improved location S_f' of the free surface defined by h'. The iteration is repeated until h^n agree closely as required; Fig. 3.9b shows an example of the free surface and equipotentials obtained in this way by Niwa and his colleagues. Generally, for moving boundary problems the BEM is a very effective tool since the discretized boundary itself offers a great deal of control in the developed algorithm. This fact has been utilized by Banerjee, Liu, Liggett and others.[16, 17]

3.9.2 Shaft Torsion

The application of the BEM to the shaft torsion problem was discussed in detail by Mendleson,[5] using both warping functions and stress functions. He also extended his analysis from the purely elastic shaft to an elastoplastic one. Earlier Jaswon and Ponter[4] had presented solutions to the elastic torsion problem for a variety of solid and hollow shafts with different regular cross-sections, again using a warping function in a DBEM formulation.

The problem we shall consider here is the twisting of a uniform elastic shaft, of arbitrary cross-section, by a torque applied as a specifically distributed shear stress system on its ends, which are free to warp. One approach would be to treat the problem as one of elasticity, which it obviously is, and use the algorithms detailed in Chapter 4. However, shaft torsion, as one of the simpler problems in elasticity, can be described by harmonic equations, as shown by St Venant. We shall develop the BEM solution in terms of a harmonic warping function $p(x)$, defined below, for an elastic shaft of cross-sectional area A, periphery S (Fig. 3.10). A torque (τ) acts on all cross-sections to produce a twist α per unit shaft length, where $\alpha = \tau/GJ$ with G the shear modulus of the shaft material and J the polar second moment of area A about a longitudinal axis through its centroid. The only non-zero stresses in the shaft are the shear stresses $\sigma_{13} = \sigma_{31}$ in the (x_1, x_3) plane and $\sigma_{23} = \sigma_{32}$ in the (x_2, x_3) plane (Fig. 3.10), where x_3 is directed along the shaft axis. If these stress components are specified in terms of $p(x)$ as

$$\sigma_{13} = \alpha G\left(\frac{\partial p}{\partial x_1} - x_2\right) \qquad \sigma_{23} = \alpha G\left(\frac{\partial p}{\partial x_2} + x_1\right) \tag{3.90}$$

then it can be shown that $p(x_i)$ ($i = 1, 2$) has to satisfy

$$\frac{\partial^2 p(x)}{\partial x_i \, \partial x_i} = 0 \tag{3.91a}$$

If we define $v_i(x) = -\partial p/\partial x_i$ and $q(x) = v_i(x)n_i(x)$, where $n_i(x)$ are the components of a specified unit vector at x_i, the value of $q(x_0)$, x_0 being on S, has to satisfy

$$q(x) = n_2 x_1 - n_1 x_2 \tag{3.91b}$$

When formulated in this way we see immediately that Eqs (3.91a,b) correspond

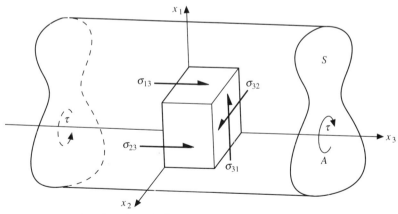

Figure 3.10 Torsion of a prismatic shape.

exactly to the two-dimensional potential flow problem with specified boundary fluxes and $\alpha = \tau/GJ$ being known. The solution for $p(x)$ then follows from either the direct or indirect BEM algorithms developed in this chapter. The stresses at any point involve $(\partial p/\partial x_i)$, the components of $v_i(x)$, which are again provided by the standard BEM analysis. However, it is important to note that in order to solve the final system equations it is necessary to assume $p(x) = 0$ at a boundary node.

It can also be shown that the torsional stiffness K_θ can be expressed as

$$K_\theta = \int_S \left[-p \frac{\partial p}{\partial n} \right] dS + \int_V (x_1^2 + x_2^2)\, dV \tag{3.92}$$

where a volume integral is involved. However, this volume integral can be converted into a surface integral by using the divergence theorem to yield

$$K_\theta = \int_S \left[-p \frac{\partial p}{\partial n} + \frac{r^3 n_r}{3} \right] dS \tag{3.93}$$

where $r^2 = x_i x_i\, (i = 1, 2)$ and n_r is the component of the normal in the direction r.

3.10 EXAMPLES OF SOLVED PROBLEMS

To demonstrate the general efficiency and precision of the BEM for potential flow problems we conclude Chapter 3 with a number of examples illustrating various features.

The first of these is a test problem with a well-known analytical solution—that of the flow under an impermeable dam sitting on the surface of an isotropic material, across which there is a hydraulic potential difference of 100 units. The streamlines for this solution are elliptical and if we distort one of them by a 5/2 scale transformation, as shown in Fig. 3.11, and use it as an outer impermeable

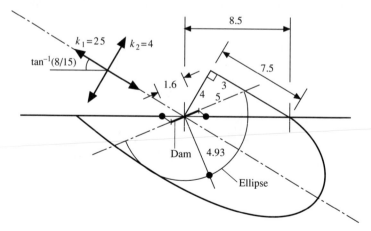

Figure 3.11 Geometric transformation.

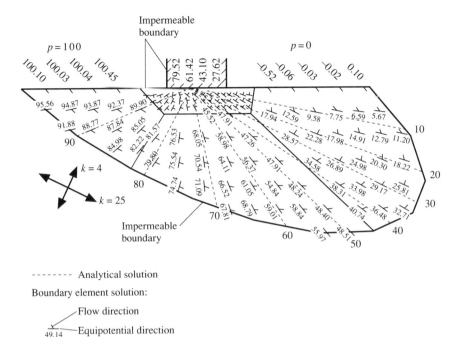

Figure 3.12 BEM results for flow beneath a dam.

boundary, we have a very convenient anisotropic test problem ($k_1 = 4$, $k_2 = 25$, say) with a known solution. If, further, we divide this region up arbitrarily into, say, five zones we can simultaneously check the precision of the zoned media algorithm. Tomlin and Butterfield[6, 7] did exactly this and obtained the solutions shown in Fig. 3.12 using the IBEM with a constant variation along each element.

Throughout most of the region the discrepancy between the calculated and analytical solutions for the potential is less than 1 per cent of the total head across the dam with a maximum error of about 2 per cent under the dam base. They used eight boundary elements under the dam and a total number of 74 elements, of which 21 were on interfaces.

Next, steady state heat conduction in a hollow cylinder is examined. For the specific case under investigation,[24, 25] the inner surface at $R_i = 30$ mm is maintained at 100°C, while the outer surface at $R_o = 80$ mm is kept at 0°C. The remaining top and bottom faces, at $Z = \pm 50$ mm, are insulated. An exact solution for the temperature at any position within the cylinder wall is well known and can be written as

$$\theta(R) = \frac{1}{\ln(R_o/R_i)}[(\theta_o - \theta_i)\ln R + \theta_i \ln R_o - \theta_o \ln R_i]$$

in which θ_i and θ_o respectively are the inner and outer surface temperatures. For the problem detailed above, this relationship produces a temperature of 38.20°C at the mid-surface of the wall.

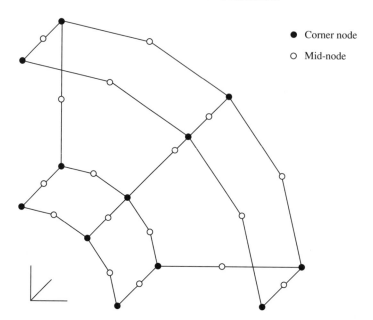

● Corner node

○ Mid-node

Figure 3.13 Potential flow in a hollow cylinder: boundary element model.

A coarse, three-dimensional boundary element idealization of the hollow cylinder was undertaken. This model employed three quadratic surface elements and 18 nodes, along with octahedral symmetry. Since the octahedral symmetry option was invoked, only the first octant was modelled and no surface elements were needed in the coordinate planes. A second, slightly more refined six-element, 29-node model is depicted in Fig. 3.13. For all BEMs described in this book we utilize a conforming element approach. Therefore, in both cases, adjacent elements share common nodes, resulting in a minimum number of degrees of freedom for a given element pattern; this of course requires a little more skill in programming.

Values for the mid-surface temperature are presented versus the total number of degrees of freedom in Fig. 3.14. It is seen that results converge quickly and, for the 29-node model, are within 0.05 per cent of the exact solution. Also, in the graph are results from a convergence study of this same problem conducted by others using a non-conforming (or discontinuous) element approach.[26] Non-conforming elements have the functional nodes repositioned interior to the edges, so that no interelement nodal sharing occurs. This alleviates some of the headaches associated with corners and, in general, simplifies the programming. However, a severe penalty is paid for this simplicity in terms of accuracy and increased computing costs. In fact, as can be seen in Fig. 3.14, even a 324-node non-conforming model does not provide the level of accuracy attained with the 29-node conforming element analysis.

As a final potential flow example, a quadratic DBEM is applied to examine steady state heat transfer in a printed circuit board (PCB).[24,25] In this problem,

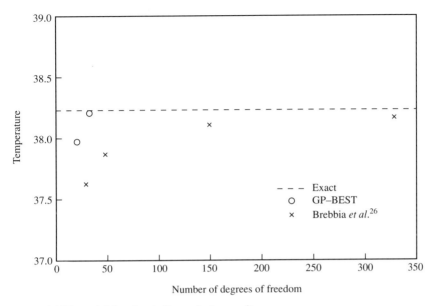

Figure 3.14 Potential flow in a hollow cylinder: results.

a very thin (0.0684 mm) copper layer resides on top of a thicker (1.9316 mm) dielectric layer composed of fibreglass cloth and epoxy resin. Of interest is the maximum temperature occurring in the PCB subjected to a localized steady input of heat along with convection on all exposed surfaces. Figure 3.15 shows a two-

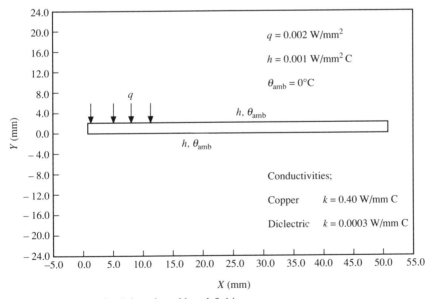

Figure 3.15 Printed circuit board: problem definition.

dimensional slice through the PCB, and includes the material properties and boundary conditions. The vertical edges are insulated to simulate symmetry. Interestingly, the copper layer is so thin that it becomes merely the heavy upper line on the cross-sectional diagram.

After magnifying the vertical scale, a two-dimensional, two-region boundary element mesh is plotted in Fig. 3.16. Each region consists of 18 elements and 36 nodes, with compatibility enforced at the copper/dielectric interface. Meanwhile, Fig. 3.17 displays the corresponding unscaled three-dimensional idealization.

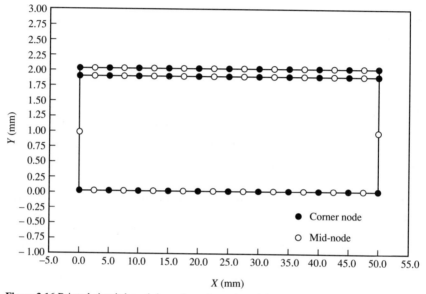

Figure 3.16 Printed circuit board: boundary element model.

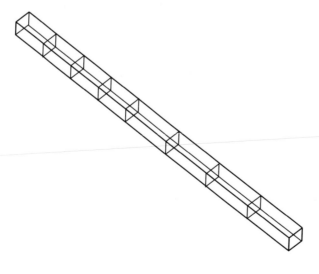

Figure 3.17 Potential flow in a hollow cylinder: boundary element model.

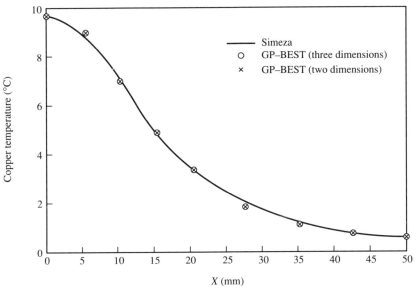

Figure 3.18 Printed circuit board: results.

Thus, once again the edges of the copper layer reduce to thick lines. Additionally, for the three-dimensional case, the front and back surfaces are insulated to simulate a two-dimensional heat flow.

A comparison of DBEM results is presented in Fig. 3.18. Notice that the temperature profile correlates quite well, even though the aspect ratio of the copper region is nearly 750, thus demonstrating the precision of the integration schemes and solutions using the DBEM. These results have been obtained using the general purpose BEM program GP-BEST.

3.11 CONCLUDING REMARKS

In this chapter we have extended the ideas behind the BEM from the one-dimensional problems of Chapter 1 to deal with potential flow in two and three dimensions. One of the more remarkable features of the analyses is that the steps in the solution procedures remain virtually unchanged by the increase in dimensionality of the problems.

It is worth emphasizing again that the necessary 'free space' unit solutions are well known for all the classical field equations, as are the relevant integral identities, which do not therefore have to be established before BEMs can be used. In fact, once the technique has been thoroughly understood the solution procedure involves merely the systematic assembly of matrix equations, their solution and backsubstitution into similar sets of equations to obtain values of the solution variables at any subsequently selected points.

We have taken particular care to bring out the 'two-point' nature of the singular solutions introduced and, at the expense of some of the equations looking rather cumbersome, persisted in tracing the roles of the two arguments through both the indirect and direct BEM procedures. Once the significance of the orderings of the argument has been fully appreciated then the compact matrix form of the discretized integral equations, which is very simple indeed, can be used.

Finally. we would strongly recommend careful study of Chapters 1 to 3 before embarking upon Chapter 4 so that the underlying operations are quite clearly understood, since they carry through identically into the solution of problems in elasticity where slight additional algebraic complexity arises in connection with the higher-rank kernel functions.

REFERENCES

1. Kellog, O.D. (1929) *Foundations of Potential Theory*, Julius Springer, Berlin.
2. Jaswon, M.A. (1963) Integral equation methods in potential theory, I, *Proceedings of Royal Society, London*, Vol. 275(A), pp. 23–32.
3. Symm, G.T. (1963) Integral equation methods in potential theory, II, *Proceedings of Royal Society, London*, Vol. 275(A), pp. 33–46.
4. Jaswon, M.A. and Ponter, A.R. (1963) An integral equation solution of the torsion problem, *Proceedings of Royal Society, London*, Vol. 275(A), pp. 237–246.
5. Mendleson, A. (1973) Boundary integral methods in elasticity and plasticity, NASA Technical Note TND-7418, 36 pp.
6. Butterfield, R. (1972) The application of integral equation methods to continuum problems in soil mechanics, in *Stress Strain Behaviour of Soils*, Roscoe Memorial Symposium, Cambridge, Foulis, pp. 573–587.
7. Butterfield, R. and Tomlin, G.R. (1972) Integral techniques for solving zoned anisotropic continuum problems, International Conference on *Various Methods in Engineering*, Southampton University, pp. 9/31–9/53.
8. Tomlin, G.R. (1972) Numerical analysis of continuum problems in zoned anisotropic media, PhD Thesis, Southampton University.
9. Niwa, Y., Kobayashi, S. and Fukui, T. (1974) An application of the integral equation method to seepage problems, Proceedings of Twenty-fourth Japanese National Conference on *Applied Mechanics*, pp. 470–486.
10. Chang, Y.P., Kang, C.S. and Chen, D.J. (1973) The use of fundamental Greens functions for the solution of problems of heat conduction in anisotropic media, *International Journal of Heat and Mass Transfer*, Vol. 16, pp. 1905–1918.
11. Banerjee, P.K. and Butterfield, R. (1981) *Boundary Element Methods in Engineering Science*, McGraw-Hill, London.
12. Liggett, J. (1977) Location of free surface in porous media, *Journal of Hydraulic Division, ASCE*, Vol. HY4, 353–365.
13. Jaswon, M.A. and Symm, G.T. (1977) *Integral Equation Methods in Potential Theory and Elastostatics*, Academic Press, London.
14. Hess, J.L. and Smith, A.M.O. (1964) Calculations of nonlifting potential flow about arbitrary three-dimensional bodies, *Journal of Ship Research*, Vol. 8(2), pp. 22–44.
15. Hess, J.L. and Smith, A.M.O. (1966) Calculations of potential flow about arbitrary bodies, in *Progress in Aeronautical Sciences*, Vol. 8, pp. 1–138, Pergamon Press, New York.
16. Banerjee, P.K. (1979) Nonlinear problems of potential flow, Chapter 2, in *Developments in BEM*, Vol. 1, Eds. P.K. Banerjee and R. Butterfield, Elsevier Applied Science, London.

17. Liggett, J.A. and Liu, P.L.F. (1983) *The Boundary Integral Equation Method for Porous Media Flow*, Allen and Unwin, London.
18. Fukui, T., Fukuhara, T. and Furuichi, T. (1988) Three-dimensional analysis of a fresh water lens in an island, *Proceedings of U.S. Japan Seminar on BEM, Boundary Element Methods in Applied Mechanics*, Eds. T. Tanaka and T.A. Cruse, Pergamon Press, New York.
19. Watson, J.O. (1979) Two and three-dimensional problems of elasticity, Chapter 3 in *Developments in Boundary Element Methods*, Vol. 1, Eds. P.K. Banerjee and R. Butterfield, Elsevier Applied Science, London.
20. Mustoe, G.G.W. (1984) Advanced integration schemes over boundary elements and volume cells for two and three-dimensional nonlinear analysis, Chapter 9 in *Developments in Boundary Element Methods*, Vol. 4, Eds. P.K. Banerjee and S. Mukherjee, Elsevier Applied Science, London.
21. Telles, J.C.F. (1987) A self adaptive coordinate transformation for efficient numerical evaluation of general boundary element integrals, *International Journal of Numerical Methods in Engineering*, Vol. 24, pp. 959–973.
22. Butenschon, H.J., Mohrmann, W. and Bauer, W. (1989), Advanced stress analysis by a commerical BEM code, Chapter 8 in *Industrial Application of BEM*, Eds P.K. Banerjee and R.B. Wilson, Elsevier Applied Science, London.
23. Lamb, H. (1932) *Hydrodynamics*, 6th edn, Dover, New York.
24. Dargush, G. (1987) Boundary element methods for the analogous problems of thermomechanics and soil consolidation, PhD Dissertation, State University of New York at Buffalo, Buffalo, NY.
25. Dargush, G. and Banerjee, P.K. (1989) Advanced development of BEM for steady state heat conduction, *International Journal of Numerical Methods in Engineering*, Vol. 28, pp. 2123–2142.
26. Brebbia, C., Telles, J.C.F. and Wrobel, L.C. (1984) *Boundary Element Techniques*, Springer-Verlag, Berlin.

FOUR

TWO- AND THREE-DIMENSIONAL PROBLEMS OF ELASTOSTATICS

4.1 INTRODUCTION

This chapter describes the development and application of the BEM to the numerical solution of two- and three-dimensional problems of small-strain elastostatics. Most of the theoretical background behind the derivations presented in this chapter was explored in Chapter 3. The development of the BEM for elasticity problems follows[1-4] that of potential theory very closely. However, unlike the integral equations of potential theory, which are scalar equations, the resulting integral equations in elasticity are vector equations. The basic singular solutions for elasticity, as would be expected, are more complex than those of potential theory. Therefore, in order to introduce them in a compact manner we have made use of some elementary operations involving indicial notation. The reader unfamiliar with this notation is advised to read Chapter 2, which explains the various symbolic operations that have been used.

4.2 GOVERNING EQUATIONS

Consider an isotropic elastic body referred to a Cartesian coordinate system as shown in Fig. 4.1, with axes x_i (with $i = 1, 2$ for two dimensions). The governing differential equation of equilibrium for an element of the body can be written as

$$\frac{\partial \sigma_{ij}}{\partial x_j} + \psi_i = 0 \tag{4.1}$$

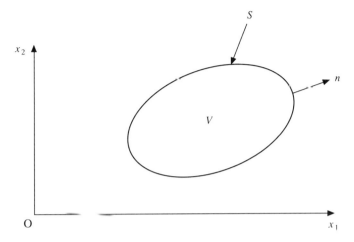

Figure 4.1 A solid V bounded by boundary S.

in which σ_{ij} are the stress components and ψ_i the components of the body forces per unit volume. Hooke's law relating the stress and strain components in an isotropic elastic solid can be written as

$$\sigma_{ij} = \lambda \delta_{ij} \varepsilon_{kk} + 2\mu \varepsilon_{ij} \tag{4.2}$$

where λ and μ are elastic constants and δ_{ij} is the Kronecker delta symbol. The strains and displacements are related by

$$\varepsilon_{ij} = \frac{1}{2} \left(\frac{\partial u_i}{\partial x_j} + \frac{\partial u_j}{\partial x_i} \right) \tag{4.3}$$

Substituting Eq. (4.2) in Eq. (4.1) and using Eq. (4.3) we obtain Navier's equations for equilibrium expressed in terms of the displacement components:

$$(\lambda + \mu) \frac{\partial^2 u_j}{\partial x_i \partial x_j} + \mu \frac{\partial^2 u_i}{\partial x_j \partial x_j} + \psi_i = 0 \tag{4.4}$$

Equation (4.4) is the governing differential equation for our problem which has to be solved subject to certain boundary conditions. For example, over a part of the boundary S displacements are assumed to be known, i.e.

$$u_i(x) = g_i(x) \tag{4.5}$$

while over the remaining part of S the surface tractions are prescribed as

$$t_i(x) = \sigma_{ij}(x) n_j(x) = h_i(x) \tag{4.6}$$

where $g_i(x)$ and $h_i(x)$ are the specified conditions on the boundary.

4.3 SINGULAR SOLUTIONS

The fundamental singular solutions play the same important role in the BEM algorithms as did their counterparts in the previous potential flow problems. The classical Kelvin solution which evaluates the displacement field $u_i(x)$ due to a concentrated unit force $e_j(\xi)$ within an elastic body forms the basis of all the subsequent analysis. For plane strain problems[3,5]

$$u_i(x) = G_{ij}(x, \xi)e_j(\xi) \tag{4.7}$$

where

$$G_{ij}(x, \xi) = C_1\left(C_2\delta_{ij}\ln r - \frac{y_i y_j}{r^2}\right)$$

$$C_1 = -\frac{1}{8\pi\mu(1-\nu)}$$

$$C_2 = 3 - 4\nu$$

$$y_i = x_i - \xi_i$$

$$r^2 = y_i y_i$$

The strains corresponding to the above displacement field can be obtained by substituting Eq. (4.7) into the strain displacement relations (4.3) as

$$\varepsilon_{ij}(x) = B_{ijk}(x, \xi)e_k(\xi) \tag{4.8}$$

where

$$B_{ijk}(x, \xi) = \frac{C_1}{r^2}\left[(1 - 2\nu)(\delta_{ik}y_j + \delta_{jk}y_i) + 2\frac{y_i y_j y_k}{r^2} - \delta_{ij}y_k\right]$$

The corresponding stresses can be deduced from the stress–strain relations as

$$\sigma_{ij}(x) = T_{ijk}(x, \xi)e_k(\xi) \tag{4.9}$$

where

$$T_{ijk}(x, \xi) = \frac{C_3}{r^2}\left[C_4(\delta_{ik}y_j + \delta_{jk}y_i - \delta_{ij}y_k) + \frac{2y_i y_j y_k}{r^2}\right]$$

$$C_3 = \frac{1}{4\pi(1-\nu)}$$

$$C_4 = 1 - 2\nu$$

We shall also require the surface tractions $t_i(x)$ at a point (x_i) on a surface with outward normal $n_j(x)$ which are calculated from

$$t_i(x) = \sigma_{ij}(x)n_j(x)$$
$$= F_{ik}(x, \xi)e_k(\xi) \tag{4.10}$$

where, in our case,

$$F_{ik} = \left(\frac{C_3}{r^2}\right)\left[C_4(n_k y_i - n_i y_k) + \left(C_4 \delta_{ik} + \frac{2 y_i y_k}{r^2}\right) y_j n_j\right]$$

Equations (4.9) and (4.10) provide all the displacement, stress, strain and surface traction components of the unit solution that we shall require. Solutions for the corresponding plane stress problem can be obtained from these for the plane strain case given above by using an effective Poisson ratio $\bar{\nu} = \nu/(1 + \nu)$.

It is worth noting here that the various functions G_{ij}, B_{ijk}, T_{ijk}, F_{ik} are singular when the load point and the field point coincide ($x_i = \xi_i$). The singularity in G_{ij} is in the term $\ln r$ (weakly singular) and that in others in $1/r$ (strongly singular). Integrals involving weakly singular functions will always exist in the normal sense of integration, even when $x_i = \xi_i$, but those involving strongly singular terms must be interpreted in the sense of a limiting value of the integral as the field point approaches the load point on the boundary.

Some readers, particularly those who prefer matrix notation, might be dismayed by the apparently unnecessary complexity of Eq. (4.7) which could, after all, be equally well written in matrix notation as $\mathbf{u} = \mathbf{Ge}$, and since ε, a strain vector, can also be expressed symbolically as $\varepsilon = \mathbf{Lu}$, Eq. (4.8) can be stated simply as

$$\varepsilon = \begin{Bmatrix} \varepsilon_{11} \\ \varepsilon_{22} \\ \gamma_{12} \end{Bmatrix} = \begin{bmatrix} \dfrac{\partial}{\partial x_1} & 0 \\ 0 & \dfrac{\partial}{\partial x_2} \\ \dfrac{\partial}{\partial x_2} & \dfrac{\partial}{\partial x_1} \end{bmatrix} \begin{Bmatrix} u_1 \\ u_2 \end{Bmatrix} = \mathbf{Lu} = \mathbf{LGe}$$

However, in order to use this equation one has to evaluate separately each of the differential operations on every term in G—not a simple matter when G is a quite complicated function. On the other hand, a brief study of the indicial notation in Chapter 2 will enable the reader to appreciate how products like $B_{ijk}e_k$ are, in fact, a concise way of presenting the functions in a form convenient for computer coding.

For three-dimensional homogeneous, isotropic, elastic media, the corresponding components of the point force solution required are also that due to Kelvin:[3, 5]

$$G_{ij}(x, \xi) = \frac{1}{16\pi\mu(1 - \nu)} \frac{1}{r}\left[(3 - 4\nu)\delta_{ij} + \frac{y_i y_j}{r^2}\right] \tag{4.7a}$$

$$B_{ijk}(x, \xi) = -\frac{1}{16\pi\mu(1 - \nu)} \frac{1}{r^2}\left[(1 - 2\nu)\left(\delta_{ik}\frac{y_j}{r} + \delta_{jk}\frac{y_i}{r}\right) - \delta_{ij}\frac{y_k}{r} + \frac{3 y_i y_j y_k}{r^3}\right] \tag{4.8a}$$

$$T_{ijk}(x, \xi) = \frac{1}{8\pi(1 - \nu)} \frac{1}{r^2}\left[\left(\delta_{ik}\frac{y_j}{r} + \delta_{jk}\frac{y_i}{r} - \delta_{ij}\frac{y_k}{r}\right)(1 - 2\nu) + \frac{3 y_i y_j y_k}{r^3}\right] \tag{4.9a}$$

and $$F_{ij}(x, \xi) = -\frac{1}{8\pi(1 - \nu)} \frac{1}{r^2}\left\{(1 - 2\nu)\left(n_j\frac{y_i}{r} - n_i\frac{y_j}{r}\right) + \left[\frac{3 y_i y_j}{r^2} + (1 - 2\nu)\delta_{ij}\right]\frac{y_k n_k}{r}\right\}$$

$$\tag{4.10a}$$

It should be noted that for some problems with an unloaded half-space the use of Mindlin's solution results in substantial savings in computational efforts.[6-8]

4.4 DIRECT BOUNDARY ELEMENT FORMULATION

The direct formulation of the BEM is most conveniently approached from the reciprocal work theorem. This theorem, which we derived in Chapter 2 using integration by parts, can be simply stated as follows: if two distinct elastic equilibrium states (ψ_i^*, t_i^*, u_i^*), (ψ_i, t_i, u_i) exist in a region V bounded by the surface S then the work done by the forces of the first system $(*)$ on the displacements of the second is equal to the work done by the forces of the second system on the displacements of the first $(*)$. Thus

$$\int_S t_i^*(x)u_i(x)\,dS(x) + \int_V \psi_i^*(z)u_i(z)\,dV(z) = \int_S t_i(x)u_i^*(x)\,dS(x)$$

$$+ \int_V \psi_i(z)u_i^*(z)\,dV(z) \qquad (4.11)$$

where x is a point on S and z is a point in V.

If we choose the actual state of displacements, tractions and body forces as u_i, t_i and ψ_i respectively and the $(*)$ system as those corresponding to a unit force system in an infinite solid, as described above, we can write, from Eq. (4.11)

$$\int_S F_{ij}(x,\xi)u_i(x)\,dS(x) + \int_V \delta_{ij}\delta(z,\xi)u_i(z)\,dV(z) = \int_S t_i(x)G_{ij}(x,\xi)\,dS(x)$$

$$+ \int_V \psi_i(z)G_{ij}(z,\xi)\,dV(z) \qquad (4.12)$$

In writing this equation we have made use of Eq. (4.10), etc., $t_i^*(x) = F_{ij}(x,\xi)$ and the following manipulation of the unit force $\psi_i^*(z) = e_i^*(z)$ in the second left-hand side term: namely,

$$\int_V e_i^*(z)u_i(z)\,dV(z) = \int_V e_i^*(\xi)\delta(z,\xi)u_i(z)\,dV(z) = \int_V e_j^*(\xi)\delta_{ij}\delta(z,\xi)u_i(z)\,dV(z)$$

The $e_j^*(\xi)$ term is now common to all the integrals and may be removed. This particular term can be simplified further by noting that

$$\int_V \delta_{ij}\delta(z,\xi)u_i(z)\,dV(z) = \int_V u_j(z)\delta(z,\xi)\,dV(z) = cu_j(\xi)$$

where $c = 1$ within V and $c = 0$ outside S. Hence from Eq. (4.12) we arrive at

$$u_j(\xi) = \int_S [t_i(x)G_{ij}(x,\xi) - F_{ij}(x,\xi)u_i(x)]\,dS(x) + \int_V \psi_i(z)G_{ij}(z,\xi)\,dV(z) \qquad (4.12a)$$

Equation (4.12a) now provides the displacement $u_j(\xi)$ at an interior point (ξ) due to any admissible combinations of t_i and u_i over S and a given distribution of ψ_i

within the volume—this equation is, in fact, well known as Somigliana's identity for the displacement vector. The functions G_{ij} and F_{ij} are precisely those defined in Eqs (4.7) and (4.10) but, stemming from the use of the reciprocal theorem, three rather subtle changes have occurred in the significance of (x, ξ) and (i, j) compared to their role in Sec. 4.3. A careful study will show that Eq. (4.12a) embodies an extension of the features discussed in connection with potential flow, namely:

1. The first argument (x) of the kernel function is now the load point (not ξ, which is now the field point coordinate) and the integration is carried through with respect to x.
2. Summation is over i, not j (i.e. the roles of i and j are reversed).
3. The normal vector n_i in F_{ij} is now through the integration point x on the surface.

In functions such as $G_{ij}(x, \xi)$ which are symmetrical with respect to both (i, j) and (x, ξ) these changes have no effect. However, for $F_{ij}(x, \xi)$, which is antisymmetric in both (i, j) and (x, ξ), the above changes must be observed very carefully.

By bringing the field point (ξ) on to a point x_0 on the surface of the domain and constructing a local coordinate system as shown in Fig. 4.2, we obtain the following results (for $\lim \xi \to \xi_0$):

$$\int_S t_i(x) G_{ij}(x, \xi) \, dS(x) = \int_S t_i(x) G_{ij}(x, \xi_0) \, dS(x)$$

$$\oint_S F_{ij}(x, \xi) u_i(x) \, dS(x) = \alpha_{ij} u_i(\xi_0) + \int_S F_{ij}(x, \xi_0) u_i(x) \, dS(x)$$

where $\alpha_{ij} = -\frac{1}{2} \delta_{ij}$ for the 'interior' problem with V inside a smooth boundary S.

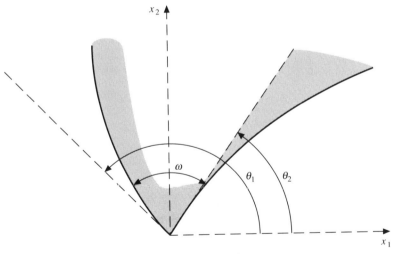

Figure 4.2 Geometry of a non-smooth boundary point.

When the field point ξ_0 is at a corner with a subtended angle ω the discontinuity term $\alpha_{ij}(\xi_0)$ becomes the function of geometry at the corner.

Substituting the above in Eq. (4.12a) for a general field point ξ we can express the integral statement as

$$c_{ij}(\xi_0)u_i(\xi_0) = \int_S [t_i(x)G_{ij}(x, \xi_0) - F_{ij}(x, \xi_0)u_i(x)] \, dS$$
$$+ \int_V \psi_i(z)G_{ij}(z, \xi_0) \, dV \tag{4.13}$$

where $c_{ij}(\xi_0) = \delta_{ij} - \alpha_{ij}(\xi_0)$.

For a two-dimensional problem the $c_{ij}(\xi_0)$ term can be determined as

$$c_{ij} = -a_3 \begin{bmatrix} a_4\omega + \frac{1}{2}(\sin 2\theta_1 - \sin 2\theta_2) & \sin^2 \theta_1 - \sin^2 \theta_2 \\ \sin^2 \theta_1 - \sin^2 \theta_2 & a_4\omega - \frac{1}{2}(\sin 2\theta_1 - \sin 2\theta_2) \end{bmatrix} \tag{4.14}$$

where $a_3 = 1/[4\pi(1 - \nu)]$ and $a_4 = 2(1 - \nu)$.

When the point ξ_0 is on a smooth boundary we have $\omega = \pi, \theta_1 = \pi, \theta_2 = 0$, leading to $c_{ij} = \frac{1}{2}\delta_{ij}$. Similar calculations can also be done for a three-dimensional situation which unfortunately leads to much more complex algebra. Fortunately, however, we shall later see that it is not really necessary to evaluate these terms because we can always determine them indirectly by using a physically meaningful solution.

Hence for a point ξ_0 on a the smooth surface we can write Eq. (4.13), assuming $\psi_i(z) = 0$ for convenience, as

$$\tfrac{1}{2}u_j(\xi_0) = \int_S [t_i(x)G_{ij}(x, \xi_0) - F_{ij}(x, \xi_0)u_i(x)] \, dS \tag{4.15}$$

Equation (4.15) is the required boundary integral representation we require for the solution of any well-posed boundary value problem. We can now use Eq. (4.12a) and the strain displacement relationship to obtain

$$\varepsilon_{ij} = \int_S [t_k(x)G^\varepsilon_{kij}(x, \xi) - F^\varepsilon_{kij}(x, \xi)u_k(x)] \, dS \tag{4.16}$$

where

$$G^\varepsilon_{kij} = \frac{1}{2}\left(\frac{\partial G_{ki}}{\partial \xi_j} + \frac{\partial G_{kj}}{\partial \xi_i}\right)$$

$$F^\varepsilon_{kij} = \frac{1}{2}\left(\frac{\partial F_{ki}}{\partial \xi_j} + \frac{\partial F_{kj}}{\partial \xi_i}\right)$$

and the stresses by using the stress–strain relationship as

$$\sigma_{ij}(\xi) = \int_S [t_k(x)G^\sigma_{kij}(x, \xi) - F^\sigma_{kij}(x, \xi)u_k(x)] \, dS \tag{4.17}$$

where

$$G_{kij}^{\sigma} = \frac{1}{4\pi\alpha(1-\nu)} \frac{1}{r^{\alpha}} \left[(1-2\nu)(\delta_{ki}r_{,j} + \delta_{kj}r_{,i} - \delta_{ij}r_{,k}) + \beta r_{,i}r_{,j}r_{,k} \right]$$

$$F_{kij}^{\sigma} = \frac{\mu}{2\pi\alpha(1-\nu)} \frac{1}{r^{\beta}} \left\{ \beta \frac{\partial r}{\partial n} \left[(1-2\nu)\delta_{ij}r_{,k} + \nu(\delta_{ik}r_{,j} + \delta_{jk}r_{,i}) - \phi r_{,i}r_{,j}r_{,k} \right] \right.$$

$$\left. + \beta\nu(n_i r_{,j}r_{,k} + n_j r_{,i}r_{,k}) + (1-2\nu)(\beta n_k r_{,i}r_{,j} + n_j \delta_{ik} + n_i \delta_{jk}) - (1-4\nu)n_k \delta_{ij} \right\}$$

where $\alpha = 2$, $\beta = 3$ and $\phi = 5$ for the three-dimensional problems and $\alpha = 1$, $\beta = 2$ and $\phi = 4$ apply to two-dimensional problems.

4.5 DISCRETIZATION OF THE BOUNDARY INTEGRALS

As in Chapter 3 (Fig. 3.3), if we divide the boundary of the region into N straight line segments, we can write Eq. (4.15) for the displacement vector of a representative nodal point on the pth boundary element with the assumption of constants t_i, u_i as

$$\tfrac{1}{2}u_j(\xi_0^p) = \sum_{q=1}^{N} \left[t_i(x^q) \int_{\Delta S} G_{ij}(x^q, \xi_0^p) \, dS(x^q) - u_i(x^q) \int_{\Delta S} F_{ij}(x^q, \xi_0^p) \, dS(x^q) \right] \quad (4.18)$$

This equation can be rewritten in matrix notation as

$$\tfrac{1}{2}\mathbf{I}\mathbf{u}^p = \sum_{q=1}^{N} \left\{ \left[\int_{\Delta S} \mathbf{G}^{pq} \, dS \right] \mathbf{t}^q - \left[\int_{\Delta S} \mathbf{F}^{pq} \, dS \right] \mathbf{u}^q \right\} \quad (4.19)$$

On the other hand, if we assume both the unknown and known values of tractions and displacements to vary linearly or quadratically over the boundary segments the equivalent equation becomes

$$\mathbf{c}\mathbf{u}^p = \sum_{q=1}^{N} \left\{ \left[\int_{\Delta S} \mathbf{G}^{pq}\mathbf{N}^q \, dS \right] \mathbf{t}_n - \left[\int_{\Delta S} \mathbf{F}^{pq}\mathbf{N}^q \, dS \right] \mathbf{u}_n \right\} \quad (4.20)$$

where $\mathbf{t}_n, \mathbf{u}_n$ are nodal values of tractions and displacements respectively, $\mathbf{u}^p$ is the displacement vector of a representative point on the pth boundary element $\mathbf{N}^q$ is the shape function for the qth boundary element and $\mathbf{c}$ is a 2×2 matrix in two dimensions and a 3×3 matrix in three dimensions, arising from the fact that if the physical problem has edges or corners, we may have a field point in such locations.

By taking the field point successively to all the nodal points on the boundary and absorbing the $\mathbf{c}$ matrix with the corresponding leading diagonal blocks of coefficients of $[\int \mathbf{F}^{pq}\mathbf{N}^q \, dS]$ we can write

$$\sum_{q=1}^{N} \left\{ \left[\int_{\Delta S} \mathbf{G}^{pq} N^q \, dS \right] \mathbf{t}_n - \left[\int_{\Delta S} \mathbf{F}^{pq} N^q \, dS \right] \mathbf{u}_n \right\} = 0 \quad (4.21)$$

or more compactly

$$\mathbf{Gt} - \mathbf{Fu} = 0 \tag{4.22}$$

Note here that with constant elements with the field point ξ at the centre of each element both G and F matrices are square. On the other hand, with higher-order boundary elements on a non-smooth surface, only the F matrix would be square and the matrix G must remain an unassembled rectangular matrix (see Sec. 3.5). After incorporating all known tractions and by eliminating any remaining extraneous (multi-valued) tractions in any directions either by simply making them equal at a node or more satisfactory by eliminating them using a suitable extrapolation procedure, the total number of equations and unknowns can be made the same.

If we now apply a rigid body displacement to the body (we can do this to any region of finite size), this will generate no traction on the boundary (that is $\mathbf{t} = 0$). Therefore we can write Eq. (4.22) as

$$\mathbf{Fu} = 0 \tag{4.23}$$

It is easy to see that for the validity of the above equation for any system of arbitrary rigid body displacements each coefficient of the 2×2 or 3×3 on-diagonal blocks must be numerically equal to the sum of the corresponding coefficients of all off-diagonal blocks with a change in sign. Since these diagonal blocks contained the terms involving $\mathbf{c}$ as well as strongly singular integrals, the evaluation of which require considerable effort, this method for determining the diagonal block components by utilizing the values of the off-diagonal blocks is a very attractive feature of the direct boundary element method.

We see immediately that the above process is virtually identical to the detailed numerical treatment described for potential flow in Chapter 3, except that here the various boundary quantities $\mathbf{t}$ and $\mathbf{u}$ are vectors (each with two components in two dimensions and three components in three dimensions) as opposed to the scalar quantities q and p respectively that we became accustomed to in Chapter 3. Correspondingly, the kernel functions $\mathbf{G}$ and $\mathbf{F}$ become 2×2 or 3×3 entries, depending on space dimensions, in the system matrices, whereas in the potential flow these were single coefficients.

In addition, the behaviour of these kernels $\mathbf{G}$ and $\mathbf{F}$ are almost identical to those of potential flow and therefore the numerical integration strategy would also be identical, as would be the scaling of coefficients (here, we scale by shear modulus) as well as the multi-region assembly techniques described in Chapter 3 where the interface conditions between two regions $1, 2$ would now be $\mathbf{u}_{12} - \mathbf{u}_{21} = 0$ and $\mathbf{t}_{12} + \mathbf{t}_{21} = 0$.

There are, however, some differences. In elasticity the boundary loading can be quite complex, which results in a much more complex deformation field, and therefore the simple constant variations of functions over each boundary element will usually not be sufficient even for the simplest of problems (bending of a beam, for example). Hence, the lowest order of boundary element that we have to consider would be the one with a linear variation of $\mathbf{t}$ and $\mathbf{u}$ over each element. This, of course, means that our $\mathbf{G}$ matrix in the final discretized system will not be square and we must utilize one of the more complex assemblies described in Sec. 3.6. Additionally, in elasticity problems local boundary conditions as well as sliding interfaces require the variables over these elements to be transformed to a local coordinate system. This creates a considerable housekeeping effort in programming.

We shall, however, try to keep the numerical implementation simple here and consider essential elements of the BEM in elasticity using linear variations of t and u over straight lines (in a two-dimensional analysis) and flat triangles or rectangles (in a three-dimensional problem). Although for all such cases the integrals involved in a BEM analysis can be calculated analytically, the algebra involved is quite formidable. We shall therefore, for demonstration purposes, use one involving less formidable algebra.

4.5.1 Integrations over Line Elements

If we closely examine the shape functions $\mathbf{N}^q$ we can see that we can split it into two parts: where N_1^q are the terms that are constant within the element and $\mathbf{N}_2^q$ are the terms that contain the variable of integration. Hence we can write the surface integrals in Eq. (4.21) as

$$\int_{\Delta S} \mathbf{G}^{pq} \mathbf{N}^q \, \mathrm{d}S = \left[\int_{\Delta S} \mathbf{G}^{pq} \, \mathrm{d}S \right] \mathbf{N}_1^q + \int_{\Delta S} \mathbf{G}^{pq} \mathbf{N}_2^q \, \mathrm{d}S$$

and

$$\int_{\Delta S} \mathbf{F}^{pq} \mathbf{N}^q \, \mathrm{d}S = \left[\int_{\Delta S} \mathbf{F}^{pq} \, \mathrm{d}S \right] \mathbf{N}_1^q + \int_{\Delta S} \mathbf{F}^{pq} \mathbf{N}_2^q \, \mathrm{d}S$$

The second integrals on the right-hand side of the above equations can be calculated quite accurately by using Gaussian integration formulae. The first integrals can be evaluated analytically by constructing a local coordinate system (y_i) on the loaded element such that y_1 is normal to the element and y_2 is the positive tangential direction. Thus if the direction cosines of the local axes y_1 and y_2 with respect to the global axes are given by e_{ij} then

$$\int_{\Delta S} G_{ij} \, \mathrm{d}S = e_{ri} e_{sj} \int_{\Delta S} G'_{rs} \, \mathrm{d}y_2 \tag{4.24}$$

$$\int_{\Delta S} F_{ik} \, \mathrm{d}S = e_{ri} e_{sk} \int_{\Delta S} F'_{rs} \, \mathrm{d}y_2 \tag{4.25}$$

where G'_{rs} and F'_{rs} are the expressions for G_{rs} and F_{rs} referred to the y_1, y_2 axes.

The integrals on the right-hand side of Eqs (4.24) and (4.25) can be calculated exactly by working with a polar coordinate system as shown in Fig. 4.3. Equations (4.24) and (4.25) can be written as

$$\int_{\Delta S} G_{ij} \, \mathrm{d}S = e_{1i} (e_{1j} \Delta G'_{11} + e_{2j} \Delta G'_{12}) + e_{2i} (e_{1j} \Delta G'_{21} + e_{2j} \Delta G'_{22})$$

$$\int_{\Delta S} F_{ik} \, \mathrm{d}S = e_{1i} (e_{1j} \Delta F'_{11} + e_{2j} \Delta F'_{12}) + e_{2i} (e_{1j} \Delta F'_{21} + e_{2j} \Delta F'_{22})$$

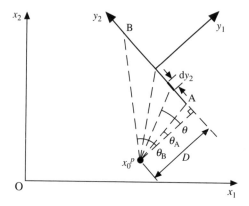

Figure 4.3 A local axes on a three-dimensional element.

where $\Delta G'_{11}$, $\Delta F'_{11}$, $\Delta G'_{12}$, $\Delta F'_{12}$, etc., are the results of integrating expressions like $\int_{\Delta S} G'_{11} \, dy_2$, $\int_{\Delta S} F'_{11} \, dy_2$, etc. Thus, for a field point inside the region (the results for a field point indefinitely close to the boundary node can be obtained from these, as shown in Fig. 4.4, see also Sec. 3.4.3) we have[5]

$$\int_{\Delta S} G_{ij} \, dS = C_1 D \left[C_2 \delta_{ij} \{ \tan\theta(\ln r - 1) + \theta \} - e_{1i} e_{1j} \theta \right.$$
$$\left. - (e_{1i} e_{2j} + e_{2i} e_{1j}) \ln r - e_{2i} e_{2j} (\tan\theta - \theta) \right]_{\theta_a, r_a}^{\theta_b, r_b} \tag{4.26}$$

and

$$\int_{\Delta S} F_{ij} \, ds = C_3 \left[C_4 \delta_{ij} \theta + e_{1i} e_{1j} (\theta + \sin\theta\cos\theta) + (e_{1i} e_{2j} + e_{2i} e_{1j}) \sin^2\theta \right.$$
$$\left. + e_{2i} e_{2j} (\theta - \sin\theta\cos\theta) + C_4 (e_{1i} e_{2j} - e_{2i} e_{1j}) \ln r \right]_{\theta_a, r_a}^{\theta_b, r_b} \tag{4.27}$$

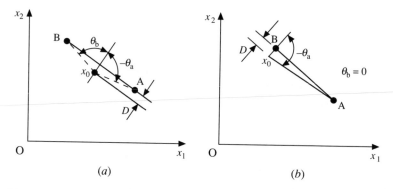

(a)

(b)

Figure 4.4 The limit of a field point x_0 as it approaches the boundary element.

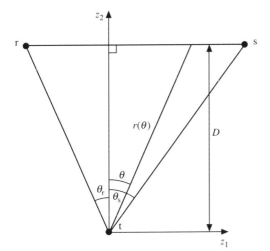

Figure 4.5 The local axes on a three-dimensional element.

4.5.2 Integration over Three-Dimensional Surface Elements

The integration schemes for the singular and non-singular parts will be identical to those discussed in earlier sections; namely when the kernel shape function products remain finite throughout the interval of integration we use numerical integration but the singular case (i.e. when the field point is at a node where the shape function tends to unity) is isolated and integrated exactly.

Evaluations of $\int_{\Delta S} G_{ij}(x, \xi_0) \, dS(x)$ By considering a local coordinate system z_1 and z_2 in the plane of the element and z_3 in the direction normal through the node under consideration (see Fig. 4.5), we have

$$\frac{y_i}{r} = \frac{\partial r}{\partial x_i} = \sin\theta \frac{\partial z_1}{\partial x_i} + \cos\theta \frac{\partial z_2}{\partial x_i} = \sin\theta e_{1i} + \cos\theta e_{2i}$$

where e_{1i} and e_{2i} are the direction cosines of the z_1 and z_2 axes in the global coordinate system.

We can then recast the integral

$$\int_{\Delta S} G_{ij}(x, \xi) \, dS(x)$$

$$= \int_{-\theta_r}^{\theta_s} \int_0^{r(\theta)} A \frac{1}{r} \left[\delta_{ij}(3 - 4\nu) + \frac{y_i y_j}{r} \right] r \, dr \, d\theta$$

$$= \int_{-\theta_r}^{\theta_s} A r(\theta) [\delta_{ij}(3 - 4\nu) + (\sin\theta e_{1i} + \cos\theta e_{2i})(\sin\theta e_{1j} + \cos\theta e_{2j})] \, d\theta$$

where $A = 1/[16\pi\mu(1 - \nu)]$ and $r(\theta) = D/\cos\theta$. Thus

$$\int_{\Delta S} G_{ij}(x, \xi) \, dS(x) = DA \left[\delta_{ij}(3 - 4\nu) \ln \left(\tan \theta + \sec \theta \right) \right.$$

$$\left. + a_{ij}\{\ln \left(\tan \theta + \sec \theta \right) - \sin \theta\} + b_{ij} \cos \theta + c_{ij} \sin \theta \right]_{-\theta_r}^{\theta_s} \quad (4.28)$$

where

$$a_{ij} = e_{1i}e_{1j}$$
$$b_{ij} = -(e_{2i}e_{1j} + e_{1i}e_{2j})$$
$$c_{ij} = e_{2i}e_{2j}$$

Evaluations of $\int_{\Delta S} F_{ij}(x, \xi) \, dS(x)$ Cruse[9] presented a method for evaluating these integrals exactly which we follow here. For a flat element $y_k n_k = 0$; therefore the function $F_{ij}(x, \xi)$ reduces to

$$F_{ij}(x, \xi) = B \frac{1}{r^2} \left[n_j \frac{y_i}{r} - n_i \frac{y_j}{r} \right]$$

where

$$B = -\frac{1 - 2\nu}{8\pi(1 - \nu)}$$

By using the identities

$$\varepsilon_{ijk}\varepsilon_{rsk}n_r \frac{\partial}{\partial x_s} \frac{1}{r} = \frac{1}{r^2} \left(n_j \frac{\partial r}{\partial x_i} - \frac{\partial r}{\partial x_j} n_i \right) = \frac{1}{r^2} \left(n_j \frac{y_i}{r} - n_i \frac{y_j}{r} \right)$$

where
ε_{ijk} is the permutation tensor (see Chapter 2) $= 0$ when $i = j$ or $j = k$ or $k = i$
$\qquad\qquad\qquad\qquad\qquad\qquad\qquad\quad = 1$ when i, j, k are cyclic
$\qquad\qquad\qquad\qquad\qquad\qquad\qquad\quad = -1$ when i, j, k are anticyclic
we can express the integral $\int_{\Delta S} F_{ij}(x, \xi) \, dS(x)$ as

$$\Delta F_{ij} = \int_{\Delta S} F_{ij}(x, \xi) \, dS(x) = B\varepsilon_{ijk} \int_{\Delta S} \varepsilon_{rsk}n_r \frac{\partial}{\partial x_s} \frac{1}{r} \, dS(x)$$

Stoke's theorem can be used to convert this into a line integral, i.e.

$$\Delta F_{ij} = B\varepsilon_{ijk} \int \frac{1}{r} \, dx_k \quad (4.29)$$

In terms of the local variables

$$dx_k = e_{1k} \, dz_1 + e_{2k} \, dz_2$$

Along the side rs we have

$$\Delta F_{ij} = B\varepsilon_{ijk}e_{1k}[\ln (z_1 + r)]_r^s \quad (4.30)$$

The total result is then obtained reorienting the local coordinate system such that the z_1 axis coincides, in turn, with the st and tr sides of the triangle. The results are then expressed in terms of the coordinates of the field point and those of the corners. Clearly Eq. (4.30) can also be used to evaluate the integral over a flat quadrilateral element in a similar manner.

4.5.3 Calculations at Interior Points

The unknown values of displacement and tractions, together with the specified values of tractions and displacements, can be used to calculate the interior values of displacement, strains and stresses from

$$\mathbf{u}(\xi) = \sum_{q=1}^{N} \left\{ \left[\int_{\Delta S} \mathbf{G}(x^q, \xi) \mathbf{N}^q \, dS \right] \mathbf{t}_n - \left[\int_{\Delta S} \mathbf{F}(x^q, \xi) \mathbf{N}^q \, dS \right] \mathbf{u}_n \right\} \tag{4.31a}$$

$$\varepsilon(\xi) = \sum_{q=1}^{N} \left\{ \left[\int_{\Delta S} \mathbf{G}^\epsilon(x^q, \xi) \mathbf{N}^q \, dS \right] \mathbf{t}_n - \left[\int_{\Delta S} \mathbf{F}^\epsilon(x^q, \xi) \mathbf{N}^q \, dS \right] \mathbf{u}_n \right\} \tag{4.31b}$$

$$\sigma(\xi) = \sum_{q=1}^{N} \left\{ \left[\int_{\Delta S} \mathbf{G}^\sigma(x^q, \xi) \mathbf{N}^q \, dS \right] \mathbf{t}_n - \left[\int_{\Delta S} \mathbf{F}^\sigma(x^q, \xi) \mathbf{N}^q \, dS \right] \mathbf{u}_n \right\} \tag{4.31c}$$

It is evident that the calculations for the stresses and strains at interior points involve much more computational effort than is necessary in the finite element method. In cases where the strains and stresses are required at a large number of interior points, it is more efficient to calculate accurate values of displacements at a sufficient number of interior nodes and then obtain the strains from them by using the numerical differentiations and the stresses by using the constitutive relations.

The expressions for strains and stresses cannot be easily constructed at the boundary points by taking the field point (ξ) in Eqs (4.31b,c) to the surface point due to the strongly singular nature of the integrals involved. The evaluation of strains and stresses at boundary points can be accomplished by considering the equilibrium of the boundary segment and utilizing constitutive and kinematic equations. The stresses and the global derivatives of displacements that lead to strains at point ξ can be obtained by coupling the following set of equations:

$$\sigma_{ij}(\xi) - \{\lambda \delta_{ij} u_{k,k}(\xi) + \mu[u_{i,j}(\xi) + u_{j,i}(\xi)]\} = 0 \tag{4.32a}$$

$$\sigma_{ij} n_j(\xi) = t_i(\xi) \tag{4.32b}$$

and
$$\frac{\partial u_i}{\partial \xi_k} \frac{\partial \xi_k}{\partial \eta_l} = \frac{\partial u_i}{\partial \eta_l} \tag{4.32c}$$

where η_l are a set of local axes in the normal and tangential directions at the field point ξ and λ and μ are Lame's constants.

Equations (4.32a,b,c) can be cast into a matrix form to determine the stress and the derivatives of displacements at a boundary point as shown in Chapter 17.

The derivatives of displacements with respect to the local axes in Eq. (4.32c) can be obtained easily using local interpolation functions for the surface element.

The calculations of stresses and strains at a point close to the surface are once again extremely difficult because of the strongly singular kernel functions involved in Eqs (4.31b,c). These difficulties can be avoided either by using an admissible rigid body motion and calculating the most difficult block of coefficients indirectly as the sum of all other blocks of coefficients along the same row with a sign change or by using integration by parts rule to transfer the derivatives from the kernels $(\mathbf{G}, \mathbf{F})$ to the functions $(\mathbf{u}, \mathbf{t})$.

4.6 BODY FORCES

4.6.1 Volume Integral Conversion Method

A large class of boundary value problems in elastostatics involve body forces generated by either a steady state temperature, seepage gradient or a gravitational potential. For all such problems the body force terms ψ_i can be expressed as[10, 11]

$$\psi_i = -\frac{\partial p}{\partial x_i} \tag{4.33a}$$

where p is a scalar function that has to satisfy the following differential equation in V:

$$\frac{\partial^2 p}{\partial x_i \partial x_i} + q = 0 \tag{4.33b}$$

For bodies subjected to a distribution of temperature or a hydraulic potential which satisfies the steady state potential flow equation (4.33b) with $q = 0$, the total stresses can then be written as

$$\sigma_{ij} = \sigma'_{ij} - \gamma \delta_{ij} p \tag{4.34}$$

where

$$\sigma'_{ij} = \frac{2\mu\nu}{1.2\nu} \delta_{ij} \frac{\partial u_m}{\partial x_m} + \mu\left(\frac{\partial u_i}{\partial x_j} + \frac{\partial u_j}{\partial x_i}\right)$$

and μ and ν are again the shear modulus and Poisson's ratio of the solid (in the case of a porous body these are the equivalent properties of the skeleton and σ'_{ij} is the effective stress tensor), $\gamma = 1$ for the seepage problem and $\alpha E/(1 - 2\nu)$ for thermal stress analysis problems, α being the coefficient of thermal expansion and E Young's modulus.

The total surface traction can then be written as

$$t_i = \sigma_{ij} n_j = \sigma'_{ij} n_j - \gamma p n_i$$

$$= t'_i - \gamma p n_i \tag{4.35}$$

Equilibrium requires

$$\frac{\partial \sigma_{ij}}{\partial x_j} = \frac{\partial \sigma'_{ij}}{\partial x_j} - \gamma \frac{\partial p}{\partial x_i} = 0 \qquad (4.36)$$

We can now use these equations to construct BEM formulations for the solution of boundary value problems. If we note that the second term in Eq. (4.36) is an equivalent body force we can write the direct integral representation for the displacements at any interior point as

$$u_j(\xi) = \int_S \left[t'_i(x) G_{ij}(x, \xi) - F_{ij}(x, \xi) u_i(x) \right] dS - \int_V \gamma \frac{\partial p(x)}{\partial x_i} G_{ij}(x, \xi) dV \qquad (4.37)$$

where

$$t'_i(x) = \sigma'_{ij}(x) n_j(x) = t_i + \gamma p n_i$$

Using the divergence theorem the volume integral in Eq. (4.37) can be rewritten as

$$-\int_V \gamma \frac{\partial p(x)}{\partial x_i} G_{ij}(x, \xi) dV = -\int_S \gamma p n_i(x) G_{ij}(x, \xi) dV + \int_V \gamma p(x) \frac{\partial G_{ij}(x, \xi)}{\partial x_i} dV \qquad (4.38)$$

Substituting Eq. (4.38) into Eq. (4.37) and using Eq. (4.35) to eliminate t'_i (the modified traction) we arrive at

$$u_j(\xi) = \int_S \left[t_i(x) G_{ij}(x, \xi) - F_{ij}(x, \xi) u_i(x) \right] dS + \int_V \gamma p(x) \frac{\partial G_{ij}(x, \xi)}{\partial x_i} dV \qquad (4.39)$$

Here we note the following well-known[3] property of the function G_{ij}:

$$G_{ij}(x, \xi) = \delta_{ij} \frac{\partial^2 \phi_s}{\partial x_k \partial x_k} + \frac{\partial^2}{\partial x_i \partial x_j} (\phi_p - \phi_s) \qquad (4.40)$$

where

$$\phi_\eta = \frac{r}{8\pi C_\eta} \qquad \text{in three dimensions}$$

$$= \frac{r^2(1 - \log r)}{8\pi C_\eta} \qquad \text{in two dimensions}$$

with η taking values p or s, while $C_p = (\lambda + 2\mu)$ and $C_s = \mu$.

With the above form of G_{ij} we can easily see that

$$\frac{\partial G_{ij}}{\partial x_i} = \nabla^2 \left(\frac{\partial \phi_p}{\partial x_j} \right) \qquad (4.41)$$

Since the potential p in Eq. (4.39) satisfies the following integral relation (see Chapter 3),

$$\int_V (p \nabla^2 p^* - p^* \nabla^2 p) dV = \int_S \left(p \frac{\partial p^*}{\partial n} - p^* \frac{\partial p}{\partial n} \right) dS \qquad (4.42)$$

we can easily see that by using $p^* = \partial \phi_p / \partial x_j$ together with the fact that $\nabla^2 p = 0$,

Eq. (4.42) provides directly the required conversion of the volume integral into an equivalent surface integral.

Thus, the required boundary integral is

$$u_j(\xi) = \int_S \left[t_i(x) G_{ij}(x, \xi) - F_{ij}(x, \xi) u_i(x) \right] \mathrm{d}S + \int_S \left[F_j(x, \xi) p(x) - G_j(x, \xi) \frac{\partial p(x)}{\partial n} \right] \mathrm{d}S$$

(4.43)

where

$$G_j = \gamma \frac{\partial \phi_p}{\partial x_j} \qquad F_j = \frac{\partial G_j}{\partial n}$$

(4.44)

In earlier work Rizzo and coworkers[12-14] derived Eq. (4.43) using properties of functions such as r and $\log r$ when they are operated on by ∇^2.

Equation (4.43) can now be written for a point ξ on the boundary in the usual manner. Its solution requires knowledge of both the potential $p(x)$ and its normal derivative $\partial p(x)/\partial n$ over the boundaries. These can be obtained from a prior solution of the underlying potential flow problem as discussed in Chapter 3.

4.6.2 Theory of Particular Integrals

Unfortunately, the above procedure is not sufficiently general for all classes of problems. Moreover, for many problems it is difficult for an engineer to configure a physical problem into one of potential flow with appropriate boundary conditions involving a potential and a flux. A far simpler as well as general approach based on the theory of particular integrals is outlined here.

The theory of particular integrals is well known and was explored earlier in Chapter 3. In this section, recall briefly the method in relation to elasticity. We can express the governing differential equation of the problem (4.4) as

$$L(u_i) + \psi_i = 0$$

(4.45a)

the solution of which can be constructed as

$$u_i = u_i^c + u_i^p, \qquad t_i = t_i^c + t_i^p, \qquad \sigma_{ij} = \sigma_{ij}^c + \sigma_{ij}^p$$

(4.45b)

where the complimentary part of the solution satisfies

$$L(u_i^c) = 0$$

(4.45c)

and the particular solution is any solution satisfying

$$L(u_i^p) + \psi_i = 0$$

(4.45d)

Self-weight If we consider the case of gravitational acceleration directed along the x_2 axis in a two-dimensional problem, the body force components are given by

$$\psi_1 = 0, \qquad \psi_2 = -\rho g$$

where ρ is the mass density and g is the acceleration. One set of particular integrals corresponding to this body force set are given, for instance, by Sokolnikoff:[3]

$$u_1^P(x) = \frac{-\rho g \lambda}{4\mu(\lambda + \mu)} x_1 x_2 \tag{4.46a}$$

$$u_2^P(x) = \frac{\rho g}{8\mu(\lambda + \mu)} [(\lambda + 2\mu)x_2^2 + \lambda x_1^2] \tag{4.46b}$$

By using the strain displacement and the stress–strain relations we can easily see that the associated stresses are

$$\sigma_{11}^P = \sigma_{12}^P = 0$$

$$\sigma_{22}^P = \rho g x_2$$

which can be used to calculate the particular traction solution $t_i^P = \sigma_{ij}^P n_j$ (for this case $t_1^P = 0, t_2^P = \rho g x_2 n_2$).

Similarly, for a three-dimensional problem, with $\psi_1 = \psi_2 = 0$ and $\psi_3 = -\rho g$, i.e. the gravitational acceleration is directed along the x_3 axis, the necessary particular integrals that satisfy Eq. (4.45d) are given by

$$u_1^P = \frac{-\lambda \rho g}{2\mu(3\lambda + 2\mu)} x_1 x_2 \tag{4.47a}$$

$$u_2^P = \frac{-\lambda \rho g}{2\mu(3\lambda + 2\mu)} x_2 x_3 \tag{4.47b}$$

$$u_3^P = \frac{(\lambda + \mu)\rho g}{2\mu(3\lambda + 2\mu)} x_3^2 + \frac{\lambda \rho g}{4\mu(3\lambda + 2\mu)} (x_1^2 + x_2^2) \tag{4.47c}$$

These can then be used to calculate the traction particular solution once again.

Centrifugal loading   For a two-dimensional body rotating about a fixed axis perpendicular to the plane of the body we have $\psi_i = \rho \omega^2 x_i$. The particular integral for u_i satisfying Eq. (4.45d) was given in Sokolnikoff[3] as

$$u_i^P = \frac{-\rho \omega^2}{8(\lambda + 2\mu)} (x_k x_k) x_i \tag{4.48}$$

and similarly for a three-dimensional body rotating about the x_3 axis the displacement field satisfying Eq. (4.45d) was given by Banerjee and coworkers as:[15–17]

$$u_\alpha^P = \frac{-\rho \omega^2}{8(\lambda + 2\mu)} \left[\frac{5\lambda + 4\mu}{4(\lambda + \mu)} x_\beta x_\beta + \frac{\mu}{\lambda + \mu} x_3^2 \right] x_\alpha \tag{4.49a}$$

$$u_3^P = \frac{\rho \omega^2}{8(\lambda + 2\mu)} x_\beta x_\beta x_3 \tag{4.49b}$$

where α, β range from 1 to 2.

These equations can, once again, be used to calculate the particular solutions for the stresses and hence tractions.

Simple thermal body forces   While the particular integrals discussed above could be chosen somewhat intuitively in the form of a polynomial whose coefficients are determined by substituting the chosen displacement fields into Eq. (4.45d), for thermal problems the body force is a little more complex because the gradients of the temperature field are involved, i.e.

$$\psi_i = -(3\lambda + 2\mu)\alpha T_{,i} \tag{4.50}$$

where α is the coefficient of thermal expansion.

As mentioned earlier, an exactly similar situation arises in the case of a porous elastic body where the body force is generated by the gradients of the pore fluid pressure, p, i.e.

$$\psi_i = -\gamma p_{,i} \tag{4.50a}$$

where γ is a material parameter which is equal to unity for soil but less than unity for rocks.

A thermoelastic particular integral formulation can then be developed by observing the fact that the particular solution can be expressed as a gradient of a thermoelastic displacement potential ϕ in accordance with the linear thermoelastic deformation theory, i.e.[10, 18]

$$u_i^P = \phi_{,i} \tag{4.51a}$$

Substituting Eq. (4.51a) into Eq. (4.45d) and simplifying yields

$$\phi_{,ii} = \beta T \tag{4.51b}$$

where

$$\beta = \left(\frac{1+\nu}{1-\nu}\right)\alpha = \frac{(3\lambda + 2\mu)\alpha}{\lambda + 2\mu}$$

Thus, an implicit relation between u_i^P and T is derived with the use of the potential function ϕ. In most engineering applications of these particular integrals the body could be divided into several regions and the temperature field could be represented by simple functions. For example, a two-dimensional particular integral for a uniform temperature c_1 can be found assuming

$$\phi(x) = \frac{c_1}{4}(x_1^2 + x_2^2)$$

Substituting this equation in (4.51b) gives

$$\phi_{,ii} = c_1$$

Therefore our choice of $\phi(x)$ was correct to give a uniform temperature. The corresponding particular integrals for displacements from Eq. (4.51a) are[18]

$$u_1^P(x) = \tfrac{1}{2}c_1\beta x_1 \quad \text{and} \quad u_2^P(x) = \tfrac{1}{2}c_1\beta x_2$$

We can now generalize the above procedure to both two- and three-dimensional problems. Let us assume

$$\phi = a_1 x_j x_j$$

where a_1 is a constant and j ranges from 1 to 2 or 1 to 3 depending on the dimensionality (d) of the problem. We easily see that

$$\phi_{,ii} = 2da_1 = \beta c_1$$

Therefore $a_1 = \beta c_1 / 2d$, leading to

$$u_i^p = \frac{\beta c_1}{d} x_i$$

In a similar manner, particular integrals for higher-order temperature distributions can be developed. Specifically, we let

$$T(x) = \sum_{m=1}^{10} y_m c_m \qquad (4.52)$$

where $y_m = \{1, x_1, x_2, x_3, x_1 x_2, x_1 x_3, x_2 x_3, x_1^2, x_2^2, x_3^2\}$. We note that this represents a quadratic variation of temperature in a region. Moreover, by letting the x_3 terms disappear we generate from Eq. (4.52) a quadratic temperature distribution in a two-dimensional region. Now by considering Eq. (4.52) term by term we can use the above procedure to generate the following particular integrals corresponding to the temperature distribution specified by Eq. (4.52):

$$u_i^p(x) = \sum_{m=1}^{10} U_{im} c_m \qquad (4.53)$$

where

$$U_{i1} = \frac{\beta}{d} x_i \qquad \text{for } m = 1$$

$$U_{im} = \frac{\beta}{2(d+2)} (\delta_{ik} x_j x_j + 2 x_i x_k) \qquad \text{for } m = 2, 3 \text{ and } 4; k = m - 1$$

$$U_{im} = \frac{\beta}{2(d+4)} [(\delta_{ik} x_\ell + \delta_{i\ell} x_m) x_j x_j + 2 x_k x_\ell x_i]$$
$$\text{for } m = 5(k = 1, \ell = 2), m = 6(k = 1, \ell = 3), m = 7(k = 2, \ell = 3)$$

$$U_{im} = \frac{\beta}{3} \delta_{ik} x_k^3 \qquad \text{for } m = 8, 9 \text{ and } 10; k = m - 1$$

Equation (4.53) can then be substituted in the strain displacement equation (4.3) and stress-strain relation for thermoelasticity (4.34) to obtain the particular solution for the stresses and hence surface tractions.

In many problems the temperature field (either steady state or transient) may not conform to these simple distributions. In such cases the region could be subdivided into a multi-zone problem and the distributions over each sub-region could be approximated by Eq. (4.52) in a least square sense.[19] Thus, if Eq. (4.52) is written for all known temperature data in a given region as

$$\{T\} = [Y]\{C\} \tag{4.54}$$

where the order of the matrix Y is $N \times M$, with N as the number of sampling points and M the order of the polynomial, a least square estimate of the coefficient matrix C can then be obtained as

$$\{C\} = ([Y]^{\mathrm{T}}[Y])^{-1}\{T\} \tag{4.55}$$

Hotspots Obviously the above polynomial distributions are not adequate for high thermal gradients that can exist around a hot spot. For any general body force system $\psi_i(z)$ the particular integral is

$$u_j^{\mathrm{P}}(\xi) = \int_V \psi_i(z) G_{ij}(z, \xi) \, \mathrm{d}V \tag{4.56}$$

G_{ij} being the fundamental point force solution for the infinite space consisting of the same material.

Obviously Eq. (4.56) does satisfy the governing differential equation (4.45d) and therefore is a valid particular integral. If the distribution of $\psi_i(z)$ is known the integral in Eq. (4.56) can be evaluated at field point ξ and the corresponding particular solutions for stresses and surface tractions also follow from it.

For a hot spot over a region D enclosed by a surface B, within our region V bounded by a surface S, we can write Eq. (4.56) as

$$u_j^{\mathrm{P}}(\xi) = -\int_D (3\lambda + 2\mu)\alpha T_{,i}(z) G_{ij}(z, \xi) \, \mathrm{d}V \tag{4.57}$$

By applying the divergence theorem we can express this as

$$u_j^{\mathrm{P}}(\xi) = -\int_B (3\lambda + 2\mu)\alpha T(x) G_{ij}(x, \xi) n_i(x) \, \mathrm{d}S$$

$$+ \int_D (3\lambda + 2\mu)\alpha T(z) \frac{\partial G_{ij}}{\partial z_k}(z, \xi) \, \mathrm{d}V \tag{4.58}$$

in which both surface and volume integral over B and D respectively can be evaluated if the distribution $T(z)$ is known. These operations could be done easily for a region D which is regular so that the geometry involved could be described analytically (e.g. circular and spherical regions).

4.7 ANISOTROPY

4.7.1 Introduction

In anisotropy the number of independent constants describing the stress–strain relation may be considerably larger (up to a maximum of 21 based on thermodynamic principles) as opposed to only two (i.e. the Lame parameters or the engineering constants—Young's modulus and Poisson's ratio) in isotropy. This becomes obvious if the following Duhamel–Neumann constitutive relation for anisotropic thermal problems comprising 21 independent elastic compliances b_{ij} is laid in a Cartesian frame of reference:

$$\epsilon_i = b_{ij}\sigma_j + \alpha_i T \quad (i,j = 1,2,\ldots,6) \tag{4.59}$$

where the convenient contracted notation is followed according to which:

σ_j represents the stress vector:

$$\begin{Bmatrix} \sigma_1 \\ \sigma_2 \\ \sigma_3 \\ \sigma_4 \\ \sigma_5 \\ \sigma_6 \end{Bmatrix} = \begin{Bmatrix} \sigma_{11} \\ \sigma_{22} \\ \sigma_{33} \\ \sigma_{23} \\ \sigma_{13} \\ \sigma_{12} \end{Bmatrix} \tag{4.60a}$$

ϵ_i represents the engineering strain vector:

$$\begin{Bmatrix} \epsilon_1 \\ \epsilon_2 \\ \epsilon_3 \\ \epsilon_4 \\ \epsilon_5 \\ \epsilon_6 \end{Bmatrix} = \begin{Bmatrix} \epsilon_{11} \\ \epsilon_{22} \\ \epsilon_{33} \\ 2\epsilon_{23} \\ 2\epsilon_{13} \\ 2\epsilon_{12} \end{Bmatrix} \tag{4.60b}$$

b_{ij} represents the matrix of compliance coefficients:

$$\begin{bmatrix} b_{11} & b_{12} & b_{13} & b_{14} & b_{15} & b_{16} \\ & b_{22} & b_{23} & b_{24} & b_{25} & b_{26} \\ & & b_{33} & b_{34} & b_{35} & b_{36} \\ \text{Symmetric} & & & b_{44} & b_{45} & b_{46} \\ & & & & b_{55} & b_{56} \\ & & & & & b_{66} \end{bmatrix} \tag{4.60c}$$

α_i represents the array of at most six coefficients of linear thermal expansion:

$$\begin{Bmatrix} \alpha_1 \\ \alpha_2 \\ \alpha_3 \\ \alpha_4 \\ \alpha_5 \\ \alpha_6 \end{Bmatrix} = \begin{Bmatrix} \alpha_{11} \\ \alpha_{22} \\ \alpha_{33} \\ 2\alpha_{23} \\ 2\alpha_{13} \\ 2\alpha_{12} \end{Bmatrix} \tag{4.60d}$$

and T denotes the net temperature above the stress free state of temperature.

Alternatively, the constitutive relation may be written in terms of stiffnesses c_{ij} as

$$\sigma_i = c_{ij}\epsilon_j - \beta_i T \quad (i,j = 1,2,\ldots,6) \tag{4.61}$$

where

$$[c_{ij}] = [b_{ij}]^{-1} \tag{4.61a}$$

$$\beta_i = c_{ij}\alpha_j \qquad (4.61b)$$

The constitutive relations given above for general anisotropy preclude purely in-plane behaviour. However, by restricting the material type to monoclinic, ordinary plane strain and stress behaviour can be defined on the plane of material symmetry. For a monoclinic material with the plane of symmetry parallel to the $x_1 x_2$ plane, the compliance matrix a_{ij} is of the following form with 13 independent constants:

$$\begin{bmatrix} a_{11} & a_{12} & a_{13} & 0 & 0 & a_{16} \\ & a_{22} & a_{23} & 0 & 0 & a_{26} \\ & & a_{33} & 0 & 0 & a_{36} \\ & \text{Symmetric} & & a_{44} & a_{45} & 0 \\ & & & & a_{55} & 0 \\ & & & & & a_{66} \end{bmatrix}$$

Thus, for two-dimensional elasticity of anisotropic media, the material constitutive relation is assumed to be made up of 'reduced' compliances b_{ij} and 'reduced' coefficients of linear thermal expansion α_i in the following form:

$$\begin{Bmatrix} \epsilon_1 \\ \epsilon_2 \\ \epsilon_6 \end{Bmatrix} = \begin{bmatrix} b_{11} & b_{12} & b_{16} \\ & b_{22} & b_{26} \\ \text{Symmetric} & & b_{66} \end{bmatrix} \begin{Bmatrix} \sigma_1 \\ \sigma_2 \\ \sigma_6 \end{Bmatrix} + \begin{Bmatrix} \alpha_1 \\ \alpha_2 \\ \alpha_6 \end{Bmatrix} T \qquad (4.62)$$

where

$$b_{ij} = a_{ij} \qquad \text{for plane stress} \qquad (4.62a)$$

$$= a_{ij} - \frac{a_{i3}a_{j3}}{a_{33}} \qquad \text{for plane strain} \qquad (4.62b)$$

$$(i,j = 1,2,6)$$

$$\alpha_i = \alpha_i^0 \qquad \text{for plane stress} \qquad (4.62c)$$

$$= \alpha_i^0 - \frac{a_{i3}\alpha_3^0}{a_{33}} \qquad \text{for plane strain} \qquad (4.62d)$$

$$(i = 1,2,6)$$

α_i^0 being the actual or unreduced coefficients of linear thermal expansion. Also, the following relations are imperative in the above reductions:

$$\epsilon_3 = a_{13}\sigma_1 + a_{23}\sigma_2 + a_{36}\sigma_6 + \alpha_3^0 T \qquad \text{for plane stress} \qquad (4.63a)$$

$$\sigma_3 = -\frac{1}{a_{33}}(a_{13}\sigma_1 + a_{23}\sigma_2 + a_{36}\sigma_6 + \alpha_3^0 T) \qquad \text{for plane strain} \qquad (4.63b)$$

In the context of two-dimensional analysis, the alternative constitutive relation (4.62) can be written by interpreting c_{ij} ($i,j = 1,2,6$) as the 'reduced' stiffness coefficients obtained from the inversion of the reduced compliance matrix $[b_{ij}]$ in Eq. (4.62).

Real materials rarely display a form of anisotropy more general than orthogonal anisotropy or orthotropy, in which case three mutually perpendicular

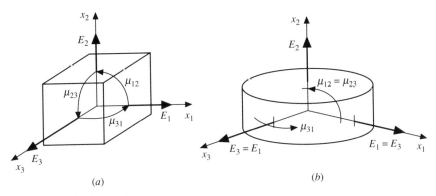

Figure 4.6 Definition of elastic constants.

planes of material symmetry exist and more commonly the cross-anisotropy (see Fig. 4.6a,b). It is usually thus sufficient to limit the study of anisotropic mechanics to orthotropic elasticity. For orthotropy, the constitutive matrix of compliances (or stiffnesses) contain at most nine independent coefficients resulting from the vanishing of 12 terms in $[b_{ij}]$ (or $[c_{ij}]$) when the constitutive relation is written in a set of principal axes aligned with the planes of material symmetry. For an orthotropic material it is possible to cast the constitutive matrix in a principal axes system in terms of 'technical or engineering constants'[20] quite similar to Young's modulus and Poisson's ratio for isotropy (Fig. 4.6). Following the notation of Lekhnitskii,[20] the compliance matrix $[b_{ij}]$ in terms of technical constants E_i, ν_{ij} and μ_{ij} can be written as given below:

$$[b_{ij}] = \begin{bmatrix} 1/E_1 & -\nu_{21}/E_2 & -\nu_{31}/E_3 & 0 & 0 & 0 \\ -\nu_{12}/E_1 & 1/E_2 & -\nu_{32}/E_3 & 0 & 0 & 0 \\ -\nu_{13}/E_1 & -\nu_{23}/E_2 & 1/E_3 & 0 & 0 & 0 \\ 0 & 0 & 0 & 1/\mu_{23} & 0 & 0 \\ 0 & 0 & 0 & 0 & 1/\mu_{13} & 0 \\ 0 & 0 & 0 & 0 & 0 & 1/\mu_{12} \end{bmatrix} \tag{4.64}$$

where

$$\frac{\nu_{ij}}{E_i} = \frac{\nu_{ji}}{E_j} \qquad (i,j = 1,2,3) \tag{4.64a}$$

Thus, there are nine independent technical constants defined by Eqs (4.64) above for the determination of which nine separate tests are necessary.

4.7.2 Fundamental Solutions

For the two-dimensional analysis of a generally anisotropic medium, the singular solutions are given by the following closed forms:[21,22]

$$G_{kj} = 2 \, \mathrm{Re}[A_{k1}B_{j1}\ln Z_1 + A_{k2}B_{j2}\ln Z_2] \qquad (k,j = 1,2) \tag{4.65}$$

where the quantities in the expressions for G_{kj} above are defined below.

The complex constants A_{kl} in Eq. (4.65) are given as follows:

$$\left\{\begin{array}{c} A_{1l} \\ A_{2l} \end{array}\right\} = \left\{\begin{array}{c} b_{11}\mu_l^2 + b_{12} - b_{16}\mu_l \\ b_{12}\mu_l + b_{22}/\mu_l - b_{26} \end{array}\right\} \qquad (l = 1, 2) \qquad (4.66)$$

The complex constants B_{kl} in Eq. (4.65) are solutions of the following equations (where δ_{jk} is the Kronecker delta, $j = 1, 2$ and $i = \sqrt{-1}$):

$$\mu_1 B_{j1} - \bar{\mu}_1 \bar{B}_{j1} + \mu_2 B_{j2} - \bar{\mu}_2 \bar{B}_{j2} = \frac{\delta_{j1}}{2\pi i}$$

$$B_{j1} - \bar{B}_{j1} + B_{j2} - \bar{B}_{j2} = -\frac{\delta_{j2}}{2\pi i} \qquad (4.67)$$

$$A_{11} B_{j1} - \bar{A}_{11} \bar{B}_{j1} + A_{12} B_{j2} - \bar{A}_{12} \bar{B}_{j2} = 0$$

$$A_{21} B_{j1} - \bar{A}_{21} \bar{B}_{j1} + A_{22} B_{j2} - \bar{A}_{22} \bar{B}_{j2} = 0$$

The complex variables Z_k in Eq. (4.65) are defined as

$$Z_k = (x_1 - \xi_1) + \mu_k(x_2 - \xi_2) \qquad (4.68)$$

μ_k (with positive imaginary parts) and $\bar{\mu}_k$ in the foregoing are the complex conjugate roots of the following fourth-order characteristic equation:

$$b_{11}\mu^4 - 2b_{16}\mu^3 + (2b_{12} + b_{66})\mu^2 - 2b_{26}\mu + b_{22} = 0 \qquad (4.69)$$

Equations (4.65) to (4.69) thus completely define G_{kj}.

The components of singular solutions F_{kj} for surface traction are given as follows:

$$\left\{\begin{array}{c} F_{1j} \\ F_{2j} \end{array}\right\} = \begin{bmatrix} (c_{11}n_1 + c_{16}n_2) & (c_{12}n_1 + c_{26}n_2) & (c_{16}n_1 + c_{66}n_2) \\ (c_{16}n_1 + c_{12}n_2) & (c_{26}n_1 + c_{22}n_2) & (c_{66}n_1 + c_{26}n_2) \end{bmatrix} \left\{\begin{array}{c} \partial G_{1j}/\partial x_1 \\ \partial G_{2j}/\partial x_2 \\ \partial G_{1j}/\partial x_2 \\ +\partial G_{2j}/\partial x_1 \end{array}\right\}$$

$$(4.70)$$

where $j = 1, 2$ and n_1 and n_2 are direction cosines of the boundary normal.

In the previous applications of the boundary element method for anisotropy in three dimensions[23] Green's functions were used in the following form from Synge[24] $(i, j = 1, 2, 3)$:

$$G_{ij}(x, \xi) = \frac{1}{8\pi^2 |x - \xi|} \oint_{|\gamma|=1} K_{ij}^{-1}(\gamma) \, dS \qquad (4.71)$$

where the line integral is taken on the unit circle in the plane normal to the vector $(x - \xi)$ and passing through x, and the function

$$K_{ij}^{-1}(\gamma) = [c_{ijkm}\gamma_k\gamma_m]^{-1} \qquad (4.72)$$

where c_{ijkl} are elastic constants.

Since it is not possible to evaluate the line integral of Eq. (4.72) in closed form for the general case, numerical integration was employed. The boundary element procedure based on the above numerically computed Green functions proved to be extremely time consuming, despite improvements in which interpolation schemes are used to avoid actually using relationship (4.71) at every Gauss point. A more efficient approach is adopted in the work of Deb and coworkers[25, 26] in which the Green function is calculated using a finite series representation at the grid points and bivariate cubic splines are then fitted through these points for the purpose of interpolation. The Green function adopted in these works is essentially based on works of Malen[27] and can be expressed in terms of spherical coordinates as

$$G_{ij}(r, \theta, \phi) = -\frac{1}{4\pi r} \sum_{\nu=1}^{6} \text{sign}(\text{Im}(\nu))\eta_i(\nu)\eta_j(\nu) \tag{4.73}$$

where, in the real variable formulation, the ν and $\eta_i(\nu)$ are the eigenvalues and eigenvectors of a 6×6 real matrix.[27] The eigenvalues and eigenvectors occur as complex conjugates. The ϕ derivative of G_{ij} can be expressed as

$$G_{ij,\phi} = -\frac{1}{4\pi r} \sum_{\nu=1}^{6} \text{sign}(\text{Im}(\nu)) \left[\frac{\partial \eta_i(\nu)}{\partial \phi} \eta_j(\nu) + \eta_i(\nu) \frac{\partial \eta_j(\nu)}{\partial \phi} \right] \tag{4.74}$$

with a similar expression for $G_{ij,\theta}$. These derivatives are required for the computation of the traction kernels F_{ij}. For the purpose of calculating interior stresses and strains, the necessary second derivatives of the kernel functions are computed at grid points using finite difference expressions. For a cross-anisotropic material the fundamental solution can be recovered in an analytical form as was shown by Pan and Chou.[28]

Armed with these fundamental solutions the entire BEM formulations outlined in Sec. 4.3 can be used for any anisotropic elasticity problem.

4.7.3 Particular Integrals

Self-weight The particular integrals for stresses are given by the following simple expressions for a two-dimensional analysis:

$$\sigma_{11}^P = \sigma_1^P = -\Gamma_1 x_1 \tag{4.75a}$$

$$\sigma_{22}^P = \sigma_2^P = -\Gamma_2 x_2 \tag{4.75b}$$

$$\sigma_{12}^P = \sigma_6^P = 0 \tag{4.75c}$$

where $\Gamma_i = \rho g_i$, g_i being the gravitational acceleration.

The corresponding particular integrals for displacements are[25, 29]

$$u_1^P = -\tfrac{1}{2}\Gamma_1 b_{11} x_1^2 - \Gamma_2 b_{12} x_1 x_2 + \tfrac{1}{2}(\Gamma_1 b_{12} - \Gamma_2 b_{26}) x_2^2 \tag{4.76a}$$

$$u_2^P = \tfrac{1}{2}(\Gamma_2 b_{12} - \Gamma_1 b_{16}) x_1^2 - \Gamma_1 b_{12} x_1 x_2 - \tfrac{1}{2}\Gamma_2 b_{22} x_2^2 \tag{4.76b}$$

It may be noted here that for the special case of an isotropic medium, with body force components $\Gamma_1 = 0$ and $\Gamma_2 - \rho g$, the above particular integrals will reduce to the ones mentioned previously.

Centrifugal loading   The particular integrals for stresses for the case of two-dimensional general anisotropy are given below:

$$\sigma^P_{11} = \sigma^P_1 = c_1(x_1^2 + x_2^2) + c_2 x_1^2 \tag{4.77a}$$

$$\sigma^P_{22} = \sigma^P_2 = c_1(x_1^2 + x_2^2) + c_2 x_2^2 \tag{4.77b}$$

$$\sigma^P_{12} = \sigma^P_6 = c_2 x_1 x_2 \tag{4.77c}$$

The quantities c_1 and c_2 in the above particular solutions are actually invariants under coordinate transformation and are given as follows:

$$c_1 = \frac{4b_{12} - b_{66}}{2(3b_{11} + 2b_{12} + 3b_{22} + b_{66})} \rho\omega^2 \tag{4.78a}$$

$$c_2 = -\frac{b_{11} + 2b_{12} + b_{22}}{3b_{11} + 2b_{12} + 3b_{22} + b_{66}} \rho\omega^2 \tag{4.78b}$$

The corresponding displacement particular integrals are[25, 29]

$$u^P_1 = \tfrac{1}{3}[(b_{11} + b_{12})c_1 + b_{11}c_2]x_1^3 + \tfrac{1}{2}b_{16}c_2 x_1^2 x_2 + [(b_{11} + b_{12})c_1 + b_{12}c_2]x_1 x_2^2 + \tfrac{1}{3}[(b_{16} + b_{26})c_1 + \tfrac{1}{2}b_{26}c_2]x_2^3 \tag{4.79a}$$

$$u^P_2 = \tfrac{1}{3}[(b_{16} + b_{26})c_1 + \tfrac{1}{2}b_{26}c_2]x_1^3 + [(b_{22} + b_{12})c_1 + b_{12}c_2]x_1^2 x_2 + \tfrac{1}{2}b_{26}c_2 x_1 x_2^2 + \tfrac{1}{3}[(b_{22} + b_{12})c_1 + b_{22}c_2]x_2^3 \tag{4.79b}$$

By substituting the appropriate elastic constants b_{ij} in the above it can be shown that these reduce to those of the isotropic case described earlier.

4.8 GENERALIZED PLANE STRESS AND PLANE STRAIN

It may sometimes be instructive to carry out an approximate two-dimensional analysis of an essentially three-dimensional problem for the purposes of a preliminary analysis. Retaining the assumptions of ordinary plane strain or ordinary (sometimes called 'generalized' to mean more consistent averaged-through-the-thickness) plane stress, we allow here the possibility of including a non-zero strain (in the case of plane strain) or a non-zero stress (in the case of plane stress) in the x_3 direction.[29] The non-zero strain (ϵ_{33}) or stress (σ_{33}) is assumed to be distributed at most linearly on the $x_1 x_2$ plane and constant in the x_3 direction. The effect of ϵ_{33} or σ_{33}, as the case may be, is taken into account at the time of reduction of the three-dimensional stress–strain constitutive relation to the two-dimensional (plane strain or stress) version. This extension of ordinary two-dimensional behaviour is here termed as 'generalized' plane strain or stress. In literature, an analysis of this kind is also referred to as quasi-three-dimensional analysis. Symbolically, the assumption of linearly varying ϵ_{33} or σ_{33} on the x_1x_2 plane is stated as follows:

$$\epsilon_{33}(\text{or}, \sigma_{33}) = A_0 + A_1x_1 + A_2x_2 \tag{4.80}$$

where A_0, A_1, and A_2 are given constants with appropriate units.

Thus, for the generalized plane strain condition in a generally anisotropic material (the isotropic case follows from all of this by substituting appropriate values of a_{ij} and $\alpha_3^0 = \alpha$),

$$\sigma_{33} = -\frac{1}{a_{33}}[a_{13}\sigma_1 + a_{23}\sigma_2 + a_{33}(A_0 + A_1x_1 + A_2x_2) + a_{36}\sigma_6 + \alpha_3^0 T] \tag{4.81}$$

and, for the generalized plane stress condition,

$$\epsilon_{33} = a_{13}\sigma_1 + a_{23}\sigma_2 + a_{33}(A_0 + A_1x_1 + A_2x_2) + a_{36}\sigma_6 + \alpha_3^0 T \tag{4.82}$$

Taking into account the above relations in the reduced constitutive law for two dimensions, it may be shown that additional body force type terms h_l are introduced into the governing Navier equation. Thus, the Navier equation in this case takes the form

$$M_{lk}u_k + h_l = 0 \tag{4.83}$$

where M_{lk} is a differential operator.

The generalized plane strain (or stress) loading terms $h_l(l = 1, 2)$ are defined below:

$$\begin{align} h_1 &= -A_1P - A_2Q \\ h_2 &= -A_1Q - A_2R \end{align} \tag{4.84}$$

In the above equations the quantities P, Q and R are constants, given as follows:

$$\begin{align} P &= c_{11}b_{13} + c_{12}b_{23} + c_{16}b_{36} \\ Q &= c_{16}b_{13} + c_{26}b_{23} + c_{66}b_{36} \\ R &= c_{12}b_{13} + c_{22}b_{23} + c_{26}b_{36} \end{align} \tag{4.85}$$

We immediately construct by a trial-and-error process a set of particular integrals for the stresses due to this known body force term h_l:

$$\begin{align} \sigma_{11}^p &= \sigma_1^p = -h_1x_1 - P(A_0 + A_1x_1 + A_2x_2) \\ \sigma_{22}^p &= \sigma_2^p = -h_2x_2 - R(A_0 + A_1x_1 + A_2x_2) \\ \sigma_{12}^p &= \sigma_6^p = -Q(A_0 + A_1x_1 + A_2x_2) \end{align} \tag{4.86}$$

The corresponding displacement particular integrals are:[25, 29]

$$\begin{align} u_1^p &= -\tfrac{1}{2}h_1b_{11}x_1^2 - h_2b_{12}x_1x_2 + \tfrac{1}{2}(h_1b_{12} - h_2b_{26})x_2^2 \\ u_2^p &= \tfrac{1}{2}(h_2b_{12} - h_1b_{16})x_1^2 - h_1b_{12}x_1x_2 - \tfrac{1}{2}h_2b_{22}x_2^2 \end{align} \tag{4.87}$$

The following interpretations must be noted for relations in Eqs (4.85) above:

For generalized plane stress:

$$b_{13} = a_{13}, \qquad b_{23} = a_{23}, \qquad b_{36} = a_{36}$$

For generalized plane strain:

$$b_{13} = \frac{a_{13}}{a_{33}}, \qquad b_{23} = \frac{a_{23}}{a_{33}}, \qquad b_{36} = \frac{a_{36}}{a_{33}}$$

4.9 EXAMPLES OF APPLICATION

4.9.1 A Perforated Plate in Tension[30]

The problem of stress concentration due to a circular hole in an infinite plate under uniform tension S has well-known solutions for stresses. Due to symmetry, a quarter of the plate under plane stress has been analysed using quadratic (QD), cubic (CUB) and quartic (QRT) elements. Details of boundary conditions and problem geometry are given in Fig. 4.7. The following material properties are assumed: Young's modulus $E = 100$ and Poisson's ratio $\nu = 0.25$. The total number of unknown degrees of freedom (DOF), equal to twice the total number of boundary

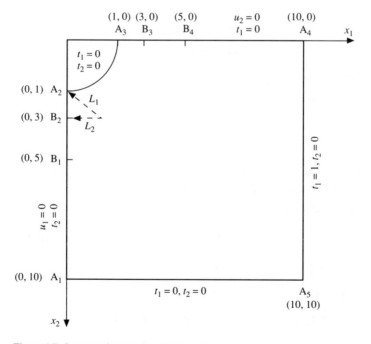

Figure 4.7 Quarter plate used in BEM analysis.

nodes, has been taken as 56 for QD and QRT elements and 57 for CUB elements. The details of meshes are as follows:

1. Arc A_2A_3 is divided into four QD/three CUB/two QRT elements.
2. A_3B_3 is divided into four QD/three CUB/two QRT elements.
3. B_3B_4 is divided into two QD/two CUB/one QRT elements.
4. B_4A_4 is divided into two QD/one CUB/one QRT elements.
5. A_4A_5 is divided into four QD/two CUB/two QRT elements.
6. Divisions of A_1A_2 and A_1A_5 are analogous to those of A_3A_4 and A_4A_5 just described.

The results for the dominant tangential stress σ_θ (for $S = 1.0$) along the boundary of interest are plotted in Fig. 4.8. On account of the diminutive nature of the hole, it is reasonable to compare the results obtained from the present example with the well-known closed form solutions[10] for the infinite plate. Interior stresses are computed along lines L_1 (inclined at $45°$ to A_2A_1) and plotted against non-dimensional distances R/A_1A_2, R being the distance of an interior point along L_1 or L_2 from the boundary. As expected, the BEM results for the finite plate are slightly higher (as they should be) as compared to the infinite plate solutions. It is of interest to note the near identicalness of the results obtained via QD, CUB and QRT elements for almost the same DOF. Since interior stresses involve a difficult surface integral, it is obvious that there will be a physical limit until which the boundary can be approached for a given adaptive integration scheme. In general, the criteria for accurate numerical integration are more stringent for points close to the boundary. In Figs 4.9 and 4.10, even at points about 3.5 per cent of the smallest element (QD) length (for $R/A_1A_2 = 0.0022$) away from the approaching boundary, the interior stresses showed no signs of deterioration. Of course, stresses at

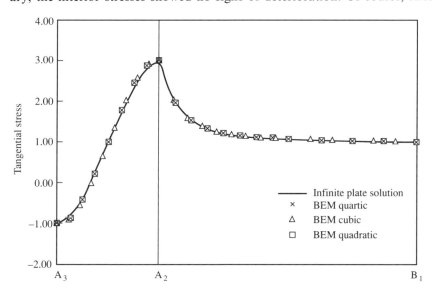

Figure 4.8 Tangential stress variation along $A_3A_2B_1$.

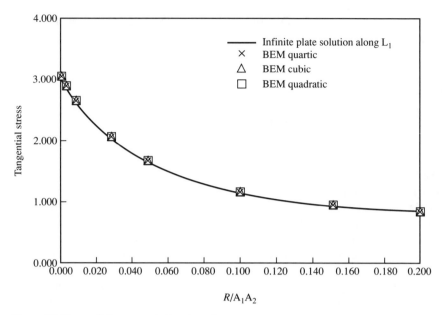

Figure 4.9 Tangential stress variation along L_1.

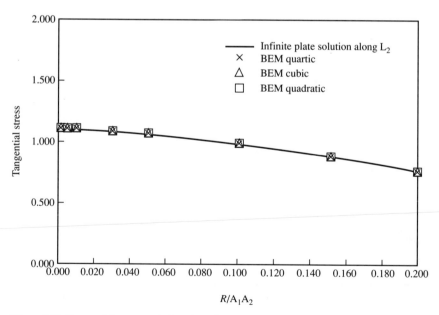

Figure 4.10 Tangential stress variation along L_2.

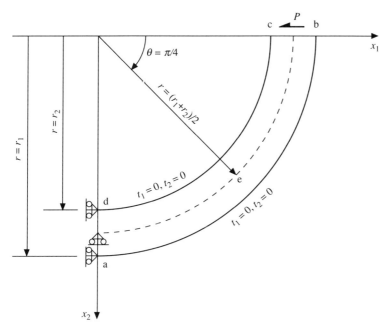

Figure 4.11 Curved beam problem with boundary conditions.

a boundary node are calculated accurately using equations in terms of boundary tractions, displacements and its tangential derivatives (see Chapter 17).

4.9.2 Bending of a Curved Beam[30]

The bending under plane stress conditions of a circular beam (Fig. 4.11) constrained at the lower end and bent by a traction-resultant P applied at the upper end in the radial direction was considered in Timoshenko and Goodier.[10]

With material properties $E = 10^6$ and $\nu = 0.25$ and curved beam geometry given by $r_1 = 9$ and $r_2 = 10$ as well as $P = 100$, BEM results of characteristic boundary displacements and tractions as well as interior stresses together with their absolute percentage errors with respect to the exact values are given in Tables 4.1 and 4.2. The quantities considered in these tables are (1) the deflection

Table 4.1 Comparison for δ_{bc} between QRT, CUB and QD elements

BEM QRT			BEM CUB			BEM QD		
DOF	δ_{bc}	|Error| (%)	DOF	δ_{bc}	|Error| (%)	DOF	δ_{bc}	|Error| (%)
48	−0.84	3.26	48	−1.29	59.88	48	−0.62	23.02
64	−0.81	0.52	60	−0.96	18.91	64	−0.74	8.5
80	−0.81	0.13	84	−0.86	5.71	80	−0.78	3.76
96	−0.81	0.05	96	−0.83	2.71	96	−0.79	1.94

Table 4.2 Comparison for $t_1(a)$ between QRT, CUB and QD elements

BEM QRT			BEM CUB			BEM QD		
DOF	$t_1(a)$ $(\times 10^{-2})$	|Error| (%)	DOF	$t_1(a)$ $(\times 10^{-2})$	|Error| (%)	DOF	$t_1(a)$ $(\times 10^{-2})$	|Error| (%)
48	−57.3	5.90	48	−118.5	119.08	48	−36.4	32.78
64	−54.6	0.92	60	−70.1	29.58	64	−47.8	11.68
80	−54.3	0.31	84	−57.4	6.16	80	−51.3	5.25
96	−54.2	0.18	96	−56.0	3.44	96	−52.6	2.79

δ_{bc} at the free end in the x_1-direction, and (2) traction $t_1(a)$ at point a. For the data used in this problem the exact solutions yield the following values:

$$\delta_{bc} = -0.8094 \qquad t_1(a) = -5409.0$$

In the boundary element meshes for this problem, two QD/two CUB/one QRT elements along sides ad or cb are taken, the rest of the elements in a mesh being mostly equally (or nearly so) distributed between the curved sides. A plot of the results on convergence of displacements corresponding to Table 4.1 is given in Fig. 4.12 where a clear superiority of the QRT elements over QD and CUB elements can be seen. As a matter of fact, QRT elements with just 64 DOF (i.e. eight elements) yield sufficiently accurate results for the chosen quantities (within about 1 per cent) while the results from a comparable number of DOF, e.g. 60 for CUB elements, are substantially poorer.

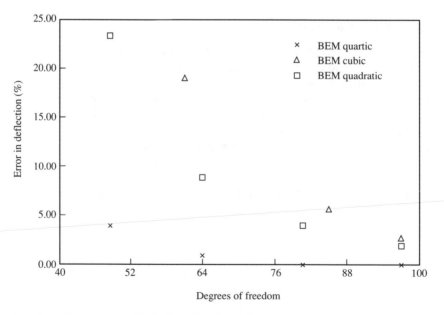

Figure 4.12 Convergence of deflection at the free end.

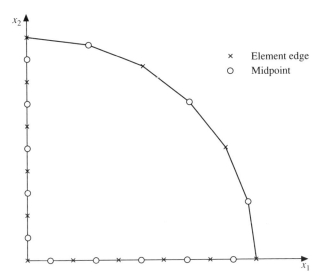

×   Element edge
○   Midpoint

Figure 4.13 Quarter disc two-dimensional boundary element model.

4.9.3 Uniformly Rotating Disc[25, 29]

An orthotropic circular disc rotating with a uniform speed exhibits quadrantal symmetry about its material principal axes. Thus making the material principal axes coincide with the global or natural axes, boundary element analysis of one quadrant of the disk needs to be carried out. The boundary element mesh for the two-dimensional analysis is shown in Fig. 4.13. The orthogonal axes of reference which are also the material principal axes pass through the centre of the disc. The major principal axis, that is the direction of alignment of fibres, is in the x_1 direction. There are five quadratic elements on each of the radial sides and three elements along the circumferential arc. The disc has a radius of 10 inches and is made of a boron/aluminium composite. The side of the quadrantal disc aligned with the x_1 axis is supported on rollers in the x_1 direction and the side aligned with the x_2 axis is roller-supported in the x_1 direction.

The deformed shape of the disc on the basis of the two-dimensional analysis is given in Fig. 4.14 where it is clear that the higher stiffness in the direction of fibres indicated by horizontal lines makes the disc undergo less deformation in that direction. The boundary element model using six-noded triangular and eight-noded quadrilateral elements for the three-dimensional analysis is given in Fig. 4.15.

In order to enforce a condition of plane strain so as to enable comparison with the two-dimensional analysis, faces perpendicular to the x_3 direction are supported on vertical rollers in addition to restricting movement in the normal direction of the faces parallel to the x_1 and x_2 axes.

As will be seen from Fig. 4.16, (on page 129), the agreement of results for tangential and normal stresses along the radial direction yielded by the two- and three-dimensional approaches is quite satisfactory.

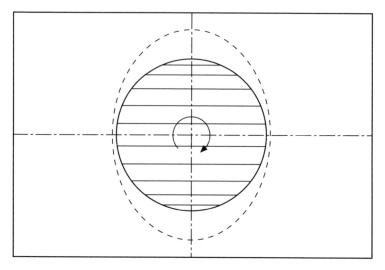

Figure 4.14 Deformed shape of rotating orthotropic disc.

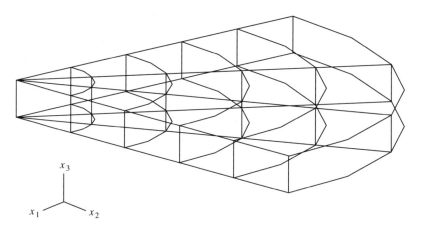

Figure 4.15 Quarter disc three-dimensional boundary element model.

4.9.4 Generalized Plane Strain Analysis of a Cube Subject to a Uniform Strain in the x_3 Direction[25, 29]

The problem of generalized plane strain, like the case of ordinary plane strain, may sometimes be an exact formulation of linear elasticity. To illustrate this point, a cube of an orthotropic material is considered with sides parallel to the x_3 axis restrained against movement in the x_1x_2 plane by rollers. If a constant strain ϵ_{33} is applied in the x_3 direction, the following solutions for stresses will be obtained:

$$\sigma_{11} = C_1\epsilon_{33} \qquad \sigma_{22} = C_2\epsilon_{33} \qquad \sigma_{33} = C_3\epsilon_{33} \qquad \sigma_{23} = \sigma_{13} = \sigma_{12} = 0$$

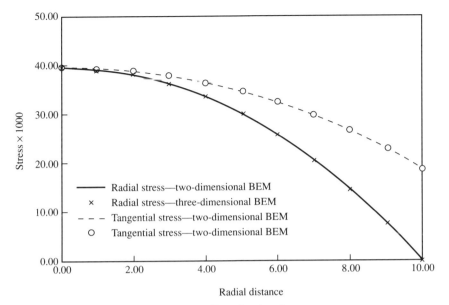

Figure 4.16 Comparison of two- and three-dimensional analyses.

In the preceding equation, C_1, C_2 and C_3 are constants and are given in terms of engineering material constants as follows:

$$C_1 = \frac{E_1 E_3}{F} [(\nu_{12} E_2 + \nu_{13} \nu_{23} E_3)\nu_{23} + (E_2 - \nu_{23}^2 E_3)\nu_{13}]$$

$$C_2 = \frac{E_2 E_3}{F} [(\nu_{12} E_2 + \nu_{13} \nu_{23} E_3)\nu_{13} + (E_1 - \nu_{13}^2 E_3)\nu_{23}]$$

$$C_3 = E_3 + \nu_{13} \frac{E_3}{E_1} C_1 + \nu_{23} \frac{E_3}{E_2} C_2$$

where

$$F = (E_1 - \nu_{13}^2 E_3)(E_2 - \nu_{23}^2 E_3) - (\nu_{12} E_2 + \nu_{13} \nu_{23} E_3)^2$$

For a fibre-reinforced composite with the properties $E_1 = 21.0 \times 10^6$, $E_2 = 1.55 \times 10^6$, $E_3 = 1.55 \times 10^6$, $\mu_{12} = 0.65 \times 10^6$, $\nu_{12} = 0.31$, $\nu_{13} = 0.31$ and $\nu_{23} = 0.49$, the non-zero stresses on the $x_1 x_2$ plane due to $\epsilon_{33} = 0.01$ will have the following numerical values:

$$\sigma_{11} = 9713.5 \qquad \sigma_{22} = 10\,429.53$$

This problem has been solved by the boundary element method using the generalized plane strain particular integrals discussed earlier. The geometry of the problem is a projection of a unit cube on the $x_1 x_2$ plane, i.e. a square of unit sides. The boundary discretization consists of one quadratic element, each along the sides of the square. Each side is roller-supported to restrain movement in a direction perpendicular to it, leading to the following results:

$$\sigma_{11} = 9691.4 \qquad \sigma_{22} = 10\,429.5$$

Obviously, the boundary element results have agreed to within 1 per cent of the exact solutions, indicating the correctness of the generalized plane strain results.

4.10 INDUSTRIAL APPLICATIONS

Elastic analysis by the BEM has benefited from about 30 years (since Banaugh and Goldsmith[31] in 1963) of continuous developments,[13–15, 28–45] its primary attraction being the ability to analyse three-dimensional problems with a boundary-only discretization which results in substantial cost savings in the data preparation. There are many large-scale industrial applications employing the state-of-the-art BEM technologies.[17, 42–45]

4.11 CONCLUDING REMARKS

Examples presented in this chapter will, we hope, help to convince the reader of the general utility and versatility of the BEM as a problem-solving tool. Since the numerical schemes outlined here are those developed for the first- or the second-generation BEM computer programs, the computational efficiencies will certainly be improved upon. Nevertheless, the algorithms presented in Chapters 3 and 4 do illustrate all the essential steps involved in the solution process that need to be mastered for static linear problems.

BEM solutions are algebraically complicated; there is a probability that in the future more efficient implementation will be achieved by using new representations of both geometries and functions so that the number of elements required to describe the boundary data, and over which the integrations are performed, is minimized. The future success of the BEM will ultimately lie in accurate modelling of boundary information and developments in more efficient surface integration techniques by using analytical methods.

REFERENCES

1. Lauricella, C. (1906) *Sull'integrazione delle equazioni dei corpi elastici isopropi*, Rendiconto Academia dei licei, Vol. 16(1), pp. 426–432.
2. Fredholm, I. (1905) Solution of fundamental problems in theory of elasticity, *Ark. Mat. Astr. och Fys.*, Vol. 2(28), pp. 1–8.
3. Sokolnikoff, I. (1956) *Mathematical Theory of Elasticity*, McGraw-Hill, New York.
4. Betti, E. (1872) Teoria dell elasticita, 11 nuovo Ciemento, 1872, pp. 7–10.
5. Banerjee, P.K. and Butterfield, R. (1981) *Boundary Element Methods in Engineering Science*, McGraw-Hill, London.
6. Mindlin, R.D. (1935) A point force in the interior of a semi-infinite solid, *Physics*, Vol. 7, pp. 195–202.
7. Butterfield, R. and Banerjee, P.K. (1971) The elastic analysis of compressible piles and pile groups, *Geotechnique*, Vol. 21(1), pp. 43–60.
8. Butterfield, R. and Banerjee, P.K. (1971) The problem of pile cap, pile group interaction, *Geotechnique*, Vol. 21(2), pp. 135–141.

9. Cruse, T.A. (1969) Numerical solutions in three-dimensional elastostatics, *International Journal of Solids and Structures*, Vol. 5, pp. 1259–1274.

10. Timoshenko, S. and Goodier, J.N. (1951) *Theory of Elasticity*, 2nd edn, McGraw-Hill, New York.

11. Terzaghi, R. (1943) *Theoretical Soil Mechanics*, John Wiley, New York.

12. Stippes, M. and Rizzo, F.J. (1977) A note on the body force integral of classical elastostatics, *ZAMP*, Vol. 28, pp. 339–341.

13. Rizzo, F.J. and Shippy, D.J. (1977) An advanced boundary integral equation method for three-dimensional thermo-elasticity, *International Journal of Numerical Methods in Engineering.*, Vol. 11, p. 1753.

14. Rizzo, F.J. and Shippy, D.J. (1979) Recent advances of the boundary element method in thermo-elasticity, Chapter VI in *Developments in Boundary Element Methods*, Vol. 1, Eds P.K. Banerjee and R. Butterfield, Elsevier Applied Science Publishers, London.

15. Henry, D.P., Pape, D.A. and Banerjee, P.K. (1987) A new axisymmetric BEM for body forces using particular integrals, *Journal of Engineering Mechanics Division, ASCE*, Vol. 113, No. 12, pp. 1880–1900.

16. Pape, D.A. and Banerjee, P.K. (1987) Treatment of body forces in a 2-D elastostatic BEM using particular integrals, *Journal of Applied Mechanics, ASME*, Vol. 54, pp. 866–871.

17. Banerjee, P.K., Wilson, R.B. and Miller, N.M. (1988) Advanced elastic and inelastic three-dimensional analysis of gas turbine engine structures by BEM, *International Journal of Numerical Methods in Engineering*, Vol. 26, pp. 393–411.

18. Henry, D. and Banerjee, P.K. (1988) A new BEM formulation for two and three-dimensional thermoelasticity using particular integrals, *International Journal of Numerical Methods in Engineering*, Vol. 26, pp. 2061–2077.

19. Deb, A. and Banerjee, P.K. (1991) Multi-domain two and three-dimensional thermoelasticity by BEM, *International Journal of Numerical Methods in Engineering*, Vol. 32, pp. 991–1008.

20. Lekhnitskii, S.G. (1963) *Theory of Elasticity of an Anisotropic Elastic Body*, Trans. P. Fern, Holden Day, San Francisco.

21. Tomlin, G.R. and Butterfield, R. (1974) Elastic analysis of zoned orthotropic continua, *Proceedings of ASCE*, Vol. EM3, pp. 511–529.

22. Snyder, M.D. and Cruse, T.A. (1975) Boundary integral equation analysis of cracked anisotropic plates, *International Journal of Fracture Mechanics*, Vol. 11, pp. 315–328.

23. Wilson, R.B. and Cruse, T.A. (1978) Efficient implementation of anisotropic three-dimensional boundary integral equation stress analysis, *International Journal of Numerical Methods in Engineering*, Vol. 12, pp. 1383–1397.

24. Synge, J.L. (1957) *The Hypercircle in Mathematical Physics*, Cambridge University Press, pp. 411–413.

25. Deb, A. (1991) Advanced Development of BEM for Linear and Nonlinear Analysis of Anisotropic Solids, PhD Dissertation, State University of New York at Buffalo, New York.

26. Deb, A., Henry, D. and Wilson, R.B. (1991) An alternate BEM for 2D and 3D anisotropic thermoelasticity, *International Journal of Solids and Structures*, Vol. 27, No. 13, pp. 1721–1738.

27. Malen, K. (1971) A unified six-dimensional treatment of elastic Green's functions and dislocations, *Physics of Statistical Solids*, Vol. 6, pp. 661–672.

28. Pan, Y.C. and Chou, T.W. (1976) Point force solution for an infinite transversely isotropic solid, *Journal of Applied Mechanics, Transactions of ASME*, Vol. 98(E), pp. 608–612.

29. Deb, A. and Banerjee, P.K. (1990) BEM for general anisotropic 2D elasticity using particular integrals, *Communications in Applied Numerical Methods*, Vol. 6, pp. 111–119.

30. Deb, A. and Banerjee, P.K. (1989) A comparison between isoparametric Lagrangian elements in 2D BEM, *International Journal of Numerical Methods in Engineering*, Vol. 28, pp. 1539–1555.

31. Banaugh, R.P. and Goldsmith, W. (1963) Diffractions of steady elastic waves by surfaces of arbitrary shape, *Journal of Applied Mechanics*, Vol. 30, pp. 589–597.

32. Rizzo, F.J. (1967) An integral equation approach to boundary value problems of classical elastostatics, *Journal of Applied Mathematics*, Vol. 25, p. 83.

33. Tomlin, G.R. (1972) Numerical analysis of continuum problems of zoned anisotropic media, PhD Thesis, Southampton University.

34. Ricardella, P. (1973) An implementation of the boundary integral techniques for plane problems in elasticity and elasto-plasticity, PhD Thesis, Carnegie Mellon University, Pittsburgh.

35. Cruse, T.A. (1974) An improved boundary integral equation method for three dimensional stress analysis, *International Journal of Computers and Structures*, Vol. 4, pp. 741–757.

36. Banerjee, P.K. and Butterfield. R. (1977) Boundary element method in geomechanics, Chapter 16 in *Finite Element in Geomechanics*, Ed. G. Gudehus, John Wiley, London.

37. Wardle, L.J. and Crotti, J.M. (1978) Two-dimensional boundary integral equation analysis for nonhomogeneous mining applications, *Proceedings of Recent Developments in Boundary Element Methods*, Southampton University.

38. Lachat, J.C. and Watson, J.O. (1976) Effective numerical treatment of boundary integral equations: a formulation for three-dimensional elasto-statics, *International Journal of Numerical Methods in Engineering*, Vol. 10, pp. 991–1005.

39. Watson, J.O. (1979) Advanced implementation of the boundary element method in two- and three-dimensional elasto-statics, Chapter 3 in *Developments in Boundary Element Methods*, Eds P.K. Banerjee and R. Butterfield, Elsevier Applied Science Publishers, London.

40. Banerjee, P.K. (1976) Integral equation methods for analysis of piece-wise non-homogeneous three-dimensional elastic solids of arbitrary shape, *International Journal of Mechanical Science*, Vol. 18, pp. 293–303.

41. Cruse, T.A. and Van Bauren, W. (1971) Three-dimensional elastic stress analysis of fracture specimen with an edge crack, *International Journal of Fracture Mechanics*, Vol. 7, pp. 1–15.

42. Henry (Jr), D. P. (1987) Advanced development of BEM for elastic and inelastic thermal stress analysis, PhD Dissertation, State University of New York at Buffalo.

43. Dargush, G.F. (1987) Boundary element methods for the analogous problems of thermomechanics and soil consolidation, PhD Dissertation, State University of New York at Buffalo.

44. Butenschön, H.J., Möhrmann, W. and Bauer W. (1988) Advanced stress analysis by a commercial BEM code, Chapter 8 in *Industrial Applications of BEM*, Eds. P.K. Banerjee and R.B. Wilson, Elsevier Applied Science, London.

45. Graf, G.L. and Gebre-Giorgis,Y. (1988) Thermoelastic analysis for design of machine components, Chapter 9 in *Industrial Applications of BEM*, Eds. P.K. Banerjee and R.B. Wilson, Elsevier Applied Science, London.

FIVE

AXISYMMETRIC PROBLEMS

5.1 INTRODUCTION

In the previous two chapters both two- and three-dimensional boundary element formulations have been described. However, there is a large class of practical problems that involve axisymmetric geometry with axisymmetric or non-axisymmetric loading. While, in these situations, a full three-dimensional analysis is valid, a reduction in the dimensionality of the problem is often quite beneficial. Consequently, this chapter details the development of axisymmetric boundary element formulations in which most of the numerical schemes for two-dimensional problems are directly usable. Thus, the main task, described below, involves the development of the axisymmetric kernels.

5.2 AXISYMMETRIC POTENTIAL FLOW

Many axisymmetric problems of potential flow can best be represented by a cylindrical coordinate system r, z, as shown in Fig. 5.1. Such axisymmetric potential flow problems have been considered by Jaswon and Symm[1] and Liggett and Liu[2] in earlier publications related to electrostatic fields and groundwater flow respectively. Dargush[3] describes an advanced multi-region axisymmetric heat transfer analysis by the BEM.

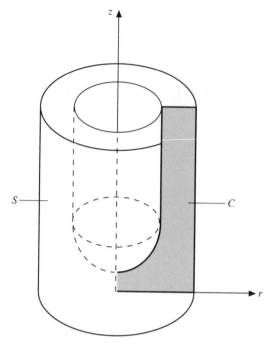

Figure 5.1 Axisymmetric geometry.

5.2.1 Axisymmetric Singular Solution

The fundamental singular solution due to an axisymmetric source e acting at point Q may be written as (see Fig. 5.2)

$$p(P) = G(P, Q)e(Q) \qquad (5.1)$$

where the function $G(P, Q)$ is obtained by expressing the three-dimensional solution

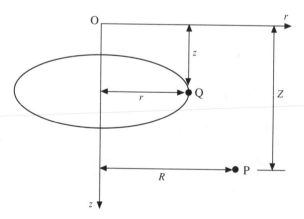

Figure 5.2 The cylindrical coordinate system.

of Chapter 3 in (r, θ, z) coordinates and integrating the result with respect to θ between the limits 0 and 2π. Thus we can write

$$G(P, Q) = \frac{r}{2\pi\kappa} \int_0^\pi \frac{d\theta}{\sqrt{a - b\cos\theta}} \tag{5.2}$$

where $a = r^2 + R^2 + (z - Z)^2$ and $b = 2rR$. The integral in Eq. (5.2) cannot be evaluated exactly and has to be dealt with as an elliptic integral, i.e.

$$G(P, Q) = \frac{r}{\pi\kappa} \left(\frac{1}{\sqrt{a + b}}\right) K(m) \tag{5.3}$$

where $K(m)$ is a complete elliptic integral of the first kind of modulus m and complementary modulus $m_1 = 1 - m$. Thus

$$K(m) = \int_0^\pi \frac{d\theta}{\sqrt{1 - m\sin^2\theta}}$$

$$m = \frac{4rR}{(r + R)^2 + (z - Z)^2} \qquad 0 \le m < 1 \tag{5.4}$$

$$m_1 = \frac{(r - R)^2 + (z - R)^2}{(r + R)^2 + (z - Z)^2}$$

It is rather inconvenient to work with the elliptic integral in (5.4) and therefore it is usually necessary to obtain a suitably close polynomial approximation (see Abramowicz and Stegan[4]). Jaswon and Symm[1] discuss a number of possible approximations for different values of the modulus m. A typical one for $K(m)$ might be[4]

$$K(m) = \sum_{i=0}^{n} [a_i m_1^i + b_i m_1^i \ln(1/m_1)] + \varepsilon(m) \tag{5.5}$$

where, for $n = 4$, the error term $\varepsilon(m)$ is $\le 2 \times 10^{-8}$ and a_i and b_i are constants.

Equation (5.5) shows that the singularity at $P = Q$ is similar to that of the two-dimensional problem (i.e. a weak logarithmic one). The normal 'velocity' in the direction n at a point P can be obtained from Eq. (5.3) as

$$q(P) = F(P, Q)e(Q) \tag{5.6}$$

where

$$F(P, Q) = -\kappa \left(\frac{\partial G}{\partial R} n_r + \frac{\partial G}{\partial Z} n_z\right)$$

and n_r and n_z are the components of the vector $\mathbf{n}$ in the direction of r and z axes respectively on a surface through $P(R, Z)$.

By using the identity

$$E(m) = m_1 \left[2m \frac{dK(m)}{dm} + K(m)\right]$$

where $E(m)$ is the complete elliptic integral of the second kind, we can express $F(P, Q)$ as

$$F(P, Q) = \left[-\frac{n_r[E(m) - K(m)]}{\pi r \sqrt{a + b}} + \frac{2n_r(r - R) + n_z(z - Z)}{\pi(a - b)\sqrt{a + b}} E(m) \right] \frac{r}{2} \quad (5.7)$$

and $E(m)$ in Eq. (5.7) may be written as the polynomial approximation

$$E(m) = 1 + \sum_{i=1}^{n} [c_i m_1^i + d_i m_1^i \ln(1/m_1)] \quad (5.8)$$

where $0 \le m < 1$, for $n = 4$, $\varepsilon(m)$ is $< 1 \times 10^{-8}$ and c_i and d_i are constants.

5.2.2 Direct BEM Formulation

To develop the direct BEM, we can either take the three-dimensional formulation of Chapter 3 and convert it to our cylindrical coordinate system or substitute the above fundamental solutions into the reciprocal identity and develop the necessary integral representation. In either case the following identity is obtained for an axisymmetric analysis:

$$c(\xi)p(\xi) = \int_S [G(x, \xi)q(x) - F(x, \xi)p(x)] \, dC(x) \quad (5.9)$$

where

ξ is the point $Q(r, z)$ and x is the point $P(R, Z)$

$c(\xi) = 1, 0$ and $\frac{1}{2}$ depending on the point ξ being in the interior, outside or on a smooth boundary point respectively

For convenience we now redefine the above $G(P, Q)$ and $F(P, Q)$ in more convenient form, i.e.

$$G(x, \xi) = \frac{2a_3}{H} K(m) \quad (5.10)$$

and
$$F(x, \xi) = \left[\left(\frac{1}{RH} \right) a_3 k K(m) + \left(\frac{2(R - r)}{\rho^2 H} - \frac{1}{RH} \right) a_3 k E(m) \right] n_r$$

$$+ \left[\left(\frac{2\bar{Z}}{\rho^2 H} \right) a_3 k E(m) \right] n_z \quad (5.11)$$

where

$n_r(R, Z)$, $n_z(R, Z)$ are normals at integration points

$\bar{Z} = (Z - z)$, $H^2 = (R + r)^2 + \bar{Z}^2$

$\rho^2 = (R - r)^2 + \bar{Z}^2$, $m = \dfrac{4Rr}{H^2}$ and $a_3 = -\dfrac{R}{2\pi k}$

Note once again, as a result of integration by parts, the load point field point combination in the fundamental solution and that in Eq. (5.9) becomes reversed.

The order of singularity in G and F obtained via the expansions of $K(m)$ and $E(m)$ becomes identical to those of two-dimensional potential flow. Eq. (5.9) can therefore be treated using the detailed treatment for a two-dimensional problem described in Chapter 3.

5.3 AXISYMMETRIC ELASTICITY

By taking the three-dimensional elastostatic formulation and by transforming both displacements and tractions as well as the kernel functions G_{ij}, F_{ij} into a cylindrical coordinate system we can easily obtain the axisymmetric elastostatic formulation as[9]

$$\begin{bmatrix} u_r(r,z) \\ u_z(r,z) \end{bmatrix} = \int_C \left\{ \begin{bmatrix} G_{rr} & G_{zr} \\ G_{rz} & G_{zz} \end{bmatrix} \begin{Bmatrix} t_r \\ t_z \end{Bmatrix} - \begin{bmatrix} F_{rr} & F_{zr} \\ F_{rz} & F_{zz} \end{bmatrix} \begin{Bmatrix} u_r \\ u_z \end{Bmatrix} \right\} dC(R,Z) \qquad (5.12)$$

where G_{ij} is the displacement vector u_i due to a ring load intensity e_j and expressed as[5]

$$G_{ij} = A_{ij}K + B_{ij}E \qquad i,j = r,z \qquad (5.13a)$$

$K = K(m)$ is the complete elliptic integral of the first kind and $E = E(m)$ is the complete elliptic integral of the second kind. The components of F_{ij} can then be calculated from

$$F_{ri} = \left[c_1 \frac{\partial G_{ri}}{\partial R} + c_2 \left(\frac{G_{ri}}{R} + \frac{\partial G_{zi}}{\partial Z} \right) \right] n_r + \mu \left[\frac{\partial G_{ri}}{\partial Z} + \frac{\partial G_{zi}}{\partial R} \right] n_z \qquad i = r,z \qquad (5.13b)$$

$$F_{zi} = \left[c_1 \frac{\partial G_{zi}}{\partial Z} + c_2 \left(\frac{G_{ri}}{R} + \frac{\partial G_{ri}}{\partial R} \right) \right] n_z + \mu \left[\frac{\partial G_{ri}}{\partial Z} + \frac{\partial G_{zi}}{\partial R} \right] n_r \qquad i = r,z \qquad (5.13c)$$

where

$$\frac{\partial G_{ij}}{\partial R} = \frac{\partial A_{ij}}{\partial R} K + \frac{\partial B_{ij}}{\partial R} E + A_{ij} \frac{\partial K}{\partial R} + B_{ij} \frac{\partial E}{\partial R}$$

$$\frac{\partial G_{ij}}{\partial Z} = \frac{\partial A_{ij}}{\partial Z} K + \frac{\partial B_{ij}}{\partial Z} E + A_{ij} \frac{\partial K}{\partial Z} + B_{ij} \frac{\partial E}{\partial Z}$$
$$i,j = r,z$$

$$n_r = n_r(R,Z) \qquad n_z = n_z(R,Z)$$

$$A_{rr} = \frac{\alpha}{rRH}(c_3 M + \bar{Z}^2) \quad B_{rr} = -\frac{\alpha}{rRH}\left(c_3 H^2 + \frac{M\bar{Z}^2}{\rho^2}\right)$$

$$A_{rz} = \frac{\alpha\bar{Z}}{RH} \qquad\qquad B_{rz} = -\frac{\alpha\bar{Z}}{R\rho^2 H}(M - 2R^2)$$

$$A_{zr} = -\frac{\alpha\bar{Z}}{rH} \qquad\qquad B_{zr} = \frac{\alpha\bar{Z}}{r\rho^2 H}N$$

$$A_{zz} = \frac{2\alpha}{H}c_3 \quad \text{and} \quad B_{zz} = \frac{2\alpha\bar{Z}^2}{\rho^2 H}$$

The required derivatives with respect to Z and R are relatively straightforward. In addition,

$$\bar{Z} = (Z - z), \quad H^2 = [(R + r)^2 + \bar{Z}^2] \qquad \rho^2 = [(R - r)^2 + \bar{Z}^2]$$

$$M = R^2 + r^2 + \bar{Z}^2 \qquad\qquad N = R^2 - r^2 + \bar{Z}^2$$

$$c_1 = \frac{2\mu(1 - \nu)}{1 - 2\nu} \qquad\qquad c_3 = 3 - 4\nu$$

$$c_2 = \frac{2\mu\nu}{1 - 2\nu} \quad \text{and} \qquad \alpha = \frac{R}{8\pi\mu(1 - \nu)}$$

where μ = shear modulus and ν = Poisson's ratio.

Note that the G_{ij} kernel of Eq. (5.12) is the ring source solution first derived by Kermanidis[6] and later again by Cruse *et al.*[7] A $2\pi R$ term, which appears after integration in the θ direction, has been absorbed in the kernel; therefore, the numerical integration is performed over a curve dC.

Once again the behaviour of these G and F kernels are somewhat analogous to those of two-dimensional elastostatics and accordingly most two-dimensional BEM solution techniques could once again be applied to Eq. (5.12).

5.3.1 Some Special Features

Treatment of kernel singularities As previously mentioned in Chapters 3 and 4, special consideration must be given to the element integration near a singular point. The axisymmetric kernels become singular under two circumstances: (1) when the radial coordinate of the field point falls on the origin ($r = 0$) and (2) when the field point and integration point coincide ($r = R$ and $z = Z$). For $r = 0$ the kernel functions related to the radial displacement are not well-behaved. Asymptotic forms for these kernels can be easily derived[8] where some of these radial displacement terms become zero. Alternatively, the computational problem involved could be easily circumvented by taking the field point at a small non-zero value when these expressions reach a finite limit.[5,9]

In the second case, when the field point coincides with the integration point, special care must be taken. The **G** (displacement) kernel is weakly singular and can be integrated numerically with element subdivision near the singular point as discussed in Chapter 3. The **F** (traction) kernel, on the other hand, is strongly singular and must be integrated with special care over the singular element. This can be circumvented by using a method analogous to the rigid body displacement technique, which is known as the 'inflation mode' for axisymmetry. A brief description of the method is given below.[5,9-13]

For every node of the system there is a 2×2 block of coefficients on the diagonal of the $[F]$ matrix corresponding to the singular node, which cannot be easily determined by numerical integration. Each of these four terms can be determined independently by assuming two different admissible displacement fields. The first is the standard rigid body displacement in the z direction ($u_r = 0$, $u_z = 1$). Here all tractions are zero and the two unknown coefficients in the 2×2 block can be determined.

Since a rigid body displacement is not admissible in the radial direction, the other two unknown coefficients in the 2×2 block are determined by an inflation mode (i.e. a linear displacement in the radial direction), $u_r = r$ and $u_z = 0$. The tractions related to this displacement are $t_r = 2(\lambda + \mu)n_r$ and $t_z = 2\lambda n_z$, in which n_r and n_z are normals on the boundary and λ and μ are Lame's constant. The unknown 2×2 blocks can be determined such that the discretized boundary system equations satisfy these conditions at nodal points.

Behaviour of kernels for interior quantities   After the boundary value problem has been solved and all unknown quantities have been determined, the displacement, stresses and strains can be calculated at interior points.

When a field point (r, z) is located in the interior of the domain and does not fall on the origin the integrals involved are well behaved. However, if the interior point falls on the origin or on the surface of the boundary, the kernels become difficult to handle numerically.

The problem of a field point falling on the origin again can be resolved by moving the point a small radial distance from the origin. Alternatively, a separate limiting form for the displacement at the origin can be derived.

When a field point lies on the boundary the displacement can be found via interpolation of the nodal values. The stresses can be obtained from the boundary tractions and displacements without any integration (see Chapter 17). A slight modification must be made to the usual two-dimensional equations to incorporate the strain in the θ direction. For stresses on the surface at the origin the relation $\epsilon_\theta - \epsilon_r = \partial u_r / \partial r$ proves useful instead of the usual $\epsilon_\theta = u_r / r$.

Particular Integrals   The particular integrals for the three-dimensional analyses under self-weight and centrifugal loading (see Chapter 4) can be converted into our cylindrical coordinate system. Thus for gravitational loading directed along the z axis (that is $\psi_z = -\rho g$, $\psi_r = 0$) the particular integrals for displacements are[5]

$$u_r^p = \frac{-\lambda \rho g}{2\mu(3\lambda + 2\mu)} rz$$

$$u_z^p = \frac{(\lambda + \mu)\rho g}{2\mu(3\lambda + 2\mu)} z^2 + \frac{\lambda \rho g}{4\mu(3\lambda + 2\mu)} r^2 \tag{5.14}$$

Using the strain–displacement and stress–strain relations with these displacements, we find

$$\sigma_r^P = \sigma_\theta^P = \sigma_{rz}^P = 0$$

$$\sigma_z^P = \rho g z \tag{5.15}$$

as the particular stresses corresponding to this gravitational loading.

For a centrifugal loading about the z axis ($\psi_r = \rho \omega^2 r$, $\psi_z = 0$) the particular integrals for displacements are given by[5]

$$u_r^p = \frac{-\rho \omega^2}{8(\lambda + 2\mu)} \left[\frac{5\lambda + 4\mu}{4(\lambda + \mu)} r^2 + \frac{\mu}{\lambda + \mu} z^2 \right] r$$

$$u_z^p = \frac{\rho \omega^2}{8(\lambda + 2\mu)} r^2 z \tag{5.16}$$

Using the strain–displacement and stress–strain relations, we can determine the σ_{ij}^P and the particular solution for surface traction $t_i^p = \sigma_{ij}^P n_j$.

5.4 AXISYMMETRIC THERMOELASTIC ANALYSIS

Once again the three-dimensional thermoelastic formulation defined in Chapter 4 (see also Chapter 9) could be converted into our cylindrical coordinate system to yield[3]

$$\begin{Bmatrix} u_r \\ u_z \\ T \end{Bmatrix} = \int_C \left(\begin{bmatrix} G_{rr} & G_{rz} & G_{rt} \\ G_{zr} & G_{zz} & G_{zt} \\ 0 & 0 & G \end{bmatrix} \begin{Bmatrix} t_r \\ t_z \\ q \end{Bmatrix} - \begin{bmatrix} F_{rr} & F_{rz} & F_{rt} \\ F_{zr} & F_{zz} & F_{zt} \\ 0 & 0 & F \end{bmatrix} \begin{Bmatrix} u_r \\ u_z \\ T \end{Bmatrix} \right) dC \tag{5.17}$$

where G_{rr}, G_{zr}, G_{rz} and G_{zz} (and similarly F_{rr}, etc.) are given in Eq. (5.13) above, while G and F have been defined in Eqs (5.10) and (5.11) and

$$G_{rt} = \frac{a_2 N}{rH} K(m) - \frac{a_2 H}{r} E(m) \tag{5.18}$$

$$G_{zt} = \frac{2 a_2 \bar{Z}}{H} K(m) \tag{5.19}$$

where $a_1 = R/[8\pi\mu(1 - \nu)]$, $a_2 = R\beta/[4\pi k(\lambda + 2\mu)]$, β being the thermoelastic constant.

Similarly,

$$F_{rt} = \left[\left(\frac{r^2 + R^2}{rRH} \right) a_2 k K(m) + \left(\frac{M\bar{Z}^2}{rRH\rho^2} - \frac{H}{rR} \right) a_2 k E(m) \right] n_r$$

$$+ \left[\left(\frac{\bar{Z}^2}{rH} \right) a_2 k K(m) - \left(\frac{N\bar{Z}}{rH\rho^2} \right) a_2 k E(m) \right] n_z \qquad (5.20)$$

$$F_{zt} = \left[-\left(\frac{\bar{Z}}{RH} \right) a_2 k K(m) - \left(\frac{\bar{H}\bar{Z}}{RH\rho^2} \right) a_2 k E(m) \right] n_r$$

$$+ \left[\left(\frac{2}{H} \right) a_2 k K(m) - \left(\frac{2\bar{Z}^2}{H\rho^2} \right) E(m) \right] n_z \qquad (5.21)$$

Equation (5.17) was first derived by Bakr and Fenner,[8] and can be either solved concurrently for the combined thermal and stress analysis problem, in which each boundary node will have three degrees of freedom, or the thermal analysis is done separately and the temperature-dependent quantities in the stress analysis problem are then separated out from the integral formulation and transferred as known quantities on the right-hand side of the system equations.[3] Numerical treatment of these additional thermoelastic terms does not pose any special difficulty.

5.5 NON-AXISYMMETRIC BOUNDARY CONDITIONS

If the problem geometry is axisymmetric but the applied boundary condition is not purely axisymmetric substantial savings can still be achieved by converting the problem into a cylindrical coordinate system (r, θ, z). In this section we consider the elastic stress analysis only; the corresponding analysis for potential flow follow from these in an obvious manner.

Let us consider the three-dimensional elastostatic equations involving zero body forces:

$$c_{ij}(\xi)u_i(\xi) = \int_S [G_{ij}(x, \xi)t_i(x) - F_{ij}(x, \xi)u_i(x)] \, dS(x) \qquad (5.22)$$

Considering the transformation of displacement and traction in cylindrical coordinates, we have

$$\{u\} = \begin{Bmatrix} u_1 \\ u_2 \\ u_3 \end{Bmatrix} = \begin{bmatrix} \cos\theta & -\sin\theta & 0 \\ \sin\theta & \cos\theta & 0 \\ 0 & 0 & 1 \end{bmatrix} \begin{Bmatrix} u_r \\ u_\theta \\ u_z \end{Bmatrix} = [T]\{u'\}$$

$$\{t\} = \begin{Bmatrix} t_1 \\ t_2 \\ t_3 \end{Bmatrix} = \begin{bmatrix} \cos\theta & -\sin\theta & 0 \\ \sin\theta & \cos\theta & 0 \\ 0 & 0 & 1 \end{bmatrix} \begin{Bmatrix} t_r \\ t_\theta \\ t_z \end{Bmatrix} = [T]\{t'\}$$

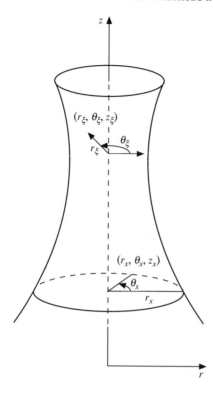

Figure 5.3 Axisymmetric body in cylindrical coordinates.

or

$$u_i = \mathcal{T}_{ik} u'_k, \qquad t_i = \mathcal{T}_{ik} t'_k$$

Introducing these transformations into Eq. (5.22), it can be written in cylindrical coordinates (Fig. 5.3) as[10-12,14,15]

$$c_{ij}(\xi) u'_i(\xi) = \int_S [G'_{ij}(x,\xi) t'_i(x) - F'_{ij}(x,\xi) u'_i(x)] \, \mathrm{d}S(x) \tag{5.23}$$

where

$$G'_{ij}(x,\xi) = \begin{bmatrix} G_{rr} & G_{r\theta} & G_{rz} \\ G_{\theta r} & G_{\theta\theta} & G_{\theta z} \\ G_{zr} & G_{z\theta} & G_{zz} \end{bmatrix} = \mathcal{T}_{ki}(\xi) G_{kl}(x,\xi) \mathcal{T}_{lj}(x) \tag{5.24a}$$

$$F'_{ij}(x,\xi) = \begin{bmatrix} F_{rr} & F_{r\theta} & F_{rz} \\ F_{\theta r} & F_{\theta\theta} & F_{\theta z} \\ F_{zr} & F_{z\theta} & F_{zz} \end{bmatrix} = \mathcal{T}_{ki}(\xi) F_{kl}(x,\xi) \mathcal{T}_{lj}(x) \tag{5.24b}$$

Symmetric   Antisymmetric

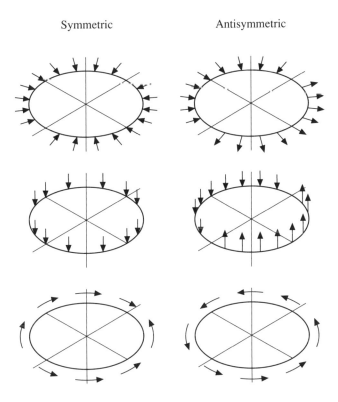

Figure 5.4 Symmetric and antisymmetric fields in axisymmetric solids.

Note that $G'_{ij}(x, \xi)$ and $F'_{ij}(x, \xi)$ are the functions of θ ($= \theta_x - \theta_\xi$). For a linear problem having symmetry in geometry, the asymmetric displacement and traction can be always decomposed into a symmetric part and an antisymmetric part[14-21] (see Fig. 5.4). This can be accomplished for the present problem if we use Fourier series expansion for the displacement field:

$$u_r(r, z, \theta) = \sum_n u_r^n(r, z) \cos n\theta + \sum_n u_r^{-n}(r, z) \sin n\theta \qquad (5.25a)$$

$$u_z(r, z, \theta) = \sum_n u_z^n(r, z) \cos n\theta + \sum_n u_z^{-n}(r, z) \sin n\theta \qquad (5.25b)$$

$$u_\theta(r, z, \theta) = \sum_n u_\theta^n(r, z) \sin n\theta + \sum_n u_\theta^{-n}(r, z) \cos n\theta \qquad (5.25c)$$

Since a symmetric displacement field yields a symmetric traction field while an antisymmetric displacement field yields an antisymmetric traction field (i.e. the symmetric field and antisymmetric field are decoupled) we can use similar expansions for tractions. Thus considering the symmetric Fourier series expansion for variables of $u'_i(\xi)$, $u'_i(x)$, $t'_i(x)$, $G'_{ij}(x,\xi)$ and $F'_{ij}(x,\xi)$ and the orthogonality of trigonometric functions $\cos n\theta$ and $\sin n\theta$, Eq. (5.22) can now be decomposed into decoupled boundary integral equations, for symmetric modes:[16-21]

$$c_{ij}(\xi)u_i^n(\xi) = \int_C [G'^n_{ij}(x,\xi)t_i^n(x) - F'^n_{ij}(x,\xi)u_i^n(x)]r_x \, dC(x) \qquad (5.26)$$

where C is the generator of an axisymmetric body and

$$G'^n_{ij}(x,\xi) = \int_0^{2\pi} \begin{bmatrix} G_{rr}\cos n\theta & G_{r\theta}\sin n\theta & G_{rz}\cos n\theta \\ -G_{\theta r}\sin n\theta & G_{\theta\theta}\cos n\theta & -G_{\theta z}\sin n\theta \\ G_{zr}\cos n\theta & G_{z\theta}\sin n\theta & G_{zz}\cos n\theta \end{bmatrix} d\theta \qquad (5.27a)$$

$$F'^n_{ij}(x,\xi) = \int_0^{2\pi} \begin{bmatrix} F_{rr}\cos n\theta & F_{r\theta}\sin n\theta & F_{rz}\cos n\theta \\ -F_{\theta r}\sin n\theta & F_{\theta\theta}\cos n\theta & -F_{\theta z}\sin n\theta \\ F_{zr}\cos n\theta & F_{z\theta}\sin n\theta & F_{zz}\cos n\theta \end{bmatrix} d\theta \qquad (5.27b)$$

Similarly, the following decoupled equation can be derived for the antisymmetric fields:[16-21]

$$c_{ij}(\xi)u_i^{-n}(\xi) = \int_L [G'^{-n}_{ij}(x,\xi)t_i^{-n} - F'^{-n}_{ij}(x,\xi)u_i^{-n}]r_x \, dL(x) \qquad (5.28)$$

where

$$G'^{-n}_{ij}(x,\xi) = \int_0^{2\pi} \begin{bmatrix} G_{rr}\cos n\theta & -G_{r\theta}\sin n\theta & G_{rz}\cos n\theta \\ G_{\theta r}\sin n\theta & G_{\theta\theta}\cos n\theta & G_{\theta z}\sin n\theta \\ G_{zr}\cos n\theta & -G_{z\theta}\sin n\theta & G_{zz}\cos n\theta \end{bmatrix} d\theta \qquad (5.29a)$$

$$F'^{-n}_{ij}(x,\xi) = \int_0^{2\pi} \begin{bmatrix} F_{rr}\cos n\theta & -F_{r\theta}\sin n\theta & F_{rz}\cos n\theta \\ F_{\theta r}\sin n\theta & F_{\theta\theta}\cos n\theta & F_{\theta z}\sin n\theta \\ F_{zr}\cos n\theta & -F_{z\theta}\sin n\theta & F_{zz}\cos n\theta \end{bmatrix} d\theta \qquad (5.29b)$$

Thus equations for symmetric and antisymmetric modes are now decoupled and the number of degrees of freedom is three for each of the symmetric and anti-symmetric cases.[16-21] It is of considerable interest to note that the zero-order symmetric field is the axisymmetric case described in the early part of this chapter and the zero-order antisymmetric mode is torsion. Note also that the symmetric and antisymmetric equations differ only in sign of a few terms in the G and F matrices which can lead to a major saving in numerical computation.[17] If the body force exists, the solution is separated into the complementary part and a particular integral which must be converted into the cylindrical coordinate system, as shown in Wang and Banerjee.[18,21]

5.5.1 Numerical Implementation

Boundary discretization   Only the discretization along the generator, C, is needed for solving Eqs (5.26) and (5.28). For a discretization of N segments, that is

$C = C_1 + C_2 + \cdots + C_N$, Eqs (5.26) and (5.28) can be written as

$$c_{ij}(\xi)u_i^n(\xi) = \sum_{m=1}^{N} \int_{C_m} [G'^n_{ij}(x,\xi)t_i^n - F'^n_{ij}(x,\xi)u_i^n]r_x \, dC_m(x) \qquad (5.30)$$

$$c_{ij}(\xi)u_i^{-n}(\xi) = \sum_{m=1}^{N} \int_{C_m} [G'^{-n}_{ij}(x,\xi)t_i^{-n} - F'^{-n}_{ij}(x,\xi)u_i^{\ n}]r_x \, dC_m(x) \qquad (5.31)$$

in which no approximation has yet been made and the equations are still exact. For a description of geometry and interpolation of boundary variables for each circumferential order, quadratic, cubic or quartic conforming boundary elements can be used. Thus

$$x_i = \sum_k N^k x_i^k \qquad (5.32a)$$

$$u_i^n = \sum_k N^k u_i^{\pm n}(x^k) \qquad (5.32b)$$

$$t_i^n = \sum_k N^k t_i^{\pm n}(x^k) \qquad (5.32c)$$

where N^k denotes the shape function (or interpolation function) in the kth node on a discrete boundary element and x_i^k, $u_i^{\pm n}(x^k)$ and $t_i^{\pm n}(x^k)$ are the coordinates, displacement and tractions at this node. Then Eqs (5.30) and (5.31) are rewritten in a discrete form as

$$c_{ij}(\xi)u_i^n(\xi) = \sum_{m=1}^{N}\sum_k \int_{C_m} G'^n_{ij}(x,\xi)N^k r_x \, dC_m(x)t_i^n(x^k)$$

$$- \sum_{m=1}^{N}\sum_k \int_{C_m} F'^n_{ij}(x,\xi)N^k r_x \, dL_m(x)u_i^n(x^k) \qquad (5.33)$$

$$c_{ij}(\xi)u_i^{-n}(\xi) = \sum_{m=1}^{N}\sum_k \int_{C_m} G'^{-n}_{ij}(x,\xi)N^k r_x \, dC_m(x)t_i^{-n}(x^k)$$

$$- \sum_{m=1}^{N}\sum_k \int_{C_m} F'^{-n}_{ij}(x,\xi)N^k r_x \, dC_m(x)u_i^{-n}(x^k) \qquad (5.34)$$

If we choose a finite number of collocation points on the boundary and let point ξ approach these points, then two decoupled system equations are formed for circumferential order n as[17,21]

$$[G^n]\{t^n\} - [F^n]\{u^n\} = \{0\} \qquad (5.35)$$

$$[G^{-n}]\{t^{-n}\} - [F^{-n}]\{u^{-n}\} = \{0\} \qquad (5.36)$$

If boundary conditions in circumferential order n are applied, then

$$[A^n]\{X^n\} = [B^n]\{Y^n\} \qquad (5.37)$$

$$[A^{-n}]\{X^{-n}\} = [B^{-n}]\{Y^{-n}\} \qquad (5.38)$$

where $\{Y^{\pm n}\}$ denotes the known boundary conditions in circumferential order n and $\{X^{\pm n}\}$ the unknown boundary conditions. Note that the original boundary conditions have to be transformed for circumferential order n according to[17,21]

$$u_r^n = \frac{1}{\alpha\pi} \int_0^{2\pi} u_r \cos n\theta \quad t_r^n = \frac{1}{\alpha\pi} \int_0^{2\pi} t_r \cos n\theta$$

$$u_\theta^n = \frac{1}{\alpha\pi} \int_0^{2\pi} u_\theta \sin n\theta \quad t_\theta^n = \frac{1}{\alpha\pi} \int_0^{2\pi} t_\theta \sin n\theta$$

$$u_z^n = \frac{1}{\alpha\pi} \int_0^{2\pi} u_z \cos n\theta \quad t_z^n = \frac{1}{\alpha\pi} \int_0^{2\pi} t_z \cos n\theta$$

$$u_r^{-n} = \frac{1}{\alpha\pi} \int_0^{2\pi} u_r \sin n\theta \quad t_r^{-n} = \frac{1}{\alpha\pi} \int_0^{2\pi} t_r \sin n\theta$$

$$u_\theta^{-n} = \frac{1}{\alpha\pi} \int_0^{2\pi} u_\theta \cos n\theta \quad t_\theta^{-n} = \frac{1}{\alpha\pi} \int_0^{2\pi} t_\theta \cos n\theta$$

$$u_z^{-n} = \frac{1}{\alpha\pi} \int_0^{2\pi} u_z \sin n\theta \quad t_z^{-n} = \frac{1}{\alpha\pi} \int_0^{2\pi} t_z \sin n\theta$$

The evaluation of circumferential integrals Since the integration is a major cost of the BEM analysis, it is very important, if possible, to attempt to evaluate the integrals with respect to circumferential variables analytically. All of the circumferential integrals for order n can be expressed in the form

$$\int_0^{2\pi} \frac{\cos^i \theta \cos n\theta}{r^b} \, d\theta \quad \text{and} \quad \int_0^{2\pi} \frac{\cos^i \theta \sin \theta \sin n\theta}{r^b} \, d\theta \quad \text{with } b = 1, 3 \text{ and } 5$$

By using the formula for trigonometric functions, these integrals can be expressed in terms of a sum of the following integrals:[10]

$$\int_0^{2\pi} \frac{\cos a\theta}{r^b} \qquad \int_0^{2\pi} \frac{\sin a\theta}{r^b} \qquad \text{where } a = 1, 2, 3, \ldots$$

Since

$$r(x, \xi) = R\left[1 - \frac{k^2}{2}(1 + \cos\theta)\right]^{1/2}$$

$$= R\left(1 - k^2 \sin^2 \frac{\theta}{2}\right)^{1/2}$$

where

$$R = [\,(r_x + r_\xi)^2 + (z_x - z_\xi)^2\,]^{1/2}$$

$$k^2 = \frac{4r_x r_\xi}{R^2}$$

by using the transformation of variables $\phi = \theta/2$, it is possible to cast the circumferential integrals into the forms[10,21]

$$\int_0^{\pi/2} \frac{\cos^a \phi \, \mathrm{d}\phi}{(1 - k^2 \sin^2 \phi)^{b/2}} \quad \text{or} \quad \int_0^{\pi/2} \frac{\cos^a \theta \, \mathrm{d}\theta}{(1 - k^2 \sin^2 \theta)^{b/2}}$$

which can be expressed once again in terms of elliptic integrals of the first kind, K, and the second kind, E. A complete list of such integrals up to the order $n = 3$ can be found in Wang.[21]

Although theoretically, these integrals can be evaluated by using rational approximations for elliptic integrals due to the limitation of precision in a computer, some numerical difficulty occurs. It has been found that although the error in the calculations of E and K is controlled to be smaller than 10^{-8}, the precisions of some integrals are lost for $k^2 < 0.2$, especially in the integrals for circumferential order $n = 2$ or greater. This problem occurs because of subtraction of large numbers compared to the final value of the integrand. To resolve this difficulty, Wang and Banerjee[18] in their implementation proposed a direct polynomial approximation for these integrals over the range where the accuracy is lost.

Boundary integration and the calculation of diagonal terms of F^n matrix   Once the integration along the circumferential direction is completed, the integration on generator C has to be done next. When ξ is away from C_m, the kernels are not singular and usual Gaussian integration can be applied. When ξ is closer to the integrated element C_m, the segmentation with more integration points is needed due to the large variation of kernel functions. When ξ is the q-collocation point (x^k) on the integrated element C_p, Eq. (5.33) can be written as

$$c_{ij}(\xi) u_i^n(x^q) + \int_{C_p} F_{ij}'^n(x, x^q) N^q r_x \, \mathrm{d}C_p(x) u_i^n(x^q) =$$

$$\sum_{m=1}^{N} \sum_k \int_{C_m} G_{ij}'^n(x, x^q) N^k r_x \, \mathrm{d}C_m(x) t_i^n$$

$$- \sum_{\substack{k \\ k \neq q}} \int_{C_p} F_{ij}'^n(x, x^q) N^k r_x \, \mathrm{d}C_m(x) u_i^n(x^q)$$

$$- \sum_{\substack{m=1 \\ m \neq p}}^{N} \sum_k \int_{C_m} F_{ij}'^n(x, x^q) N^k r_x \, \mathrm{d}C_m(x) u_i^{nk} \qquad (5.39)$$

All the integrals in the above are integrable except the one at the left-hand side in which the singularity of the kernel F occurs. By investigating the integrals in the equation, one finds that only five of nine integrals (those containing the trigonometric function $\cos n\theta$) are singular. The sums of these integrals and their corresponding constant c_{ij} which form five of nine terms of a 3×3 diagonal block in the F matrix are found by applying the elastic solutions that satisfy the governing differential equation. The adopted elastic solutions and their corresponding tractions at point (r, θ, z) for circumferential order $n = 0, 1, 2$ and 3 are given as follows:[17,21]

1. Circumferential order $n = 0$:
   (a) Inflation deformation:

$$\begin{Bmatrix} u_r \\ u_\theta \\ u_z \end{Bmatrix} = \begin{Bmatrix} r_\xi \\ 0 \\ 0 \end{Bmatrix} \quad \text{and} \quad \begin{Bmatrix} t_r \\ t_\theta \\ t_z \end{Bmatrix} = \begin{Bmatrix} 2(\lambda + \mu)n_r \\ 0 \\ 2\lambda n_z \end{Bmatrix}$$

   (b) Rigid body motion in the θ direction:

$$\begin{Bmatrix} u_r \\ u_\theta \\ u_z \end{Bmatrix} = \begin{Bmatrix} 0 \\ r_\xi \\ 0 \end{Bmatrix} \quad \text{and} \quad \begin{Bmatrix} t_r \\ t_\theta \\ t_z \end{Bmatrix} = \begin{Bmatrix} 0 \\ 0 \\ 0 \end{Bmatrix}$$

   (c) Rigid body motion in the z direction:

$$\begin{Bmatrix} u_r \\ u_\theta \\ u_z \end{Bmatrix} = \begin{Bmatrix} 0 \\ 0 \\ 1 \end{Bmatrix} \quad \text{and} \quad \begin{Bmatrix} t_r \\ t_\theta \\ t_z \end{Bmatrix} = \begin{Bmatrix} 0 \\ 0 \\ 0 \end{Bmatrix}$$

2. Circumferential order $n = 1$
   (a) Rigid body motion along the X axis ($\theta = 0$):

$$\begin{Bmatrix} u_r \\ u_\theta \\ u_z \end{Bmatrix} = \begin{Bmatrix} \cos\theta \\ -\sin\theta \\ 0 \end{Bmatrix} \quad \text{and} \quad \begin{Bmatrix} t_r \\ t_\theta \\ t_z \end{Bmatrix} = \begin{Bmatrix} 0 \\ 0 \\ 0 \end{Bmatrix}$$

   (b) Rigid body rotation about the Y axis ($\theta = \pi/2$):

$$\begin{Bmatrix} u_r \\ u_\theta \\ u_z \end{Bmatrix} = \begin{Bmatrix} z\cos\theta \\ -z\sin\theta \\ -r\cos\theta \end{Bmatrix} \quad \text{and} \quad \begin{Bmatrix} t_r \\ t_\theta \\ t_z \end{Bmatrix} = \begin{Bmatrix} 0 \\ 0 \\ 0 \end{Bmatrix}$$

3. Circumferential order $n = 2$:
   (a) Plane deformation:

$$\begin{Bmatrix} u_r \\ u_\theta \\ u_z \end{Bmatrix} = \begin{Bmatrix} r\cos 2\theta \\ -r\sin 2\theta \\ 0 \end{Bmatrix} \quad \text{and} \quad \begin{Bmatrix} t_r \\ t_\theta \\ t_z \end{Bmatrix} = \begin{Bmatrix} 2\mu n_r \cos 2\theta \\ -2\mu n_r \sin 2\theta \\ 0 \end{Bmatrix}$$

   (b) Vertical deformation only:

$$\begin{Bmatrix} u_r \\ u_\theta \\ u_z \end{Bmatrix} = \begin{Bmatrix} 0 \\ 0 \\ r^2 \cos 2\theta \end{Bmatrix} \quad \text{and} \quad \begin{Bmatrix} t_r \\ t_\theta \\ t_z \end{Bmatrix} = \begin{Bmatrix} 2\mu n_z r \cos 2\theta \\ -2\mu n_z r \sin 2\theta \\ 2\mu n_r r \cos 2\theta \end{Bmatrix}$$

4. Circumferential order $n = 3$:
   (a) Plane deformation:

$$\begin{Bmatrix} u_r \\ u_\theta \\ u_z \end{Bmatrix} = \begin{Bmatrix} r^2 \cos 3\theta \\ -r^2 \sin 3\theta \\ 0 \end{Bmatrix} \quad \text{and} \quad \begin{Bmatrix} t_r \\ t_\theta \\ t_z \end{Bmatrix} = \begin{Bmatrix} 4\mu n_r r \cos 3\theta \\ -4\mu n_r r \sin 3\theta \\ 0 \end{Bmatrix}$$

   (b) Vertical deformation only:

$$\begin{Bmatrix} u_r \\ u_\theta \\ u_z \end{Bmatrix} = \begin{Bmatrix} 0 \\ 0 \\ r^3 \cos 3\theta \end{Bmatrix} \quad \text{and} \quad \begin{Bmatrix} t_r \\ t_\theta \\ t_z \end{Bmatrix} = \begin{Bmatrix} 3\mu n_z r^2 \cos 3\theta \\ -3\mu n_z r \sin 3\theta \\ 3\mu n_r r \cos 3\theta \end{Bmatrix}$$

For other higher circumferential orders, the needed elastic solutions can be easily found.

5.6 EXAMPLES

5.6.1 Conical Water Tank[5,9]

An analysis of a water tank, shown in Fig. 5.5a, is carried out using the BEM. The hoop stress on the inside surface of the tank subjected to hydrostatic pressure is shown in Fig. 5.5b along with the result from Chapter 9 of Zienkiewicz.[22] The two results are similar although an exact comparison is not possible since the exact geometry data and material constants are not given for the later case. In Fig. 5.5b, a comparison of BEM solutions is presented for the hoop stress on both the inner and outer surfaces (along **AB**) of the tank for two different loadings. The first loading is hydrostatic pressure only and the second is a combination of the hydrostatic pressure and self-weight. Note that in the lower section the hoop stress at the outer surface is compressive, i.e. the bending effect is dominant there.

One hundred isoparametric quadratic boundary elements are used to model the body. The tank is assumed to be constructed of concrete with $\rho = 4.65$ lbm/ft^3, $E = 3.472 \times 10^6$ lb/in^2, and $\nu = 0.18$. The thickness of the uniform section is taken to be 1 ft. The time required for integration is reduced by dividing the tank into six sub-regions. Two or more elements are used at every interface to insure an accurate representation of the variables across the interface. The computation time for a single region mesh is 50 per cent greater than that of the six sub-region model, although the computed results are almost identical. Solution to within 5 per cent of the one shown is achievable with half the number of elements.

5.6.2 Concrete Reactor Pressure Vessel[3]

Next, in a somewhat more practical example a steady state thermal stress analysis is conducted for the concrete reactor pressure vessel shown in Zienkiewicz.[22] The

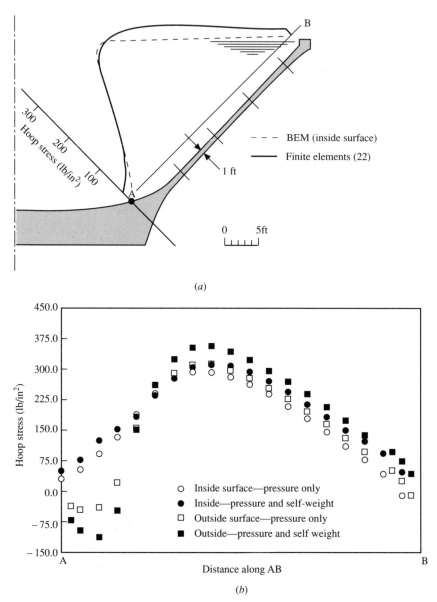

Figure 5.5 Conical water tank subjected to (*a*) hydrostatic pressure and (*b*) hydrostatic pressure and self-weight.

axisymmetric geometry of one half of the vessel, shown in Fig. 5.6, was digitized from the above reference. The inner spherical surface is maintained at 400°C, while the temperature of the outer surface is 0°C. Symmetry conditions are applied along the edge of $Z = 0$. The concrete has an elastic modulus of

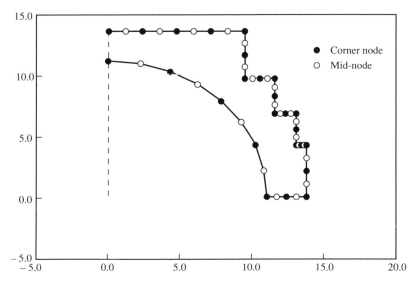

Figure 5.6 Concrete reactor pressure vessel: boundary element model.

2.58×10^6 lb/in^2, a Poisson ratio of 0.15, a coefficient of thermal expansion equal to $5 \times 10^{-6}/°C$ and uniform thermal conductivity. Interestingly, the thermal stresses for this problem are independent of the actual size of the vessel. Only the shape is important.

A 24-element, 49-node boundary element mesh is also depicted in Fig. 5.6; on the other hand, the finite element model involved 206 elements and 235 nodes. Figure 5.7 contains the maximum principal stress contours obtained by Zienkiewicz,[22] along with the corresponding boundary values calculated with the GP-BEST code. Good agreement is evident between the two analysis methods for this problem. Additionally, it should be noted that Bakr and Fenner[8] also examined this problem via their boundary element formulation and achieved correlation with Zienkiewicz.

For the present GP-BEST analysis, contour plots are, perhaps, conspicuously missing. In boundary elements, the generation of these plots requires the computation of stresses at a large number of interior points, which can become an expensive undertaking. Fortunately, however, for the majority of practical problems, including the reactor pressure vessel, it is only the surface stresses that are of major interest.

5.6.3 Toriodal Shell under Uniform Internal Pressure[17,21]

This problem is a purely axisymmetric one which involves the circumferential order $n = 0$ only. The thickness of the toriodal shell is 0.5 inch and its average radius is 15 inches. The material properties are: Poisson's ratio 0.3 and Young's

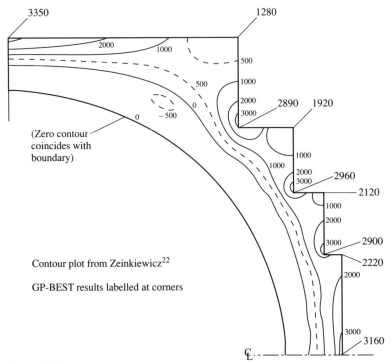

Figure 5.7 Concrete reactor pressure vessel.

modulus 10^7 lb/in². The only applied loading is the uniform internal pressure $p = 1.0$ lb/in². A six-region mesh with 12 elements in each region is used (Fig. 5.8) for the upper half of the toroidal shell and a roller boundary condition is applied along the line of symmetry of the shell. The average radial displacement on the mid-surface of the toroidal shell obtained by the present

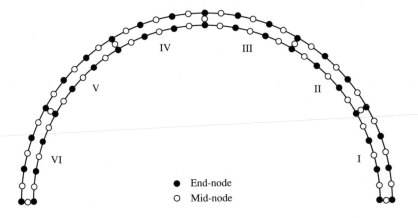

Figure 5.8 Six-region boundary element model of toroidal shell.

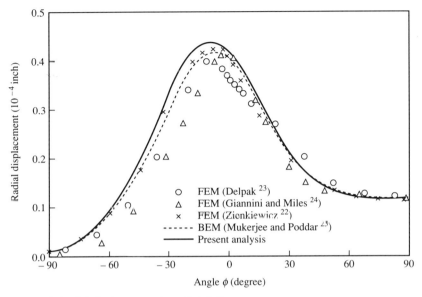

Figure 5.9 Radial displacement of toroidal shell.

analysis is plotted in Fig. 5.9 and compared with the results by Zienkiewicz,[22] Delpak,[23] Giannini and Miles[24] and Mukherjee and Poddar.[25] The maximum radial displacement from the present analysis is slightly higher than the previously published results by shell theory. It is interesting to note that the finite element results differ by as much as 10 per cent depending on the type of shell elements used. The previous BEM results for this problem are due to Mukherjee and Poddar[25] who specialized the three-dimensional BEM formulation for solids into a thin shell theory. In the present analysis, the toriodal shell is treated as a three-dimensional solid and a very good result is seen to have been obtained.

5.6.4 A Cylindrical Shell under Lateral Loads[17,21]

It is generally very difficult to analyse thin three-dimensional solids under bending by the BEM without using shell theory because the numerical integration on the boundary is very difficult. Here the problem is resolved by a very precise numerical implementation without using any approximated theory of structural analyses, such as shell theory. The configuration of the cylindrical shell is shown in Fig. 5.10. The internal radius is $r = 120$ ft the heights $h = 200$ ft and its thickness $t = 1$ ft. Poisson's ratio is assumed to be $1/6$ for a concrete structure. The applied external loading is expressed as

$$t_r = \sum t_r^n \cos n\theta$$

According to membrane theory the average hoop stress and average vertical

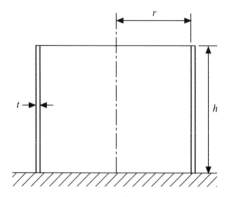

Figure 5.10 The configuration of cylindrical shell.

stress resultants per unit length are given as[26]

$$N_\theta = \sum_n N_\theta^n \cos n\theta = t_r r_{avg} = \sum_n t_r^n r_{avg} \cos n\theta$$

$$N_z = \sum_n N_z^n \cos n\theta N_z = \sum_n \frac{(h-z)^2}{2r_{avg}} n^2 t_r^n \cos n\theta$$

respectively. For the analysis of this problem, an eight-region (with 12 quadratic elements in each region) BE mesh is used. The average vertical stress resultant at different heights is shown in Fig. 5.11. Also, the result according to membrane

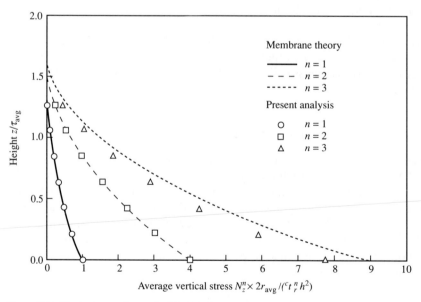

Figure 5.11 Vertical stress in a cylindrical shell (at $\theta = 0$) under lateral loads of circumferential orders $n = 1, 2$ and 3.

theory is computed for comparison. Good agreement can be seen for $n = 0, 1, 2$. The discrepancy for $n = 3$ is probably due to neglect of the circumferential flexural effect in the membrane theory. This example problem is probably close to the limit of the three-dimensional BEM formulation unless, of course, the integrals are specialized into thin shell theory by a suitable approximation.

5.7 CONCLUDING REMARKS

The axisymmetric BEM has been described for elasticity, potential flow and thermoelastic analyses under steady state conditions. In addition, the axisymmetric analysis under non-axisymmetric loading was also considered. In these analyses, a method of applying separate elastic solutions for different circumferential orders for the determination of singular diagonal blocks was described and the analysis was applied to long and thin structures. The use of Fourier series is not inherently restricted to the present analysis; one can in many situations convert a three-dimensional problem into a pseudo-two-dimensional one by analytically representing the variations in the third direction using a Fourier series and completing the integrations using either an analytical or a numerical procedure.

REFERENCES

1. Jaswon, M.A. and Symm, G.T. (1977) *Integral Equations in Potential Theory and Elastostatics,* Academic Press.
2. Liggett, J.A. and Liu, P.L.F. (1983) *The Boundary Integral Equation Method in Porous Media Flow,* Allen and Unwin, London.
3. Dargush, G. (1987) Boundary element method for the analogous problems of thermomechanics and soil consolidation, PhD Dissertation, State University of New York at Buffalo.
4. Abramowicz, M. and Stegan, I.A. (1974) *Handbook of Mathematical Functions,* Dover, New York.
5. Henry, D.P., Pape, D.A. and Banerjee, P.K. (1987) A new axisymmetric BEM formulation for body forces using particular integrals, *Journal of Engineering, Mechanical Division, ASCE,* Vol. 113, pp. 671–688.
6. Kermanidis, T. (1975) A numerical solution for axially symmetrical elasticity problems, *International Journal of Solids and Structures,* Vol. 11, pp. 493–500.
7. Cruse, T.A., Snow, D.W. and Wilson, R.B. (1977) Numerical solutions in axisymmetric elasticity, *Computers and Structures,* Vol. 7, pp. 445–451.
8. Bakr, A.A. and Fenner, R.T. (1983) Boundary integral equation analysis of axisymmetric thermoelastic problems, *Journal of Strain Analysis,* Vol. 18, No. 4, pp. 239–251.
9. Henry (Jr), D.P. (1987) Advanced development of boundary element method for elastic and inelastic thermal stress analysis, PhD Dissertation, State University of New York at Buffalo.
10. Nigam, R.K. (1979) The boundary integral equation method for elastostatic problems involving axisymmetric geometry and arbitrary boundary conditions, thesis presented to the University of Kentucky, in partial fulfilment of the requirements for the degree of Master of Science.
11. Rizzo, F.J. and Shippy, D.J (1979) A boundary integral approach to potential and elasticity problems for axisymmetric bodies with arbitrary boundary conditions, *Mechanics Research Communications,* Vol. 6, pp. 99–103.
12. Rizzo, F.J. and Shippy, D.J. (1986) A boundary element method for axisymmetric elastic bodies,

Chapter 3 in *Developments in Boundary Element Methods*, Vol. 4, Eds P.K. Banerjee and J.O. Watson, pp. 67–90, Applied Science Publishers, London, UK.

13. Sarihan, V. and Mukherjee, S. (1982) Axisymmetric viscoplastic deformation by the boundary element method, *International Journal of Solids and Structures*, Vol. 18, pp. 1113–1128.

14. Mayr, M., Drexler, W. and Kuhn, G. (1980) A semianalytical boundary integral approach for axisymmetric elastic bodies with arbitrary boundary conditions, *International Journal of Solids and Structures*, Vol. 16, pp. 863-871.

15. Mayr, M. (1975) Integral equation method for rotationally symmetric problems of elasticity, PhD. Dissertation, Technical University of Munich, Germany.

16. Le, M.F., Tong, Y.W. and Wie, J.F. (1986) On the BEM solution of rotationally symmetrical body under nonsymmetrical surface load, in *Proceedings of the International Conference*, Ed. Qinghua Du, Beijing, October 1986, pp. 14–17, Pergamon Press, Oxford.

17. Wang, H.C. and Banerjee, P.K. (1988) Free vibration of axisymmetric bodies by BEM, *ASME Journal Applied Mechanics*, Vol. 110, pp. 435–442.

18. Wang, H.C. and Banerjee, P.K. (1989) Multi-domain general axisymmetric stress analysis by BEM, *International Journal of Numerical Methods in Engineering*, Vol. 28, pp. 2065–2083.

19. Wang, H.C. and Banerjee, P.K. (1990) Free-vibration of axisymmetric solids by BEM using particular integrals, *International Journal of Numerical Methods in Engineering*, Vol. 29, pp. 985–1001.

20. Wang, H.C. and Banerjee, P.K. (1990) Advanced periodic dynamic analysis of three-dimensional solids by BEM, *International Journal of Numerical Methods in Engineering*, Vol. 30, pp. 115–131.

21. Wang, H.C. (1989) A general purpose development of BEM for axisymmetric solids, PhD dissertation, State University of New York at Buffalo.

22. Zienkiewicz, O.C. (1977) *The Finite Element Method*, McGraw-Hill, London.

23. Delpak, R. (1975) Role of the curved parametric element in linear analysis of thin rotational shells, PhD Thesis, Department of Civil Engineering and Building, The Polytechnic of Wales.

24. Giannini, M. and Miles, G.A. (1970) A curved element approximation in the analysis of axisymmetric thin shells, *International Journal of Numerical Methods in Engineering*, Vol. 2, pp. 459–476.

25. Mukherjee, S. and Poddar, B. (1986) An integral equation formulation for elastic and inelastic shell analysis, *Proceedings of International Conference*, Beijing, October 1986., Pergamon Press, Oxford.

26. Billington, D.P. (1982) *Thin Shell Concrete Structures*, 2nd edn, McGraw-Hill, New York.

ELASTIC ANALYSIS OF SOLIDS WITH HOLES AND INCLUSIONS

6.1 INTRODUCTION

In the design of components from all fields of engineering, it is not uncommon to find bodies that contain many small holes which are used for cooling passages, electrical wiring passages, etc. These small holes can significantly change the entire stress distribution in a body and therefore create major difficulties for the design engineer.

Analysing components containing holes by modern computer-based analyses, such as the boundary element method, or the finite element method, is very time consuming and cumbersome, particularly when many holes are present. In particular, the finite element method would require a very fine mesh to accurately capture the local stress concentration created by the holes. The labour cost alone for model generation of a body with many holes by either method could be staggering. Moreover, in design situations, several models with changes in the locations of the holes may need to be generated and analysed.

Recently a number of researchers have developed BEMs for applications to problems of potential flow involving regions with circular holes[1-5] (see Chapter 8). In this chapter a boundary element approach is introduced to readily analyse three-dimensional elastic structures containing tubular holes and inclusions of circular cross-section.[6,7] The holes and inclusions are modelled in three-dimensional space using curvilinear line elements or 'hole elements' or 'inclusion elements' with a prescribed radius. These assume a variation in the displacement field about the circumference defined by a trigonometric function, and a linear or quadratic variation is assumed along its length. The two-dimensional integration over the surface of the hole is thus reduced to a one-dimensional integration along its length by carrying out an analytic integration in the circumferential direction. A boundary element model employed using the present approach requires fewer nodes and elements than a traditional BEM analysis in which surface elements are used to model the hole. Moreover, in a design analysis the location of these holes and inclusions can easily be changed without altering the discretization of the outer boundary. Further, the numerically integrated coefficients of the outer boundary may be reused when the body is reanalysed with the hole or the inclusion relocated. The analysis for inclusions developed here can be used for the micromechanical analysis of composites.

6.2 BEM FORMULATION FOR EMBEDDED HOLES

6.2.1 Integral Formulation

The boundary integral equation for displacement of a three-dimensional elastic body is applicable in an analysis of a body that contains embedded holes. This equation can be expressed as[6,7]

$$c_{ij}(\xi)u_i(\xi) = \int_S \left[G_{ij}(x,\xi)t_i(x) - F_{ij}(x,\xi)u_i(x) \right] \mathrm{d}S(x)$$

$$+ \sum_{m=1}^{M} \int_{S^m} \left[G_{ij}(x,\xi)t_i(x) - F_{ij}(x,\xi)u_i(x) \right] \mathrm{d}S^m(x) \qquad (6.1)$$

where

S = outer surface of the body
S^m = surface of the mth hole
M = number of embedded holes in the body

The boundary element discretization of Eq. (6.1) in the traditional manner (outlined in earlier chapters) requires a very fine discretization about the hole. Alternatively, a new formulation is introduced in this chapter for the efficient modelling and analysis of holes using 'hole elements'. A hole is defined with hole elements by describing the centreline of the (curvilinear, tubular) hole with nodal points, defining the connectivity of the nodal points and specifying the radius of the hole at each of these nodal points. A long hole (which may be allowed to vary in diameter) can be described by a number of hole elements connected end to end, and any hole element not connected to another is assumed to be closed at the end by a circular disc.

Using the concept of the hole element, the essential part of the formulation is the conversion of the two-dimensional surface integration of the hole to a one-dimensional integration. To facilitate an analytic integration in the circumferential direction, the three-dimensional kernel functions are first expressed in local coordinates with the centre of the coordinate system coinciding with the centre of the hole and the z axis aligned with the centreline of the tube (see Fig. 6.1).

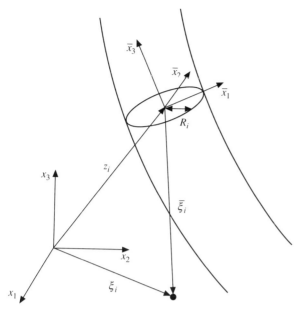

Figure 6.1 Orientation of the local coordinate system on the centre-line of the hole.

The relative translation ξ_i' is added to the field coordinate ξ_i and the rotation is applied using the appropriate vector transformation:

$$\xi_i = a_{ij}\bar{\xi}_j + \xi_i' \tag{6.2}$$

where a_{ij} are the direction cosines between the axis of the local and global coordinate systems and the bar indicates a local variable.

The integration point x_i for a ring can now be expressed in cylindrical coordinates relative to the centre of the hole as

$$x_1 = R\cos\theta \qquad x_2 = R\sin\theta \qquad x_3 = 0 \tag{6.3}$$

where R represents the radius of the hole, that is $R = (x_1^2 + x_2^2)^{1/2}$.

The normal vectors are transformed by

$$n_1 = n_r\cos\theta \qquad n_2 = n_r\sin\theta \qquad n_3 = n_z \tag{6.4}$$

where n_r and n_z represents the normals of the side of the hole in local coordinates and are dependent on the change in the radius of the hole. On the side of a straight hole $n_r = 1$ and $n_z = 0$, and on the flat surface closing the end of the hole $n_r = 0$ and $n_z = 1$.

Next a circular shape function is employed to approximate the variation in the displacement about the circumference of the hole. The circular shape function is multiplied and integrated with the three-dimensional F_{ij} kernel, allowing the nodal values of displacement to be brought outside the integral. The shape function is expressed as (see Fig. 6.2)[6,7]

$$u_i = M^\gamma U_i^\gamma \quad \text{(summation over } \gamma \text{ is implied, } \gamma = 1, 2, 3) \tag{6.5}$$

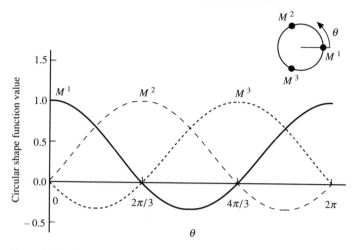

Figure 6.2 Value of the circular shape functions about the hole.

where

$$M^1(\theta) = \frac{1}{3} + \frac{2}{3}\cos\theta$$

$$M^2(\theta) = \frac{1}{3} + \frac{\sqrt{3}}{3}\sin\theta - \frac{1}{3}\cos\theta$$

$$M^3(\theta) = \frac{1}{3} - \frac{\sqrt{3}}{3}\sin\theta - \frac{1}{3}\cos\theta$$

and U_i^γ are the nodal displacements.

A modified circular shape function is used in the integration over the end of the hole to insure continuity of displacement at the centre of the end surface. The modified shape function is expressed as

$$\bar{M}^\gamma = aM^\gamma + \frac{b}{3} \qquad \gamma = 1, 2, 3$$

$$a = \frac{r}{R} \qquad\qquad b = \frac{R-r}{R}$$

where:

R = radius of the hole at the end

r = location of the integration point as it sweeps from $r = 0$ to $r = R$

M^γ = circular shape function defined above

The displacement must also be transformed between the local and the global systems by

$$u_i = a_{ik}\bar{u}_k \tag{6.6a}$$

or

$$\bar{u}_j = a_{mj}u_m \tag{6.6b}$$

If internal pressure, p, is present in the hole, the pressure is assumed to be constant about the circumference. The traction vector in Eq. (6.1), however, is written as a global vector and must be converted to a local scalar variable. This transformation vector is multiplied and integrated along with the G_{ij} kernel, allowing the constant pressure value to be brought outside the integration. The load transformation is

$$t_i = N_i p \tag{6.7}$$

where N_i is the transformation vector:

$$N_1 = n_r \cos \theta \qquad N_2 = n_r \sin \theta \qquad N_3 = n_z \tag{6.8}$$

The last two terms in Eq. (6.1) can now be analytically integrated in the circumferential direction. For the mth hole the two integrals involved can be expressed as[6,7]

$$\int_{S^m} G_{ij}(x, \xi) t_i(x) \, ds^m(x) = \int_{C^m} a_{jk} \int_0^{2\pi} \bar{G}_{ik}(R, \theta, z, \bar{\xi}) N_i R \, d\theta p \, dC^m$$

$$= \int_{C^m} G_j^{\mathrm{H}}(R, z, \bar{\xi}) p \, dC^m(z) \tag{6.9}$$

$$\int_{S^m} F_{ij}(x, \xi) u_i(x) \, ds^m(x) = \int_{C^m} a_{jk} \int_0^{2\pi} \bar{F}_{\ell k}(R, \theta, z, \bar{\xi}) M^\gamma R \, d\theta a_{i\ell} U_i^\gamma \, dC^m$$

$$= \int_{C^m} F_{ij}^{\mathrm{H}\gamma}(R, z, \bar{\xi}) U_i^\gamma \, dC^m(z) \tag{6.10}$$

where the indicated integration over C^m is now a one-dimensional curvilinear integration along the hole and G_j^{H} and F_{ij}^{H} represent the analytically integrated hole kernels. Note, since the transformation vector a_{ik} is independent of θ, it may be taken outside the $d\theta$ integration.

The kernel functions G_j^{H} and F_{ij}^{H} obtained from the analytical integration are very lengthy and therefore are not presented here. They contain functions of elliptical integrals[8,9] which in general are expressed numerically by common series[9,10] approximations (see Chapter 5). For a range of input values (coordinate locations), several higher-order elliptic integrals were found to produce incorrect numerical results. To overcome this problem, several new series were derived using a best-fit polynomial approximation (as a function of the modulus of elliptic integrals) using values of the integrals calculated by an accurate numerical integration in the circumferential direction.[8–11]

'Hole kernels' corresponding to the strain equation are derived from the displacement equation (6.1) by differentiation and application of the strain–displacement equations. The stress equation is then found using Hooke's law.

6.2.2 Numerical Integrations

After the analytical integration in the circumferential direction is complete, a hole in a three-dimensional solid can be modelled as a two-dimensional curvilinear line element with a prescribed radius at each longitudinal node. Linear and quadratic shape functions may be utilized in modelling the geometry and field variables along the hole element as well as for the boundary elements on the outer surface of the three-dimensional body. In the discretized form the displacement boundary integral equation for an elastic body containing tubular holes can be expressed in the following manner:

$$c_{ij}u_i(\xi) = \sum_{n=1}^{N} \int_{S^m} G_{ij}(x,\xi)L^\beta(\eta_1,\eta_2) \, dS^n(x)T_i^{\beta n}$$

$$- \sum_{n=1}^{N} \int_{S^m} F_{ij}(x,\xi)L^\beta(\eta_1,\eta_2) \, dS^n(x)U_i^{\beta n}$$

$$+ \sum_{m=1}^{M} \int_{C^m} G_j^H(R,z,\bar{\xi})N^\alpha(\eta) \, dC^m(z)p^{\alpha m}$$

$$- \sum_{m=1}^{M} \int_{C^m} F_{ij}^{H\gamma}(R,z,\bar{\xi})N^\alpha(\eta) \, dC^m(z)U_i^{\gamma \alpha m} \tag{6.11}$$

where

$T_i^{\beta n}$ = nodal tractions
$U_i^{\beta n}$ = nodal displacements
$p^{\gamma m}$ = nodal pressures inside holes
$L^\beta(\eta_1,\eta_2)$ = two-dimensional shape function; summation over β is implied where $\beta = 1$ to the number of nodes in the nth boundary element
$N^\alpha(\eta)$ = one-dimensional shape function; summation over α is implied where $\alpha = 1$ to the number of nodes in the mth hole element

It is important to note that the displacement of the hole varies in the longitudinal as well as the circumferential direction, i.e.

$$u_j = M^\gamma N^\alpha U_j^{\gamma \alpha} \tag{6.12}$$

Numerical evaluation of the integrals in Eq. (6.11) can be carried out using integration schemes outlined in Chapter 3 (see also Refs 12 and 13). The ends of the holes are assumed to be flat surfaces and one-dimensional numerical integration is carried out in the radial direction. The coefficients obtained from the integration over the end are lumped with their respective coefficients from the integration of the side of the hole.

Equations for stress or strain at points in the interior of a body can be discretized and integrated in a similar manner.

6.2.3 System Equations

In constructing a boundary element system, Eq. (6.11) is written for each nodal point used in the representation of the field variables over the surfaces of the outer boundary and holes. For every node along the axis a hole element (three equations, each with three degrees of freedom) must be written corresponding to each of the three nodes in the circular shape function. Therefore, a single quadratic hole element would require a total of nine nodes each with three degrees of freedom.

Equations for the nodes on the outer boundary are written at the nodes of the boundary elements in the usual manner. The jump terms and the principal value integral associated with the singular elements are calculated using the rigid body translation technique. In a rigid body translation, the summation of the coefficients of the traction kernel corresponding to the holes should be zero for the equations written at the nodes on the outer boundary (due to numerical inaccuracies, some residual value remains). It is therefore recommended to use only the coefficients of the outer boundary in this calculation in order to prevent any residual error due to the holes from entering in this calculation.

The equations written for the nodes on the surface of the hole may be handled in a similar manner. However, the rigid body technique is not recommended for the calculations of coefficients at a singular point on a hole element, since imprecise numerical integration or approximations associated with the integration of a curvilinear hole element could introduce errors in the system that may have some adverse influence on the results. Therefore, the equations corresponding to the singular nodes of the holes are taken at a point slightly inside the hole. Note that the nodal values of the displacement and pressure associated with shape functions of the hole elements still represent the actual values for the nodes on the surface of the hole. By moving the equation point inside the hole, the singular integration is circumvented and all the values of the tensor c_{ij} in Eq. (6.11) are zeros. The numerical integration of a hole element, however, for a point within the hole element must still be performed with utmost precision to maintain accuracy. This requires the splitting of an integration interval in at least two parts so that the near singular point lies exactly between two integration intervals.

When the numerical integration is complete, the system equations are assembled incorporating interface conditions (perfect bond, springs, sliding, etc.) in the usual manner for multi-region applications. All nodal displacements at the holes will be contained in the system as unknowns and any specified internal pressure within the hole will be multiplied by their respective coefficients and added to the right-hand side of the system equation. Specified and unknown boundary conditions on the outer boundary are collected and assembled with their coefficients in the usual manner. The system is then solved, yielding results for both the unknown conditions on the outer boundary and for the displacements on the surface of the holes.

Results for displacements, stresses and strains at interior points in the body may be calculated using the discretized integral equations. Results for points on the outer boundary or on the surface of the hole are most readily calculated using shape functions and the boundary stress–strain calculations discussed in Chapter 17.

6.3 INTEGRAL FORMULATION FOR INCLUSIONS

6.3.1 Boundary Integrals

The conventional boundary integral equations for elastostatic analyses are used to derive a boundary element formulation for the analysis of composite structures. The boundary integral equation written for a point in the interior of the composite matrix is modified by adding to it the boundary integral equations of each inclusion written at the same point in the composite matrix. This eventually eliminates the displacement variables at the inclusion–matrix interface from the system, and therefore reduces the total number of equations required for a solution of the system.

The direct boundary integral equation for the displacement at a point ξ inside an elastic composite matrix is[6]

$$c_{ij}(\xi)u_i(\xi) = \int_S \left[G_{ij}^O(x,\xi)t_i^O(x) - F_{ij}^O(x,\xi)u_i^O(x) \right] \mathrm{d}S(x)$$
$$+ \sum_{n=1}^N \int_{S^n} \left[G_{ij}^H(x,\xi)t_i^H(x) - F_{ij}^H(x,\xi)u_i^H(x) \right] \mathrm{d}S^n(x) \quad (6.13)$$

where

S, S^n = surfaces of the outer boundary of the matrix and the nth hole (left for fibre) respectively

N = number of individual fibres

Superscripts O, H and I (below) identify the quantities on the outer surface of the matrix, the surface of the hole and inclusions respectively.

The conventional boundary integral equation for displacement can also be written for each of the N fibre inclusions. For the displacement at a point ξ in the nth fibre inclusion we can write[6]

$$c_{ij}^I(\xi)u_i(\xi) = \int_{S^n} \left[G_{ij}^I(x,\xi)t_i^I(x) - F_{ij}^I(x,\xi)u_i^I(x) \right] \mathrm{d}S^n(x) \quad (6.14)$$

We next examine the interface conditions between the composite matrix and the fibre. For a perfect bond the displacement of the matrix and that of the fibre inclusion are equal and the tractions along the interface are equal and opposite:

$$u_i^H(x) = u_i^I(x) \quad (6.15a)$$
$$t_i^H(x) = -t_i^I(x) \quad (6.15b)$$

When the elastic modulus of the fibre is much greater than the modulus of the composite matrix, the Poisson ratio of the inclusion can be assumed equal to that of the matrix with little error. Although this assumption can easily be relaxed, it does lead to considerable efficiency without much error in the analysis. Therefore, upon consideration of the surface normals at the interface and examination of the F_{ij} kernels, we can write the following relation for the nth inclusion:

$$F_{ij}^I(x,\xi) = -F_{ij}^H(x,\xi) \quad (6.15c)$$

Substitution of Eqs (6.15) into Eq. (6.14) yields the following modified boundary integral equation for fibre inclusion n:

$$c_{ij}^I(\xi)u_i(\xi) = \int_{S^n} \left[-G_{ij}^I(x,\xi)t_i^H(x) + F_{ij}^H(x,\xi)u_i^H(x) \right] dS^n(x) \qquad (6.16)$$

Finally, adding the N fibre inclusion Eq. (6.14) to Eq. (6.13) and cancelling terms yields the modified boundary integral equation for the composite matrix:[6]

$$\bar{c}_{ij}(\xi)u_i(\xi) = \int_S \left[G_{ij}^O(x,\xi)t_i^O(x) - F_{ij}^O(x,\xi)u_i^O(x) \right] dS(x)$$

$$+ \sum_{n=1}^N \int_{S^n} \bar{G}_{ij}^n(x,\xi)t_i^H(x) \, dS^n(x) \qquad (6.17)$$

where

$\bar{G}_{ij}^n(x,\xi) = G_{ij}^H(x,\xi) - \left[G_{ij}^I(x,\xi) \right]^n$

$\quad \bar{c}_{ij}(\xi) =$ constants dependent on the geometry for a point ξ on the outer boundary

$\quad \bar{c}_{ij}(\xi) = \delta_{ij}$ for a point ξ in the interior of the matrix

6.3.2 Numerical Implementation

By using the shape functions given in Eq. (6.5) and tranforming the traction between the local and the global axes by $t_i = a_{ik}\bar{t}_k$ or $\bar{t}_j = a_{mj}t_m$, the last term in Eq. (6.17) can now be analytically integrated in the circumferential direction. For the *mth* hole the two integrals involved can be expressed as, for example:

$$\int_{S^m} \bar{G}_{ij}^n(x,\xi)t_i^H(x) \, dS^m(x) = \int_{C^m} a_{jk} \int_0^{2\pi} \bar{G}_{\ell k}^{local}(R,\theta,z,\bar{\xi})M^\gamma R \, d\theta a_{i\ell} t_i^\gamma \, dC^m$$

$$= \int_{C^m} \bar{G}_{ij}^\gamma(R,z,\bar{\xi})t_i^\gamma \, dC^m(z) \qquad (6.18)$$

where the indicated integration over C^m is now a one-dimensional curvilinear integration along the hole and $\bar{G}_{ij}^\gamma$ represent the semi-analytically integrated kernel functions. Note, since the transformation vector a_{ik} is independent of angle θ, it may be taken outside the $d\theta$ integration. Similar semi-analytic integration is also performed in Eq. (6.16).

In the discretized form the displacement boundary integral equation for an elastic body containing inclusion [Eq. (6.16)] can be expressed for a single inclusion as

$$c_{ij}^I(\xi)u_i(\xi) = -\sum_{p=1}^P \left[\int_{C^p} G_{ij}^{I\gamma}(x,\xi)N^\alpha(\eta) \, dC^p \right] t_i^{\alpha\gamma}$$

$$+ \sum_{p=1}^P \left[\int_{C^p} F_{ij}^{I\gamma}(x,\xi)N^\alpha(\eta) \, dC^p \right] u_i^{\alpha\gamma} \qquad (6.19)$$

where

$P =$ number of line elements

$N^\gamma(\eta)$ represents a shape function over the curvilinear line element

In a similar manner, Eq. (6.17) can be discretized using one-dimensional shape functions over the surface of inclusion and two-dimensional shape functions over the outer surface of the body in the following manner:

$$
\bar{c}_{ij}(\xi)u_i(\xi) = \sum_{q=1}^{Q} \left[\int_{S^q} G_{ij}^O(x,\xi)L^\beta(\eta_1,\eta_2)\, \mathrm{d}S^q \right] t_i^\beta
$$

$$
- \sum_{q=1}^{Q} \left[\int_{S^q} F_{ij}^O(x,\xi)L^\beta(\eta_1,\eta_2)\, \mathrm{d}S^q \right] u_i^\beta
$$

$$
+ \sum_{p=1}^{P} \left[\int_{C^p} \bar{G}_{ij}^\gamma(x,\xi)N^\alpha(\eta)\, \mathrm{d}C^p \right] t_i^{\alpha\gamma} \tag{6.20}
$$

After the derivation of the modified boundary integral equations and the analytical circumferential integration of the kernel functions, the next critical step in the formulation is the assembly of the inclusion in the system equations. Here, efficiency is of utmost importance. The strategy is to retain in the system only traction variables on the matrix–fibre interface. This is in contrast to a general multi-region problem where both displacement and tractions are retained on an interface. The elimination of the displacement on the interface is achieved through a backsubstitution of the inclusion equations in the system equations, which are made up exclusively from equations written for the composite matrix.

Equation (6.20) is used to generate a system of equations for nodes on the outer surface of the composite matrix and for nodes on the surface of the holes containing the inclusions. Written in matrix form we have:

On the matrix outer surface: $\mathbf{G}^O\mathbf{t}^O - \mathbf{F}^O\mathbf{u}^O + \bar{\mathbf{G}}\mathbf{t}^H = \mathbf{0}$ $\qquad$ (6.21a)

On the matrix hole surface: $\mathbf{G}^O\mathbf{t}^O - \mathbf{F}^O\mathbf{u}^O + \bar{\mathbf{G}}\mathbf{t}^H = \mathbf{I}\mathbf{u}^H$ $\qquad$ (6.21b)

where

$\mathbf{t}^O$ and $\mathbf{u}^O$ = traction and displacement vectors on the outer surface of the composite matrix

$\mathbf{t}^H$ and $\mathbf{u}^H$ = traction and displacement vectors on the surface of the inclusion

$\mathbf{I}$ = the identity matrix

$\mathbf{G}^O$ and $\mathbf{F}^O$ matrices contain coefficients from the integration over the outer boundary

$\bar{\mathbf{G}}$ matrix contains coefficients integrated about the hole/inclusion

Our goal is to eliminate $\mathbf{u}^H$ from the system. To this end, Eq. (6.19) is written for every node on an inclusion, collocating slightly outside the boundary of the inclusion, where $c_{ij}^I(\xi) = 0$. This results in

$$
\mathbf{F}^{I2}\mathbf{u}^I = \mathbf{G}^{I2}\mathbf{t}^I
$$

Superscript **I2** identifies the equations written at points located slightly outside the boundary of the inclusions.

Noting that $\mathbf{u}^H = \mathbf{u}^I$ and $\mathbf{t}^H = -\mathbf{t}^I$ we have, from above,

$$
\mathbf{F}^{I2}\mathbf{u}^H = -\mathbf{G}^{I2}\mathbf{t}^H \tag{6.22}
$$

Pre-multiplying Eq. (6.21*b*) by the $\mathbf{F^{12}}$ matrix of Eq. (6.22) yields

$$\mathbf{F^{12}G^Ot^O} - \mathbf{F^{12}F^Ou^O} + \mathbf{F^{12}\bar{G}t^H} = \mathbf{F^{12}u^H} \tag{6.23}$$

Equation (6.22) can now be set equal to Eq. (6.23) and the final form of the system is derived:

On the outer surface: $\qquad\qquad \mathbf{G^Ot^O} - \mathbf{F^Ou^O} + \mathbf{\bar{G}t^H} = \mathbf{0}$

On the hole: $\quad \mathbf{F^{12}G^Ot^O} - \mathbf{F^{12}F^Ou^O} + (\mathbf{F^{12}\bar{G} + G^{12}})\mathbf{t^H} = \mathbf{0} \tag{6.24}$

At every point on the outer surface, either the traction or the displacement is specified and on the surface of the inclusion only the tractions are retained. Therefore, the number of equations in the system is equal to the final number of unknowns. After the solution of Eqs (6.24), Eq. (6.21b) is used to determine the displacement at the matrix–inclusion interface. It should be noted that since the displacement $\mathbf{u}^I$ for a particular inclusion is present only in the inclusion equation corresponding to that matrix–inclusion interface, backsubstitution can be performed one inclusion at a time.

6.4 NUMERICAL APPLICATIONS

6.4.1 Examples of Holes in Solids

A cube with a hole[7] A test problem of a cube of dimensions *a* with a cylindrical hole of diameter *b* was subjected to simple tension such that the deformed shape of the hole is elliptical (Fig. 6.3). Figure 6.4 shows the discretization used which

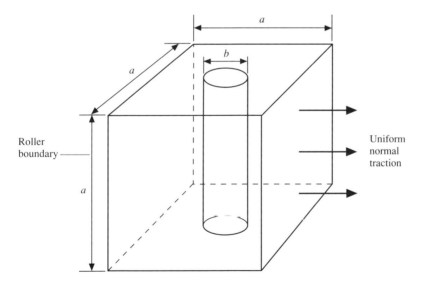

Figure 6.3 Three-dimensional cube (of length *a*) with a hole (of diameter *b*) at the centre.

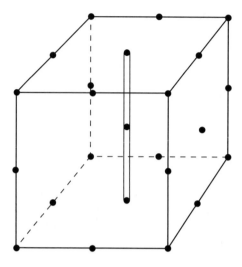

Figure 6.4 Discretization of the cube using one quadratic hole element to model the hole.

includes a mixture of five eight-noded and one nine-noded surface elements. The hole, which is of constant radius, was modelled using the hole element with three nodes in the longitudinal direction. A double line is used as the convention to indicate the location of the centreline of a hole element. Young's modulus $= 100\,\text{lb/in}^2$, $\nu = 0.3$ and an applied traction of $100\,\text{lb/in}^2$ were assumed.

Figure 6.5 shows the results of the current analysis for various dimensions of the hole compared with three-dimensional results in which the hole was modelled with traditional boundary elements (shown in Fig. 6.6). The diameter of the hole

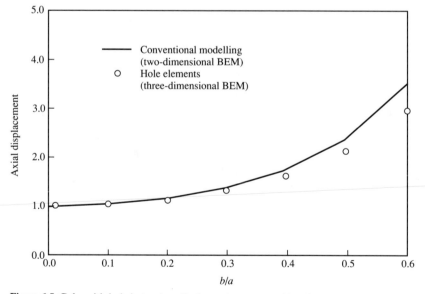

Figure 6.5 Cube with hole in tension: displacement versus void ratio.

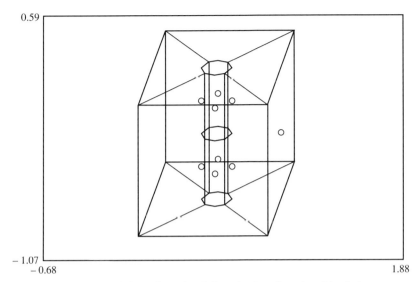

0.59

−1.07
−0.68   1.88

Figure 6.6 Conventional three-dimensional discretization of a cube with a hole.

was varied from $b/a = 0.01$ to $b/a = 0.6$. It can be seen that the displacement at the middle node of the loaded face of the cube is in good agreement with the three-dimensional results. The discrepancy at the larger void ratio is to be expected since the displacement field around the hole is too complex to be modelled by the coarse surface mesh.

It was originally thought that the approximate nature of the hole element, while perfectly adequate to quantify the loss of stiffness due to their presence, would not be suitable for the determination of the local stress concentration factors. Figure 6.7 on page 170 shows that the actual stress concentration up to a point close to the hole surface is in excellent agreement with those obtained by a full three-dimensional modelling by the BEM with the mesh shown in Fig. 6.6. Even at points close to the hole surface (up to 0.15 times the radius) there is a reasonable agreement between the results. For the full three-dimensional modelling, the boundary stress calculation is employed at surface points on the hole to complete the stress plot.

As expected, the hole element allows for a much more efficient analysis, as indicated in Table 6.1. The advantage of the hole element would, of course, increase with the number of holes.

Table 6.1 Relative computing efficiencies for hole models

Model type	Relative CPU time
Cube only with eleven interior points	1.0
Cube with one hole element and eleven interior points	1.67
Traditional three-dimensional model of cube with hole (Fig. 6.6) and eleven interior points	4.60

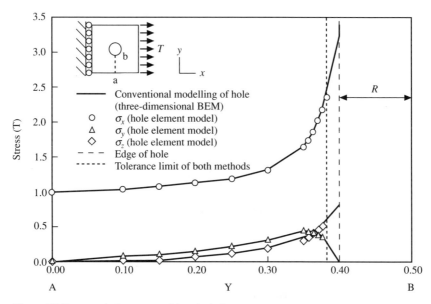

Figure 6.7 Stress variation approaching the hole.

Analysis of a thick cylinder with four holes[7]   A more complex problem demonstrating the capabilities of hole elements is shown in Fig. 6.8, where a 15° segment

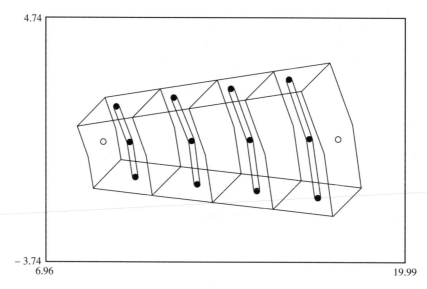

Figure 6.8 Three dimensional model of a slice of a thick cylinder with four holes modelled with hole elements.

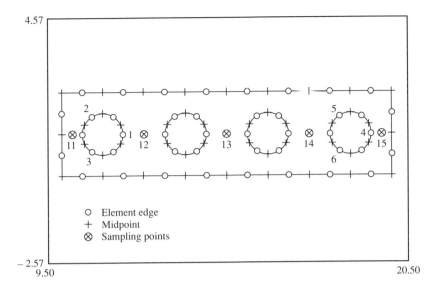

Figure 6.9 Axisymmetric discretization of a thick cylinder with four holes.

of a thick cylinder (inner radius = 10 inches, outer radius = 20 inches, height of the cylinder = 2 inches) has four 1 inch diameter holes evenly spaced at radii of $r = 11.25$, 13.75, 16.25 and 18.75. Young's modulus of the material $E = 30 \times 10^6$ lb/in^2 and Poisson's ratio $\nu = 0.3$. The problem is first modelled with eighteen quadratic surface elements and four quadratic hole elements. Sixteen of the quadratic surface elements are eight-noded and two are nine-noded elements. The cylindrical segment is loaded with an internal pressure of 100 000 lb/in^2 under the conditions of plane strain, which was simulated by applying rollers on all plane surfaces.

The problem is also analysed using an axisymmetric analysis under plane strain conditions using twenty quadratic surface elements to model the cylinder and six quadratic surface elements for each hole, as shown in Fig. 6.9. Tables 6.2 and 6.3 show the variations of the displacements (radial and axial) of the inner and outer holes respectively. Tables 6.4 and 6.5 show the radial displacements and the hoop stresses of five interior points located in between holes.

Table 6.2 Deformation of the inner hole

Node	Radial displacement		Axial displacement	
	Axisymmetric	Hole element	Axisymmetric	Hole element
1	0.0657	0.0692	0.0	0.0
2	0.0725	0.0714	0.0006	0.0005
3	0.0725	0.0714	−0.0006	−0.0005

Table 6.3 Deformation of the outer hole

Node	Radial displacement		Axial displacement	
	Axisymmetric	Hole element	Axisymmetric	Hole element
1	0.0483	0.0496	0.0	0.0
2	0.0494	0.0489	0.0003	0.0003
3	0.0494	0.0489	−0.0003	−0.0003

Table 6.4 Radial displacement at interior points

Point	Component	Axisymmetric	Hole element
11	u_r	0.0773	0.0775
12	u_r	0.0640	0.0641
13	u_r	0.0558	0.0561
14	u_r	0.0508	0.0509
15	u_r	0.0477	0.0478

Table 6.5 Hoop stress at interior points

Point	Component	Axisymmetric (ksi)	Hole element
11	$\sigma_{\theta\theta}$	207.8	214.1
12	$\sigma_{\theta\theta}$	150.7	152.0
13	$\sigma_{\theta\theta}$	114.2	115.9
14	$\sigma_{\theta\theta}$	92.9	93.9
15	$\sigma_{\theta\theta}$	79.5	81.0

Once again the hole algorithm is found to be capable of reproducing the local displacement as well as the stresses in this model.

6.4.2 Examples of Fibre Composite Analysis

Two examples are presented to demonstrate the applications of the composite formulation. In the discretization diagrams of models containing the inclusion, a double line is used to indicate the centreline of the inclusion elements. The lengths of these elements are shown in proper proportion for the three-dimensional views; however, the radii of the inclusions are not indicated on these diagrams.

Thick cylinder with circumferential inclusions[6] The strength of a cylinder under internal pressure can be increased by adding stiff circumferential inclusions. The cylinder shown in Fig. 6.8 is again considered which includes four fully bonded inclusions of radius 0.5. By using roller boundary conditions on the faces of symmetry, only a 15° slice of the thick cylinder is modelled. The elastic modulus of the cylinder is assumed to be 100, and the effect of inclusions with five different moduli

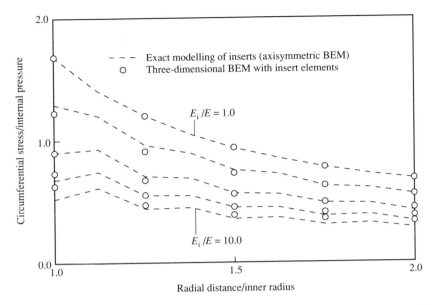

Figure 6.10 Circumferential stress through a pressurized thick cylinder with circumferential inclusions for $E_i/E = 1.0$, 2.5, 5.0, 7.5 and 10.0.

of 100, 250, 500, 750 and 1000 is studied. The Poisson ratio is 0.3 for both the composite matrix and inclusion, and the internal pressure in the cylinder is 100.

Results from a multi-region, axisymmetric BEM analysis were used for comparison with the three-dimensional inclusion results. The axisymmetric model consists of twenty quadratic boundary elements on the outer surface and six boundary elements per hole and inclusion (Fig. 6.9). Figure 6.10 shows circumferential stress for points along the top edge. This stress is smooth for the homogeneous case ($E_i/E = 1.0$) and exhibits increasing fluctuations as the E_i/E ratio increases and the fibres take on more of the load. The circumferential stress of the three-dimensional inclusion model is in good agreement with the axisymmetric results for all cases.

A beam with reinforcement in bending[6]   Reinforced concrete can now be modelled exactly as a three-dimensional body and studied in detail. The present example considers a reinforced concrete beam. Here the concrete plays the role of the composite matrix and the reinforcement bars play the role of the fibre. In Fig. 6.11, a $4 \times 1 \times 1$ beam with four fibres is modelled using twenty-eight quadratic boundary elements. The effect of the ratio of fibre modulus to matrix modulus (E_i/E) is studied for a range of values between 1 and 100. The Poisson ratio is 0.3 for both the beam and reinforcement.

The beam is completely fixed at one end and a downward shear traction of 100 is applied to the other end. The non-dimensional vertical displacement of the end obtained from the present analysis is shown in Fig. 6.12 as a function of E_i/E.

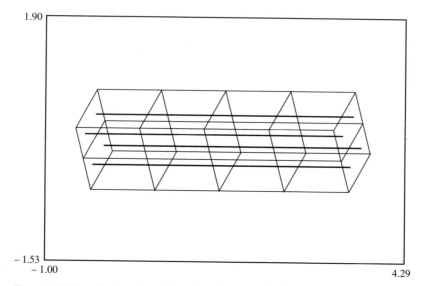

Figure 6.11 Discretization of a reinforced beam utilizing quadratic inclusion elements to model the four inclusions.

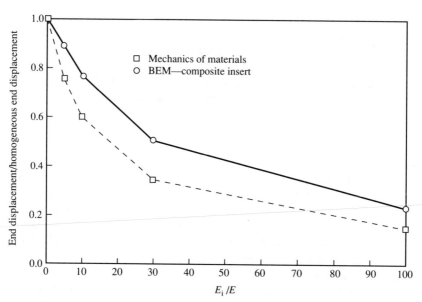

Figure 6.12 Non-dimensional vertical end displacement of a reinforced beam in bending versus the modulus of the inclusion over modulus of the beam.

The non-dimensional displacement is defined as the end displacement of the reinforced beam divided by the displacement of a homogeneous beam under similar boundary conditions.

The end displacement obtained from the mechanics of material solution is also displayed in Fig. 6.12 in a non-dimensional form. The curvature of the two plots are very similar but differ in magnitude. This difference is attributed to the fact that, although the mechanics of material solution accounts for the stiffening due to the fibres, it does not include the effect of interaction between individual fibres.

6.5 CONCLUDING REMARKS

An approximate BEM formulation for the efficient three-dimensional elastic analysis of solids with holes and inclusions is presented. The analysis is efficient as well as reasonably accurate for a range of problems when compared with conventional modelling. One of the major advantages of this formulation lies in the fact that the locations and arrangements of these holes and inclusions can be altered without rebuilding the BEM discretization of the outer boundary. It is also possible to store the major part of the boundary element coefficients and supplement them with new integrated coefficients for the altered design for reassembly and solution at a fraction of the original cost.

REFERENCES

1. Barone, M.R. and Caulk, D.A. (1981) Special boundary integral equations for approximate solution of Laplace's equation in two-dimensional regions with circular holes, *Quarterly Journal of Mechanical Applied Mathematics*, Vol. 34, pp. 265–268.
2. Caulk, D.A. (1983) Analysis of elastic torsion in a bar with circular holes by a special boundary integral method, *Journal of Applied Mechanics*, Vol. 50, pp. 101–108.
3. Barone, M.R. and Caulk, D.A. (1985) Special boundary integral equations for approximate solution of potential problems in three-dimensional regions with slender cavities of circular cross-section, *IMA Journal of Applied Mathematics*, Vol. 35, pp. 311–325.
4. Dargush, G.F. and Banerjee, P.K. (1989) Advanced development of the boundary element method for steady-state heat conduction, *International Journal of Numerical Methods in Engineering*, Vol. 28, pp. 2123–2142.
5. Dargush, G.F. and Banerjee, P.K. (1991) Application of BEM to transient heat conduction, *International Journal of Numerical Methods in Engineering*, Vol. 31, pp. 1231–1247.
6. Banerjee, P.K. and Henry, D.P. (1992) Elastic analysis of three-dimensional solids with fiber inclusions by BEM, *International Journal of Solids and Structures*, Vol. 29, No. 10, pp. 2423–2440.
7. Henry, D.P. and Banerjee, P.K. (1991) Elastic analysis of three-dimensional solids with small holes by BEM, *International Journal of Numerical Methods in Engineering*, Vol. 31, No. 2, pp. 369–384.
8. Wang, H.C. and Banerjee, P.K. (1989) Multi-domain general axisymmetric analysis by BEM, *International Journal of Numerical Methods in Engineering*, Vol. 28, pp. 2065–2083.
9. Byrd, P.F. and Friedman, M.D. (1954) *Handbook of Elliptic Integrals for Engineers and Physicists*, Springer-Verlag, Berlin.
10. Abramowitz, M. and Stegan, I.A. (1974) *Handbook of Mathematical Functions*, Dover, New York.
11. Hastings (Jr), C. (1955) *Approximations for Digital Computers*, Princeton University Press, Princeton.

12. Watson, J.O. (1979) Advanced implementation of the BEM for two and three-dimensional elastostatics, Chapter 3 in *Developments in Boundary Element Methods*, Vol. 1, Eds P.K. Banerjee and R. Butterfield, Elsevier Applied Science, London.
13. Banerjee, P.K., Wilson, R.B. and Miller, N. (1988) Advanced elastic and inelastic stress analysis of gas turbine engine structures by BEM, *International Journal of Numerical Methods in Engineering*, Vol. 26, pp. 393–411.

GENERAL BODY FORCES

7.1 INTRODUCTION

Body forces can occur in a medium due to events that can be either conservative or non-conservative. Conservative body forces such as those occurring due to self-weight, centrifugal loading, seepage forces, thermal fields, etc., can be reduced completely to a boundary integral formulation either via the use of the divergence theorem and utilizing the condition that the body forces are gradients of some potential or alternatively by developing particular solutions for the inhomogeneous differential equation. Both of these methods were discussed in Chapters 3, 4 and 5.

For a non-conservative body force system that can either be real or a pseudo body force system resulting from splitting the governing differential equations of the problem into a linear operator, for which a fundamental solution can be constructed, and a part that has to be treated as a body force, one must either retain the volume integral or develop a different particular integral formulation.

7.2 INTEGRAL FORMULATIONS AND SOLUTION

For the general body force system ψ_i, the integral formulation, following from Chapter 4, is given by

$$c_{ij}u_i(\xi) = \int_S \left[G_{ij}(x,\xi)t_i(x) - F_{ij}(x,\xi)u_i(x) \right] dS(x) + \int_V G_{ij}(x,\xi)\psi_i(x)\,dV(x) \qquad (7.1)$$

where G_{ij} and F_{ij} have been defined in Chapter 4. In order to solve such a problem we must discretize not only the surface of the body into boundary elements but also the volume into several volume cells so that integrals in Eq. (7.1) could be evaluated as

$$c_{ij}u_i(\xi) = \sum_{n=1}^{N} \int_{\Delta S^n} (G_{ij}t_i - F_{ij}u_i)\,\mathrm{d}S^n + \sum_{m=1}^{M} \int_{\Delta V^m} G_{ij}\psi_i\,\mathrm{d}V^m \qquad (7.2)$$

By utilizing the shape functions $N^k(x)$ over boundary elements and $M^k(x)$ over volume cells to account for the variations of t_i, u_i over boundary elements and of ψ_i over volume cells respectively, we can express Eq. (7.2) as

$$c_{ij}u_j(\xi) = \sum_{n=1}^{N} \left[\left(\int_{\Delta S^n} G_{ij}N^k\,\mathrm{d}S^n \right)t_i^k - \left(\int_{\Delta S^n} F_{ij}N^k\,\mathrm{d}S^n \right)u_i^k \right]$$

$$+ \sum_{m=1}^{M} \left(\int_{\Delta V^m} G_{ij}M^k\,\mathrm{d}V^n \right)\psi_i^k \qquad (7.3)$$

By taking the point ξ once again to all boundary nodes we can obtain the following matrix equations

$$\mathbf{G}^s\mathbf{t}^s - \mathbf{F}^s\mathbf{u}^s + \mathbf{G}^v\boldsymbol{\psi}^v = 0 \qquad (7.4)$$

where the superscripts s and v denote quantities obtained from surface and volume integrations respectively. By incorporating appropriate boundary conditions we can express Eq. (7.4) as

$$\mathbf{Ax} = \mathbf{b} - \mathbf{G}^v\boldsymbol{\psi}^v \qquad (7.5)$$

where the system equation is identical to that for a medium without body forces except that the right-hand side is augmented with an additional matrix product term $\mathbf{G}^v\boldsymbol{\psi}^v$. Therefore, we see that the introduction of any general body force system does not lead to additional complexity, although some integration efforts are required over volume cells.

7.3 INITIAL STRESSES AND INITIAL STRAINS

The body force system discussed above can arise from thermal stresses or lack of fit stresses or strains[1,2] existing in the body. Such stresses are often described as 'initial stresses' or 'initial strains'.[1] For some problems these are known a priori, while for others (non-linear problems) they have to be determined as a part of the solution process.

7.3.1 Boundary Integral Formulations

In an 'initial stress' formulation the initial stress σ_{ij}^0 is defined as

$$\sigma_{ij}^0 = \sigma_{ij}^e - \sigma_{ij} \qquad (7.6)$$

where $\sigma^e_{ij} = D^e_{ijkl}\varepsilon_{kl}$. D^e_{ijkl} is the elastic constitutive tensor and σ_{ij} is the correct stress.

The governing equilibrium equations then become[2]

$$\frac{\partial \sigma_{ij}}{\partial x_j} = 0 \quad \text{or} \quad \frac{\partial \sigma^e_{ij}}{\partial x_j} \frac{\partial \sigma^0_{ij}}{\partial x_j} - 0 \tag{7.7}$$

It is immediately evident that the term $-(\partial \sigma^0_{ij}/\partial x_j)$ is equivalent to a body force ψ_i and Eq. (7.7) may be written as

$$\frac{\partial \sigma^e_{ij}}{\partial x_j} + \psi_i = 0 \tag{7.8}$$

Over the boundaries the displacement vector u_i remains unchanged but the boundary traction becomes[1,2]

$$t_i = \sigma_{ij}n_j = \sigma^e_{ij}n_j - \sigma^0_{ij}n_j = t'_i - t^0_i \tag{7.9}$$

where t_i is the correct boundary traction, t'_i is the modified boundary traction and t^0_i is the traction due to the initial stress.

The direct BEM formulation for this problem is clearly that given by Eq. (7.1) with a body force ψ_i and a modified traction t'_i. Using the divergence theorem the volume integral in Eq. (7.1) can then be manipulated as follows:[2,3]

$$\int_V \psi_i(x)G_{ij}(x,\xi)\,\mathrm{d}V = -\int_V \frac{\partial \sigma^0_{ik}(x)}{\partial x_k}G_{ij}(x,\xi)\,\mathrm{d}V$$

$$= \int_V \frac{\partial G_{ij}(x,\xi)}{\partial x_k}\sigma^0_{ik}(x)\,\mathrm{d}V - \int_S G_{ij}(x,\xi)\sigma^0_{ik}(x)n_k(x)\,\mathrm{d}S$$

$$= \int_V B_{ikj}(x,\xi)\sigma^0_{ik}(x)\,\mathrm{d}V - \int_S G_{ij}(x,\xi)t^0_i(x)\,\mathrm{d}S \tag{7.10}$$

where $B_{ikj}(x,\xi) = \partial G_{ij}(x,\xi)/\partial x_k$. Substituting Eq. (7.10) into Eq. (7.1) and making use of Eq. (7.9) to eliminate the modified traction t'_i we obtain[2-6]

$$c_{ij}u_i(\xi) = \int_S [t_i(x)G_{ij}(x,\xi) - F_{ij}(x,\xi)u_i(x)]\,\mathrm{d}S + \int_V B_{ikj}(x,\xi)\sigma^0_{ik}(x)\,\mathrm{d}V \tag{7.11}$$

as the necessary boundary integral formulation for our problem.

Since σ^0_{ik} is symmetric we can write

$$B_{ikj}\sigma^0_{ik} = \frac{1}{2}\left(\frac{\partial G_{ij}}{\partial x_k} + \frac{\partial G_{kj}}{\partial x_i}\right)\sigma^0_{ik} \tag{7.12}$$

Therefore B_{ikj} is none other than the strain ε_{ik} at x due to a unit force vector e_j acting at ξ.

If the initial strains ε_{ij}^0 are known, the correct stresses σ_{ij} are given by

$$\sigma_{ij} = D_{ijkl}^e \varepsilon_{kl}^e = D_{ijkl}^e (\varepsilon_{kl} - \varepsilon_{kl}^0)$$

which can be recast as

$$\sigma_{ij} = D_{ijkl}^e \varepsilon_{kl} - D_{ijkl}^e \varepsilon_{kl}^0 = \sigma_{ij}^e - \sigma_{ij}^0 \tag{7.13}$$

Using Eqs (7.13) and (7.11) we write the initial strain formulation for our problem as[3]

$$c_{ij} u_i(\xi) = \int_S [t_i(x) G_{ij}(x, \xi) - F_{ij}(x, \xi) u_i(x)] \, dS + \int_V T_{ikj}(x, \xi) \varepsilon_{ik}^0(x) \, dV \tag{7.14}$$

where $T_{ikj} = D_{ikmn}^e B_{mnj}$ gives the stresses σ_{ik} at a point x due to a unit force $e_j(\xi)$ within an infinite solid. Quantities G_{ij}, F_{ij} B_{ikj} and T_{ikj} have been defined in Eqs (4.7) to (4.9).

7.3.2 Interior Stresses

We can differentiate Eqs (7.11) and (7.14) for a field point ξ in the interior (with $c_{ij} = \delta_{ij}$) of the body. Thus we have, from Eq. (7.11),

$$\frac{\partial u_j(\xi)}{\partial \xi_k} = \int_S \left[t_i(x) \frac{\partial G_{ij}(x, \xi)}{\partial \xi_k} - \frac{\partial F_{ij}(x, \xi)}{\partial \xi_k} u_i(x) \right] dS$$

$$+ \frac{\partial}{\partial \xi_k} \int_V B_{ipj}(x, \xi) \sigma_{ip}^0(x) \, dV(x) \tag{7.15}$$

and from Eq. (7.14),

$$\frac{\partial u_j(\xi)}{\partial \xi_k} = \int_S \left[t_i(x) \frac{\partial G_{ij}(x, \xi)}{\partial \xi_k} - \frac{\partial F_{ij}(x, \xi)}{\partial \xi_k} u_i(x) \right] dS$$

$$+ \frac{\partial}{\partial \xi_k} \int_V T_{ipj}(x, \xi) \varepsilon_{ip}^0(x) \, dV \tag{7.16}$$

The reason for keeping the derivatives outside the volume integrals becomes clear if we look at functions B_{ipj} and T_{ipj} of Eqs (7.11) and (7.14). The order of singularity of these functions is $(1/r^{n-1})$ which is integrable in the normal sense, where n is the number of space dimensions. If we differentiate under the integral sign the singularity would then be $(1/r^n)$, which does not exist in the ordinary sense. In earlier works on two-dimensional problems these were avoided by first integrating analytically and then taking derivatives. The surface integrals of course are never singular at any interior point (that is $x \neq \xi$) and therefore pose no difficulty.

These volume integrals can be evaluated by considering a small exclusion D around the field point ξ and thus we can express the volume integral in Eq. (7.16) as[7,8]

$$
\frac{\partial}{\partial \xi_k} \int_V T_{ipj}(x,\xi)\varepsilon_{ip}^0(x)\,\mathrm{d}V(x) = \int_{V-D} \frac{\partial}{\partial \zeta_k} T_{ipj}(x,\xi)\varepsilon_{ip}^0(x)\,\mathrm{d}V(x)
$$

$$
+ \frac{\partial}{\partial \xi_k} \int_D T_{ipj}(x,\xi)\varepsilon_{ip}^0(x)\,\mathrm{d}V(x) \qquad (7.17)
$$

If we now assume that D is small enough such that $\varepsilon_{ip}^0(x) = \varepsilon_{ip}(\xi)$, i.e. the initial strain is constant within this small exclusion, we can write the second integral on the right-hand side of Eq. (7.17) as[7,8]

$$
-\varepsilon_{ip}^0(\xi) \int_D \frac{\partial}{\partial x_k} T_{ipj}(x,\xi)\,\mathrm{d}V(x) = -\varepsilon_{ip}^0(\xi) \int_{\partial D} T_{ipj}(x,\xi)n_k(x)\,\mathrm{d}S(x) \qquad (7.18)
$$

where the integral now needs to be evaluated over the surface of the exclusion D with a field point ξ at its centre. Thus, evaluating this surface integral analytically Eqs (7.15) and (7.16) can be expressed as[8]

$$
u_{j,k} = \int_S [t_i(x)G_{ij,k}(x,\xi) - F_{ij,k}(x,\xi)u_i(x)]\,\mathrm{d}S
$$

$$
+ \int_{V-D} B_{ipj,k}(x,\xi)\sigma_{ip}^0(x)\,\mathrm{d}V + \frac{\delta_{jk}\sigma_{mm}^0 - (8-10\nu)\sigma_{jk}^0}{30\mu(1-\nu)} \qquad (7.19)
$$

and $\quad u_{j,k} = \int_S [t_i(x)G_{ij,k}(x,\xi) - F_{ij,k}(x,\xi)u_i(x)]\,\mathrm{d}S$

$$
+ \int_{V-D} T_{ipj,k}(x,\xi)\varepsilon_{ip}^0(x)\,\mathrm{d}V + \frac{(1-5\nu)\varepsilon_{mm}^0\delta_{jk} - (8-10\nu)\varepsilon_{jk}^0}{15(1-\nu)} \qquad (7.20)
$$

respectively, for a three-dimensional problem.

It is of considerable interest to note that the last terms on the right-hand side of Eqs (7.19) and (7.20) which resulted from surface integrations over the exclusion D are independent of the size of exclusion as long as the initial stress or initial strain is locally homogeneous[7-9] (constant) over the volume D. This is, of course, a well-known result and form the basis of the 'Eshelby tensor' for inhomogeneous inclusions in an elastic body. It is also possible to treat these terms using different interpretations.[10-14]

Thus, the strain at the point ξ can be expressed as[9,15]

$$
\varepsilon_{jk}(\xi) = \int_S [G_{ijk}^\epsilon(x,\xi)t_i(x) - F_{ijk}^\epsilon(x,\xi)u_i(x)]\,\mathrm{d}S
$$

$$
+ \int_{V-D} B_{ipjk}^\epsilon(x,\xi)\sigma_{ip}^0(x)\,\mathrm{d}V + J_{ipjk}^\epsilon\sigma_{ip}^0(\xi) \qquad (7.21)
$$

where

$$G_{ijk}^{\varepsilon}(x,\xi) = -B_{kji}(x,\xi)$$

$$F_{ijk}^{\varepsilon} = \frac{C_3}{(\alpha-1)r^{\alpha}}\{C_4 n_i(\delta_{jk} - \alpha z_j z_k) - n_j(C_4\delta_{ik} + \alpha\nu z_i z_k)$$

$$- n_k(C_4\delta_{ij} + \alpha\nu z_i z_j) - \alpha z_m n_m[\nu\delta_{ij}z_k + \nu\delta_{ik}z_j + \delta_{jk}z_i - (\alpha+2)z_i z_j z_k]\}$$

$$B_{ipjk}^{\varepsilon}(x,\xi) = \frac{C_1}{(\alpha-1)r^{\alpha}}[C_4(\delta_{ij}\delta_{kp} + \delta_{ik}\delta_{jp}) - \delta_{ip}\delta_{jk}$$

$$+ \alpha\nu(\delta_{ij}z_k z_p + \delta_{ik}z_j z_p + \delta_{jp}z_i z_k + \delta_{kp}z_i z_j)$$

$$+ \alpha(\delta_{jk}z_i z_p + \delta_{ip}z_j z_k) - \alpha(\alpha+2)z_i z_j z_k z_p]$$

$$J_{ipjk}^{\varepsilon} = \frac{C_7}{\alpha(\alpha+2)}\{2[(\alpha+1) - \nu(\alpha+2)]\delta_{ij}\delta_{kp} - \delta_{ip}\delta_{jk}\}$$

The stresses at an interior point ξ can now be obtained from Eq. (7.21) by using the constitutive relationships ($\sigma_{ij} = D_{ijkl}^{e}\varepsilon_{kl} - \sigma_{ij}^{0}$) as[9,15]

$$\sigma_{jk}(\xi) = \int_S [G_{ijk}^{\sigma}(x,\xi)t_i(x) - F_{ijk}^{\sigma}(x,\xi)u_i(x)]\,dS$$

$$+ \int_{V-D} B_{ipjk}^{\sigma}(x,\xi)\sigma_{ip}^{0}(x)\,dV + J_{ipjk}^{\sigma}\sigma_{ip}^{0}(\xi) \qquad (7.22)$$

where G_{ijk}^{σ} and F_{ijk}^{σ} are defined by Eq. (4.17) and

$$B_{ipjk}^{\sigma}(x,\xi) = \frac{C_3}{(\alpha-1)r^{\alpha}}[C_4(\delta_{ip}\delta_{jk} - \delta_{ij}\delta_{kp} - \delta_{ik}\delta_{jp} - \alpha\delta_{jk}z_i z_p) - \alpha\delta_{ip}z_j z_k$$

$$- \alpha\nu(\delta_{ij}z_k z_p + \delta_{ik}z_j z_p + \delta_{jp}z_i z_k + \delta_{kp}z_i z_j) + \alpha(\alpha+2)z_i z_j z_k z_p]$$

$$J_{ipjk}^{\sigma} = \frac{C_8}{\alpha(\alpha+2)}\{[(\alpha^2-2) - \nu(\alpha^2-4)]\delta_{ij}\delta_{kp} + [1 - \nu(\alpha+2)\delta_{ip}\delta_{jk}]\}$$

and

$$C_1 = \frac{1}{8\pi\mu(1-\nu)}$$

$$C_3 = \frac{-1}{4\pi(1-\nu)}$$

$$C_4 = 1 - 2\nu$$

$$C_7 = \frac{1}{2\mu(1-\nu)}$$

$$C_8 = \frac{-1}{1 - \nu}$$

$$r^2 = y_i y_i$$

$$y_i = x_i - \zeta_i$$

$$z_i = \frac{y_i}{r}$$

$$n_m = \text{the normal on the boundary at } x$$

All of the expressions given above are valid for two-dimensional plane strain problems ($\alpha = 2$) as well as the three-dimensional problems ($\alpha = 3$). The corresponding expressions for the two-dimensional plane stress problems can be obtained from these by replacing ν by $\nu/(1 + \nu)$ and using $\alpha = 2$. It is also possible to obtain similar expressions for the initial strain formulations by invoking Eq. (7.13). Unfortunately, during the process of contraction from three to two dimensions this formulation requires separate expressions for B^ϵ and B^σ kernels for two-dimensional plane stress and plane strain problems.[11,16] The equivalent axisymmetric forms are defined in Henry and Banerjee.[17,18]

The expressions for strains and stresses cannot be easily constructed at the boundary points by taking the field point (ξ) in Eqs (7.21) and (7.22) to the surface point due to the strongly singular nature of the integrals involved. The evaluation of strains and stresses at boundary points can be accomplished by considering the equilibrium of the boundary segment and utilizing constitutive and kinematic equations. The stresses and the global derivatives of displacements which lead to strains at point ξ can be obtained by coupling the following set of equations:[9,15]

$$\sigma_{ij}(\xi) - \{\lambda \delta_{ij} u_{k,k}(\xi) + \mu[u_{i,j}(\xi) + u_{j,i}(\xi)]\} = -\sigma_{ij}^0(\xi)$$

$$\sigma_{ij} n_j(\xi) = t_i(\xi) \tag{7.23}$$

and
$$\frac{\partial u_i}{\partial \xi_k} \frac{\partial \xi_k}{\partial \eta_l} = \frac{\partial u_i}{\partial \eta_l}$$

where η_l are a set of local axes in the normal and tangential directions at the field point ξ and λ and μ are Lame's constants.

By making use of the shape function in the third equation of (7.23) and rearranging the equation into known and unknown components, the above sets of equations can be inverted and rearranged to form (see Chapter 17)[9,15]

$$\varepsilon_{jk} = \bar{G}_{ijk}^\varepsilon t_i - \bar{F}_{ijk}^\varepsilon u_i + \bar{B}_{ipjk}^\varepsilon \sigma_{ip}^0 \tag{7.24a}$$

$$\sigma_{jk} = \bar{G}_{ijk}^\sigma t_i - \bar{F}_{ijk}^\sigma u_i + \bar{B}_{ipjk}^\sigma \sigma_{ip}^0 \tag{7.24b}$$

7.3.3 Numerical Integration over Volume Cells

For a problem with a known distribution of initial stresses, σ_{ij}^0, within the volume, Eqs (7.11), (7.14), (7.21) and (7.22) provide the exact formulation; the numerical implementation can therefore be made as sophisticated as the analyst desires. Assuming that the boundary and the volume are divided into N isoparametric boundary elements and M isoparametric volume cells (see Chapter 2) respectively, the boundary integral equations for the displacement can then be represented as[9,15,17–20]

$$
c_{ij}u_i(\xi_0^p) = \sum_{n=1}^{N}\left\{\int_{\Delta S^n} G_{ij}[x^n(\eta),\xi_0^p]N^\alpha(\eta)\,\mathrm{d}S^n \bar{t}_i^{\alpha n} - \int_{\Delta S^n} F_{ij}[x^n(\eta),\xi_0^p]N^\alpha(\eta)\,\mathrm{d}S^n \bar{u}_i^{\alpha n}\right\}
$$

$$
+ \sum_{m=1}^{M}\int_{\Delta V^m} B_{ipj}^\sigma[x^m(\eta),\xi_0^p]M^\beta(\eta)\,\mathrm{d}V^m \bar{\sigma}_{ip}^{\beta m} \tag{7.25}
$$

where ξ_0^p is the field point in the pth boundary node; $x^n(\eta)$ is the load point in element n; $x^m(\eta)$ is the load point in cell m; $\bar{u}^{\alpha n}$, $\bar{t}_i^{\alpha n}$ are the nodal values of displacements and tractions of the nth element; $\bar{\sigma}_{ip}^{\beta m}$ are the nodal values of initial stresses of the mth cell; ΔS^n is the nth boundary element; and ΔV^m is the mth interior cell.

Similarly, the stress (or strain) boundary integral equations can be expressed as

$$
\sigma_{jk}(\xi) = \sum_{n=1}^{N}\left\{\int_{\Delta S^n} G_{ijk}^\sigma[x^n(\eta),\xi]N^\alpha(\eta)\,\mathrm{d}S^n \bar{t}_i^{\alpha n} - \int_{\Delta S^n} F_{ijk}^\sigma[x^n(\eta),\xi]N^\alpha(\eta)\,\mathrm{d}S^n \bar{u}_i^{\alpha n}\right\}
$$

$$
+ \sum_{m=1}^{N}\int_{\Delta V^m} B_{ipjk}^\sigma[x^m(\eta),\xi]M^\alpha(\eta)\,\mathrm{d}V^m \bar{\sigma}_{ip}^{\beta m} + J_{ipjk}\sigma_{ip}^0(\xi) \tag{7.26}
$$

where ξ is the coordinate of the interior field point.

In order to evaluate the integrals over each boundary element and volume cell we need to examine the behaviour of the integrand (i.e. the kernel, shape function and Jacobian products). If the integrand is bounded it can be evaluated using the repeated use of Gaussian integration formulae. One of the factors that determine the accuracy of the integration is the order of the quadrature rule which can be selected automatically by considering the error bound of the quadrature. Watson[21] and, more recently, Mustoe[20] developed the strategy of setting the order of integration to a prescribed value based on error considerations. When the error is exceeded, the element or the cell is subdivided and the prescribed quadrature rule is applied to each subdivision. For simple elemental geometries it is of course possible and desirable to evaluate all of these integrals analytically.[3,5,6,16,22]

When the field point and load point coincide on the boundary element or the volume cell over which the integrals are being evaluated, the above-mentioned numerical procedure breaks down completely. The behaviour of these singular or weakly singular integrals depends on the strength of the singularity. A singular integral $\int_S dS/r^m$, where r is the distance between the field and load points, is only integrable in the normal sense if $m \leq (n-1)$, where n is the dimensionality of S. When $m > (n-1)$, the principal value of the integral is supplemented by a jump term resulting from a non-zero path followed by the integrand, and such operations are difficult to handle numerically since they are particularly sensitive to mesh distortion.

All the surface integrals involved in the above equations have been discussed in Chapter 3. The primary focus of this section is therefore the treatment of the volume integrals involved.

For a two-dimensional analysis involving plane stress, plane strain or the axisymmetric cases, the volume integral is transformed and the quadrature rule is applied as (the three-dimensional cases follow in an analogous manner)[9,15,20]

$$\int_{\Delta V_m} B_{ijk}[x^n(\eta), \xi] M^\beta(\eta) \, d\Omega_m = \int_{-1}^{1} \int_{-1}^{1} B_{ipj}[x^n(\eta), \xi]$$

$$M^\beta(\eta) J(\eta) \, d\eta_1 \eta_2 = \sum_{a=1}^{A} \sum_{b=1}^{B} w^a w^b B_{ipj}[x^n(\eta^{ab}), \xi] M^\beta(\eta^{ab}) J(\eta^{ab}) \qquad (7.27)$$

where the Jacobian is defined by

$$dV_m = J(\eta_1, \eta_2) \, d\eta_1 \, d\eta_2$$

and

$$J(\eta_1, \eta_2) = \left| \frac{\partial x_i}{\partial \eta_j} \right| \qquad i,j = 1, 2$$

The non-singular integration is therefore similar to the corresponding evaluation of the surface integrals in three dimensions. The singular integration of kernels involving $1/r$ behaviour can be developed by carefully studying the behaviour of the kernel, shape functions and Jacobian products. The procedure of integration is depicted in Fig. 7.1 and is as follows:[9,15,20,21]

1. The curvilinear cell is transformed to a unit cell by using the Jacobian transformation employed in the non-singular integration.
2. Depending upon the location of the singular point, the cell is subdivided into two (or multiple of two), when the singularity is at a corner node, or three (or multiple of three), when the singular node is on the cell side.
3. Each triangular sub-cell is then transformed to a unit triangular cell with the origin of coordinates through the singular point.
4. A rectangular system of coordinates is constructed through the field point with one axis perpendicular to the side opposite to the singular point.

By defining a polar coordinate system through the field point as shown in

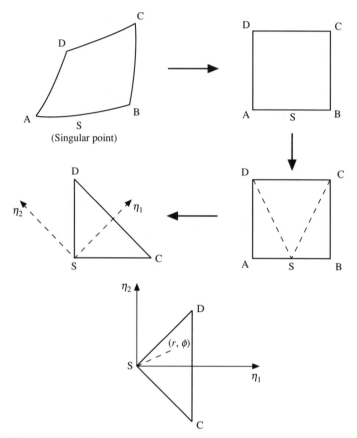

Figure 7.1 Mapping procedure for two-dimensional singular volume cell.

Fig. 7.1, any sampling point within the triangle can be expressed as

$$\eta_1 = r \cos \phi$$

$$\eta_2 = r \sin \phi$$

This transformation of the area integral from a rectangular to a polar coordinate system provides the Jacobian

$$dA = J \, dr \, d\phi = r \, dr \, d\phi$$

Since the integrand is singular of the order $1/r$, the integrand is bounded in the transformed domain and therefore it can be evaluated accurately by selecting a suitable order of integration rule.

The singular integrals involving function B_{ipjk}^{σ} (or B_{ipjk}^{ε}) are strongly singular of order $1/r^2$ and the integrand can be made bounded by excluding a continuous contour in the form of a circle and mapping the remaining curvilinear cell to a unit cell to be adoptable for quadrature formulae. The transformed integral in these cases can be expressed as[9,15,20]

$$\int_{V^m} \mathbf{B}^\sigma[x(\eta), \xi] M^\beta(\eta) J(\eta) \, \mathrm{d}V^m = \int_{-1}^1 \int_{-1}^1 \mathbf{B}^\sigma[x(\eta), \xi]$$

$$M^\beta(\eta) J(\eta) \, \mathrm{d}\eta_1 \, \mathrm{d}\eta_2 = \sum_{k=1}^K \int_{\Omega k} \mathbf{B}^\sigma[x(\eta), \xi] M^\beta(\eta) J(\eta) \, \mathrm{d}\Omega_k$$

$$= \sum_{k=1}^K \int_0^1 \int_0^1 \mathbf{B}^\sigma[x(\bar\eta), \xi] M^\beta(\bar\eta) J_1(\bar\eta) \, \mathrm{d}\bar\eta \qquad (7.28)$$

where $\bar\eta$ is a function of r, ϕ and K is the number of sub-cells.

For singular three-dimensional volume integrals, the volume cell can be transformed to a unit cube and the cube is subdivided into tetrahedra through the field point, similar to the two-dimensional version, as shown in Fig. 7.1. Using a local spherical polar coordinate system (r, θ, ϕ) with its origin at the field point, the integral of the sub-cell can be transformed by the Jacobian as[9,11,15,20]

$$\mathrm{d}V = J \, \mathrm{d}r \, \mathrm{d}\phi \, \mathrm{d}\theta = r^2 \sin\theta \, \mathrm{d}r \, \mathrm{d}\phi \, \mathrm{d}\theta$$

The integrand involving the B_{ipj} kernel is singular of the order $1/r^2$ and therefore is bounded in the transformed domain. However, the volume integral B^σ_{ipjk} is singular of the order $1/r^3$ and in the transformed domain the behaviour is approximately of the order $1/r$. The integral, however, is made bounded by excluding a sphere and mapping the remainder of the tetrahedra to a unit cube as shown in Fig. 7.2. The

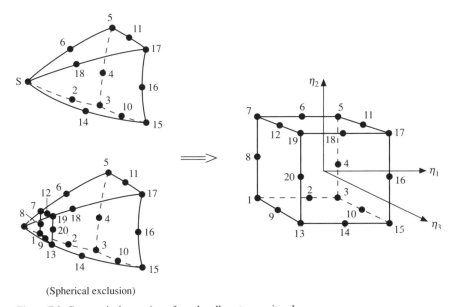

(Spherical exclusion)

Figure 7.2 Geometrical mapping of a sub-cell on to a unit cube.

integral is then computed by applying the Gaussian quadrature together with extensive subdivision in the mapped space.

The above integration procedure becomes quite expensive due to the large number of sub-divisions necessary for the accurate evaluation of the singular and near singular integrals. It can therefore be beneficial to apply the integration by parts rule to reduce the order of singularity. However, this will require the evaluation of an additional integral over the boundary of the cell involved.

7.3.4 Initial Stress Expansion Technique

As mentioned above, the B^σ_{ijkl} and B^ε_{ijkl} kernels are strongly singular when integrated over a volume containing the field point, and accurate numerical integration at this singular point for the most general case as outlined above is costly and difficult. The difficulties, however, can be alleviated by determining the coefficient of the singular node in an indirect manner.[12,17-19]

In this procedure, the coefficients of the stress equations related to the non-singular nodes are integrated in the usual manner using Gaussian integration with appropriate element subdivisions. In each stress equation there remain three undetermined coefficients in two dimensions, or six in three dimensions, or four in axisymmetry, corresponding to the initial stress at the singular node, which cannot be evaluated easily. In a manner analogous to the 'rigid body' motion technique for the surface equations, each of these coefficients is calculated by assuming one (each row) of the three (six or four) admissible initial stress states and compatible displacement fields given in Tables 7.1, 7.2 and 7.3.

In an unrestrained body the resulting stresses are either zero (in two and three dimensions) or can be determined (in axisymmetry) for an assumed initial stress field. Thus, for each assumed initial stress state, one unknown coefficient in each stress equation can be calculated by using all the other coefficients. It should be noted that in order to apply this method the entire region must be covered with cells, which appears inefficient. Although this is true in a single-region problem, such is not the case in a multi-region problem, since the technique is applied to each region independently and is only used in regions where volume cells exist, which happens to be the most efficient BEM analysis. Telles[12] tentatively

Table 7.1 Stress states for initial stress expansion technique in two-dimensional plane strain (plane stress) analysis

Coefficient to be determined corresponds to	Nodal values of assumed stress state				
	σ^o_{xx}	σ^o_{yy}	σ^o_{xy}	u_x	u_y
σ^o_{xx}	E	0	0	$(1-\nu^2)x$	$-\nu(1+\nu)y$
σ^o_{yy}	0	E	0	$-\nu(1+\nu)x$	$(1-\nu^2)y$
σ^o_{xy}	0	0	E	$(1+\nu)y$	$(1+\nu)x$

The stress states for two-dimensional plane strain analysis given here can be applied to the plane stress case, if the following modified material parameters are used: $\bar{E} = E(1 + 2\nu)/(1 + \nu)^2$; $\bar{\nu} = \nu/(1 + \nu)$.

Table 7.2 Stress states for initial stress expansion technique in three-dimensional analysis

Coefficient to be determined corresponds to u_z	Nodal values of assumed stress state								
	σ^o_{xx}	σ^o_{yy}	σ^o_{zz}	σ^o_{xy}	σ^o_{xz}	σ^o_{yz}	u_x	u_Y	u_y
σ^o_{xx}	E	0	0	0	0	0	x	$-\nu y$	$-\nu z$
σ^o_{yy}	0	E	0	0	0	0	$-\nu x$	y	$-\nu z$
σ^o_{zz}	0	0	E	0	0	0	$-\nu x$	$-\nu y$	z
σ^o_{xy}	0	0	0	E	0	0	$(1+\nu)y$	$(1+\nu)x$	0
σ^o_{xz}	0	0	0	0	E	0	$(1+\nu)z$	0	$(1+\nu)x$
σ^o_{yz}	0	0	0	0	0	E	0	$(1+\nu)z$	$(1+\nu)y$

Table 7.3 Stress states for initial stress expansion technique in axisymmetric analysis

Coefficient to be determined corresponds to	Nodal values of assumed stress state											
	σ_{rr}	σ_{zz}	$\sigma_{\theta\theta}$	σ_{rz}	σ^o_{rr}	σ^o_{zz}	$\sigma^o_{\theta\theta}$	σ^o_{rz}	u_r	u_z	t_r	t_z
σ^o_{rr}	$\dfrac{(2-\nu)E}{3(1-\nu)C}r$	$\dfrac{\nu E}{(1-\nu)C}r$	$2\sigma_{rr}$	0	0	0	$\dfrac{-E}{C}r$	0	$\dfrac{(1+\nu)(1-2\nu)}{3(1-\nu)C}r^2$	0	$\sigma_{rr}n_r$	$\sigma_{zz}n_z$
σ^o_{zz}	0	0	0	0	E	0	E	0	$(1-\nu)r$	$-2\nu z$	0	0
$\sigma^o_{\theta\theta}$	0	0	0	0	0	E	0	0	$-\nu r$	z	0	0
σ^o_{rz}	0	0	0	0	0	0	0	μ	0	r	0	0

$C =$ arbitrary parameter with dimensions of length, added to ensure dimensional homogeneity. The value can be taken as unity for simplicity.

discussed this idea of using simple initial strain fields for determining these co-efficients for a two-dimensional elastoplastic problem.

7.4 PARTICULAR INTEGRALS FOR INITIAL STRESSES

Particular solutions of the inhomogeneous differential equation (7.7) written in operator notation as

$$L(u^p_i) = \sigma^0_{ij,j} \qquad (7.29)$$

can be used once again to construct the total solution as the sum of the complementary solution and particular integral, as was shown in Chapters 3 and 4.

7.4.1 Particular Integrals

In the present formulation, the two- and three-dimensional particular integrals for displacement are related to the Galerkin vector F_i via[17,23,24]

$$u^p_i = \frac{1-\nu}{\mu}F_{i,ki} - \frac{1}{2\mu}F_{k,ki} \qquad (7.30)$$

Substituting this into Eq. (7.29) renders a relationship between the Galerkin vector and the initial stress:

$$F_{i,kkjj} = \frac{1}{1-\nu}\sigma^0_{ij,j} \tag{7.31}$$

In subsequent steps of this derivation, it will be advantageous for the implicit expression of Eqs (7.30) and (7.31) to be related by a second-order tensor, rather than a vector. Therefore, a tensor function h_{ij} is introduced such that[17,24]

$$h_{ij,mmnn} = \sigma^0_{ij} \tag{7.32}$$

Substitution of this equation into Eq. (7.31) and simplifying yields an expression for the Galerkin vector in terms of this new function:

$$F_i = \frac{1}{1-\nu}h_{ij,j} \tag{7.33}$$

Finally, substituting this expression into Eq. (7.30) yields the desired particular integral for displacement:[17,24]

$$u^p_i = \frac{1}{\mu}h_{il,lkk} - \frac{1}{2\mu(1-\nu)}h_{lm,ilm} \tag{7.34}$$

The particular integrals for strain, stress and traction are then found using the strain–displacements and stress–strain relations (7.13), etc.

In passing, we note similar relations for particular integrals for a known initial strain ε^0_{ij} are possible, assuming[17,24]

$$h_{ij,mmnn} = \varepsilon^0_{ij} \tag{7.35}$$

where the associate particular integral for the displacement is

$$u^p_i = \frac{\nu}{1-\nu}h_{kk,ijj} + 2h_{ij,jkk} - \frac{1}{1-\nu}h_{kj,ikj} \tag{7.36}$$

The above particular integrals have little practical use in this implicit form. However, by applying the global shape function, a concept first introduced by Rayleigh,[25] an explicit formulation can be developed for the general initial stress or initial strain problems.

7.4.2 Global Shape Function

The tensor $h_{ij}(x)$ can be expressed in terms of a fictitious tensor density $\phi_{ij}(\xi)$ as an infinite series using a suitable global shape function $C(x,\xi)$:[25]

$$h_{ml}(x) = \sum_{n=1}^{\infty} C(x,\xi_n)\phi_{ml}(\xi_n) \begin{cases} m,l = 1,2 \text{ for two dimensions} \\ m,l = 1,2,3 \text{ for three dimensions} \end{cases} \tag{7.37}$$

Several functions could be considered; however, the best results have been obtained with the following expression:[17,24]

$$C(x,\xi_n) = A^4_0(\rho^4 - b_n\rho^5) \tag{7.38}$$

where

A_0 = characteristic length
ρ = Euclidean distance between the field point x and the source point ξ_n
b_n = suitable constant for the present discussion (b_n can assume the value of unity)

All distances are non-dimensionalized by a characteristic length A_0.

The unknown fictitious densities are related to the initial stress rates through Eq. (7.32). Substituting Eq. (7.37) into Eq. (7.32) leads to

$$\sigma^0_{lm} = \sum_{n=1}^{\infty} K(x, \xi_n)\phi_{lm}(\xi_n) \tag{7.39}$$

where
$K(x, \xi_n) = C_{,mmnn}(x, \xi_n) = a - b\rho$
$\qquad a = 8d(d + 2)$
$\qquad b = b_n \times 15(d + 3)(d + 1)$
$\qquad d = 3$ for three-dimensional analysis
$\qquad d = 2$ for two-dimensional (plane strain) analysis

The particular integral for the displacement can now be found as a function of $\phi_{lm}(\xi_n)$ by substituting Eq. (7.37) into Eq. (7.34):[17,24]

$$u_i^p(x) = \sum_{n=1}^{\infty} D_{iml}(x, \xi_n)\phi_{lm}(\xi_n) \tag{7.40}$$

where

$$D_{iml}(x, \xi_n) = A_0 \left[(c_1 + d_1\rho)(y_i\delta_{lm} + y_m\delta_{il}) + (c_2 + d_2\rho)y_l\delta_{im} + \frac{d_1}{\rho}y_iy_ly_m \right]$$

$$y_i = [x_i - (\xi_n)_i]$$

$$c_1 = \frac{-8}{2\mu(1 - \nu)}$$

$$c_2 = c_1 + \frac{8(d + 2)}{\mu}$$

$$d_1 = \frac{15b_n}{2\mu(1 - \nu)}$$

$$d_2 = d_1 - \frac{15(d + 3)b_n}{\mu}$$

The particular integrals for strain, stress and traction are found by substituting Eq. (7.40) into the strain–displacement and stress–strain relations (7.13).

The above equations are equally valid for two and three dimensions, but for two-dimensional analysis the formulation assumes the plane strain condition. The corresponding plane stress formulation can be obtained from the plane strain case by substituting the modified material constant $\bar{\nu} = \nu/(1 + \nu)$.

3 Numerical Implementation

e boundary integral equation representing the complementary function is dis-
tized and numerically integrated in the usual manner for a system of boundary
nodes. The final equations can be expressed for each region in matrix form as (see
Chapters 3 and 4)

$$\mathbf{Gt} - \mathbf{Fu} = [\mathbf{Gt}^P - \mathbf{Fu}^P] \tag{7.41a}$$

$$\sigma^c = \mathbf{G}^\sigma \mathbf{t} - \mathbf{F}^\sigma \mathbf{u} + [\mathbf{F}^\sigma \mathbf{u}^P - \mathbf{G}^\sigma \mathbf{t}^P + \sigma^P] \tag{7.41b}$$

Essentially, the particular integral solution procedure consists of evaluating
the relevant particular integrals at nodal points and solving the above equation
for these values and a set of appropriate boundary conditions. However, here
the particular integrals are functions of initial stress or initial strain.

For the purpose of numerical evaluation, the infinite series representations of
the particular integrals are truncated at a finite number of N terms. Thus for $u_i^P(x)$
we can write

$$u_i^P(x) = \sum_{n=1}^{N} D_{iml}(x, \xi_n)\phi_{lm}(\xi_n) \tag{7.42}$$

and similar equations for $t_i^P(x)$, $\sigma_{ij}^P(x)$ and $\sigma_{ij}^0(x)$.

The choice of N is dictated by the complexity of the problem. Generally, fictitious nodes ξ_n should be introduced at all boundary collocation nodes, and additional nodes should be added through the interior, consistent in fineness with the
boundary mesh. It should be noted that the particular integrals are derived for each
region, independent of the other regions in the multi-region problem.

The above equations for u_i^P and similar ones for t_i^P are written at each boundary node and expressed in matrix form for each region as[17,24,26]

$$\mathbf{u}^P = \mathbf{D}\phi \tag{7.43a}$$

$$\mathbf{t}^P = \mathbf{T}\phi \tag{7.43b}$$

Equations for the initial stresses are also written at the N nodal points corresponding to the N nodes and expressed in matrix form as

$$\sigma^0 = \mathbf{K}\phi \tag{7.44}$$

or

$$\phi = \mathbf{K}^{-1}\sigma^0 \tag{7.45}$$

If the initial stresses are known Eq. (7.45) can be used to determine ϕ and hence the
particular solutions for displacements u_i^P and t_i^P. In a more general situation, however, we can backsubstitute Eq. (7.45) into Eqs (7.43a,b) to render

$$\mathbf{u}^P = \mathbf{DK}^{-1}\sigma^0 \tag{7.46a}$$

$$\mathbf{t}^P = \mathbf{TK}^{-1}\sigma^0 \tag{7.46b}$$

Similar particular integral expressions can be written for displacements and stresses at interior points of interest. Thus we can now express Eqs (7.41a,b) as[17,24,26]

$$\mathbf{Gt} - \mathbf{Fu} + \mathbf{B}\sigma^0 = \mathbf{0} \tag{7.47a}$$

and

$$\sigma = \mathbf{G}^\sigma \mathbf{t} - \mathbf{F}^\sigma \mathbf{u} + \mathbf{B}^\sigma \sigma^0 \tag{7.47b}$$

where

$$\mathbf{B} = [\mathbf{GT} - \mathbf{FD}]\mathbf{K}^{-1} \quad \text{and} \quad \mathbf{B}^\sigma = -[\mathbf{G}^\sigma \mathbf{T} - \mathbf{F}^\sigma \mathbf{D}]\mathbf{K}^{-1}$$

In a multi-region problem, the boundary integral equations and particular integrals are generated independently for each region, leading to a set of equations in each region similar in form to Eqs (7.47a,b). Interface conditions expressing the interaction of real quantities between regions are applied and, after assembly, the unknown boundary quantities and corresponding coefficients are brought on the left-hand side and the known boundary conditions to the right. The final system can then be written as[17,24,26]

$$\mathbf{A}^b \mathbf{x} = \mathbf{B}^b \mathbf{y} + \mathbf{C}^b \sigma^0 \tag{7.48a}$$

$$\sigma = \mathbf{A}^\sigma \mathbf{x} + \mathbf{B}^\sigma \mathbf{y} + \mathbf{C}^\sigma \sigma^0 \tag{7.48b}$$

Finally, careful observation of Eq. (7.39) reveals that the components of the initial stress σ_{lm}^0 are independently related to the components of the nodal density ϕ_{lm}. Moreover, each component has the same relation. Therefore, the tensor quantities of Eq. (7.39) are reduced to a scalar quantity before the inversion in Eq. (7.44) is performed.

7.4.4 Thermal Fields

The initial stress or initial strains for an arbitrary temperature field $T(x)$ is simpler since the initial strain is simply $\delta_{ij}\alpha T$, where α is the coefficient of thermal expansion. Thus, the particular solution for the displacement is given by

$$u_i^P(x) = \sum_{n=1}^{\infty} D_i(x, \xi_n)\phi(\xi_n) \tag{7.49a}$$

where

$$D_i(x, \xi_n) = kC_{,i}(x, \xi_n) = kA_0(2 - 3b_n\rho)y_i$$

and

$$k = \frac{\alpha(3\lambda + 2\mu)}{\lambda + 2\mu}$$

The temperature distribution can be related to a scalar ϕ via[17,27]

$$T(x) = \sum_{n=1}^{\infty} K(x, \xi_n)\phi(\xi_n) \tag{7.49b}$$

where

$$K(x, \xi_n) = C_{,ii}(x, \xi_n) = [2d - 3(1 + d)b_n\rho]$$

and $d = 3$ for a three-dimensional or $d = 2$ for a two-dimensional (plane strain) analysis.

A particular integral for strain can be found by substituting Eq. (7.49a) into the strain–displacement relation; thus:[17,27]

$$\varepsilon_{kl}^p(x) = \sum_{n=1}^{\infty} E_{kl}(x, \xi_n)\phi(\xi_n) \tag{7.50}$$

where

$$E_{kl}(x, \xi_n) = kC_{,kl}(x, \xi_n) = k\left[2\delta_{kl} - 3b_n\left(\delta_{kl}\rho + \frac{y_k y_l}{\rho}\right)\right]$$

This result introduced into the stress–strain law for thermoelasticity produces the particular integral for stress:[17,27]

$$\sigma_{ij}^p(x) = \sum_{n=1}^{\infty} S_{ij}(x, \xi_n)\phi(\xi_n) \tag{7.51}$$

where

$$S_{ij}(x, \xi_n) = D_{ijkl}^e E_{kl}(x, \xi_n) - \delta_{ij}\beta K(x, \xi_n)$$

$$D_{ijkl}^e = \lambda\delta_{ij}\delta_{kl} + 2\mu\delta_{ik}\delta_{jl}$$

and

$$\beta = \alpha(3\lambda + 2\mu)$$

The above equations are once again equally valid for two and three dimensions, but for two-dimensional analysis, the formulation assumes the plane strain condition. The corresponding plane stress formulation can be obtained from the plane strain case by substituting the modified material constants $\bar{\alpha}$ and $\bar{\lambda}$ into the plane strain equations in place of α and λ respectively, where

$$\bar{\alpha} = \frac{\alpha(3\lambda + 2\mu)}{2(2\lambda + \mu)}$$

$$\bar{\lambda} = \frac{2\mu\lambda}{\lambda + 2\mu}$$

7.4.5 General Remarks

If a distribution of initial stress or initial strain can be described by a simple polynomial in a manner similar to the temperature distribution approximations described in Chapter 4, the above analysis simplifies very considerably. The particular integrals for such cases can be developed in terms of explicit expressions involving only the constants of the polynomial (see Chapter 4) rather than the fictitious function ϕ, as in global shape function.

The use of the global shape function is an area that has not been fully explored in the BEM. It is possible that there may lie a family of functions that are better suited to the BEM than the distance-based approximations used here.

7.5 EIGENVALUE ANALYSIS

It is perhaps appropriate at this stage to use the global shape function with the elasticity and potential flow formulations to investigate whether such rational approximations could be introduced in an eigenvalue analysis. Eigenvalue extraction in the BEM has always been a rather unreliable and difficult procedure, arising from the fact that the frequency parameters for the eigenvalues are always non-linearly embedded into the fundamental solutions, as will be seen later in Chapter 10. However, global shape functions and the theory of particular integrals[28-33] provide us with an opportunity to investigate this problem in which all the essential equations of elasticity and potential flow can be utilized.

7.5.1 Boundary Integral Equations

For a homogeneous medium subjected to harmonic excitation in the absence of sound sources within the domain, the governing differential equation is

$$p_{,ii}(x) + \left(\frac{\omega}{c}\right)^2 p(x) = 0 \qquad (7.52a)$$

where p is the acoustic pressure, ω is the circular frequency in radians per second and c is the speed of sound. For a homogeneous elastic body subjected to harmonic excitation in the absence of body forces, the equilibrium equation is

$$(\lambda + \mu)u_{j,ij}(x) + \mu u_{i,jj}(x) + \rho\omega^2 u_i(x) = 0 \qquad (7.52b)$$

where λ, μ are Lame material constants, ρ is the density and u_i is the displacement.

We can construct particular solutions of the governing differential equations (7.52a,b) by assuming the inhomogeneous terms such as $(\omega/c)^2 p(x)$ and $\rho\omega^2 u_i(x)$ over the domain to be approximated by some other set of functions $(\omega/c)^2 \hat{p}(x)$ and $\rho\omega^2 \hat{u}_i(x)$. The above governing equations (7.52a,b) can then be rewritten as:

$$L(p) + \left(\frac{\omega}{c}\right)^2 \hat{p} = 0 \qquad (7.53a)$$

$$\mathbf{L}(u_i) + \rho\omega^2 \hat{u}_i = 0 \qquad (7.53b)$$

respectively.

Although infinitely many functions may be selected for the approximation of the forcing term, two general classes of functions are discussed in this chapter. The first type of function, which was originally used by Lord Rayleigh[25] in 1896, is based on expressing the field variable in the forcing function by an arbitrary variation in positive powers of the distance between the origin and the point ξ_m. In particular, a function that varies linearly with the distance r is constructed as

$$\hat{p}(x) = \sum_{m=1}^{M} [R - r(x, \xi_m)] \phi(\xi_m) \tag{7.54a}$$

$$\hat{u}_i(x) = \sum_{m=1}^{M} [R - r(x, \xi_m)] \phi_i(\xi_m) \tag{7.54b}$$

where $r(x, \xi_m)$ is the distance between the source point ξ_m to the field point x at which the particular solutions are to be evaluated, R is an arbitrary constant, such as the largest dimension of the problem, and $\phi(\xi_m)$ and $\phi_i(\xi_m)$ are unknown coefficients associated with the source point ξ_m.

Assuming complete polynomials of order 2, a polynomial function approximation can also be constructed as (see Chapters 3 and 4)

$$\hat{p}(x) = \sum_{m=1}^{M} K_m(x) \phi_m \tag{7.55a}$$

$$\hat{u}_i(x) = \sum_{m=1}^{M} K_m(x) \phi_{im} \tag{7.55b}$$

The functions used for the approximation of the field variable in the forcing term can be used in two different ways:

1. A suitable combination of functions may be used to approximate the forcing term throughout a sub-region or an entire region.
2. A relatively small set of functions can be associated at specific geometric locations in the region and the overall approximation is built up as a sum of these point-based functions.

7.5.2 Global Interpolation

Acoustic analysis　Particular solutions for the acoustic analysis are obtained by using the approximation given by Eq. (7.54a). The particular solution for acoustic pressure is easily determined as[30]

$$p^0(x) = \left(\frac{\omega^2}{c^2} \right) \sum_{m=1}^{M} P(x, \xi_m) \phi(\xi_m) \tag{7.56}$$

where

$$P(x, \xi_m) = (C_1 r - C_2 R) r^2$$

$$r = r(x, \xi_m) = \sqrt{y_i y_i}$$

$$y_i = y_i(x, \xi_m) = x_i - \xi_{im}$$

$$C_1 = \frac{1}{3(d+1)}$$

$$C_2 = \frac{1}{2d}$$

and d is the dimensionality of the problem; the normal derivative solution q^0 can then be easily calculated from Eq. (7.56).

Free vibration analysis   The determination of particular solutions for the free vibration analysis is facilitated by the use of the Galerkin vector g_i, which is related to the displacements by[23,34]

$$u_i^p = a g_{i,jj} - b g_{j,ij} \tag{7.57}$$

where, in terms of shear modulus μ and Poisson's ratio ν, $a = (1-\nu)/\mu$ and $b = 1/(2\mu)$. Substituting the above Galerkin vector in the governing differential equation (7.52b), we obtain

$$(1-\nu) g_{i,jjkk} + \rho \omega^2 \hat{u}_i = 0 \tag{7.58}$$

The displacement particular solution is then determined by approximating the displacement in the inertia term by Eq. (7.54b) to arrive at[29,34]

$$u_i^p(x) = \rho \omega^2 \sum_{m=1}^{M} U_{ij}(x, \xi_m) \phi_j(\xi_m) \tag{7.59}$$

where

$$U_{ij}(x, \xi_m) = (D_1 + D_2 r) \delta_{ij} r^2 + (D_3 + D_4 r) y_i y_j$$

$$D_1 = -\frac{2(d+2)(1-\nu) - 1}{60\mu(1-\nu)} R$$

$$D_2 = \frac{2(d+3)(1-\nu) - 1}{144\mu(1-\nu)}$$

$$D_3 = \frac{1}{30\mu(1-\nu)} R$$

$$D_4 = -\frac{1}{48\mu(1-\nu)}$$

Equation (7.59) can be used to determine the particular solutions for stresses and tractions in the usual manner.

7.5.3 Piece-wise Polynomial Function

Acoustic analysis   A particular solution for the acoustic pressure is easily obtained using the polynomial approximation (7.55a) in Eq. (7.53a) as[34,35]

$$p^0(x) = \left(\frac{\omega^2}{c^2}\right) \sum_{m=1}^{10} P_m(x)\phi_m \qquad (7.60)$$

where

$$
\begin{aligned}
&P_1 = A_1 x_i x_i \\
&P_m = A_2 x_i x_i x_j && m = 2, 3, 4, \quad j = m - 1 \\
&P_m = A_3 x_j^4 && m = 5, 6, 7, \quad j = m - 4 \\
&P_m = A_4 x_i x_i x_j x_k && m = 8(j = 1, k = 2), 9(j = 1, k = 3), 10(j = 2, k = 3)
\end{aligned}
$$

The constants A_k depend on the problem dimensionality d as

$$A_1 = -\frac{1}{2d} \qquad A_2 = -\frac{1}{2(d+2)} \qquad A_3 = -\frac{1}{12} \qquad A_4 = -\frac{1}{2(d+4)}$$

The normal derivative particular solution can then easily be constructed.

Free vibration analysis   Here again, the task of deriving particular solutions is simplified by the use of the Galerkin vector equation (7.57). For the representation given by Eq. (7.55b) a displacement particular solution is obtained as (the stress and traction equations are obtained in the usual manner)[34,35]

$$u_i^p(x) = \rho\omega^2 \sum_{m=1}^{M} U_{ijm}(x)\phi_{jm} \qquad (7.61)$$

where

$$
\begin{aligned}
&U_{ij1} = 12B_1[(a\delta_{ij} - b\delta_{i1}\delta_{j1})x_1^2 + (a\delta_{ij} - b\delta_{i2}\delta_{j2})x_2^2 + (a\delta_{ij} - b\delta_{i3}\delta_{j3})x_3^2] \\
&U_{ijm} = 20B_2(a\delta_{ij} - b\delta_{ik}\delta_{jk})x_k^3 && m = 2, 3, 4, \quad k = m - 1 \\
&U_{ijm} = 30B_3(a\delta_{ij} - b\delta_{ik}\delta_{jk})x_k^4 && m = 5, 6, 7, \quad k = m - 4 \\
&U_{ijm} = 3B_4\{2a\delta_{ij}(x_k^2 + x_l^2) - b[2\delta_{il}\delta_{jl}x_k^2 + 3(\delta_{ik}\delta_{jl} + \delta_{il}\delta_{jk})x_k x_l + 2\delta_{ik}\delta_{jk}x_l^2]\}x_k x_l \\
&\qquad\qquad m = 8(k = 1, l = 2), m = 9(k = 1, l = 3), m = 10(k = 2, l = 3)
\end{aligned}
$$

The constants B_k are given by

$$B_1 = -\frac{1}{24d(1-\nu)} \qquad B_2 = -\frac{1}{120(1-\nu)} \qquad B_3 = -\frac{1}{360(1-\nu)} \qquad B_4 = -\frac{1}{72(1-\nu)}$$

In both (acoustic and elastic) cases, the two-dimensional primitive function is a sub-set of the three-dimensional function and therefore particular solutions for the two-dimensional problem are obtained from the above three-dimensional solutions by setting components in the third direction to zero.

7.5.4 Formulation of the Algebraic Eigenvalue Problem

In the present analysis, the geometrical, physical and particular field variables can be approximated by isoparametric quadratic shape functions and the resulting discretized integrals are evaluated numerically. Evaluation of these boundary integrals at all surface nodes leads to the algebraic equivalent:

$$[G]\{q\} - [F]\{p\} = [G]\{q^0\} - [F]\{p^0\} \qquad (7.62a)$$

for the acoustic eigenfrequency problem, and

$$[\mathbf{G}]\{\mathbf{t}\} - [\mathbf{F}]\{\mathbf{u}\} = [\mathbf{G}]\{\mathbf{t}^0\} - [\mathbf{F}]\{\mathbf{u}^0\} \qquad (7.62b)$$

for the free vibration problem.

The particular solutions required in Eqs (7.62a,b) are given earlier; therefore, using these particular solutions, the above equations are reduced to the form (for the acoustic case, as an example)

$$[G]\{q\} - [F]\{p\} = \left(\frac{\omega}{c}\right)^2 ([G][Q] - [F][P])\{\phi\} \qquad (7.63)$$

The unknown coefficients $\{\phi\} = [\hat{K}]^{-1}\{p\}$ can now be eliminated from Eq. (7.63) and expressed in terms of physical variables as

$$[G]\{q\} - [F]\{p\} = \left(\frac{\omega}{c}\right)^2 [M]\{p\} \qquad (7.64)$$

where

$$[M] = ([G][Q] - [F][P])[\hat{K}]^{-1}$$

The matrix $[\hat{K}]$ above is obtained either by a collocation process based on Eq. (7.54a) or by a least square minimization process based on the polynomial approximation (7.55a), as described in Chapter 4.

Separation of the physical variables in terms of unknown $\{X\}$ and known $\{Y\}$ vectors leads to[29-35]

$$\left([A] - \frac{\omega^2}{c^2}[\bar{M}]\right)\{X\} = \left([B] - \frac{\omega^2}{c^2}[\hat{M}]\right)\{Y\} \qquad (7.65)$$

where the mass matrices $[\bar{M}]$ and $[\hat{M}]$ are augmented to the same dimensions as $[A]$ and $[B]$ by adding zeros appropriately. Similar equations can also be derived for the free vibration problem.

Since the applied boundary loading $\{Y\}$ is zero in the eigenfrequency analysis, Eq. (7.65), for a single-region problem, reduces to a standard generalized eigenvalue problem.[36] In a multi-region analysis both the system matrices $[A]$, $[B]$ and the mass matrices $[\bar{M}]$, $[\hat{M}]$ are assembled in exactly the same way as that used in the static analysis. In this case physical field variables at the interface between sub-regions will generally be non-zero, but the interface quantities at adjacent sub-regions are related through continuity and compatibility conditions. The eigenvalues can then be extracted as described in Wilson et al.[33] and Raveendra and Banerjee.[34,35]

7.6 EXAMPLES OF APPLICATIONS

7.6.1 Volume Integration Based on Initial Stress[3,8]

Chiu[37] has published analytical solutions for the displacement and stress fields due
to uniform initial strains in a cuboidal region within an infinite and semi-infinite
elastic solid. These solutions, which were derived by an integral transform
method, have been used to verify the accuracy of the numerical solution to the
equivalent initial stress problems. Any initial stress system is related to an initial
strain system via $\sigma_{ij}^0 = D_{ijkl}^e \varepsilon_{kl}^0$.

In order to reproduce Chiu's results the equivalent initial stresses were first
obtained from the above equation, in which Poisson's ratio of the medium was
assigned the value 0.3. The results are non-dimensionalized with respect to
Young's modulus.

The solutions obtained to the problem of the vertical displacement of a surface
point vertically above a cube embedded in the interior of an elastic half-space are
plotted against the depth of embedment in Fig. 7.3. Three cases are considered:

1. Vertical initial strains, ε_{33}^0 only.
2. Horizontal initial strains, ε_{11}^0 only.
3. Uniform initial strains in all directions $\varepsilon_{11}^0 = \varepsilon_{22}^0 = \varepsilon_{33}^0$.

This problem was analysed earlier by Banerjee *et al.*[3] using a very simple implementation. It can be seen that the earlier numerical results are in excellent agreement with Chiu's results except for the case when the cube is very close to the

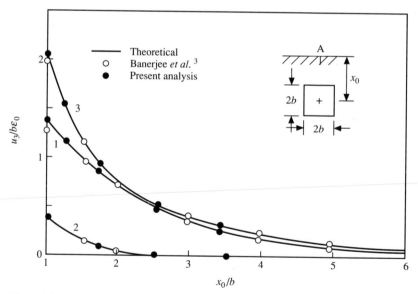

Figure 7.3 Surface displacements due to initial strains (ϵ_0) in a cube.

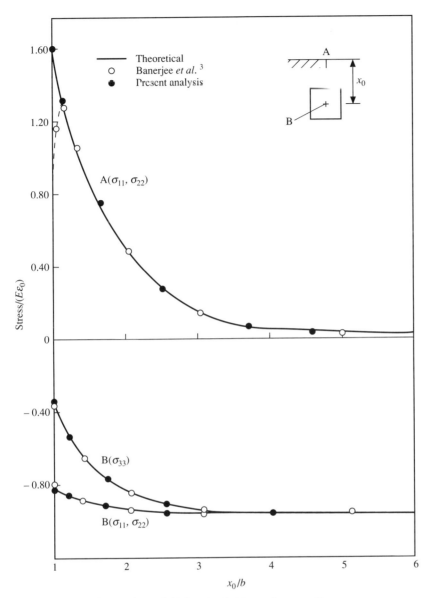

Figure 7.4 Normal stress due to initial strains (ϵ_0) in a cube—case 3.

surface. Nevertheless, the maximum error is only about 4 per cent. The results of the newer analysis[8] (see Fig. 7.4) are almost identical to those of Chiu.

For our purposes we need only consider the stresses induced at the centre of the cell and at ground level. Again, the agreement with earlier results is excellent when the cube is remote from ground level but deteriorates rapidly when the cube is just below the surface. The results are meaningless when the cube is near ground level. The newer analysis (Fig. 7.4) provides much more accurate values of the stresses.

7.6.2 Comparison of Results Based on Global Interpolation and Polynomial Function Approximations for Eigenvalue Analysis[34]

Rectangular acoustic cavity (two-dimensional) The accuracy of the acoustic eigen-frequency analysis is demonstrated by solving for the eigenfrequencies of a rectangular acoustic cavity of length (a) 40 m and height (b) 2 m. Single-region, two-region and four-region BEM models were created and acoustically rigid (that is $\partial p / \partial n = 0$) boundary conditions were imposed on all surfaces. When $a \gg b$, the behaviour of the cavity, at lower modes, corresponds to the behaviour of an equivalent one-dimensional problem. For the one-dimensional problem, the analytical eigenfrequency, in hertz, is given by

$$f = \frac{cn}{2a}$$

where c is the speed of sound which was taken as 340 m/s and n is the mode number. The problem was solved by using both the global interpolation approximation (GIA) and the polynomial function approximation (PFA). The results for the first four modes are given in Table 7.4. The computing times (all solution times are in seconds and these are obtained on an HP 9000 series 800 computer) indicate that the polynomial function approximation is considerably more efficient than the global interpolation approximation. However, to get accurate higher mode frequencies, the problem body needs to be substructured, particularly when the polynomial function approximation is used.

Automotive compartment with and without seat (two-dimensional)[30,34] The second problem considered is a passenger compartment of an automobile. The design of automobile compartments requires the resonant frequencies that cause discomfort to occupants. The specific problem studied is the compartment of a hatchback automobile that was investigated previously by Nefske *et al.*[38] using a two-dimensional finite element model and Banerjee *et al.*[30] using a BEM based on global interpolation of the forcing function. The dimensions of the acoustic compartment, in the present

Table 7.4 Acoustic eigenfrequencies (Hz) for rectangular cavity

Mode	Analytical	GIA			PFA	
		1 region	2 regions	4 regions	2 regions	4 regions
1	4.25	4.24	4.24	4.22	4.25	4.25
2	8.5	8.46	8.48	8.45	8.73	8.51
3	12.75	12.68	12.69	12.70	13.32	12.80
4	17.0	16.88	16.89	16.97	18.17	17.73
Time		148	133	142	80	97

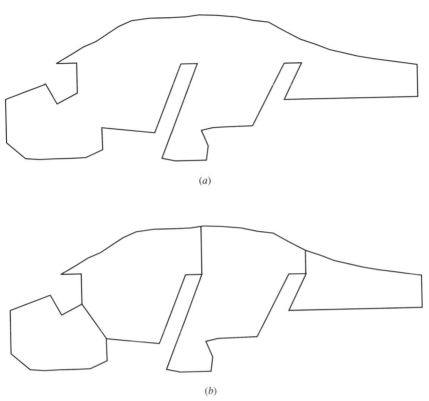

(a)

(b)

Figure 7.5 BEM model for automobile compartment with seats: (a) single-region model, (b) four-region model.

case, were taken from an automobile similar to the one used by Nefske *et al.* Both compartments without seats and with seats were considered and they were modelled by single-region and four-region boundary element models, typically, as shown in Fig. 7.5. Table 7.5 shows a comparison of finite element and experimental results with various boundary element results. According to Nefske *et al.* the discrepancy

Table 7.5 Eigenfrequencies (Hz) for a hatchback car

	Without seat					With seat				
			1 region	4 regions				1 region	4 regions	
Mode	Experiment	FEM	GIA	GIA	PFA	Experiment	FEM	GIA	GIA	PFA
1	60	68	69	72	73	53	50	48	49	49
2	110	105	104	110	111	—	79	75	78	78
3	135	152	153	155	155	—	125	116	122	129
4	—	179	190	186	193	—	163	157	159	166
Time			109	167	103			266	265	155

between the finite element results and the experimental data was due to the wall flexibility and the structural acoustic coupling. However, the boundary element results agree well with the finite element results. In all cases, the results indicate that the resonant frequencies are reduced when seats are included.

Fixed-end arch with and without opening (two-dimensional)[29,34,35] This example is concerned with the evaluation of eigenvalues of a fixed-end arch. Two different cases were studied; in the first case the fixed-end arch was considered without opening and in the second case the arch was considered to have rectangular openings. The material properties were taken as $E = 10^8$ units, $\rho = 1.0$ unit and $\nu = 0.2$. Single- and multi-region boundary element models were used; the six-region models of the arch without the openings and with the openings are shown in Fig. 7.6a,b respectively. The first four modes based on global interpolation and

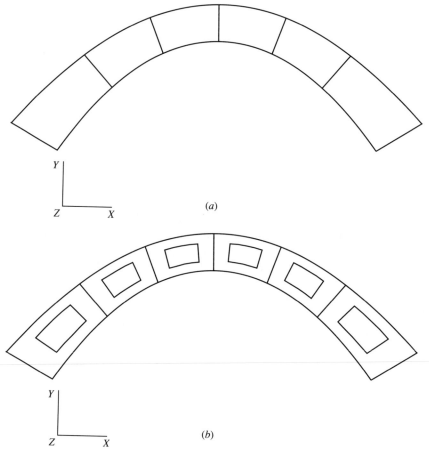

Figure 7.6 Multi-region BEM model for fixed-end arch: (*a*) without opening, (*b*) with opening.

Table 7.6 Natural frequencies (Hz) for fixed-end arch

(a) Without opening

Mode	1 region GIA	3 regions GIA	3 regions PFA	6 regions GIA	6 regions PFA
1	89.3	89.0	90.0	87.6	89.8
2	127.6	126.8	127.8	124.4	127.9
3	178.4	179.5	177.1	176.8	180.4
4	236.2	235.9	239.4	231.8	237.6
Time	155	165	122	255	219

(b) With opening

Mode	1 region GIA	3 regions GIA	3 regions PFA	6 regions GIA	6 regions PFA
1	58.4	56.7	57.6	56.1	56.8
2	100.0	97.2	101.9	97.9	100.1
3	115.1	114.9	116.4	114.2	115.3
4	145.2	144.7	159.0	144.6	150.4
Time	1214	1460	544	946	703

polynomial function approximations of the inertia force, for various boundary element models, are shown in Table 7.6a and b. The global interpolation solutions and the polynomial function results using sub-structured models agree well with each other.

7.7 CONCLUDING REMARKS

Integral formulations for volume effects resulting from initial stresses and initial strains have been treated using both volume integration and particular integral methods. It was shown that general volume effects which could not be completely converted into a surface integral could still be treated by the BEM. For eigenfrequency and free vibration problems, by treating the forcing function in the governing differential equation as an initially unknown distributed source, approximating the forcing function by global interpolation and polynomial function representations and utilizing the particular solutions of the governing inhomogeneous differential equations, surface-only integral equations are derived. The numerical examples presented indicate the applicability of the BEM techniques for the solution of realistic engineering problems. The results further indicate that the solution technique based on a polynomial function approximation of the forcing function is computationally more efficient than the solution based on the global interpolation procedure; however, to obtain accurate results the problem domain needs to be sub-structured, particularly when polynomial functions are used for the approximation.

REFERENCES

1. Reisner, H. (1931) Initial stresses and sources of initial stresses (in German), *Zeitschrift Agnew Mathematik und Mechanik*, Vol. 11, pp. 1–8.
2. Banerjee, P.K. and Butterfield, R. (1981) *Boundary Element Methods in Engineering Science*, McGraw-Hill, London.
3. Banerjee, P.K., Cathie, D.N. and Davies, T.G. (1979) Two and three-dimensional problems of elastoplasticity, Chapter 4 in *Developments of Boundary Element Methods—Vol. I*, pp. 65–95, Elsevier Applied Science Publishers, London.
4. Banerjee, P.K. and Davies, T.G. (1979) Analysis of case histories of laterally loaded pile groups, in *Numerical Methods of Offshore Piling*, pp. 101–108, Institution of Civil Engineers, London.
5. Banerjee, P.K. and Cathie, D.N. (1980) A direct formulation and numerical implementation of the boundary element method for two-dimensional problems of elastoplasticity, *International Journal of Mechanical Science*, Vol. 22, pp. 233–245.
6. Cathie, D.N. and Banerjee, P.K. (1980) Numerical solutions in axisymmetric elastoplasticity by the boundary element method, in *Innovative Numerical Analysis for Engineering Science*, Eds R.P. Shaw *et al.*, pp. 331–340, University of Virginia Press, Charlottesville.
7. Bui, H.D. (1978) Some remarks about the formulation of three-dimensional thermoelastoplastic problems by integral equations, *International Journal of Solids and Structures*, Vol. 14, pp. 935–939.
8. Banerjee, P.K. and Davies, T.G. (1984) Advanced implementation of the boundary element methods for three-dimensional problems of elastoplasticity and viscoplasticity, in *Development in Boundary Element Methods*, Vol. 3, Eds P.K. Banerjee and S. Mukherjee, Elsevier Applied Science Publishers, London.
9. Raveendra, S.T. (1984) Advanced development of BEM for two and three-dimensional inelastic analysis, PhD dissertation, State University of New York at Buffalo.
10. Lu, S. and Ye, T.Q. (1991) Direct evaluation of singular integrals in elastoplastic analysis by BEM, *International Journal of Numerical Methods in Engineering*, Vol. 32, pp. 295–311.
11. Mukherjee, S. (1977) Corrected boundary integral equations in planar thermoelastoplasticity, *International Journal of Solids and Structures*, Vol. 13, pp. 331–335.
12. Telles, J.C.F. (1983) *The Boundary Element Method Applied to Inelastic Problems*, Springer-Verlag, New York.
13. Dallner, R. and Kuhn, G. (1992) Efficient evaluation of volume integrals in BEM, *Computer Methods in Applied Mechanics and Engineering* (unpublished work).
14. Okada, H. and Atluri, S.N. (1988) Some recent developments in finite strain elastoplasticity, *International Journal of Numerical Methods in Engineering*, Vol. 23, pp. 985–1002.
15. Banerjee, P.K. and Raveendra, S.T. (1986) Advanced boundary element analysis of two- and three-dimensional problems of elastoplasticity, *International Journal of Numerical Methods in Engineering*, Vol. 23, pp. 985–1002.
16. Mukherjee, S. (1982) *Boundary Element Methods in Creep and Fracture*, Applied Science Publishers, London.
17. Henry (Jr), D.P. (1987) Advanced development of the boundary element method for elastic and inelastic thermal stress analysis, PhD Dissertation, State University of New York at Buffalo, Buffalo, New York.
18. Henry (Jr), D.P. and Banerjee, P.K. (1987) A thermoplastic BEM analysis for substructured axisymmetric bodies, *Journal of Engineering, Mechanical Division ASCE*, Vol. 113, pp. 1880–1900.
19. Bakr, A.A., Abdul-Mihsein, J.J. and Fenner, R.T. (1986) Isoparametric quadratic boundary element in 2D, axisymmetric and 3D stress analysis problems, in *BEM Theory and Application*, pp. 97–110. A meeting of the Stress Analysis Group of the Institute of Physics, Techno House, Bristol.
20. Mustoe, G.G. (1984) Advanced integration schemes over boundary elements and volume cells for two and three-dimensional nonlinear analysis, in *Developments in Boundary Element Methods*, Vol. 3, Eds P.K. Banerjee and S. Mukherjee, Elsevier Applied Science Publishers, London.
21. Watson, J.O. (1979) Advanced implementation of the boundary element methods for two- and

three-dimensional elastostatics, Chapter 3 in *Developments in Boundary Element Methods*—Vol. I, pp. 31–63, Elsevier Applied Science Publishers, London.

22. Ricardella, P. (1973) An implementation of the boundary integral technique for plane problems of elasticity and elastoplasticity, PhD Thesis, Carnegie-Mellon University.

23. Fung, Y.C. (1965) *Foundation of Solid Mechanics*, pp. 192–193, Prentice-Hall, Englewood Cliffs, New Jersey.

24. Henry, D.P. and Banerjee, P.K. (1988) A new boundary element formulation for two- and three-dimensional elastoplasticity using particular integrals, *International Journal of Numerical Methods in Engineering*, Vol. 26, pp. 2079–2096.

25. Rayleigh, J.W.S. (1896) *The Theory of Sound*, Macmillan, London.

26. Banerjee, P.K. and Henry, D.P. (1988) Advanced applications of BEM to inelastic analysis of solids, Chapter 2 in *Industrial Applications of BEM*, Eds P.K. Banerjee and R.B. Wilson, pp. 39–75, Elsevier Applied Science, London.

27. Henry, D.P. and Banerjee, P.K. (1988) A new boundary element formulation for two- and three-dimensional thermoelasticity using particular integrals, *International Journal of Numerical Methods in Engineering*, Vol. 26, pp. 2061–2077.

28. Nardini, D. and Brebbia, C.A. (1982) A new approach to free vibration analysis using boundary elements, in *Proceedings of 4th International Conference on BEM*, Ed. C.A. Brebbia, pp. 313–326, Springer-Verlag, Berlin.

29. Ahmad, S. and Banerjee, P.K. (1986) Free vibration analysis of BEM using particular integrals, *Journal of Engineering Mechanics, ASCE*, Vol. 112, pp. 682–695.

30. Banerjee, P.K., Ahmad, S. and Wang, H.C. (1988) A new BEM formulation of the acoustic eigen-frequency analysis, *International Journal of Numerical Methods in Engineering*, Vol. 26, pp. 1299–1309.

31. Wang, H.C. and Banerjee, P.K. (1988) Axisymmetric free-vibration problems by the boundary element method, *Journal of Applied Mechanics, ASME*, Vol. 55, pp. 437–442.

32. Wang, H.C. and Banerjee, P.K. (1990) Free vibration of axisymmetric solids by BEM using particular integrals, *International Journal of Numerical Methods in Engineering*, Vol. 29, pp. 985–1001.

33. Wilson, R.B., Miller, N.M. and Banerjee, P.K. (1990) Free-vibration analysis of three-dimensional solids by BEM, *International Journal of Numerical Methods in Engineering*, Vol. 29, pp. 1737–1757.

34. Raveendra, S.T. and Banerjee, P.K. (1992) Eigenvalue analysis by BEM, Chapter 8 in *Advanced Dynamic Analysis by BEM*, Eds P.K. Banerjee and S. Kobayashi, pp. 282–320, Elsevier Applied Science, London.

35. Raveendra, S.T. and Banerjee, P.K. (1992), BEM analysis of free-vibration problems using polynomial particular integrals, *International Journal of Solids and Structures*, Vol. 29, No. 16, pp. 2023–2037.

36. Garbow, B.S. (1980) EISPACK—for the real generalized eigenvalue problems, Report, Applied Mathematics Division, Argonne National Laboratory, USA.

37. Chiu, Y.P. (1978) On the stress field and surface deformation in a half space with cuboidal zone in which initial strains are uniform, *Journal of Applied Mechanics*, Vol. 45, pp. 302–306.

38. Nefske, D.J., Wolf, J.A. and Howell, L.J. (1982) Structural-acoustic finite element analysis of the automobile passenger compartment: a review of current practice, *Journal of Sound and Vibration*, Vol. 80, pp. 247–266.

EIGHT

TRANSIENT POTENTIAL FLOW (DIFFUSION) PROBLEMS

8.1 INTRODUCTION

Previous chapters have all been concerned with 'steady state' systems—those in which neither the problem variables nor boundary conditions change with time. However, a great many problems of practical importance do involve 'transient' phenomena, the simplest of which is a large group governed by the linear 'diffusion' equation. In addition to the classical diffusion of gases and liquids the topics of most interest to the engineering analyst might be the heating and cooling of bodies, electrical and hydraulic diffusion phenomena and the transient flow through porous media.

The major reference source for analytical solutions, Green's functions, etc., for the diffusion equation (in the terminology of heat transfer) is Carslaw and Jaeger's well-known book.[1] There is also an extensive numerical solution literature which can be categorized by the manner in which the time-dependent terms in the equation are dealt with, irrespective of the underlying method of analysis. The two basic methods used are either (1) a 'time-marching' process in which, step by step, the solution is evaluated at successive time intervals following an initially specified state or (2) by taking the Laplace transform of the time variable under which the (parabolic) diffusion equation becomes an elliptic one, resembling Poisson's equation, which can be solved in the transform space by the techniques described in previous chapters.

It has been nearly twenty years since Rizzo and Shippy[2] first applied boundary integral methods for transient heat conduction analysis. Their approach involved a solution in the Laplace transform domain, followed by a numerical inversion.

Later, Tomlin[3] utilized a more convenient time domain indirect integral method for the analogous problem of transient groundwater flow. This work by Tomlin examined two-dimensional anisotropic, multi-zone media. Meanwhile, Chang *et al.*[4] developed the first time-domain direct boundary integral method for planar transient heat conduction. Once again anisotropy was included. Three-dimensional bodies were initially investigated by Shaw.[5] More recently, Banerjee, Butterfield and Tomlin[6-9] presented solutions of more complex two-dimensional problems, while Wrobel and Brebbia[10] addressed axisymmetric analysis.

Two-dimensional, axisymmetric and three-dimensional problems are all considered in this chapter within a consistent framework, which utilizes a time-domain direct BEM. A convolution approach described here eliminates the need for volume discretization. This not only provides a considerable reduction in the manpower requirements for modelling, but also permits more precise solutions than any of the popular domain-based methods. In addition, a number of new capabilities is introduced for problems of transient heat conduction. Included in this category is the specification of thermal resistance across an interface, an accurate method for the calculation of the strongly singular terms in the flux system matrix, and most significantly an efficient treatment of multiple cooling holes embedded in three-dimensional bodies.[11-14]

8.2 GOVERNING EQUATIONS

The flow of heat in a homogeneous, isotropic body is governed by the familiar differential equation[1,15]

$$k\theta_{,ii} - \rho c_\epsilon \dot{\theta} + \psi = 0 \tag{8.1}$$

in which θ represents the temperature, k the conductivity, ρ the mass density, c_ϵ the specific heat at constant strain and ψ the body sources. Indicial notation is employed. Thus, commas represent differentiation with respect to spatial coordinates, while a superposed dot denotes a time derivative. In addition to Eq. (8.1), which is often called simply the diffusion equation, suitable boundary and initial conditions must be specified. In particular, for all points x on S, for all times t, either

$$\theta = \Theta(x, t) \tag{8.2a}$$

or
$$q = Q(x, t) \tag{8.2b}$$

or
$$q = H(x, t)[\Theta_{amb}(x, t) - \theta] \tag{8.2c}$$

where q is the heat flux normal to the surface S, $H(x, t)$ is a convection coefficient and $\Theta_{amb}(x, t)$ is an ambient fluid temperature. Meanwhile, the initial conditions

$$\theta_0 = \theta(x, 0) \tag{8.3}$$

must be given for all points x in the domain V of the body. Complete details on the theory of transient heat conduction can be found in the classic text by Carslaw and Jaeger.[1]

8.3 BOUNDARY INTEGRAL FORMULATION

The desired boundary integral equation can be derived directly from the governing differential equation (8.1) or by utilizing the well-known reciprocal theorem developed below using integration by parts. Clearly, Eq. (8.1) must hold for all points of the body at each instant of time. Therefore, the left-hand side of Eq. (8.1) multiplied by an arbitrary function, say $\tilde{g}$, and integrated over time and space must also equal zero; i.e.[9,13]

$$\int_0^T \int_V \tilde{g}(k\theta_{,ii} - \rho c_\epsilon \dot{\theta} + \psi) \, dV \, dt = 0 \tag{8.4}$$

Applying integration by parts, repeatedly, to the applicable terms in Eq. (8.4), both the spatial and temporal derivatives can be transferred from θ to $\tilde{g}$. The result can be written as

$$\int_0^T \int_S (\tilde{g}k\theta_{,i} n_i - k\tilde{g}_{,i} n_i \theta) \, dS \, dt - \int_V (\tilde{g}\rho c_\epsilon \theta|_0^T) \, dV + \int_0^T \int_V \tilde{g}\psi \, dV \, dt$$

$$+ \int_0^T \int_V (k\tilde{g}_{,ii} + \rho c_\epsilon \dot{\tilde{g}})\theta \, dV \, dt = 0 \tag{8.5}$$

with n_i defined as the outer normal to the surface S at x. Now if $\tilde{g}$ is selected to be the fundamental solution to the adjoint heat equation then the last volume integral in Eq. (8.5) can be collapsed to a point. Thus, let $\tilde{g} = \tilde{g}(x, t; \xi, \tau)$ satisfy

$$k\tilde{g}_{,ii} + \rho c_\epsilon \dot{\tilde{g}} + \delta(x - \xi)\delta(t - \tau) = 0 \tag{8.6}$$

in which the generalized Delta function has been introduced to represent a unit pulse source applied at the point ξ at time τ. Also, imposed upon $\tilde{g}$ is the causality condition for the adjoint heat equation

$$\tilde{g}(x, t; \xi, \tau) = 0 \qquad \text{for } t > \tau \tag{8.7}$$

Figure 8.1 shows a typical two-dimensional potential flow problem which has been represented as a three-dimensional one by introducing a third coordinate axis, say $x_3 = t, \xi_3 = \tau$. The above statement of causality implies that the discretized events along this third axis are not retroactive, i.e. sources applied at time τ can only affect events at later times.

Then for ξ interior to S and for $\tau < T$, Eq. (8.5) becomes

$$\theta(\xi, \tau) = \int_0^\tau \int_S \left[-\tilde{g}(x, t; \xi, \tau) q(x, t) + \tilde{f}(x, t; \xi, \tau) \theta(x, t) \right] dS(x) \, dt$$

$$+ \int_V [\tilde{g}(x, 0; \xi, \tau)\rho c_\epsilon \theta(x, 0)] \, dV(x)$$

$$+ \int_0^\tau \int_V [\tilde{g}(x, t; \xi, \tau) \psi(x, t)] \, dV(x) \, dt \tag{8.8}$$

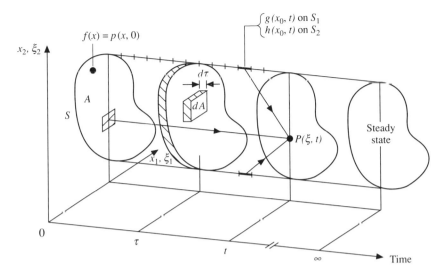

Figure 8.1 Spatial and time domains for a plane diffusion problem.

where

$$q(x, t) = -k \frac{\partial \theta(x, t)}{\partial x_i} n_i(x) \tag{8.9a}$$

$$\tilde{f}(x, t; \xi, \tau) = -k \frac{\partial \tilde{g}(x, t; \xi, \tau)}{\partial x_i} n_i(x) \tag{8.9b}$$

However, the adjoint fundamental solution $\tilde{g}$ can be related to the well-known fundamental solution g of the heat equation via

$$\tilde{g}(x, t; \xi, \tau) = g(\xi, \tau; x, t) \tag{8.10}$$

where

$$g(\xi, \tau; x, t) = \frac{\exp\{-[(x_i - \xi_i)(x_i - \xi_i)]/[4\kappa(\tau - t)]\}}{[4\pi\kappa(\tau - t)]^{n/2}} \tag{8.11}$$

with the thermal diffusivity defined by

$$\kappa = \frac{\rho c_\epsilon}{k} \tag{8.12}$$

and n equal to the number of spatial dimensions. Note that $g(\zeta, \tau; x, t)$ is symmetric with respect to x and ξ, but not in terms of t and τ. As a result, Eq. (8.8) can be rewritten as

$$\theta(\xi, \tau) = \int_S (-g * q + f * \theta) \, dS + \int_V (g_0 \rho c_\epsilon \theta_0) \, dV + \int_V (g * \psi) \, dV \tag{8.13}$$

in which f is the flux kernel, where

$$f(\xi, \tau; x, t) = -k \frac{\partial g(\xi, t; x, t)}{\partial x_i} n_i(x) \tag{8.14a}$$

while

$$g_0 \equiv g(\xi, \tau; x, 0) \tag{8.14b}$$

and the symbol $*$ represents a Riemann convolution integral that is, for example,

$$g * q \equiv \int_0^\tau g(\xi, \tau; x, t) q(x, t) \, dt \tag{8.15}$$

For a point ξ on the bounding surface S, the kernel functions $g(\xi, \tau; x, t)$ and $f(\xi, \tau; x, t)$ are singular at $x = \xi$ at $t = \tau$. Consequently, Eq. (8.13) must be modified, producing

$$c(\xi)\theta(\xi, \tau) = \int_S (-g * q + f * \theta) \, dS + \int_V (g_0 \rho c_\epsilon \theta_0 + g * \psi) \, dV \tag{8.16}$$

where $c(\xi)$ depends only upon the local geometry of the body at ξ.

Equation (8.16) is the desired boundary integral equation. For most practical problems, volume discretization is completely eliminated. In fact, from Eq. (8.16) it can be seen that volume integration is only needed for problems involving non-zero initial temperature distributions or internal body sources. The latter, if present, are quite often of a local nature which can be easily handled via analytical integration or particular integrals (see Chapter 7). These source terms are assumed absent in the remainder of this chapter. Additionally, when the prescribed initial temperature field corresponds to any steady state distribution, then Eq. (8.16) can be reformulated as

$$c(\xi)\bar{\theta}(\xi, \tau) = \int_S (-g * \bar{q} + f * \bar{\theta}) \, dS \tag{8.17}$$

where

$$\bar{\theta}(x, t) = \theta(x, t) - \theta(x, 0) \tag{8.18a}$$
$$\bar{q}(x, t) = q(x, t) - q(x, 0) \tag{8.18b}$$

Presumably, both $\theta(x, 0)$ and $q(x, 0)$ are known everywhere on S from a previous steady state analysis. Equation (8.17), once again, involves only surface quantities.

8.4 INDIRECT FORMULATION

The IBEM formulation can be deduced from the direct one, and thereby established rigorously, by using a device once again originally due to Lamb.[16]

Consider a complementary region (V^+) exterior to V with a common boundary S, no internal source $(\psi^+ = 0)$ and zero initial values (that is $\theta_0^+ = 0$) (Fig. 8.2). If the boundary distribution of (θ, q) over the common surface (S) of

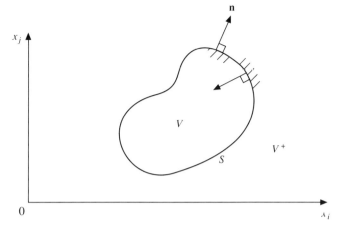

Figure 8.2 Interior and exterior regions and boundary normals.

V^+ are prescribed as (θ^+, q^+), say, then from Eq. (8.17), in V^+,

$$\alpha^+ \theta(\xi, t) = \int_S (-f * \theta^+ - g * q^+) \, dS \qquad (8.19)$$

where the sign has changed on f since the outward normal to S and V^+ is opposed to that of S and V (Fig. 8.2). Writing both Eqs (8.19) and (8.17) for any point within V leads to

$$0 = \int_S (-f * \theta^+ - g * q^+) \, dS$$

and

$$\theta(\xi, t) = \int_S (f * \theta - g * q) \, dS \qquad (8.20)$$

We may choose (θ^+, q^+) such that $\theta = \theta^+$ and define $(q + q^+) = -\phi$, a new boundary source density distribution, which produces, when Eqs (8.20) are added together,

$$\theta(\xi, t) = \int_S (g * \phi) \, dS \qquad (8.21)$$

Since g is symmetric in x and ξ we can interchange them, which means that integration is now carried out with respect to ξ and τ and Eq. (8.21), when written out fully, becomes

$$\theta(x, t) = \int_S [g(x, t; \xi, \tau) * \phi(\xi, \tau)] \, dS \qquad (8.22)$$

Differentiating Eq. (8.22) with respect to x and using Eq. (8.9a) we can calculate the n direction flux at x_i as

$$q(x, t) = \int_S (F * \phi) \, dS \qquad (8.23)$$

with arguments identical to Eq. (8.22).

Finally, the IBEM formulation is generated by taking x to x_0 on S, from inside V, as

$$p(x_0, t) = \int_S (G * \phi) \, \mathrm{d}S \tag{8.24a}$$

$$q(x_0, t) = \frac{\omega}{4\pi} \phi(x_0, t) + \int_S (F * \phi) \, \mathrm{d}S \tag{8.24b}$$

In both the direct and indirect BEM statements the only changes from the steady state formulation are:

1. One additional dimension (time) of integration is involved [expressed here by the convolutions $(F * \phi)$, etc.].
2. There is a corresponding increased dimensionality of the singular solution.
3. Additional contributions occur on the right-hand side of the final equations generated by the initially specified values of the potential $\theta(x, 0) = \theta_0(x)$ throughout the region.

8.5 NUMERICAL SOLUTION

8.5.1 Solution by Laplace Transform

The most powerful analytical methods for solving the diffusion equation (and, in fact, other classes of problem involving convolution integrals) are based on Laplace transform of the time coordinate. Some authors, principally Rizzo and Shippy,[2] have explored the use of this technique in conjunction with the BEM. The key attraction is that when the time coordinate is so transformed the dimensionality of the differential equation is effectively reduced by one to produce an equivalent elliptic equation.

The major steps involved require:

1. The Laplace transform of a function $\theta(x, t)$ is defined as $\theta^*(x, s)$ where

$$\theta^*(x, s) \equiv \mathcal{L}[\theta(x, s)] = \int_0^\infty \theta(x, t) \, \mathrm{e}^{-st} \, \mathrm{d}t \tag{8.25}$$

By operating on the diffusion equation (8.1) and the boundary conditions by $\mathcal{L}$ we obtain

$$-k \frac{\partial^2 \theta^*(x, s)}{\partial x_i \, \partial x_i} - \rho c_\varepsilon s \theta^*(x, s) = \psi^*(x, s) \tag{8.26a}$$

and

$$A\theta^* + Bq^* = C^* \tag{8.26b}$$

It may well prove to be very difficult to evaluate the transform of $\psi(x, t)$ and we only consider the $\psi = 0$ case.

2. Since Eq. (8.26a) resembles the steady state potential flow we can easily obtain the required integral representation and the fundamental solution. The

necessary integral representation can be expressed for the DBEM as

$$c(\xi)\theta^*(\xi) = \int_S [f^*(x,\xi,s)\theta^*(x,s) - g^*(x,\xi,s)q^*(x,s)]\,dS(x) \tag{8.27}$$

where

$$q^*(x,s) = -k\frac{\partial\theta^*}{\partial x_i}n_i(x) \tag{8.28}$$

$$g^*(x,\xi,s) = \frac{1}{4\pi kr}\exp\left(-r\sqrt{\frac{s}{\kappa}}\right) \qquad \text{for three-dimensions} \tag{8.29a}$$

$$g^*(x,\xi,s) = \frac{1}{2\pi k}K_0\left(r\sqrt{\frac{s}{\kappa}}\right) \qquad \text{for two dimensions} \tag{8.29b}$$

where K_0 is the modified Bessel function of the second kind.[17]

Utilizing Eqs (8.29a,b) and (8.28) we obtain the corresponding flux kernels as

$$f^*(x,\xi,s) = -k\frac{\partial g^*(x,\xi,s)}{\partial x_i}n_i(x) = \frac{1}{4\pi r^2}\frac{y_i n_i}{r}\left(1 + r\sqrt{\frac{s}{\kappa}}\right)\exp\left(-r\sqrt{\frac{s}{\kappa}}\right) \tag{8.30}$$

and $\quad f^*(x,\xi,s) = \frac{1}{2\pi r}\frac{y_i n_i}{r}\left(r\sqrt{\frac{s}{\kappa}}\right)K_1\left(r\sqrt{\frac{s}{\kappa}}\right) \tag{8.31}$

and $\qquad \kappa = k/\rho c_\epsilon$

3. By using the techniques developed in Chapter 3 we can discretize and solve Eq. (8.27) with transformed boundary conditions (8.26b) for several values of the Laplace transform parameters. The real solutions for $\theta(x,t)$ and $q(x,t)$ can then be obtained by a numerical inversion of the Laplace transform.

Unfortunately the numerical inversion is essentially a curve-fitting process which often becomes dependent on the solution. As yet there exists no universal numerical method for doing this inversion very accurately. Nevertheless, the above procedure is very attractive from a programming point of view.

8.5.2 Time-marching Processes

The essence of such processes is to 'march', from $t = 0$, step by step using time increments $\Delta\tau$ to any specified time $\tau = N\Delta\tau$ in N such steps. One fundamental question is immediately obvious. Are there limits on the magnitude of $\Delta\tau$, since, from the point of view of computational efficiency, a few large steps would be preferable to many small ones?

The published information on admissible $\Delta\tau$ (dimensionless time) relates mainly to finite difference methods (Crandall[18] gave a particularly good account of the basic problem) and finite element methods for which many provide very interesting information on the use of higher-order approximations. Using whole body discretization schemes and the conventional explicit forward difference marching process the maximum value of the ratio $\Delta\tau/(\Delta x)^2 = \frac{1}{2}$ to ensure convergence, Δx being the mesh or element dimension.

However, an examination of Eqs (8.16) and (8.17) will show that the BEM formulations are implicit in time (i.e. the various quantities at time $t = N\Delta\tau$ are calculated from the boundary integrals involving the known and unknown boundary values and known sources in the interior up to time t). Therefore, if in the discretized system we do not introduce gross approximations in the time domain the criteria governing stability should be much less stringent than that discussed above.

Two basically different time-marching processes can be used, both of which lead to systems of equations that are simultaneous in space but successive in time.

Method 1 (convolution) In both the direct and the indirect formulations established above the treatment of the time variable is virtually identical to that of the spatial variables. Therefore, as depicted in Fig. 8.1, it should be possible to treat a planar diffusion problem as one in 'three dimensions' (the third being time) and proceed step by step to a solution at time t. The basic boundary element philosophy would lead us to discretize the whole spatial and temporal boundary (Fig. 8.1) and determine all the unknown boundary information, from which any $\theta(\xi, t)$, etc., could be calculated, bearing in mind that sources introduced at $\tau > t$ have no retroactive effect at $\tau \leq t$.

We shall demonstrate all the important features of this algorithm by solving a general one-dimensional problem using the DBEM. Consider the (x, t) plane of Fig. 8.3 and a uniform one-dimensional field ($L = 1$) extending from $x = 0$ to $x = 1$, with $\kappa =$ unity and $\theta_0(x, 0)$ and the initially specified potential along it. The other boundaries of our problem are lines parallel to the time axis and, as

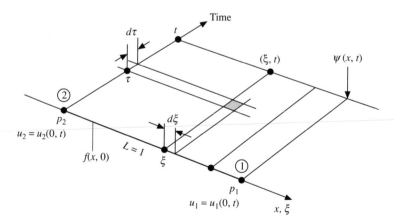

Figure 8.3 Definition of domains and boundary conditions for the 'one-dimensional' example.

an example, we shall consider the potentials (θ_1, θ_2) to be constant and specified along them. We realize immediately that the boundary fluxes (q_1, q_2) are thereby both (unknown) functions of time. By analogy with Eq. (1.8) the DBEM statement (8.16) now becomes

$$\theta(\xi, t) = -[g * q - f * \theta]_0^L + \int_0^L (\theta_0 g + g * \psi) \, dx \tag{8.32}$$

where, from Eq. (8.11),

$$g = \frac{\exp[-r^2/4(t-\tau)]}{2\sqrt{\pi}(t-\tau)^{1/2}} \qquad f = \frac{g}{2(t-\tau)} r \operatorname{sgn} r \qquad r = x - \xi \tag{8.33}$$

Since (θ_1, θ_2) are constant with time we can evaluate the $(f * \theta_1)$ and $(f * \theta_2)$ convolutions analytically as

$$(f * \theta_1) = \theta_1 r \operatorname{sgn} r \int_0^t \frac{\exp[-r^2/4(t-\tau)]}{4\sqrt{\pi}(t-\tau)^{3/2}} \, d\tau = \frac{\theta_1}{2} (\operatorname{sgn} r) \operatorname{erfc} \left| \frac{r}{2\sqrt{t}} \right| \tag{8.34}$$

where $\operatorname{erfc}(z) = 1 - \operatorname{erf}(z)$ and erf is the error function defined by

$$\operatorname{erf}(z) = \frac{2}{\sqrt{\pi}} \int_0^z e^{-\tau^2} d\tau$$

If $q_1 = $ constant in time, say, had been specified in lieu of θ_1, then convolutions such as $(g * q_1)$ can be evaluated analytically as

$$(g * q_1) = q_1 \int_0^t \frac{\exp[-r^2/4(t-\tau)]}{2\sqrt{\pi}(t-\tau)^{1/2}} = q_1 \left[\sqrt{\frac{t}{\pi}} \exp\left(-\frac{r^2}{4t}\right) - \frac{|r|}{2} \left(\operatorname{erfc} \left| \frac{r}{2\sqrt{t}} \right| \right) \right] \tag{8.35}$$

For convenience consider $\psi = 0$ and $\theta_0 = $ constant $= $ unity, say, when the final term in Eq. (8.32) becomes, using $g(x, t|\xi, 0)$ from Eq. (8.33),

$$\int_0^L g \, dx = \int_{-\xi}^{L-\xi} \frac{\exp(-r^2/4t)}{2\sqrt{\pi} t^{1/2}} \, dr = 2 \left[\operatorname{erf}\left(\frac{L-\xi}{2\sqrt{t}}\right) + \operatorname{erf}\left(\frac{\xi}{2\sqrt{t}}\right) \right] \tag{8.36}$$

Then, from Eqs (8.32), (8.34) and (8.36) we have, taking ξ to the boundaries at $(L^-, 0^+)$ in turn,

$$\begin{Bmatrix} \theta(L^-, t) \\ \theta(0, t) \end{Bmatrix} = \begin{Bmatrix} \theta_1 \\ \theta_2 \end{Bmatrix}$$

$$= \begin{bmatrix} -g_1 & g_2 \\ -g_2 & g_1 \end{bmatrix} * \begin{Bmatrix} q_1 \\ q_2 \end{Bmatrix} + \begin{bmatrix} \frac{1}{2} & \frac{1}{2} \operatorname{erf}(0.5L/\sqrt{t}) \\ \text{symmetric} & \frac{1}{2} \end{bmatrix} \begin{Bmatrix} \theta_1 \\ \theta_2 \end{Bmatrix} + 2 \begin{Bmatrix} 1 \\ 1 \end{Bmatrix} \operatorname{erf}\left(\frac{L}{2\sqrt{t}}\right) \tag{8.37}$$

where $g_1 = g(L, t; L^-, \tau)$ and $g_2 = g(0, t; L^-, \tau)$ which, from the spatial symmetry of g, are respectively identical to $g(0, t; 0^+, \tau)$ and $g(L, t; 0^+, \tau)$.

We can now see the vital point that, whereas the final terms are similar to that in the steady state problem described in Chapter 1, including the $(\frac{1}{2})$ coefficients multiplying (θ_1, θ_2), the (g, q) products are convolutions. Consequently we have to discretize the time axis, as expected, so that each of the $(g * q)$ terms can be expressed as a matrix product of, say, n discrete boundary element subdivisions of both q_1 and q_2 within the time interval $(0, t)$ of interest.

At the Nth time interval we can write the $(g * q)$ terms as

$$\int_0^{t=N\Delta\tau} g(x, t; \xi, \tau)q(x, \tau)\,d\tau = \int_0^{(N-1)\Delta\tau} g(x, t; \xi, \tau)q(x, \tau)\,d\tau$$

$$+ \int_{(n-1)\Delta\tau}^{N\Delta\tau} g(x, t; \xi, \tau)q(x, \tau)\,d\tau \qquad (8.38)$$

in which the first integral on the right-hand side involves the known information from the solution involving the $(N-1)$th time step. Therefore, if we assume that (q_1, q_2) remain constant over the time step $\Delta\tau$ we can then express Eq. (8.37) as

$$\left(\int_{(N-1)\Delta\tau}^{N\Delta\tau} \begin{bmatrix} g_1 & -g_2 \\ g_2 & -g_1 \end{bmatrix} d\tau \right) \left\{ \begin{array}{c} q_1 \\ q_2 \end{array} \right\} = \int_0^{(N-1)\Delta\tau} \left(\begin{bmatrix} -g_1 & g_2 \\ -g_2 & g_1 \end{bmatrix} \left\{ \begin{array}{c} q_1 \\ q_2 \end{array} \right\} \right) d\tau$$

$$+ \begin{bmatrix} -\frac{1}{2} & \frac{1}{2}\mathrm{erfc}(0.5L/\sqrt{t}) \\ \text{Symmetric} & -\frac{1}{2} \end{bmatrix} \left\{ \begin{array}{c} \theta_1 \\ \theta_2 \end{array} \right\} + 2 \left\{ \begin{array}{c} 1 \\ 1 \end{array} \right\} \mathrm{erf}\left(\frac{L}{2\sqrt{t}} \right) \qquad (8.39)$$

Equation (8.39) can now be solved for (θ_1, θ_2) at time $t = N\Delta t$ since the right-hand side involves only known quantities. It is important to note also that having obtained the solution for the Nth time step $(N = 1, 2, \ldots)$, the solution for the $(N+1)$th step merely requires the right-hand side of Eq. (8.39) to be supplemented with new terms involving the convolution produce $(g * q)$ over the time intervals $[N\Delta\tau, (N+1)\Delta\tau]$ and $[(N-1)\Delta\tau, N\Delta\tau]$ respectively.

Turning our attention to the general boundary value problem, within each time increment, the primary field variables θ and q can be assumed constant. As a result,

$$g * q = \sum_{n=1}^{N} q^n(x) \int_{(n-1)\Delta\tau}^{n\Delta\tau} g(\xi, N\Delta\tau; x, t)\,dt \qquad (8.40a)$$

$$f * \theta = \sum_{n=1}^{N} \theta^n(x) \int_{(n-1)\Delta\tau}^{n\Delta\tau} f(\xi, N\Delta\tau; x, t)\,dt \qquad (8.40b)$$

in which q^n and θ^n are the flux and temperature respectively during the nth time step. Since the kernel function is known in explicit form, the integrations remaining in Eqs (8.40) can be performed analytically. (Note also that these analytical integrations are possible for linear variation in time, as discussed in Chapter 10.) After letting

$$G^{N+1-n}(\xi; x) = -\int_{(n-1)\Delta\tau}^{n\Delta\tau} g(\xi, N\Delta\tau; x, t)\, dt \qquad (8.41a)$$

$$F^{N+1-n}(\xi; x) = -\int_{(n-1)\Delta\tau}^{n\Delta\tau} f(\xi, N\Delta\tau; x, t)\, dt, \qquad (8.41b)$$

the boundary integral equation becomes

$$c(\xi)\theta(\xi, \tau) = \sum_{n=1}^{N}\left\{\int_{S}\left[G^{N+1-n}(\xi; x)q^{n}(x) - F^{N+1-n}(\xi; x)\theta^{n}(x)\right]dS\right\} \qquad (8.42)$$

In writing Eq. (8.42) the following convention has been adopted:

$$
\begin{array}{ll}
G^{n}(\xi; x) = G(\xi, n\Delta\tau; x, 0) & \text{for } n = 1 \\
G^{n}(\xi; x) = G(\xi, n\Delta\tau; x, 0) - G(\xi, (n-1)\Delta\tau; X, 0) & \text{for } n > 1 \\
F^{n}(\xi; x) = F(\xi, n\Delta\tau; x, 0) & \text{for } n = 1 \\
F^{n}(\xi; x) = F(\xi, n\Delta\tau; x, 0) - F(\xi, (n-1)\Delta\tau; X, 0) & \text{for } n > 1
\end{array}
$$

With the usual definition of $y_i = x_i - \xi_i$, $r^2 = y_i y_i$, $\kappa = \rho c_e/k$ and $\eta = r/(\kappa\tau)^{1/2}$, the two-dimensional kernels G and F in Eq. (8.42) are given by

$$G(\xi, \tau; x, t) = \frac{1}{4\pi k} E_1\left(\frac{\eta^2}{4}\right) \qquad (8.42a)$$

$$F(\xi, \tau; x, t) = \frac{1}{2\pi r}\left(\frac{y_i n_i}{r}\right)e^{-\eta^2/4} \qquad (8.42b)$$

$$E_1(z) = \int_{z}^{\infty}\frac{e^{-x}}{x}\, dx$$

whereas those for the three-dimensional problems are

$$G(\xi, \tau; x, t) = \frac{1}{4\pi kr}\operatorname{erfc}\left(\frac{\eta}{2}\right) \qquad (8.42c)$$

$$F(\xi, \tau; x, t) = \frac{1}{4\pi r^2}\left(\frac{y_i n_i}{r}\right)[1 - h_1(\eta)] \qquad (8.42d)$$

$$\operatorname{erfc}(z) = 1 - \operatorname{erf}(z)$$

$$\operatorname{erf}(z) = \frac{2}{\sqrt{\pi}}\int_{0}^{z} e^{-x^2}\, dx$$

$$h_1 = \operatorname{erf}\left(\frac{\eta}{2}\right) - \frac{\eta}{\sqrt{\pi}}e^{-\eta^2/4}$$

This time-stepping convolution formulation was first employed by Tomlin[3] and Chang et al.[4] in 1973. Later, this algorithm was adopted by a number of researchers, including Banerjee and coworkers.[6-9] Wrobel and Brebbia[10] also discussed the convolution approach, but chose to utilize the volume-based recurring initial condition method (method 2, described below), which was also presented in Banerjee et al.[8] and Banerjee and Butterfield.[9]

With the subdivision of the surface into M elements, Eq. (8.42) can be rewritten as

$$c(\xi)\theta(\xi, \tau) = \sum_{n=1}^{N} \left[\sum_{m=1}^{M} \left(q_{m\omega}^n \int_{S_m} G^{N+1-n} N_\omega \, dS - \theta_{m\omega}^n \int_{S_m} F^{N+1-n} N_\omega \, dS \right) \right] \quad (8.43)$$

where $\theta_{m\omega}^n$ and $q_{m\omega}^n$ are now nodal values of temperature and flux respectively and N_ω are the shape functions.

In general, the integration required in Eq. (8.43) can only be performed analytically for very simple cases. Instead, numerical integration is utilized. Details of this spatial integration process can be found in Chapter 3; however, a discussion of the evaluation of the strongly singular integral involving the flux kernel is in order. In the present work, this integral is evaluated very accurately by the following indirect method.[11-13] First, it must be recognized that the instantaneous source flux kernel $f(\xi, \tau; x, t)$ is only singular at $\xi = x$ and $\tau = t$, and that this singularity is identical to that present in the steady-state heat conduction flux kernel, F^{ss}. Consequently, F^n for $n > 1$ is completely non-singular, so that standard Gaussian integration is applicable. Only F^1 is singular at $\xi = x$. However, this can be rewritten as

$$F^1 = F^{ss} + F^{1,tr} \quad (8.44)$$

where $F^{1,tr}$ is also non-singular. The term F^{ss} can be determined via the equipotential method normally used for steady state heat conduction. Thus, by assuming that the temperature is unity throughout the body,

$$c(\xi) + \sum_{m=1}^{M} \int_{S_m} F^{ss} \, dS = 0 \quad (8.45)$$

Writing Eq. (8.45) at any node of the boundary element model permits the calculation of $c(\xi)$ plus the contribution of $\int_s F^{ss} \, dS$ to that node. Direct integration of the strongly singular kernel is thus avoided and $F^{1,tr}$ is integrated using Gaussian quadrature. As noted above, both kernel functions are non-singular for $n > 1$. Consequently, lower-order Gaussian formulae can be employed for accurate evaluation of these integrals in a fraction of the computing time required for the initial time step.

A system of equations is formed at each time step N by employing a nodal collocation process in conjunction with the discretized integral equation defined by Eq. (8.43). This leads to the matrix equations

$$\sum_{n=1}^{N} ([G^{n+1-n}]\{q^n\} - [F^{n+1-n}]\{\theta^n\}) = \{0\} \quad (8.46)$$

The above formulation considers bodies composed of a single material. This restriction can be eliminated by introducing the concept of the geometric modelling region (GMR) or zones, as described in Chapter 3. During integration, the GMRs remain separate entities. Thus the integrals in Eq. (8.43) are evaluated only over the surface of the individual GMR which contains the point ξ. However, at the assembly stage compatibility relationships are enforced on common boundaries between adjacent regions. These relationships can model either completely bonded contacts or can include an interfacial thermal resistance. The latter is appropriate for simulating air gaps or surface coatings.

In any case, for a well-posed problem at each time $N \Delta \tau$ there will be exactly P unknowns among $\{\theta^N\}$ and $\{q^N\}$. Furthermore, all of the quantities $\{\theta^n\}$ and $\{q^n\}$ for $n < N$ are known from previous time steps. Therefore, Eq. (8.46) can be rewritten as

$$[A]\{x^N\} = [B]\{y^N\} + \sum_{n=1}^{N-1} ([G^{N+1-n}]\{q^n\} - [F^{N+1-n}]\{\theta^n\}) \qquad (8.47)$$

in which
$\{x^N\}$ = unknown components of $\{\theta^N\}$ and $\{q^N\}$
$\{y^N\}$ = known components of $\{\theta^N\}$ and $\{q^N\}$
and $[A]$ and $[B]$ are the associated coefficient matrices obtained from $[F^1]$ and $[G^1]$. All of the quantities on the right-hand side of Eq. (8.47) are known and, hence, can be combined into a load vector $\{b^N\}$.

Method 2 (recurring initial condition) The terms included within the summation appearing in Eq. (8.47) represent the contribution of past events. For the convolution approach, this contribution only involves surface quantities. In the alternate recurring initial condition approach, the solution from each time step is used as the initial conditions for the next step. Since this initial state will not, in general, correspond to a steady temperature distribution, the first volume integral appearing in Eq. (8.16) must be retained. Consequently, complete volume discretization is required for the recurring initial condition approach and, as will be seen in the examples, some precision is lost compared to numerical convolution.

It will be shown in Chapter 13 that non-linearities in the governing differential equation can be handled in a BEM formulation by modifying the value of the source term ψ in Eq. (8.1). Thus in a completely general diffusion algorithm, where both the boundary conditions and the internal sources may vary with time, the following time-marching process is more advantageous.

For simplicity of presentation both θ and q values will be assumed to remain constant throughout any $\Delta \tau$ time step and ψ will be represented by its average value over each step. The values of θ and q over any boundary element can be represented as $\theta(x, \tau) = N(x)\boldsymbol{\theta}$ and $q(x, \tau) = N(x)\mathbf{q}$ where $N(x)$ are isoparametric shape functions and $\boldsymbol{\theta}$ and $\mathbf{q}$ are vectors of nodal (θ, q) values. Clearly some form of shape function could also be introduced over the $\Delta \tau$ elements along the time axis, as discussed in Chapter 10.

We now proceed as before and write, from Eq. (8.16), for the pth boundary node,[8,9,13]

$$c\theta(\xi^p, t) = \sum_{q=1}^{n} (\mathbf{F}^{pq}\boldsymbol{\theta}^q - \mathbf{G}^{pq}\mathbf{q}^q) + \sum_{l=1}^{m} (\mathbf{C}^{pl}\boldsymbol{\psi}^l + \mathbf{D}^{pl}\theta_0^l) \qquad (8.48)$$

where

$$\mathbf{F}^{pq} = \int_0^{\Delta\tau} d\tau \int_{\Delta S_q} f(x^q, t; \xi^p, \tau)\mathbf{N}(x^q)\, dS_q$$

$$\mathbf{G}^{pq} = \int_0^{\Delta\tau} d\tau \int_{\Delta s_q} g(x^q, t; \xi^p, \tau)\mathbf{N}(x^q)\, dS_q$$

$$\mathbf{C}^{pl} = \int_0^{\Delta\tau} d\tau \int_{\Delta V_l} g(x^l, t; \xi^p, \tau)\mathbf{M}(x^l)\, dV_l \qquad (8.49)$$

$$\mathbf{D}^{pl} = \int_{\Delta V_l} g(x^l, t; \xi^p, 0)\mathbf{M}(x^l)\, dV_l$$

where $\mathbf{M}(x^l)$ is the shape function for the lth cell.

By writing Eq. (8.48) at each boundary node for time $t = \Delta t$ and absorbing the $\alpha\theta(\xi^p, t)$ coefficients into the diagonal elements of $\mathbf{F}^{pq}$, we can obtain the following matrix equation:

$$[G]\{\theta\} - [F]\{q\} = [C]\{\psi\} + [D]\{\theta_0\} \qquad (8.50)$$

which can now be solved in a recurring initial condition process.[8,9,13]

8.6 ANALYTICAL EVALUATION OF INTEGRALS

In order to construct the matrices for the final system of equations (8.47) and (8.48) it is necessary to evaluate accurately the integrals involved. Due to the introduction of the spatial as well as time integrations it may appear at first glance that the arithmetic calculations involved are prodigious. However, a closer examination of these will reveal that:

1. Sources applied at time τ can only affect events at later times (i.e. discretized events along the time axis are not retroactive) and utilizing the time-shifting properties of the kernel function we only need one time step integration for each new time step.
2. The potential and flow velocity at any point are unaffected by a source generated at any other point at the same instant of time. Therefore, if the time increments are kept small, the effect of an element i of a source applied on an element j during the time interval ΔT is negligible when the elements i and j are at some distance (and time) apart.

Nevertheless, it is necessary to evaluate a number of complicated integrals. For the vast majority of these calculations $\xi \neq x$ and $t \neq \tau$; therefore they involve only smooth functions and can be evaluated using ordinary Gaussian integration formulae. For singular cases (i.e. when the load point and the field point coincide over a boundary element or over a volume cell), it is possible to calculate all contributions due to a constant or linear spatial distribution.

By constructing a local coordinate system through the field point as shown in Fig. 8.4, the effects of the source of strength q_0 can be obtained by integrating Eq. (8.11) as

$$G(0, t) = \int_0^t d\tau \int_{\Delta S} g(0, t; \xi, \tau) q_0 \, dS$$

$$= \frac{q_0}{2\pi\sqrt{ct}} \left\{ L_1 E_1(R_1^2) + L_2 E_1(R_2^2) - 2P\left[\text{erf} \, xpc\left(\frac{L_1}{P}, P\right) + \text{erf} \, xpc\left(\frac{L_2}{P}, P\right) \right] \right.$$

$$\left. + \sqrt{\pi} e^{-P^2} (\text{erf} \, L_1 + \text{erf} \, L_2) \right\} \qquad (8.51)$$

where

$$P = \frac{h}{2\sqrt{ct}} \qquad L_1 = \frac{l_1}{2\sqrt{ct}} \qquad L_2 = \frac{l_2}{2\sqrt{ct}}$$

and

$$R_1^2 = P^2 + L_1^2 \qquad R_2 = P^2 + L_2^2$$

The corresponding $F = \partial G/\partial n$ can be obtained directly from Eq. (8.51) as

$$F = \frac{q_0}{2\pi t} \left\{ \left[\text{erf} \, xpc\left(\frac{L_1}{P}, P\right) + \text{erf} \, xpc\left(\frac{L_2}{P}, P\right) \right] \cos\theta - \frac{1}{2}[E_1(R_1^2) - E_1(R_2^2)] \sin\theta \right\}$$

$$(8.52)$$

where θ is the angle between the normal to the boundary element and the direction n, as shown in Fig. 8.4.

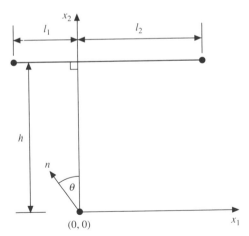

Figure 8.4 Potential at the origin due to a uniform line segment source.

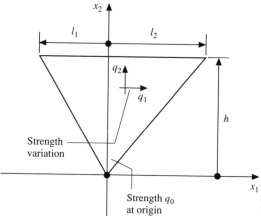

Figure 8.5 Instant triangular source: vertex at the origin.

In order to take account of the initial condition over the volume it is necessary to integrate the fundamental solutions over a triangular cell. If we define an instant triangular source by the vertices $(-l_1, h)$, (l_2, h) and $(0, 0)$, as shown in Fig. 8.5, we can integrate Eq. (8.11) and obtain the potential at the field point $(0, 0)$ at a time t due to a source varying linearly over the triangle as $q_0 + q_1 \xi_1 + q_2 \xi_2$ (where q_0, q_1 and q_2 are constants) applied at time $\tau = 0$. Thus

$$G(0, t) = \frac{q_0}{2\pi} \left[\operatorname{erf} xp\left(\frac{L_1}{P}, P\right) + \operatorname{erf} xp\left(\frac{L_2}{P}, P\right) \right]$$

$$+ \frac{1}{2} \sqrt{\frac{ct}{\pi}} \left(\frac{q_1 P + q_2 L_1}{R_1} \operatorname{erf} R_1 - \frac{q_1 P - q_2 L_2}{R_2} \operatorname{erf} R_2 \right)$$

$$- \frac{q_2}{2} \sqrt{\frac{ct}{\pi}} \, e^{-P^2} (\operatorname{erf} L_1 + \operatorname{erf} L_2) \tag{8.53}$$

For the more general case of a field point lying in the interior, potential G can be obtained by summing the effects of the three sub-triangles meeting at the point.

8.7 ADDITIONAL TOPICS

8.7.1 Anisotropy

In general anisotropic material, the discretized integral equation (8.43) remains valid provided that the coordinate axes are modified appropriately. However, this is not a suitable approach for a general purpose implementation because severe element distortion can result. Instead, explicit anisotropic kernel functions must be employed. These fundamental solutions were first defined by Chang *et al.*[4]

8.7.2 Axisymmetry

Wrobel and Brebbia[10] presented a boundary element formulation for axisymmetric heat conduction using complete volume discretization in a recurring initial condition algorithm. Although this represents an important contribution, they were not able to write the time-integrated kernel functions in explicit form. As an alternative suggestion,[11,13] here the three-dimensional kernels are split into steady state and transient components. Circumferential integration is then performed analytically for the steady portion. The resulting steady state axisymmetric kernels are well known and given in Chapter 5. However, the remaining transient portion, which is completely non-singular, is then integrated numerically in the circumferential direction in which graded sub-segmentation and variable-order Gaussian formulae are employed. With the passage of time the spatial variation of these kernels becomes much more gradual. Thus, the computational effort expended for integration can be greatly reduced for later time steps. The results remain very accurate using this approach.[11,13] Additionally, the extension to anisotropic axisymmetric bodies is relatively straightforward.

8.7.3 Embedded Holes

The introduction of small holes is often required in engineering design to enhance the heating or cooling of otherwise solid bodies. Practical examples include injection moulds and jet engine turbine blades. In both instances, the holes must be strategically sized and placed to ensure adequate cooling of the component. A powerful engineering analysis tool can be developed (see also Chapter 6) for these components by introducing the concept of hole elements within a two-dimensional boundary element formulation. This idea was first conceived by Barone and Caulk[19] for steady state heat conduction. More recently, Dargush and Banerjee[13,14] generalized the methodology to include multiple higher-order hole elements embedded in a multi-region body.

An idealization of a typical cylindrical hole of circular cross-section is depicted in Fig. 8.6. Naturally, the boundary integral equation (8.43) is applicable to a heat-conducting body containing any number of such exclusions. Let

$$c(\xi)\theta(\xi, \tau) = \sum_{n=1}^{N} \left\{ \int_{S-S^h} [G^{N+1-n}(\xi; x)q^n(x) - F^{N+1-n}(\xi; x)\theta^n(x)] \, dS \right.$$

$$\left. + \int_{S^h} [G^{N+1-n}(\xi; x)q^n(x) - F^{N+1-n}(\xi; x)\theta^n(x)] \, dS \right\} \qquad (8.54)$$

where S^h is the surface of the holes. For small-diameter holes, as a first approximation, it can be assumed that the temperature and flux do not vary in the circumferential direction. Thus, at any cross-section, the temperature and flux are constants, although both may vary along the length of the hole. Each physical hole can then be represented by a series of hole elements which connect two or three nodes positioned on the centreline. The radius of the hole is assumed

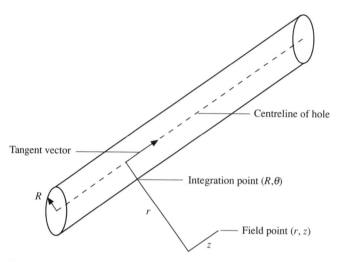

Figure 8.6 Hole element geometric definition.

constant, while the longitudinal variation of temperature and flux is either linear or quadratic. Setting up a local cylindrical coordinate system, as shown in Fig. 8.6, permits the straightforward evaluation of the circumferential integration at any cross-section. This integration is identical to that needed for axisymmetric transient heat conduction. The remaining longitudinal integration is performed numerically utilizing techniques developed for two-dimensional line integrals.

In developing the system equations, Eq. (8.54) must be written at each node of the problem, including those on the outer surface and on the hole centreline. Two types of hole nodes are possible. The first type, non-endpoints, are not on the surface of the exclusion and thus $c(\xi)$ is zero. All integration for these nodes is non-singular. The remaining endpoint nodes reside on a circular disc which bounds the hole. In this case, $c(\xi)$ equals $\frac{1}{2}$. The contribution of the temperature and the flux on this disc is also included in the formulation through the use of an analytical integration.

As noted above, the assumption of constant circumferential temperature and flux is a first-order approximation. If increased accuracy is required or if a component has a high void ratio, then this assumption may be inappropriate. Higher-order variations can be incorporated, as was recently done by Henry and Banerjee[20] for elastostatics (see Chapter 6).

8.8 EXAMPLES OF APPLICATIONS

8.8.1 Unidirectional Heat Conduction[13]

As a first example, consider the case of unidirectional transient heat conduction in a two-dimensional rectangular slab. The baseline boundary element model for

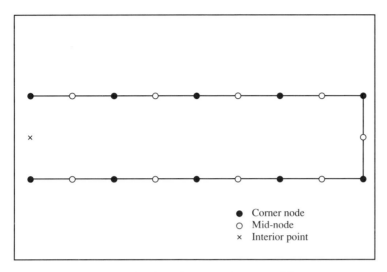

Figure 8.7 Unidirectional diffusion—boundary element model.

one-half of the slab is depicted in Fig. 8.7. A total of nine quadratic boundary elements are employed, and planar symmetry is imposed.

The problem begins with the slab at rest in thermodynamic equilibrium at zero temperature. The bottom and side edges are perfectly insulated. Then, suddenly, at time zero, the temperature of the top edge at $Y = 1$ is elevated to θ_0 and maintained at that level. Note that the unidirectional diffusion length is L while Y represents the normalized spatial coordinate in that direction. Thus, for this geometry, $Y = x_2/L$ and the exact solution can be written as (e.g. Carslaw and Jaeger[1])

$$\theta(Y, T) = \theta_0 \left\{ 1 - \frac{4}{\pi} \sum_{n=0}^{\infty} \frac{(-1)^n}{2n+1} \exp\left[-(2n+1)^2 \pi^2 T/4\right] \cos\frac{(2n+1)\pi Y}{2} \right\} \quad (8.55)$$

where T is the non-dimensional time factor defined by $T = \kappa t/L^2$. This apparently simple problem proves to be quite a test for any numerical method.

Values of the non-dimensional temperature (θ/θ_0) obtained from the general purpose boundary element program, GP-BEST, are compared with this exact solution in Fig. 8.8. Numerical results are presented for a point on the plane of symmetry at the bottom edge $(Y = 0.0)$ and at the interior point $(Y = 0.5)$ identified in Fig. 8.8. The analysis was run with several different values of the non-dimensional time step (ΔT). Not only does the boundary element solution converge to the analytical result, but even with a very large time step there are no oscillations or instabilities. The method is fully implicit and, unlike the Crank–Nicholson approach used in the finite difference and finite element methods, there are no arbitrary constants introduced. Additionally, to obtain comparable accuracy with a finite element approach many more elements are required in the vertical direction.

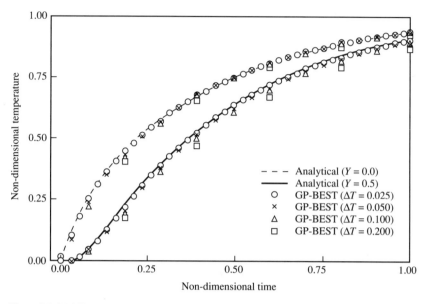

Figure 8.8 Unidirectional diffusion—transient heat conduction.

However, as shown in Fig. 8.9, there is a sensitivity to the aspect ratio of the region. Employing a constant time step of $\Delta T = 0.05$, it is apparent from the diagram that there is a degradation in accuracy as the width of the modelled region L_x is reduced. The solution for $L_x = 0.250$ can be improved by utilizing a

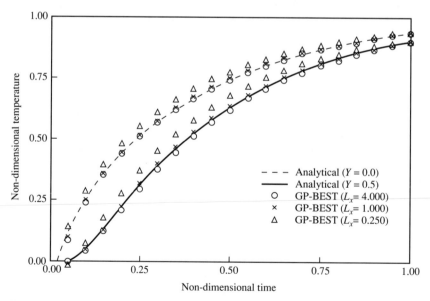

Figure 8.9 Unidirectional diffusion—transient heat conduction.

smaller time step. Therefore, even though there is no diffusion in the horizontal direction, the time step must be based upon the minimum distance between nodal points. Consequently, for thin-walled bodies, unless alternate formulations are developed, extremely small time steps must be employed.

Provided that a sufficiently small step is selected, the greatest error occurs during the initial stages of the diffusive process. This error, which was noted in the early efforts of Chang et al.[4] and Banerjee et al.,[8] has not been eliminated by the introduction of higher-order elements, adaptive integration and state-of-the-art equation solvers. The error is present because the heat flux is infinite at time zero in the mathematical problem. This behaviour cannot be accommodated with finite temporal and spatial discretization. Of course, in a physical problem the temperature will not undergo an instantaneous jump at time zero, and an accurate solution at small times is possible with a boundary element convolution approach.

A number of researchers (e.g. Banerjee et al.[8] and Wrobel and Brebbia[10]) recommend the use of a recurring initial condition approach rather than using convolution. In addition to requiring complete volume discretization, the recurring initial condition algorithm generally provides less accurate results for a given surface mesh. Four quadratic volume cells were added to the baseline model, and very precise numerical quadrature of the initial condition volume integral in Eq. (8.49) was undertaken. Results shown in Fig. 8.10 indicate that this latter approach is less accurate, especially at small values of time. Consequently, the convolution algorithm is recommended for linear problems, even though an additional surface integration is required at each time step.

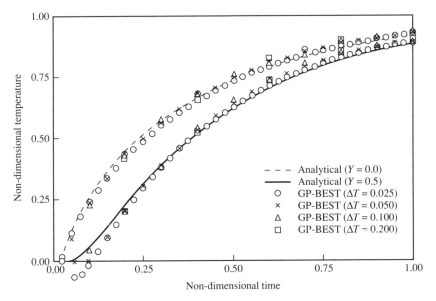

Figure 8.10 Unidirectional diffusion—recurring initial condition algorithm.

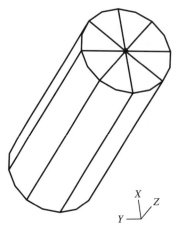

Figure 8.11 Cylinder with an embedded hole—boundary element model.

8.8.2 Cylinder with an Embedded Hole[13]

The performance of the hole element is examined in this next example. A finite circular cylinder of length 40.0 and radius 8.0 contains an embedded longitudinal hole of unit radius. The entire cylinder is at zero temperature, until time zero, when the temperature of the hole surface is elevated to unity.

For the case of a concentric hole within an infinitely long outer cylinder, an analytical solution is available from Carslaw and Jaeger.[1] The temperature away from the ends of the finite cylinder should approach this solution.

The BEM surface model is shown in Fig. 8.11. In addition, four quadratic hole elements are employed. Unit diffusivity is assumed. Results at $r = (a + b)/2$ obtained with a time step of 1.0 are compared with the analytical solution in Fig. 8.12. The correlation is excellent, thus validating the hole element approach for transient problems.

8.9 CONCLUDING REMARKS

In this chapter we have presented DBEM and IBEM formulations which provide solutions for classical diffusion problems in any number of space dimensions. One rather satisfying result is that, by using the notation for a Riemann convolution integral, the final equations are only minimally different in form from those for steady state flow problems.

Transient problems of this kind have received relatively less attention from BEM analysts than many other fields, and questions concerning optimum values of time steps and precision at early times in the diffusion process have been answered satisfactorily. It is quite clear that, whereas Laplace transform methods may not look very promising, successful time-marching algorithms have been developed and tested on very general diffusion problems.[3–13,21–30]

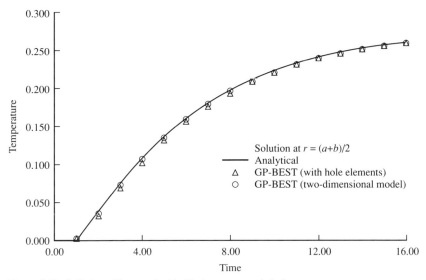

Figure 8.12 Cylinder with an embedded hole—concentric hole.

REFERENCES

1. Carslaw, H.S. and Jaeger, J.C. (1959) *Conduction of Heat in Solids*, 2nd edn, Oxford University Press.
2. Rizzo, F.J. and Shippy, D.J. (1971) A method of solution for certain problems of transient heat conduction, *AIAA Journal*, Vol. 8, p. 2004.
3. Tomlin, G.R. (1973) Numerical analysis of continuum problems in zoned anisotropic media, PhD Thesis, Southampton University.
4. Chang, Y.P., Kang, C.S. and Chen, D.J. (1973) The use of fundamental Green functions for solution of problems of heat conduction in anisotropic media, *International Journal of Heat and Mass Transfer*, Vol. 16, pp. 1905–1918.
5. Shaw, R.P. (1974) An integral equation approach to diffusion, *International Journal of Heat and Mass Transfer*, Vol. 17, pp. 693–699.
6. Banerjee, P.K. and Butterfield, R. (1976) Boundary element methods in geomechanics, Chapter 16 in *Finite Elements in Geomechanics*, Ed. G. Gudehus, John Wiley, London.
7. Butterfield, R. and Tomlin, G.R. (1972) Integral techniques for solving zoned anisotropic continuum problems, *Proceedings of International Conference on Variational Methods in Engineering*, pp. 9/31–9/51, Southampton University.
8. Banerjee, P.K., Butterfield, R. and Tomlin, G.R. (1981) Boundary element methods for two-dimensional problems of transient ground-water flow, *International Journal of Numerical Methods in Geomechanics*, Vol. 5, pp. 15–31.
9. Banerjee, P.K. and Butterfield, R. (1981) *Boundary Element Methods in Engineering Science*, McGraw-Hill, London.
10. Wrobel, L.C. and Brebbia, C.A. (1981) A formulation of the boundary element method for axisymmetric transient heat conduction, *International Journal of Heat and Mass Transfer*, Vol. 24, pp. 843–850.
11. Dargush, G.F. (1987) Boundary element methods for the analogous problems of thermomechanics and soil consolidation, PhD Dissertation, State University of New York at Buffalo.
12. Banerjee, P.K., Ahmad, S. and Manolis, G.D. (1986) Transient elastodynamic analysis of three

dimensional problems by boundary element method, *Earthquake Engineering, Structural Dynamics*, Vol. 14, pp. 933–949.

13. Dargush, G. and Banerjee, P.K. (1991) Application of boundary element method to transient heat conduction, *International Journal of Numerical Methods in Engineering*, Vol. 31, No. 6, pp. 1231–1248.

14. Dargush, G.F. and Banerjee, P.K. (1989) Advanced development of the boundary element method for steady-state heat conduction, *International Journal of Numerical Methods in Engineering*, Vol. 28, pp. 2123–2142.

15. Morse, P. and Feshbach, H. (1953) *Methods of Theoretical Physics*, Vol. II, McGraw-Hill, New York.

16. Lamb, H. (1932) *Hydrodynamics*, Cambridge University Press.

17. Abramowicz, M. and Stegan, I.A. (1965) *Handbook of Mathematical Functions*, Dover, New York.

18. Crandall, S.H. (1956) *Engineering Analysis*, McGraw-Hill, New York.

19. Barone, M.R. and Caulk, D.A. (1985) Special boundary integral equations for approximate solution of potential problems in three-dimensional regions with slender cavities of circular cross-section, *IMA Journal of Applied Mathematics*, Vol. 35, pp. 311–325.

20. Henry, D.P. and Banerjee, P.K. (1991) Elastic stress analysis of three-dimensional bodies with small holes by BEM, *International Journal of Numerical Methods in Engineering*, Vol. 31, No. 2, pp. 369–384.

21. Chuang, Y.K. and Szekely, J. (1971) On the use of Green's function for solving melting or solidification problems, *International Journal of Heat and Mass Transfer*, Vol. 14, pp. 1285–1294.

22. Chuang, Y.K. (1971) The melting and dissolution of a solid in a liquid with a strong exothermic head of solution, PhD thesis, State University of New York at Buffalo.

23. Chuang, Y.K. and Szekely, J. (1972) The use of Green's functions for solving melting or solidification problems in cylindrical coordinates system, *International Journal of Heat and Mass Transfer*, Vol. 15, pp. 1171–1174.

24. DeMey, G. (1977) Numerical solution of a drift-diffusion problem, *Comparative Physics Communications*, Vol. 13, pp. 81–88.

25. DeMey, G. (1978) An integral equation method to calculate the transient behaviour of a photovoltaic solar cell, *Solid-State Electronics*, Vol. 21, pp. 595–596.

26. DeMey, G. (1976) An integral equation approach to a.c. diffusion, *International Journal of Heat and Mass Transfer*, Vol. 19, pp. 702–704.

27. Chuang, Y.K. and Ehrich, O. (1974) On the integral technique for spherical growth problems, *International Journal of Heat and Mass Transfer*, Vol. 17, pp. 945–953.

28. O'Neill, K. (1983) Boundary integral equation solution of moving boundary phase change problems, *International Journal of Numerical Methods in Engineering*, Vol. 19, pp. 1825–1850.

29. Zabaras, N. and Mukherjee, S. (1987) An analysis of solidification problems by the BEM, *International Journal of Numerical Methods in Engineering*, Vol. 24, pp. 1879–1900.

30. Zabaras, N., Mukherjee, S. and Richmond, O. (1988) An analysis of inverse heat transfer problems with phase charge using an integral method, *Journal of Heat Transfer*, Vol. 110, pp. 554–561.

NINE

POROELASTIC AND THERMOELASTIC ANALYSES

9.1 INTRODUCTION

The development of the boundary element for two-dimensional, three-dimensional and axisymmetric quasi-static poroelasticity and thermoelasticity is discussed in detail. These formulations, for the complete Biot theory, operates directly in the time domain and requires only boundary discretization. Furthermore, as a result of a well-known thermodynamic analogy, the same BEM is equally applicable to quasi-static thermoelasticity and poroelasticity (consolidation).

In the quasi-static realm, Banerjee and Butterfield[1] and others[2-4] presented a staggered procedure for solving the coupled equations. The algorithm requires the solutions of the transient pore fluid (or heat) flow equation followed by an elastic analysis including body forces at each time step. Unfortunately, this is not a boundary-only formulation, and complete volume discretization is required.

Another alternative involves casting the problem in the Laplace transform domain. Cheng and Liggett[5] took that approach for planar quasi-static poro-elasticity. While this is a boundary-only formulation, the procedure is not, in general, very satisfactory. Transformation to the Laplace domain introduces several serious side-effects, including sensitivity to the selection of values for the transform parameter, requirements for numerical inversion of the transform and limitations to strictly linear problems. More recently, Nishimura[6] developed the first two-dimensional time-domain boundary element formulation for regions infiltrated by an incompressible fluid.

Returning to thermoelasticity, Tanaka and Tanaka[7] presented a reciprocal theorem and the corresponding boundary element formulation, but with no implementation, for the time-domain coupled problem. However, kernel functions are not discussed and no numerical results are included. In a series of papers, Sladek and Sladek[8-10] presented a collection of fundamental solutions, in both Laplace transform and time domains, under the classifications of coupled, uncoupled, transient and quasi-static thermoelasticity. Boundary integral equations were included, although in several instances these were written inappropriately in terms of displacement and traction rates. More recently, Chaudounet[11] again resorts to the volume-based approach, while Masinda[12] and Sharp and Crouch[13] have directed efforts towards converting the thermal body force volume integral into a surface integral.

This chapter discusses the development of boundary element methods for general two- and three-dimensional Biot's theory of consolidation and the corresponding theory of coupled and uncoupled thermoelasticity, and is essentially based upon the recent efforts of Banerjee, Dargush and Chopra[14-21].

9.2 FUNDAMENTAL THEORY

9.2.1 Thermomechanics

The equilibrium of a thermoelastic body is written in differential form as

$$\sigma_{ji,j} + b_i = 0 \tag{9.1}$$

where σ_{ji} is the stress tensor and b_i is the body force per unit volume. The stresses are related to the strain through a constitutive tensor D^e_{ijkl}. For a linear, isotropic and homogeneous thermoelastic material, the constitutive equation, known as the Duhamel-Neumann relation, can be expressed as

$$\sigma_{ij} = 2\mu\varepsilon_{ij} + \delta_{ij}\lambda\varepsilon_{kk} - \delta_{ij}(3\lambda + 2\mu)\alpha(T - T_0) \tag{9.2}$$

where T is the present temperature, T_0 is the temperature of a zero reference state, μ and λ are Lame's elastic constants and α is the coefficient of linear expansion. Combining the equilibrium, strain–displacement and stress–strain relations leads to the familiar Navier equation with a generalization to include the additional temperature effects:

$$(\lambda + \mu)u_{j,ij} + \mu u_{i,jj} + b_i - (3\lambda + 2\mu)\alpha T_{,i} = 0 \tag{9.3}$$

However, another equation is required to complete the theoretical development since temperature fields are unknown in Eq. (9.3). The first law of thermodynamics, which postulates the conservation of energy, is utilized for this purpose, as presented by Boley and Weiner:[22]

$$kT_{,jj} + \psi = \rho c_\varepsilon \dot{T} + (3\lambda + 2\mu)\alpha(T)\dot{u}_{j,j} \tag{9.4}$$

where k is the conductivity, assumed to be spatially invariant. This thermoelastic theory, exemplified by Eqs (9.3) and (9.4), is called coupled quasi-static thermoelasticity and is sufficiently general and applicable to a large spectrum of engineering problems. The set of equations (9.3) and (9.4) is clearly coupled since displacements and temperatures appear in both equations and, therefore, must be solved simultaneously.

An alternate way to frame the governing equations is to utilize the entropy density s as the basic parameter. Thus, in the absence of body forces and sources, the reformulated governing equations may be expressed in the following manner:[14,20]

$$\left(\lambda + \frac{T_0\beta^2}{\rho c_\epsilon} + \mu\right)u_{j,ij} + \mu u_{i,jj} - \left(\frac{T_0\beta}{\rho c_\epsilon}\right)s_{,i} = 0 \qquad (9.5a)$$

$$\left\{\frac{k}{\rho c_\epsilon}\left[\frac{\lambda + 2\mu}{\lambda + T_0\beta^2/(\rho c_\epsilon) + 2\mu}\right]\right\}s_{,jj} = \dot{s} \qquad (9.5b)$$

The above equations are not directly expressed in terms of the temperature and are thus less appealing. The advantage, however, will be evident after the discussion on poroelasticity in the next section. When expressed in the form of Eqs (9.5a,b), the governing equations of thermoelasticity can be compared with the corresponding equations of poroelasticity to illustrate the complete analogous nature of the two phenomena. Additionally, it is evident from Eq. (9.5b) that the entropy density s, a measure of the molecular disorder, undergoes a diffusive process within a thermoelastic medium. The rate of this diffusion depends upon the existing entropy density and a material parameter c. This property c is called the thermal diffusivity and is given by

$$c = \frac{k}{\rho c_\epsilon}\frac{\lambda + 2\mu}{\lambda + T_0\beta^2/(\rho c_\epsilon) + 2\mu} \qquad (9.6)$$

9.2.2 Poroelasticity

The behaviour of multi-phase porous materials such as soils is governed by the behaviour of its individual constituents as well as an interaction between the various phases. The study of the diffusive process that takes place in a soil medium, made up of solid particles (skeleton) and pore fluids, as it is loaded, is known as poroelasticity or soil consolidation. Terzaghi[23] first laid down the theoretical foundation to poroelastic analysis with the introduction of the effective stress concept. The generalized effective stress concept may now be expressed as

$$\sigma_{ij} = \sigma'_{ij} - \delta_{ij}\beta u_p \qquad (9.7)$$

where σ_{ij} is the total stress state, σ'_{ij} is the effective stress state and β is a dimensionless material parameter relating the compressibilities of the solid and fluid constituents and may be expressed as

$$\beta = 1 - \frac{K}{K'_s} \qquad (9.8)$$

where $K = \lambda + 2\mu/3$ is the bulk modulus of the soil under drained conditions and K'_s is an arbitrary constant which reduces to the bulk modulus of the solid constituents under certain specific conditions. If both the solid and fluid constituents are considered incompressible, the parameter β reduces to 1 and Eq. (9.7) becomes identical to Terzaghi's equation; λ and μ are Lame's constants under purely drained conditions.

For a linear, elastic soil medium the constitutive equation can now be expressed in terms of the effective stress related to the total strain, as

$$\sigma'_{ij} = 2\mu\varepsilon_{ij} + \lambda\delta_{ij}\varepsilon_{kk} \tag{9.9}$$

Starting again with equilibrium and making use of the strain–displacement relation and the constitutive relation, Navier's equation for quasi-static poroelasticity may be written as

$$(\lambda + \mu)u_{j,ij} + \mu u_{i,jj} - \beta u_{p,i} + b_i = 0 \tag{9.10}$$

To form a complete set with Eq. (9.10), the conservation of mass of the fluid phase is utilized:

$$q_{i,i} + \dot{m} = \bar{\psi} \tag{9.11}$$

where q_i is the fluid mass flux vector, m the fluid mass content per unit volume and $\bar{\psi}$ the fluid mass supply per unit volume.

Utilizing Darcy's law in an isotropic form and expressing the fluid mass content as a function of displacements and pore pressure, the desired form of the conservation of mass equation is obtained as[19,20,21]

$$\rho_0\kappa u_{p,jj} - \left(\frac{\rho_0\beta^2}{\lambda_u - \lambda}\right)\dot{u}_p - \rho_0\beta\dot{u}_{j,j} + \bar{\psi} = 0 \tag{9.12}$$

where ρ_0 is the fluid mass density, λ_u the undrained λ and κ the permeability of the soil. This can be simplified by substituting $\bar{\psi} = \rho_0\psi$ and the set of governing equations for coupled quasi-static poroelasticity (CQP) may be written as

$$(\lambda + \mu)u_{j,ij} + \mu u_{i,jj} - \beta u_{p,i} + b_i = 0 \tag{9.13a}$$

$$\kappa u_{p,jj} - \alpha\dot{u}_p - \beta\dot{u}_{j,j} + \psi = 0 \tag{9.13b}$$

where $\alpha = \beta^2/(\lambda_u - \lambda)$. Thus, the constants β and α can now be expressed in terms of the undrained bulk modulus K_u as

$$\beta = \frac{1}{B}\left(1 - \frac{K}{K_u}\right) \tag{9.14a}$$

$$\alpha = \frac{\beta}{K_u B} \tag{9.14b}$$

where B is the well-known Skempton's coefficient of pore pressure.

The governing equations can be reformulated, as was done in the previous section, such that the equations are written in terms of the fluid mass content instead of the pore pressure p. Thus, Eqs (9.13a,b) can be rewritten as[16,20]

$$(\lambda_u + \mu)u_{j,ij} + u_{i,jj} - \left(\frac{\lambda_u - \lambda}{\beta\rho_0}\right)m_{,i} = 0 \tag{9.15a}$$

$$\left[\kappa\left(\frac{\lambda_u - \lambda}{\beta^2}\right)\frac{\lambda + 2\mu}{\lambda_u + 2\mu}\right]m_{,jj} = 0 \tag{9.15b}$$

In a manner similar to the entropy density in thermoelasticity, the fluid mass content m obeys the law of simple diffusion in a poroelastic medium. The material parameter determining the rate at which the diffusive process takes place is called the diffusivity, c_v, and is given by

$$c_v = \kappa\left(\frac{\lambda_u - \lambda}{\beta^2}\right)\frac{\lambda + 2\mu}{\lambda_u + 2\mu} \tag{9.16}$$

The undrained Lame parameter is not suited for measurement and can be replaced by the undrained bulk modulus K_u and the undrained Poisson ratio ν_u. The diffusivity can now be expressed as

$$c_v = \frac{2\mu\kappa(1 - \nu)}{1 - 2\nu}\left[\frac{B^2}{9}\frac{(1 + \nu_u)^2(1 - 2\nu)}{(\nu_u - \nu)(1 - \nu_u)}\right] \tag{9.17}$$

For a medium with incompressible fluid and solid constituents, the relationships are considerably simpler. Since $K_u \to \infty$ and $B \to 1$, thus $\nu_u \to \frac{1}{2}$, $\beta \to 1$, $\alpha \to 0$ and $c_v \to 2\mu\kappa(1 - \nu)/(1 - 2\nu)$. Additionally, since $\beta \to 1$, the definition of effective stresses reduces to that given by Terzaghi's theory.

9.2.3 Analogous Nature of Thermoelasticity and Poroelasticity

This section summarizes the analogy between the different parameters obtained from the discussion of the two theories. Biot[24-26] was the first to identify this correspondence and noted that both phenomena satisfy the same fundamental principles of thermodynamics.

Several interesting observations can be made upon examining the two sets of equations (9.5a,b) and (9.15a,b). It is immediately evident that the entropy density s and the fluid mass content m follow the same diffusive behaviour. In addition, the undrained Lame constant λ_u in poroelasticity corresponds to the term $[\lambda + T_0\beta^2/(\rho c_\epsilon)]$ in thermoelasticity. Since this term defines a material constant obtained under isentropic conditions, it follows that isentropic behaviour in CQT corresponds to the undrained behaviour in poroelasticity. A detailed study of the term-by-term correspondence between the two theories was presented by Dargush and Banerjee[27] and is reproduced in Table 9.1.

Table 9.1 Poroelastic–thermoelastic analogy

Thermoelasticity		Poroelasticity
Entropy density	$s \leftrightarrow m$	Fluid mass content per unit volume
Temperature	$T \leftrightarrow u_p$	Pore pressure
Heat flux	$q \leftrightarrow q$	Fluid mass flux
Diffusivity	$c \leftrightarrow c$	Diffusivity
$c = \frac{k}{\rho c_\epsilon} \frac{\lambda + 2\mu}{\lambda_s + 2\mu}$		$c = \kappa \left(\frac{\lambda_u - \lambda}{\beta^2} \right) \frac{\lambda + 2\mu}{\lambda_u + 2\mu}$
Isentropic Lame constant	$\lambda_s \leftrightarrow \lambda_u$	Undrained Lame constant
$\lambda_s = \lambda + \frac{\theta_0 \beta^2}{\rho c_\epsilon}$		$\lambda_u = \lambda + \frac{B\beta K}{1 - B\beta}$
Thermoelastic constant	$\beta \leftrightarrow \beta$	Poroelastic constant
$\beta = (3\lambda + 2\mu)\alpha$		$\beta = \frac{3}{B} \left(\frac{\lambda_u - \lambda}{3\lambda_u + 2\mu} \right)$
Conductivity over reference temperature	$\frac{k}{T_0} \leftrightarrow \kappa$	Permeability

9.3 FUNDAMENTAL SOLUTIONS

9.3.1 Three-dimensional Solutions

The infinite space, unit step point force and point source fundamental solutions were first presented by Nowacki.[28] From a poroelastic standpoint, the first efforts to recognize these kernels in independent manner were those of Cleary[29] and Rudnicki.[30]

Unit step force solution   Consider the displacement and temperature (or pore pressure) response to a unit step point force in the jth direction applied at a point ξ in an infinite medium. The forcing function can, therefore, be written as

$$b_j(\xi, t) = \delta(\xi_1)\delta(\xi_2)\delta(\xi_3)H(t)e_j \tag{9.18}$$

where $\delta(\xi_i)$ is the standard Dirac delta function, $H(t)$ is the Heaviside step function and e_j is a unit vector in the jth direction. As a result of the action of Eq. (9.18), the following displacement and temperature response are obtained at any point x:

$$u_i(x, t) = \frac{1}{16\pi r} \frac{1}{\mu(1 - \nu)} \left[\left(\frac{y_i y_j}{r^2} \right) g_1(\eta) + \delta_{ij} g_2(\eta) \right] e_j \tag{9.19a}$$

$$T(x, t) = \frac{1}{4\pi} \left[\frac{\beta}{k(\lambda + 2\mu)} \right] \left[\left(\frac{y_j}{r} \right) g_3(\eta) \right] e_j \tag{9.19b}$$

where $y_i = x_i - \xi_i$, $r^2 = y_i y_i$ and $\eta = r/(ct)^{1/2}$. The evolution functions $g_1(\eta), g_2(\eta)$ and $g_3(\eta)$ are given by the following expressions:

$$g_1(\eta) = 1 + c_1[h_1(\eta) - h_2(\eta)] \tag{9.20a}$$

$$g_2(\eta) = (3 - 2\nu) - c_1 \left[h_1(\eta) - \frac{h_2(\eta)}{3} + \frac{2\eta}{3\sqrt{\pi}} e^{-\eta^2/4} \right] \tag{9.20b}$$

$$g_3(\eta) = \frac{h_1(\eta)}{\eta^2 t} \tag{9.20c}$$

where $c_1 = (\nu_s - \nu)/(1 - \nu_s) = (\nu_u - \nu)/(1 - \nu_u)$, $h_1(\eta) = \mathrm{erf}\,(\eta/2) - (\eta/\sqrt{\pi})\,e^{-\eta^2/4}$ and $h_2(\eta) = 6h_1(\eta)/\eta^2 - (\eta/\sqrt{\pi})\,e^{-\eta^2/4}$ with $\nu_s(\nu_u)$ being the isentropic (undrained) Poisson ratio. The error function, erf, in the above, is defined as

$$\mathrm{erf}\,(z) = \frac{2}{\sqrt{\pi}} \int_0^z e^{-x^2}\,\mathrm{d}x \tag{9.21}$$

From Eqs. (9.19a,b), it becomes evident that the displacement response is singular in nature as the field point x approaches load point ξ. The singularities of the different kernel functions are discussed in detail in a later section. The large time behaviour corresponds to the steady state uncoupled isothermal solutions, whereas the instantaneous behaviour may be described as the isentropic response in thermoelasticity and the undrained response in poroelasticity.

Unit step heat source solution Solution to the governing equations can be obtained when a unit step heat source is applied at ξ in an infinite medium as

$$\psi(\xi, t) = \delta(\xi_1)\delta(\xi_2)\delta(\xi_3)H(t) \tag{9.22}$$

The resulting displacement and temperature response at any point x_i can be expressed as

$$u_i(x, t) = \frac{1}{8\pi} \left[\frac{\beta}{k(\lambda + 2\mu)} \right] \left[\left(\frac{y_i}{r} \right) g_4(\eta) \right] \tag{9.23a}$$

$$T(x, t) = \frac{1}{4\pi r} \left(\frac{1}{k} \right) g_5(\eta) \tag{9.23b}$$

where $g_4(\eta) = \mathrm{erfc}(\eta/2) + 2h_1(\eta)/\eta^2$, $g_5(\eta) = \mathrm{erfc}(\eta/2)$ and the complementary error function, erfc, is defined as $\mathrm{erfc}(z) = 1 - \mathrm{erf}(z)$.

Once again, on examining Eq. (9.23b), it is seen that there exists a $1/r$ singularity which would influence a numerical scheme. The evolution functions $g_4(\eta)$ and $g_5(\eta)$ provide the transition from isentropic to steady state behaviour.

9.3.2 Two-dimensional Solutions

Two-dimensional plane strain solutions are described here. First, consider the effect of unit step forces acting in the j direction at the points (ξ_1, ξ_2, ξ_3) where the range of ξ_3 is from $-\infty$ to ∞. Thus, the total applied force is a line load and the desired solution can be obtained by integrating the three-dimensional Green function along the entire ξ_3 direction. The response at any point $(x_1, x_2, 0)$ is given by

$$u_i(x, t) = \frac{1}{8\pi} \left[\frac{1}{\mu(1 - \nu)} \right] \left[\left(\frac{y_i y_j}{r^2} \right) \bar{g}_1(\eta) + (\delta_{ij}) \bar{g}_2(\eta) \right] e_j \qquad (9.24a)$$

$$T(x, t) = \frac{r}{2\pi} \left[\frac{\beta}{k(\lambda + 2\mu)} \right] \left[\left(\frac{y_j}{r} \right) \bar{g}_3(\eta) \right] e_j \qquad (9.24b)$$

where

$$\bar{g}_1(\eta) = 1 + c_1 [1 - \bar{h}_1(\eta)] \qquad (9.25a)$$

$$\bar{g}_2(\eta) = -(3 - 4\nu) \ln r + c_1 [\bar{h}_2(\eta)] \qquad (9.25b)$$

$$\bar{g}_3(\eta) = \frac{\bar{h}_1(\eta)}{4t} \qquad (9.25c)$$

and

$$\bar{h}_1(\eta) = \frac{4}{\eta^2} (1 - e^{-\eta^2/4}) \quad \text{and} \quad \bar{h}_2(\eta) = \frac{1}{2} \left[E_1 \left(\frac{\eta^2}{4} \right) + \ln \left(\frac{\eta^2}{4} \right) + \bar{h}_1(\eta) \right]$$

The function E_1 is the exponential integral defined by

$$E_1(z) = \int_z^\infty \frac{e^{-x}}{x} \, dx \qquad (9.26)$$

Next, the continuous-line fluid mass source Green function is presented. Again, this can be obtained by integrating the three-dimensional fundamental solution expressed in Eqs (9.23a,b). The resulting displacements and temperature fields are

$$u_i(x, t) = \frac{r}{4\pi} \left[\frac{\beta}{k(\lambda + 2\mu)} \right] \left[\left(\frac{y_i}{r} \right) \bar{g}_4(\eta) \right] \qquad (9.27a)$$

$$T(x, t) = \frac{1}{2\pi} \left(\frac{1}{k} \right) [\bar{g}_5(\eta)] \qquad (9.27b)$$

where

$$\bar{g}_4(\eta) = \frac{\bar{h}_1(\eta)}{2} + \frac{E_1(\eta^2/4)}{2} \quad \text{and} \quad \bar{g}_5(\eta) = \frac{E_1(\eta^2/4)}{2}$$

Armed with these fundamental solutions, a boundary element formulation for coupled quasi-static thermoelasticity (CQT) will be developed in the following section.

9.4 BOUNDARY INTEGRAL FORMULATION

The direct boundary element method requires the use of an appropriate reciprocal theorem, which was presented by Ionescu-Cazimir[31] based upon the work of Biot.

Thus, for a body of volume V and surface S under zero initial conditions, the time-domain reciprocal theorem for CQT may be written as

$$\int_S \left[t_i^{(1)} * u_i^{(2)} + q^{(1)} * T^{(2)} - u_i^{(1)} * t_i^{(2)} - T^{(1)} * q^{(2)} \right] dS$$

$$+ \int_V \left[\dot{b}_i^{(1)} * u_i^{(2)} + \psi^{(1)} * T^{(2)} - u_i^{(1)} * b_i^{(2)} - T^{(1)} * \psi^{(2)} \right] dV = 0 \quad (9.28)$$

In the above equation, the superscripts (1) and (2) denote two independent states in the body characterized by the two sets of parameters, namely $[u_i^{(1)}, t_i^{(1)}, T^{(1)}, q^{(1)}, b_i^{(1)}, \; \psi^{(1)}]$ and $[u_i^{(2)}, t_i^{(2)}, T^{(2)}, q^{(2)}, b_i^{(2)}, \psi^{(2)}]$. The asterisk ($*$) symbol is used to indicate a Riemann convolution integral. For instance,

$$t_i^{(1)} * u_i^{(2)} = \int_0^t t_i^{(1)}(x, t - \tau) u_i^{(2)}(x, \tau) \, d\tau = \int_0^t t_i^{(1)}(x, \tau) u_i^{(2)}(x, t - \tau) \, d\tau \quad (9.29)$$

As evident from Eq. (9.28), the resulting boundary element formulations would appear in terms of the primary variables such as displacements, tractions, temperatures and fluxes.

Let one of those states, say state (2), be the, as yet, unknown solution to a given boundary value problem. Then the remaining state may be chosen arbitrarily. However, by initially selecting, for state (1), the infinite region response to a unit step force in the j direction acting at ξ within V and beginning at time zero, the volume integrals in the reciprocal theorem can be made to vanish. That is, in Eq. (9.28), at any point x in V, let the applied forces and sources equal

$$b_i^{(1)}(x, t) = \delta(x - \xi) H(t) \delta_{ij} e_j \quad (9.30a)$$

$$\psi^{(1)}(x, t) = 0 \quad (9.30b)$$

and represent the response by

$$u_i^{(1)}(x, t) = g_{ij}(x - \xi, t) e_j \quad (9.30c)$$

$$T^{(1)}(x, t) = \dot{g}_{Tj}(x - \xi, t) e_j \quad (9.30d)$$

$$t_i^{(1)}(x, t) = f_{ij}(x - \xi, t) e_j \quad (9.30e)$$

$$q^{(1)}(x, t) = \dot{f}_{Tj}(x - \xi, t) e_j \quad (9.30f)$$

In this notation, the subscript T does not vary, but instead takes the value 4 for three-dimensional regions and 3 for two-dimensional problems. Obviously, Eqs (9.30c,d) are the same fundamental solutions that were presented in the previous sections, but now with a different nomenclature. For example, in the three-dimensional case, the function $g_{ij}(x - \xi, t)$ is defined by Eq (9.19a) while $\dot{g}_{Tj}(x - \xi, t)$ is contained in Eq. (9.19b). On the other hand, the functions $f_{ij}(x - \xi, t)$ and $\dot{f}_{Tj}(x - \xi, t)$ can be obtained from g_{ij} and $\dot{g}_{Tj}$ through the constitutive relationships. The superposed dots, as before, represent time derivatives which have been introduced here only for notational convenience. Ultimately, after the time integration these unit step solutions will be required in explicit form, and these are also detailed for both two- and three-dimensions in Dargush[14] and Dargush and Banerjee.[15–17]

In the absence of body forces and sources and under zero initial conditions, simplifying the volume integrals and factoring out the common e_j term, Eq. (9.28) leads to the following set of equations at an interior point ξ:

$$u_j(\xi, t) = \int_S [\dot{g}_{ij}(x - \xi, t) * t_i(x, t) + \dot{g}_{Tj}(x - \xi, t) * q(x, t)$$

$$- \dot{f}_{ij}(x - \xi, t) * u_i(x, t) - \dot{f}_{Tj}(x - \xi, t) * T(x, t)] \, dS(x) \qquad (9.31)$$

A similar relationship is also desired for the interior temperatures. To that end, return to the reciprocal theorem (9.28) and select, instead, for state (1), the infinite space response to a unit pulse heat source, acting at time zero and at point ξ within V. That is, let

$$b_i^{(1)}(x, t) = 0 \qquad (9.32a)$$

$$\psi^{(1)}(x, t) = \delta(x - \xi)\delta(t) \qquad (9.32b)$$

and consequently

$$u_i^{(1)}(x, t) = g_{iT}(x - \xi, t) \qquad (9.32c)$$

$$T^{(1)}(x, t) = \dot{g}_{TT}(x - \xi, t) \qquad (9.32d)$$

$$t_i^{(1)}(x, t) = f_{iT}(x - \xi, t) \qquad (9.32e)$$

$$q^{(1)}(x, t) = \dot{f}_{TT}(x - \xi, t) \qquad (9.32f)$$

Since this is the solution due to a unit pulse, the functions g_{iT} and $\dot{g}_{TT}$ are the time derivatives of the unit step Green functions presented previously. In the end, again the kernels will be these unit step solutions. These may also be found in explicit form in Dargush[14] and Dargush and Banerjee.[15–18]

By selecting state (2) as the actual problem, in which all body forces and heat sources are absent, the reciprocal theorem reduces to

$$T(\xi, t) = \int_S [\dot{g}_{iT}(x - \xi, t) * t_i(x, t) + \dot{g}_{TT}(x - \xi, t) * q(x, t)$$

$$- \dot{f}_{iT}(x - \xi, t) * u_i(x, t) - \dot{f}_{TT}(x - \xi, t) * T(x, t)] \, dS(x) \qquad (9.33)$$

This, of course, is the desired statement for the interior temperature in terms of boundary quantities.

Equations (9.31) and (9.33) can now be combined and rewritten in a more convenient matrix notation as

$$\left\{ \begin{array}{c} u_j(\xi, t) \\ T(\xi, t) \end{array} \right\} = \int_S \left[\begin{array}{cc} \dot{g}_{ij} & \dot{g}_{iT} \\ \dot{g}_{Tj} & \dot{g}_{TT} \end{array} \right]^{\mathrm{T}} * \left\{ \begin{array}{c} t_i(x, t) \\ q(x, t) \end{array} \right\} - \left[\begin{array}{cc} \dot{f}_{ij} & \dot{f}_{iT} \\ \dot{f}_{1j} & \dot{f}_{TT} \end{array} \right]^{\mathrm{T}} * \left\{ \begin{array}{c} u_i(x, t) \\ T(x, t) \end{array} \right\} \mathrm{d}S(x)$$

(9.34)

However, a further compaction of these equations is possible if the displacement and traction vectors are generalized to incorporate the temperature (or pore pressure) and flux respectively as an additional component. Hence, for a three-dimensional problem, define, for example,

$$u_\alpha = \{u_1 \ u_2 \ u_3 \ T\}^{\mathrm{T}}$$

(9.35a)

$$t_\alpha = \{t_1 \ t_2 \ t_3 \ q\}^{\mathrm{T}}$$

(9.35b)

Henceforth, all Greek indices are used to denote generalized degrees of freedom varying from one to four in three dimensions and one to three in two dimensions. Utilizing this new convention in Eq. (9.34), the generalized boundary integral equations may be expressed as

$$c_{\beta\alpha}(\xi)u_\beta(\xi, \tau) = \int_S [\dot{g}_{\beta\alpha}(x; \xi, \tau) * t_\beta(x) - \dot{f}_{\beta\alpha}(x; \xi, \tau) * u_\beta(x)] \, \mathrm{d}S(x)$$

(9.36)

where the tensor $c_{\beta\alpha}$ depends upon the local geometry at ξ.

The time-dependent integrals in Eq. (9.36) can be expressed as the sum of the steady state and transient components as follows:

$$\int_S \dot{g}_{\beta\alpha} * t_\beta \, \mathrm{d}S = \int_S G_{\beta\alpha} t_\beta \, \mathrm{d}S + \int_S g_{\beta\alpha}^{\mathrm{tr}} * t_\beta \, \mathrm{d}S$$

(9.37a)

$$\int_S \dot{f}_{\beta\alpha} * u_\beta \, \mathrm{d}S = \int_S F_{\beta\alpha} u_\beta \, \mathrm{d}S + \int_S f_{\beta\alpha}^{\mathrm{tr}} * u_\beta \, \mathrm{d}S$$

(9.37b)

As a result of this decomposition, Eq. (9.36) can be rewritten as

$$c_{\beta\alpha}(\xi)u_\beta(\xi, \tau) = \int_S [G_{\beta\alpha}(x; \xi, \tau)t_\beta(x, \tau) - F_{\beta\alpha}(x; \xi, \tau)u_\beta(x, \tau)] \, \mathrm{d}S(x)$$

$$+ \int_S [g_{\beta\alpha}^{\mathrm{tr}}(x; \xi, \tau) * t_\beta(x) - f_{\beta\alpha}^{\mathrm{tr}}(x; \xi, \tau) * u_\beta(x)] \, \mathrm{d}S(x)$$

(9.38)

The advantage of this separation will be more evident in the numerical implementation of these kernels with a BEM algorithm, where it is seen that all singularities reside only in the steady state components, $G_{\beta\alpha}$ and $F_{\beta\alpha}$. All of these steady state components are available in Chapters 3 and 4 for both two- and three-dimensional problems (note that for the steady state, $G_{Tj} = 0$ and $F_{Tj} = 0$). The remaining transient portion is completely non-singular.

9.4.1 Integral Formulation for Axisymmetry

In order to model an axisymmetric geometry, adopting a cylindrical coordinate system (r, θ, z) becomes more convenient, as shown in Chapter 5. Transforming the generalized quantities, i.e. displacements, traction, fluxes and temperatures (or pore pressures), into the cylindrical system yields $\bar{u}_\alpha = T_{\beta\alpha} u_\beta$ and $\bar{t}_\alpha = T_{\beta\alpha} t_\beta$, where $\bar{u}_\alpha = \{u_r u_\theta u_z T\}^T$, $\bar{t}_\alpha = \{t_r t_\theta t_z q\}^T$ and the transformation tensor $T_{\beta\alpha}$ is written explicitly as

$$T_{\beta\alpha} = \begin{bmatrix} \cos\theta & -\sin\theta & 0 & 0 \\ \sin\theta & \cos\theta & 0 & 0 \\ 0 & 0 & 1 & 0 \\ 0 & 0 & 0 & 1 \end{bmatrix} \tag{9.39}$$

Upon introduction of the above quantities into Eq. (9.38) and integrating from 0 to 2π in the θ direction yields the integral equations in a cylindrical system:

$$\bar{c}_{\beta\alpha}(\xi)\bar{u}_\beta(\xi, \tau) = \int_C \int_0^{2\pi} [\bar{G}_{\beta\alpha}(x; \xi)\bar{t}_\beta(x, \tau) - \bar{F}_{\beta\alpha}(x; \xi)\bar{u}_\beta(x, \tau)]\, d\theta\, dC$$

$$+ \int_C \int_0^{2\pi} [\bar{g}_{\beta\alpha}^{tr}(x; \xi, \tau) * \bar{t}_\beta(x) - \bar{f}_{\beta\alpha}^{tr}(x; \xi, \tau) * \bar{u}_\beta(x)]\, d\theta\, dC \tag{9.40}$$

In the above equation, $\bar{G}_{\beta\alpha} = \tilde{T}_{\gamma\beta} G_{\gamma\delta} T_{\delta\alpha} r(x)$, $\bar{F}_{\beta\alpha} = \tilde{T}_{\gamma\beta} F_{\gamma\delta} T_{\delta\alpha} r(x)$, $\bar{g}_{\beta\alpha}^{tr} = \tilde{T}_{\gamma\beta} g_{\gamma\delta}^{tr} T_{\delta\alpha} r(x)$, $\bar{f}_{\beta\alpha}^{tr} = \tilde{T}_{\gamma\beta} f_{\gamma\delta}^{tr} T_{\delta\alpha} r(x)$ and $\bar{C}_{\beta\alpha} = \tilde{T}_{\gamma\beta} C_{\gamma\delta} T_{\delta\alpha}$, with $T_{\delta\alpha}$ as the transformation matrix (9.39) evaluated at x and $\tilde{T}_{\gamma\beta}$ as the transformation matrix evaluated at ξ, which reduces to an identity matrix when the z axes of the two systems coincide at ξ.

Equation (9.40) is applicable to the most general axisymmetric thermoelastic or poroelastic analysis in which the boundary conditions and the resulting behaviour need not be symmetric about the z axis. However, if as a specialized case only purely axisymmetric behaviour is considered, all $u_\theta = 0$ and $t_\theta = 0$ and the θ degree of freedom can be eliminated from any subsequent consideration. Since the generalized quantities are now independent of θ, the circumferential integration of the steady state kernels in Eq. (9.40) can be analytically evaluated in terms of elliptic integrals (see Chapter 5). Upon doing so, the following definition may now be used:

$$\hat{G}_{\beta\alpha} = \int_0^{2\pi} \bar{G}_{\beta\alpha}\, d\theta \tag{9.41a}$$

$$\hat{F}_{\beta\alpha} = \int_0^{2\pi} \bar{F}_{\beta\alpha}\, d\theta \tag{9.41b}$$

and Eq. (9.40) then becomes

$$\bar{c}_{\beta\alpha}(\xi)\bar{u}_\beta(\xi, \tau) = \int_C [\hat{G}_{\beta\alpha}(x; \xi)\bar{t}_\beta(x, \tau) - \hat{F}_{\beta\alpha}(x; \xi)\bar{u}_\beta(x, \tau)]\, dC$$

$$+ \int_C \int_0^{2\pi} [\bar{g}_{\beta\alpha}^{tr}(x; \xi, \tau) * \bar{t}_\beta(x) - \bar{f}_{\beta\alpha}^{tr}(x; \xi, \tau) * \bar{u}_\beta(x)]\, d\theta\, dC \tag{9.42}$$

The explicit forms of all the steady state kernels are available in current literature. For instance, G_{ij} and F_{ij} are the same as in axisymmetric elasticity, G_{TT} and F_{TT} are the same as axisymmetric potential flow and G_{Tr}, G_{Tz}, F_{Tr} and F_{Tz} are given in Bakr and Fenner[32] as well as in Dargush and Banerjee.[18,19] All of these steady state axisymmetric kernels are defined in Chapter 5.

9.5 NUMERICAL IMPLEMENTATION

This section describes a general numerical implementation of the above equations. Most of the features employed in the current development have been developed previously for analyses in earlier chapters. For the sake of continuity and completeness, the different features of the numerical algorithm are presented here. These include temporal discretization, a discussion of kernel singularities and the treatment of axisymmetric problems. The corresponding two- and three-dimensional problems follow these in an obvious manner (see Chapter 8).

9.5.1 Temporal Discretization

The time interval from zero to t can be divided into N equal increments of duration Δt without any precision loss. Thus, similar to transient potential flow, the convolution integrals in Eqs (9.40) and (9.42) with limits from zero to t can be represented as a sum of N integrals in the following manner:

$$g_{\beta\alpha}^{\text{tr}}(x;\xi,\tau) * t_\beta(x) = \int_0^t g_{\beta\alpha}^{\text{tr}}(x,t;\xi,\tau)t_\beta(x)\,\mathrm{d}\tau$$

$$= \sum_{n=1}^N \int_{(n-1)\Delta t}^{n\Delta t} g_{\beta\alpha}^{\text{tr}}(x,t;\xi,\tau)t_\beta(x)\,\mathrm{d}\tau \qquad (9.43a)$$

$$f_{\beta\alpha}^{\text{tr}}(x;\xi,\tau) * u_\beta(x) = \int_0^t f_{\beta\alpha}^{\text{tr}}(x,t;\xi,\tau)u_\beta(x)\,\mathrm{d}\tau$$

$$= \sum_{n=1}^N \int_{(n-1)\Delta t}^{n\Delta t} f_{\beta\alpha}^{\text{tr}}(x,t;\xi,\tau)u_\beta(x)\,\mathrm{d}\tau \qquad (9.43b)$$

If, as a first approximation, the primary variables u_β and t_β remain constant within each time increment Δt, these variables can be brought outside the integrals in Eq. (9.43) and the convolution terms now become

$$g_{\beta\alpha}^{\text{u}} * t_\beta = \sum_{n=1}^N t_\beta^n(x) \int_{(n-1)\Delta t}^{n\Delta t} g_{\beta\alpha}^{\text{tr}}(x,t;\xi,\tau)\,\mathrm{d}\tau \qquad (9.44a)$$

$$f_{\beta\alpha}^{\text{tr}} * u_\beta = \sum_{n=1}^N u_\beta^n(x) \int_{(n-1)\Delta t}^{n\Delta t} f_{\beta\alpha}^{\text{tr}}(x,t;\xi,\tau)\,\mathrm{d}\tau \qquad (9.44b)$$

where $t_\beta^n(x)$ and $u_\beta^n(x)$ are the generalized variables for the nth time step. The temporally discretized boundary integral equation for the Nth time step can then be written as follows:

$$c_{\beta\alpha}(\xi)u_\beta^N(\xi) = \int_C [G_{\beta\alpha}(x;\xi)t_\beta^N(x) - F_{\beta\alpha}(x;\xi)u_\beta^N(x)]\,\mathrm{d}C$$

$$+ \sum_{n=1}^N \left\{ \int_C [G_{\beta\alpha}^{N-n+1}(x;\xi)t_\beta^n(x) - F_{\beta\alpha}^{N-n+1}(x;\xi)u_\beta^n(x)]\,\mathrm{d}C \right\} \qquad (9.45)$$

The time integration in Eqs (9.44a,b) is done analytically and the resulting transient boundary kernels were developed by Dargush and Banerjee.[15–19] In the axisymmetric case, the circumferential integration required in the time-dependent parts of Eq. (9.45) can not be done analytically. Fortunately, the kernels involved are non-singular, allowing for a standard numerical quadrature scheme for integration.[18,19,21]

9.5.2 Singularities in the Kernel

As mentioned previously, all the singularities are confined exclusively to the steady-state kernels listed above, whereas all transient kernels (that is $G_{\beta\alpha}^{N-n+1}$ and $F_{\beta\alpha}^{N-n+1}$) are non-singular. It is also clear that the singular behaviour is encountered only for the first time step since the steady state kernels are computed only once. In fact, the singularities of G_{ij} and F_{ij} are identical to those for the corresponding elastostatic kernels, while G_{TT} and F_{TT} singularities are identical to those for potential flow. G_{iT}, G_{Tj} and F_{iT} are non-singular while F_{Tj} is weakly singular.

Another form of singularity exists in axisymmetric kernels when the load point falls on the origin ($\xi_r = 0$). This problem is handled by simply moving the singular point a small radial distance away from the origin; alternatively, a limiting form of the kernels involved may be derived for a point on the origin, as suggested by Bakr and Fenner.[32]

9.6 SUB-SETS OF THE GENERAL COUPLED THEORY

A number of important engineering analyses are embedded within the general coupled quasi-static theory (CQT). These sub-sets can be extracted by specializing CQT formulations under certain specific conditions. Uncoupled quasi-static thermoelasticity (UQT), steady state thermoelasticity and poroelasticity, elastostatics, diffusion analysis and potential flow are the sub-theories of CQT described in the following sections.

9.6.1 Uncoupled Quasi-static Thermoelasticity

For most engineering materials, the coupling in the energy equation between the displacement and temperature fields is found to be insignificant. By ignoring the coupling terms in the energy equation, the governing differential equations for the uncoupled quasi-static theory can be written as

$$kT_{,jj} - \rho c_\epsilon \dot{T} + \psi = 0 \tag{9.46a}$$

$$(\lambda + \mu)u_{j,ij} + \mu u_{i,jj} - (3\lambda + 2\mu)\alpha T_{,i} + b_i = 0 \tag{9.46b}$$

It is evident from Eq. (9.46a) that temperature now obeys the laws of diffusion and the diffusivity, in this case, is $c = k/(\rho c_\epsilon)$. The known temperature field is then introduced into Eq. (9.46b) to determine the deformation response.

Using the fundamental temperature and displacement solutions due to a unit point force and a unit pulse heat source in an infinite medium and their derivatives a set of boundary integral equations can be derived as[16,17]

$$c_{TT}(\xi)T(\xi, t) = \int_S [g_{TT}(x, \xi; t, \tau) * q(x, t) - f_{TT}(x, \xi; t, \tau) * T(x, t)] \, dS(x) \tag{9.47a}$$

$$c_{ij}(\xi)u_j(\xi, t) = \int_S [G_{ij}(x, \xi)t_j(x, t) - F_{ij}(x, \xi)u_j(x, t) + g^{tr}_{Ti}(x, \xi; t, \tau) * q(x, t)$$

$$- f^{tr}_{Ti}(x, \xi; t, \tau) * T(x, t)] \, dS(x) \tag{9.47b}$$

where

$$G_{ij}(x, \xi) \text{ and } F_{ij}(x, \xi) = \text{displacement and traction solutions due to a unit}$$
point force
$$g^{tr}_{Ti}(x, \xi; t, \tau) \text{ and } f^{tr}_{Ti}(x, \xi; t, \tau) = \text{displacement and traction solutions due to a unit}$$
pulse heat source
$$g_{TT}(x, \xi; t, \tau) \text{ and } f_{TT}(x, \xi; t, \tau) = \text{temperature and flux solutions due to a unit pulse}$$
heat source

The corresponding fundamental solutions are determined by simplifying the fundamental solutions for CQT. The response due to a unit step point force at ξ in the j direction is then identical to those of elasticity developed in Chapters 4 and 5 and, of course, $T(x, t) = 0$. The unit step heat source at ξ produces the following response, in three dimensions:

$$u_i(x, t) = \frac{1}{8\pi}\left[\frac{\beta}{k(\lambda + 2\mu)}\right]\left\{\left(\frac{y_i}{r}\right)\left[\text{erfc}\left(\frac{\eta}{2}\right) + \frac{2h_1(\eta)}{\eta^2}\right]\right\} \tag{9.48a}$$

$$T(x, t) = \left(\frac{1}{4\pi r}\right)\left[\left(\frac{1}{k}\right)\text{erfc}\left(\frac{\eta}{2}\right)\right] \tag{9.48b}$$

whereas in two dimensions

$$u_i(x, t) = \frac{\beta r}{2\pi\kappa(\lambda + 2\mu)}\left\{\left(\frac{y_i}{r}\right)\left[\frac{2}{\eta^2}(1 - e^{-\eta^2/4}) + \frac{1}{2}E_1\left(\frac{\eta^2}{4}\right)\right]\right\} \tag{9.49a}$$

$$T(x, t) = \frac{1}{2\pi\kappa}\left[\frac{1}{2}E_1\left(\frac{\eta^2}{4}\right)\right] \tag{9.49b}$$

Note that these unit step heat source solutions are identical to those described for the fully coupled cases earlier, except here the diffusivity c in $\eta = r/\sqrt{ct}$ is simply $c = \kappa/(\rho c_e)$.

Evidently, in this theory an applied force does not produce any temperature change, whereas a heat source produces both displacement and temperature changes. Therefore, the thermal response can be determined without involving the entire gamut of displacement equations. Thus the discretized integral equations become[14–21,33,34]

$$c_{TT}(\xi)T^N(\xi, t) = \int_S [G_{TT}(x, \xi)q(x) - F_{TT}(x, \xi)T(x)]\,dS(x)$$

$$+ \sum_{n=1}^{N} \left\{ \int_S [G_{TT}^{N-n+1}(x, \xi; t)q(x, t) - F_{TT}^{N-n+1}(x, \xi; t)T(x, t)]\,dS(x) \right\}$$

$$(9.50a)$$

and

$$c_{ij}(\xi)u_j(\xi, t) = \int_S [G_{ij}(x, \xi)t_j(x, t) - F_{ij}(x, \xi)u_j(x, t)]\,dS(x)$$

$$+ \sum_{n=1}^{N} \int_S [G_{Ti}^{N-n+1}(x, \xi; t)q(x, t) - F_{Ti}^{N-n+1}(x, \xi; t)T(x, t)]\,dS(x) \quad (9.50b)$$

9.6.2 Steady State Thermoelasticity and Poroelasticity

In the steady state case, the analysis is time independent and the governing differential equations of the coupled theory can be simplified to those discussed in Chapter 4. The fundamental solutions of the deformation equation, for this case, is identical to those of elastostatics and $T(x, \xi) = 0$. On the other hand, the response due to a unit step heat source at ξ is written, in three dimensions, as

$$u_i(x) = \frac{1}{8\pi} \left[\frac{\beta}{k(\lambda + 2\mu)} \right] \left[\left(\frac{y_i}{r} \right) \right] \quad (9.51a)$$

and in two dimensions as

$$u_i(x) = \frac{r}{8\pi} \left[\frac{\beta}{k(\lambda + 2\mu)} \right] \left[\left(\frac{y_i}{r} \right) r(1 - 2\ln r) \right] \quad (9.51b)$$

The corresponding fundamental solutions for temperature are exactly identical to those of steady state potential flow. All convolution integrals are eliminated and the resulting integral equations may be written as[14–21,33,34]

$$c_{TT}(\xi)T^N(\xi) = \int_S [G_{TT}(x, \xi)q(x) - F_{TT}(x, \xi)T(x)]\,dS(x) \quad (9.52a)$$

$$c_{ij}(\xi)u_j(\xi) = \int_S [G_{ij}(x, \xi)t_j(x) - F_{ij}(x, \xi)u_j(x)]\,dS(x)$$

$$+ \int_S [G_{Ti}(x, \xi)q(x) - F_{Ti}(x, \xi)T(x)]\,dS(x) \quad (9.52b)$$

The above thermoelastic case was considered in Chapters 3, 4 and 5.

9.7 APPLICATIONS

9.7.1 Poroelasticity

Consolidation under a strip load[15] A number of analytical, as well as finite element, solutions are available for the consolidation of a single poroelastic stratum beneath a strip load. The geometry for this problem is completely defined in the xy plane by the depth of the layer H and the total breadth of the strip load $2a$. The lower boundary remains impervious, while free drainage is permitted along the entire upper surface. For convenience, non-dimensional forms of the material parameters are utilized. In particular, let $E = 1.0$, $\nu = 0.0$, $\kappa = 1.0$, $\nu_u = 0.5$ and $B = 1.0$. Notice, for this set of properties, that the diffusivity is unity. The presentation of the results is simplified further by also assuming unit values for both the half-width a and the applied traction.

Analytical solutions for this general problem were developed by Gibson *et al.*,[35] although previous work for the special case of infinite H is contained in Schiffman *et al.*[36] The finite element results included in the following graphs are those of Yousif.[37]

The boundary element mesh for the particular case of $H/a = 5$ is depicted in Fig. 9.1. It contains 12 elements, 25 boundary nodes and 9 interior points. Symmetry is imposed along $x = 0$. As in the finite element analysis cited above, the region is truncated at a horizontal dimension of $L/a = 10$. While theoretically the vertical elements along $x = 10$ are unnecessary, closure of the region permits accurate determination of the strongly singular diagonal blocks of the **F** matrix via rigid body considerations.

Results are presented for $H/a \to \infty$. Fig. 9.2 contains the pore pressure time history for a point, on the centreline, a unit depth below the surface. The well-known

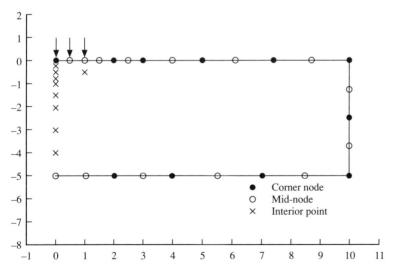

Figure 9.1 Consolidation under a strip load—boundary element model.

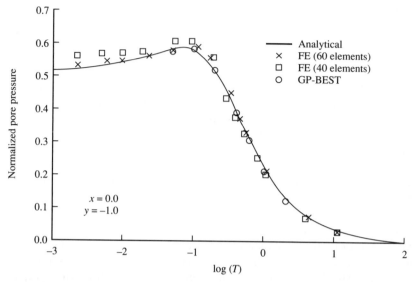

Figure 9.2 Consolidation under strip load—results.

Mandel–Cryer[38] effect, of increasing pore pressure during the early stages of the process, is evident in the graph. Actually, only the final portion of the rising pore pressure response was captured in the boundary element analysis, since a relatively large time step was selected. Even with this large step, however, the analytical solution is accurately reproduced.

Consolidation under rectangular loads[15,27]   The next poroelastic example involves the fully three-dimensional problem of the consolidation of a half-space under rectangular loads of various dimensions. The three specific cases considered involve length (*b*) to width (*a*) ratios of 1.0, 2.0 and 8.0 for the loaded area. A boundary element model corresponding to the latter case is shown in Fig. 9.3.

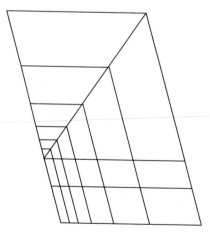

Figure 9.3 Consolidation under a load ($b/a = 8$)— boundary element model.

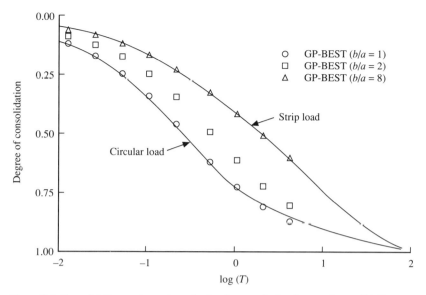

Figure 9.4 Consolidation under rectangular loads—results for degree of consolidation.

Notice that with the present formulation, discretization is required on only a small portion of the half-space. Additionally, quadrantal symmetry is utilized to reduce computational effort.

Figure 9.4 displays the results for the degree of consolidation at the centre of the loaded area versus non-dimensional time. For comparative purposes, the analytical curves of Gibson *et al.*[35] for consolidation under strip and circular loads are also shown. It is evident, from the graph, that the behaviour under a rectangular load with an aspect ratio of 8 is nearly identical to the response due to a strip load. At the other end of the spectrum, the results for a square load closely follow the analytical solution for a circular load, with a slight deviation present at large times. As expected, the boundary element results for $b/a = 2$ are between the two limiting cases.

Consolidation of a solid sphere[18,21]   Based upon the theories of Terzaghi and, subsequently, of Biot, the first quasi-static analytical solutions to the problem of a sphere subjected to hydrostatic pressure p were presented by Cryer.[38] Closed form solutions for the surface displacement and the pore pressure change at the centre of the sphere were developed in the above and these are discussed below. It is notable that Cryer's formulations are constructed in terms of non-dimensional quantities. Hence, by definition, $T = c_v t / a^2$, $U_p = u_p / P$, $R = r / a$, $\mu_c = \mu / (2\mu + \lambda)$ and $\lambda_c = \lambda / (2\mu + \lambda)$, where a is the radius of the sphere, c_v is the coefficient of consolidation, u_p is the pore pressure, p is the applied load intensity, t is the time and r is the radial distance from the centre.

Three different cases, corresponding to $\mu_c = 0, 0.25$ and 0.5, are considered

Table 9.2 Material constants

μ_c	ν	E
0.5	0.0	1.0
0.25	0.333	0.666
0.01	0.495	0.0299

using the poroelastic BEM formulation. However, since $\mu_c = 0$ implies incompressible behaviour leading to $\nu = 0.5$, a value of 0.01 is used to approximate the condition $\mu_c = 0$. The material constants for this problem corresponding to each of the above cases are shown in Table 9.2.

The fluid properties, namely the permeability, the density of the fluid and the undrained Poisson ratio, are all selected as unity. The axisymmetric BEM mesh is shown in Fig. 9.5 to consist of four three-noded elements. The size of the time step is selected as 0.006 25 based on the diffusivity and the element length. The results obtained from the present analysis are compared to the exact solutions discussed earlier. Figure 9.6 displays the BEM results for the variation of the radial displacement at the surface ($R = 1$) with time for $\mu_c = 0$, 0.25, 0.50. Cryer's solution for this case is also plotted. The agreement between the solutions is excellent.

9.7.2 Thermoelasticity

Sudden cooling of a steel sphere[19,21] As an illustrative example of axisymmetric quasi-static thermoelasticity, consider the behaviour of a rapidly cooled sphere. The radius of the solid sphere is taken as unity and the initial uniform temperature is 200°F. It is then rapidly cooled to 0°F. The exact solutions to this problem, for the variations of temperature and stresses in the sphere, have been provided by Carslaw and Jaeger.[39]

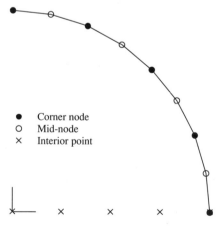

●   Corner node
○   Mid-node
×   Interior point

Figure 9.5 Consolidation of a solid sphere—boundary element model.

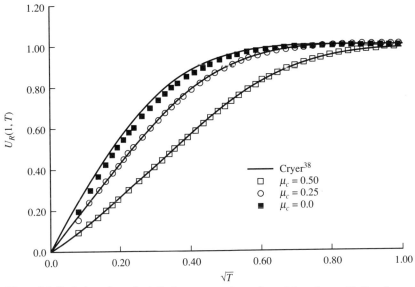

Figure 9.6 Variation of a radical displacement at the surface of the sphere with time for $\mu_c = 0.0, 0.25$ and 0.50.

The material properties for the sphere were selected to be those of carbon steel. Thus,

$$E = 30 \times 10^6 \, \text{lb/in}^2 \qquad\qquad k = 5.8 \, \text{in} \cdot \text{lb/s} \cdot \text{in} \cdot {}^{\circ}\text{F}$$

$$\nu = 0.3 \qquad\qquad\qquad\qquad \rho c_\epsilon = 283.0 \, \text{in} \cdot \text{lb/in}^3 \cdot {}^{\circ}\text{F}$$

$$\alpha = 6.0 \times 10^{-6}/{}^{\circ}\text{F}$$

The axisymmetric boundary element model for this problem consists of only three quadratic elements comprising of seven functional nodes, as shown in Fig. 9.5. Only the positive octant of the sphere is modelled utilizing the symmetry conditions. The sphere is completely unconstrained. In addition, a number of interior points are placed within the sphere to monitor the response of the temperature and stresses in the interior. Based upon the diffusivity and a representative size and using a time factor $\Delta\tau$ of 0.05, a time step value of 1.25 s is utilized. The time factor $\Delta\tau$ is given by

$$\Delta\tau = \frac{ct}{L^2}$$

where L is the element length.

However, another parallel analysis using a smaller time step ($\Delta t = 0.625$ s) is also carried out to check the convergence of the results with time. The resulting temperature variation with time is plotted in Fig. 9.7 for three r/b ratios, namely, 0.0, 0.5 and 0.75. The analytical solutions obtained above are also plotted for the temperature. Excellent correlation is obtained between the two sets of results. The next figure, Fig. 9.8, displays the results for circumferential

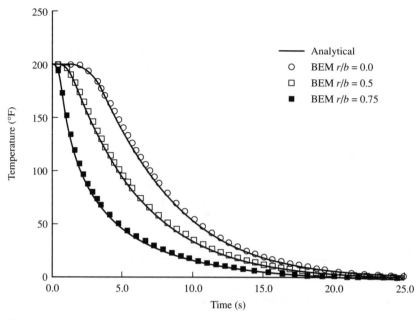

Figure 9.7 Temperature variation with time for $r/b = 0.0$, 0.5 and 0.75.

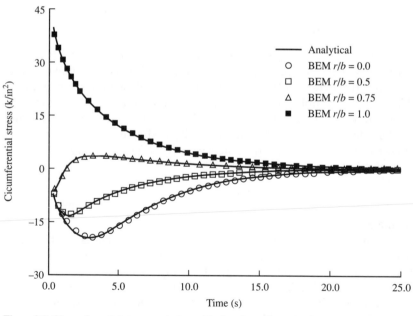

Figure 9.8 Circumferential stress variation with time for $r/b = 0.0$, 0.5, 0.75 and 1.0.

stresses from the analysis as well as the exact solutions.[21] The variation of circumferential stress at the same four radii ($r/b = 0, 0.5, 0.75$ and 1.0) are displayed with time. The highest stress is on the surface at a very short time ($t = 0^+$), following which it decays to zero. Other smaller amplitude peaks are observed for the different radii and the stress at the centre, peaks at the same instance as the radial stress.

Turbine blade   For the final application, the plane strain response of an internally cooled turbine blade is examined under the start-up thermal transients. The boundary element model of the blade is illustrated in Fig. 9.9. In this problem, the two generic modelling regions (GMR) approach is chosen solely to enhance computational efficiency. This is accomplished by reducing the aspect ratio of individual GMRs and by creating a block-banded system matrix. The leading (left-hand) GMR consists of 26 quadratic elements, while 24 elements are used to model the trailing (right-hand) region.

The blade is manufactured of stainless steel with the following thermomechanical properties:

$$E = 29 \times 10^6 \text{ lb/in}^2 \qquad\qquad k = 1.65 \text{ in} \cdot \text{lb/s} \cdot \text{in} \cdot {}^\circ\text{F}$$
$$\nu = 0.30 \qquad\qquad\qquad\qquad \rho c_\epsilon = 368.0 \text{ in} \cdot \text{lb/in}^3 \cdot {}^\circ\text{F}$$
$$\alpha = 9.6 \times 10^{-6}/\text{in} \cdot \text{in} \cdot {}^\circ\text{F}$$

During operation a hot gas flows outside the blade, while relatively cool gas passes through the internal holes. The gas temperature transients are plotted in Fig. 9.10 for a typical start-up. Convection film coefficients are specified as follows: outer surface at leading edge $h = 50$ in $\cdot$ lb/s $\cdot$ in$^2 \cdot {}^\circ$F, remainder of the outer surface $h = 20$ in $\cdot$ lb/s $\cdot$ in$^2 \cdot {}^\circ$F, inner cooling hole surfaces $h = 10$ in $\cdot$ lb/s $\cdot$ in$^2 \cdot {}^\circ$F and a time step of 0.2 s is employed for the BEM analysis.

The response at two points, A on the leading edge and B at mid-span, are displayed in Figs 9.11 and 9.12. Notice that temperatures and stresses are consistently higher on the leading edge, reaching peak values of approximately 1500 ${}^\circ$F and -60 k/in^2 respectively.

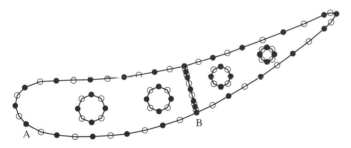

Figure 9.9 Turbine blade—boundary element model.

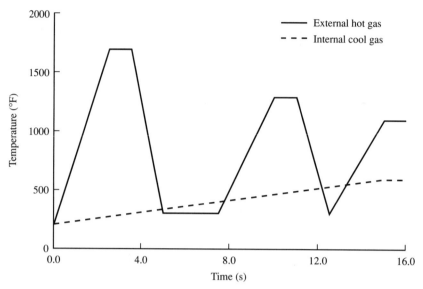

Figure 9.10 Turbine blade—start-up transient.

Also, as is evident from Fig. 9.12, significant stress reversals occur during the start-up. It should be emphasized that the stresses presented for points A and B are surface stresses, calculated by satisfying the constitutive laws, strain–displacement and equilibrium directly at the boundary point in the usual manner (see Chapters 4, 7 and 17).

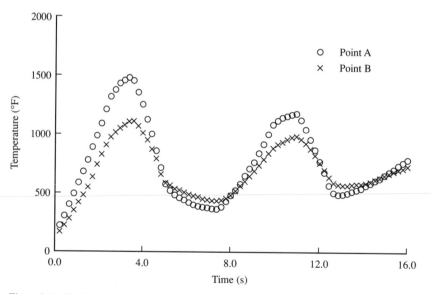

Figure 9.11 Turbine blade—GP-BEST results for temperature.

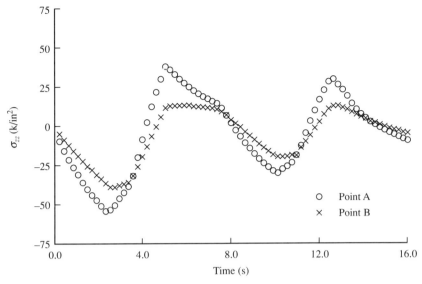

Figure 9.12 Turbine blade—GP-BEST results for surface stresses.

9.8 CONCLUDING REMARKS

This chapter was devoted to developing a linear, time-domain BEM formulation for thermomechanics and soil consolidation. The boundary integral equations were derived for two- and three-dimensional cases by utilizing an appropriate reciprocal theorem and selecting one of the independent states as the appropriate infinite space fundamental solution. The exact equations were then discretized in time and space using numerical implementation techniques discussed in earlier chapters. This allows for the analysis of a number of significant practical applications in both fields, involving complex geometries and time-dependent boundary conditions. A number of sub-sets of the general coupled theory were discussed as special cases.

REFERENCES

1. Banerjee, P.K. and Butterfield, R. (1981) *Boundary Element Methods in Engineering Science*, McGraw-Hill, London.
2. Kuroki, T. and Onishi, K. (1982) Boundary element method in Biot's linear consolidation, *Applied Mathematics Modelling*, Vol. 6, pp. 105–110.
3. Aramaki, G. and Yasuhara, K. (1985) Application of the boundary element method for axisymmetric Biot's consolidation, *Engineering Analysis*, Vol. 2, pp. 184–191.
4. Aramaki, G. (1986) Boundary elements for thin layers with high permeability in Biot's consolidation analysis, *Applied Mathematics Modelling*, Vol. 10, pp. 82–86.
5. Cheng, A. H-D. and Liggett, J.A. (1984) Boundary integral equation method for linear porous-elasticity with applications to soil consolidation, *International Journal of Numerical Methods in Engineering*, Vol. 20, pp. 255–278.

6. Nishimura, N. (1985) A BIE formulation for consolidation problems, in *Boundary Elements VIII, Proceedings of the 7th International Conference*, Eds C.A. Brebbia and G. Maier, Springer-Verlag, Berlin.

7. Tanaka, M. and Tanaka, K. (1981) Boundary element approach to dynamic problems in coupled thermoelasticity—1. Integral equation formulation, *Solid Mechanical Archives*, Vol. 6, pp. 467–491.

8. Sladek, V. and Sladek, J. (1983) Boundary integral equation method in thermoelasticity, Part I: general analysis, *Applied Mathematics Modelling*, Vol. 7, pp. 241–253.

9. Sladek, V. and Sladek, J. (1984) Boundary integral equation method in thermoelasticity, Part III: uncoupled thermoelasticity, *Applied Mathematics Modelling*, Vol. 8, pp. 413–418.

10. Sladek, V. and Sladek, J. (1984) Boundary integral equation method in two-dimensional thermoelasticity, *Engineering Analysis*, Vol. 1, pp. 135–148.

11. Chaudounet, A. (1987) Three-dimensional transient thermoelastic analysis by the BIE method, *International Journal of Numerical Methods in Engineering*, Vol. 24, pp. 25–45.

12. Masinda, J. (1984) Application of the boundary element method to 3D problems of nonstationary thermoelasticity, *Engineering Analysis*, Vol. 1, pp. 66–69.

13. Sharp, S. and Crouch, S.L. (1986) Boundary integral methods for thermoelasticity problems, *Journal of Applied Mechanics*, Vol. 53, pp. 298–302.

14. Dargush, G.F. (1987) Boundary element methods for the analogous problems of thermomechanics and soil consolidation, PhD Thesis, State University of New York at Buffalo.

15. Dargush, G.F. and Banerjee, P.K. (1989) A time domain element method for poroelasticity, *International Journal of Numerical Methods in Engineering*, Vol. 28, pp. 2423–2449.

16. Dargush, G.F. and Banerjee, P.K. (1989) The boundary element method for plane problems of thermoelasticity, *International Journal of Solids and Structures*, Vol. 25, pp. 999–1021.

17. Dargush, G.F. and Banerjee, P.K. (1990) BEM analysis for three-dimensional problems of transient thermoelasticity, *International Journal of Solids and Structures*, Vol. 26, pp. 199–216.

18. Dargush, G.F. and Banerjee, P.K. (1991) A boundary element method for axisymmetric soil consolidation, *International Journal of Solids and Structures*, Vol. 7, pp. 897–915.

19. Dargush, G.F. and Banerjee, P.K. (1992) Time dependent axisymmetric thermoelastic boundary element analysis, *International Journal of Numerical Methods in Engineering*, Vol. 33, pp. 695–717.

20. Dargush, G.F. and Banerjee, P.K. (1989) Boundary element methods for poroelastic and thermoelastic analyses, in *Developments in Boundary Element Methods*, Vol. 5, Eds P.K. Banerjee and R.B. Wilson, Elsevier Applied Science, London.

21. Chopra, M.B. (1992) Linear and nonlinear analyses of axisymmetric problems in thermomechanics and soil consolidation, PhD Thesis, State University of New York at Buffalo.

22. Boley, B.A. and Weiner, J.H. (1960) *Theory of Thermal Stresses*, John Wiley, New York.

23. Terzaghi, K. (1925) *Erdbaumechanik auf Bodenphysikalischer Grundlage*, F. Deuticke, Vienna.

24. Biot, M.A. (1941) General theory of three-dimensional consolidation, *Journal of Applied Physics*, Vol. 12, pp. 155–164.

25. Biot, M.A. (1956) Thermoelasticity and irreversible thermodynamics, *Journal of Applied Physics*, Vol. 27, pp. 240–253.

26. Biot, M.A. (1956) General solutions of the equations of elasticity and consolidation for a porous material, *Journal of Applied Mechanics*, Vol. 23, No. 1, pp. 91–96.

27. Dargush, G.F. and Banerjee, P.K. (1991) A new boundary element method for three-dimensional coupled problems of consolidation and thermoelasticity, *Journal of Applied Mechanics*, Vol. 58, pp. 28–36.

28. Nowacki, W. (1966) Green's functions for a thermoelastic medium (quasistatic problems), *Bulletin of Institute Political Jasi, Sierie Noua*, Vol. 12, Nos. 3–4, pp. 83–92.

29. Cleary, M.P. (1977) Fundamental solutions for a fluid-saturated porous solid, *International Journal of Solids and Structures*, Vol. 13, pp. 785–806.

30. Rudnicki, J.W. (1987) Fluid mass sources and point forces in linear elastic diffusive solids, *Mechanics and Materials*, Vol. 5, pp. 383–393.

31. Ionescu-Cazimir, V. (1964) Problem of linear coupled thermoelasticity. Theorems on reciprocity for the dynamic problem of coupled thermoelasticity, I, *Bulletin de l'Academie Polonaise des Sciences, Series des Sciences Techniques*, Vol. 12, No. 9, pp. 473–488.

32. Bakr, A.A. and Fenner, R.T. (1983) Boundary integral equation analysis of axisymmetric thermo-elastic problems, *Journal of Strain Analysis*, Vol. 18, pp. 239–251.

33. Raveendra, S.T. and Banerjee, P.K. (1992) Boundary element analysis of cracks in thermally loaded planar structures, *International Journal of Solids and Structures*, Vol. 29, No. 18, pp. 2301–2317.

34. Raveendra, S.T., Banerjee, P.K., and Dargush, G.F. (1992) Three-dimensional analysis of thermally loaded cracks, *International Journal of Numerical Methods in Engineering*, Vol. 36, pp. 1909–1926.

35. Gibson, R.E., Schiffman, R.L. and Pu, S.L. (1970) Plane strain and axially symmetric consolidation of a clay layer on a smooth impervious base, *Quarterly Journal of Mechanical Applied Mathematics*, Vol. 23, pp. 505–520.

36. Schiffman, R.L., Chen, A.T-F. and Jordan, J.C. (1969) An analysis of consolidation theories, *ASCE, Journal of Soil Mechanics and Foundations Division*, Vol. 95(SM1), pp. 285–312.

37. Yousif, N.B. (1984) Finite element analysis of some time independent construction problems in geotechnical engineering, PhD Thesis, State University of New York at Buffalo.

38. Cryer, C.W. (1963) A comparison of the three-dimensional consolidation theories of Biot and Terzaghi, *Quarterly Journal of Mechanical Applied Mathematics*, Vol. XVI, No. 4, pp. 401–411.

39. Carslaw, H.S. and Jaeger, J.C. (1959) *Conduction of Heat in Solids*, 2nd edn, Clarendon Press, Oxford.

DYNAMIC ANALYSIS

10.1 INTRODUCTION

Perhaps unexpectedly, the use of integral formulation for the analysis of transient phenomena in solids and fluids has a long history. In a great many of these problems a part of the boundary is at infinity and, since they are particularly attractive for such cases, boundary element methods have been used quite extensively. References 1 to 26 provide a good account of the classical work in elastodynamics and related topics. Although the basic integral formulations for elastodynamics and wave propagation have been known for well over a hundred years, their adaptation to construct numerical algorithms for the solutions of boundary value problems is relatively new and the first computer-based developments only appeared in the early sixties, mainly starting with the excellent papers of Banaugh and Goldsmith.[8,26]

10.2 GOVERNING EQUATIONS AND FUNDAMENTAL SOLUTIONS OF ELASTODYNAMICS

10.2.1 Governing Equations

The elastodynamic, small-displacement field $u_i(x, t)$ in an isotropic homogeneous elastic body is governed by Navier's equation:

$$(\lambda + \mu)\frac{\partial^2 u_i}{\partial x_i\, \partial x_j} + \mu \frac{\partial^2 u_j}{\partial x_i\, \partial x_i} + \rho(\psi_j - \ddot{u}_j) = 0 \tag{10.1}$$

where

λ, μ = elastic constants
ρ = mass density of the deformed body
ψ_j = body forces per unit mass
$\ddot{u}_j = \partial^2 u_j / \partial t^2$, the acceleration

A domain V, bounded by a surface S, is considered subject to the following initial conditions:

$$u_i(x, t) = u'_i(x) \qquad\qquad \text{at } t = 0$$

$$\frac{\partial u_i(x, t)}{\partial t} = \dot{u}_i(x, t) = V'_i(x) \qquad \text{at } t = 0 \qquad (10.2)$$

and the boundary conditions on S are, in general,

$$u_i(x, t) = f_i(x, t) \qquad\qquad \text{on } S_1$$

$$t_i(x, t) = \sigma_{ij}(x, t) n_j(x) = g_i(x, t) \qquad \text{on } S_2 \qquad (10.3)$$

where $S_1 + S_2 = S$ and the stresses σ_{ij} are given by

$$\sigma_{ij} = \rho(c_p^2 - 2c_s^2)\delta_{ij}\frac{\partial u_m}{\partial x_m} + \rho c_s^2 \left(\frac{\partial u_i}{\partial x_j} + \frac{\partial u_j}{\partial x_i} \right) \qquad (10.4)$$

where

$$c_p = \left(\frac{\lambda + 2\mu}{\rho} \right)^{1/2} \qquad c_s = \left(\frac{\mu}{\rho} \right)^{1/2}$$

Therefore,

$$\left(\frac{c_s}{c_p} \right)^2 = \frac{1 - 2\nu}{2(1 - \nu)}$$

The constants c_p and c_s, the irrotational and equivoluminal wave velocities respectively, are also called the dilatational and shear wave velocities (or simply the P-wave and S-wave velocities). We shall often use the symbols c_1 and c_2 respectively to designate these wave velocities.

10.2.2 The Singular Solution of Stokes

The fundamental point force solution for the dynamic problem was first obtained by Stokes in 1849 although certain features of the solution were anticipated by Poisson in 1829. In order to demonstrate various features of this solution most simply it is convenient to rewrite Eq. (10.1) in vector notation as

$$\mu\nabla \cdot \nabla\mathbf{u} + (\lambda + \mu)\nabla \cdot \mathbf{u} + \rho\psi = \rho\frac{\partial^2 \mathbf{u}}{\partial t^2} \qquad (10.5)$$

By using the Stokes–Helmholtz resolution theorem, which states that a sufficiently smooth vector field may be decomposed into irrotational and solenoidal parts, $\psi(x, t)$ can be written as

$$\psi = \nabla f + \nabla \times \mathbf{F} \tag{10.6}$$

Similarly, for the displacement vector we have

$$\mathbf{u} = \nabla V + \nabla \times \mathbf{V} \tag{10.7}$$

where $f(x, t)$ and $V(x, t)$ are scalars whereas $\mathbf{F}(x, t)$ and $\mathbf{V}(x, t)$ are vector functions of position and time.

Using Eqs (10.6) and (10.7) and writing $\nabla^2 = \nabla \cdot \nabla$ we can rewrite Eq. (10.5) as

$$\nabla \left(c_p^2 \nabla^2 V + f - \frac{\partial^2 V}{\partial t^2} \right) + \nabla \times \left(c_s^2 \nabla^2 \mathbf{V} + \mathbf{F} - \frac{\partial^2 \mathbf{V}}{\partial t^2} \right) = \mathbf{0} \tag{10.8}$$

This equation is satisfied identically if we choose V and $\mathbf{V}$ to be solutions of the non-homogeneous wave equations

$$c_p^2 \nabla^2 V + f = \frac{\partial^2 V}{\partial t^2} \tag{10.9a}$$

$$c_s^2 \nabla^2 \mathbf{V} + \mathbf{F} = \frac{\partial^2 \mathbf{V}}{\partial t^2} \tag{10.9b}$$

It is now possible to solve Eqs (10.9a,b) and construct the total displacement field $\mathbf{u}$, using Eq. (10.7), as

$$\mathbf{u} = \mathbf{u}^1 + \mathbf{u}^2 \tag{10.10}$$

where $\mathbf{u}^1$ is the solution of Eq. (10.9a) and $\mathbf{u}^2$ that of Eq. (10.9b). Clearly the first part, $\mathbf{u}^1$, is curl free whereas the second part, $\mathbf{u}^2$, is divergence free, which means that

$$\nabla \times \mathbf{u}^1 = \mathbf{0} \quad \text{and} \quad \nabla \cdot \mathbf{u}^2 = 0$$

The solution we require, the fundamental solution of the equations of elastodynamics, is that in an infinite region in three dimensions due to a concentrated body force acting at a point ξ in a fixed direction e_j but of time-dependent magnitude $f(t)$, i.e. we need specifically the solution of the elastodynamic equations for a body force

$$\rho \psi_i(x, t) = f(t)\delta(x - \xi) e_i \tag{10.11}$$

The displacement at a point x at time t due to a force vector of magnitude $f(t)$ acting at ξ in the direction of the x_i axis is given by[3]

$$u_i(x, t) = G_{ij}(x, t; \xi, f) e_j(\xi) \tag{10.12}$$

where

$$G_{ij} = G_{ij}^1 + G_{ij}^2$$

with

$$G_{ij}^1 = \frac{1}{4\pi\rho r}\left[-\left(\frac{3y_i y_j}{r^2} - \delta_{ij}\right)\int_0^{c_p^{-1}} \lambda f(t - \lambda r)\,d\lambda + \frac{y_i y_j}{r^2 c_p^2}f\left(t - \frac{r}{c_p}\right)\right]$$

$$G_{ij}^2 = \frac{1}{4\pi\rho r}\left[\left(\frac{3y_i y_j}{r^2} - \delta_{ij}\right)\int_0^{c_s^{-1}} \lambda f(t - \lambda r)\,d\lambda - \frac{y_i y_j}{r^2 c_s^2}f\left(t - \frac{r}{c_s}\right) + \frac{\delta_{ij}}{c_s^2}f\left(t - \frac{r}{c_s}\right)\right]$$

G_{ij}^1 and G_{ij}^2 are respectively known as the irrotational and equivoluminal parts of the Stokes tensor G_{ij}. The functions f vanish for negative arguments reflecting that the longitudinal and the shear parts of the wave due to a force $f(t)$ acting at ξ could only be felt at x after times $t = r/c_p$ and r/c_s respectively.

The stresses at (x, t) are given by

$$\sigma_{ij} = \rho[(c_p^2 - 2c_s^2)G_{mk,m}\delta_{ij} + c_s^2(G_{ik,j} + G_{jk,i})]\,e_k$$

$$= T_{ijk}(x, t; \xi, f)\,e_k(\xi) \tag{10.13}$$

The traction at a point (x, t) can be obtained from

$$t_i(x, t) = F_{ik}(x, \xi, f)\,e_k(\xi) \tag{10.14}$$

with

$$F_{ik} = T_{ijk}n_j(x)$$

The above equations, (10.12) to (10.14), are also valid for an impulsive force $\delta(t - \tau)$ applied at time τ, i.e.

$$f(t) = \delta(t - \tau)$$

We can therefore obtain the solution for a unit impulse applied at a point ξ at time τ by replacing f by δ and t by $(t - \tau)$. For example, the displacements and tractions for this case are given by

$$u_i(x, t) = g_{ij}(x, t; \xi, \tau)\,e_j(\xi) \tag{10.15a}$$

$$t_i(x, t) = f_{ik}(x, t; \xi, \tau)\,e_k(\xi) \tag{10.15b}$$

For two-dimensional problems the above solutions can be integrated in the third direction to derive the necessary components of the two-dimensional fundamental solution.[3]

10.2.3 Steady State Elastodynamics

If we choose our instant of observation of motion to be a sufficiently long time after the initiating disturbances we can assume that the physical components of the problem are harmonic in time with an angular frequency ω (i.e. we are then dealing with a steady state elastodynamic problem). The analysis is then greatly simplified since the time variable is thereby eliminated from the governing differential equations and the initial value–boundary value problem reduces to a boundary value problem only.

Thus, if we assume that

$$\psi_i(x, t) = \psi_i(x, \omega) e^{-i\omega t}$$

$$u_i(x, t) = u_i(x, \omega) e^{-i\omega t} \tag{10.16}$$

$$t_i(x, t) = t_i(x, \omega) e^{-i\omega t}$$

where $\psi_i(x, \omega)$, $u_i(x, \omega)$ and $t_i(x, \omega)$ are complex amplitudes of body forces, displacements and tractions respectively.

We can substitute Eqs (10.16) in Eq. (10.5) to obtain

$$[\mu \nabla^2 + (\lambda + \mu)\boldsymbol{\nabla}\boldsymbol{\nabla}\cdot]\mathbf{u}(x, \omega) + \rho\omega^2\mathbf{u}(x, \omega) = -\rho\psi(x, \omega) \tag{10.17}$$

which is the governing differential equation of steady state elastodynamics. The fundamental solution of Eq. (10.17) for a unit harmonic body force, i.e. the solution of

$$[\mu \nabla^2 + (\lambda + \mu)\boldsymbol{\nabla}\boldsymbol{\nabla}\cdot]\mathbf{G}(x, \xi, \omega) + \rho\omega^2\mathbf{G}(x, \xi, \omega) = -\delta(x, \xi)$$

is given by[3,6]

$$\mathbf{G}(x, \xi, \omega) \equiv G_{mn}(x, \xi, \omega)$$

$$= \frac{1}{4\pi\rho\omega^2}\left[\delta_{mn}k_s\frac{e^{ik_s r}}{r} - \frac{\partial^2}{\partial x_m \partial x_n}\left(\frac{e^{ik_p r}}{r} - \frac{e^{ik_s r}}{r}\right)\right] \tag{10.18a}$$

for three-dimensional problems

$$= \frac{i}{4\rho}\left\{\frac{1}{C_s^2}\delta_{mn}H_0(k_s r) - \frac{1}{\omega^2}\frac{\partial^2}{\partial x_m \partial x_n}[H_0(k_p r) - H_0(k_s r)]\right\} \tag{10.18b}$$

for two-dimensional problems

where $k_p = \omega/c_p$, $k_s = \omega/c_s$ and $H_0(\eta)$ is the zero-order Hankel function of the first kind with argument η [note that the two-dimensional fundamental solutions can also be expressed in terms of modified Bessel functions, as shown later in Eq. (10.30a)].

Whereas in the above we have reduced the transient problem to a steady state one by considering the time of our observation to be sufficiently far from the initial disturbance it is also possible to proceed 'in reverse' and to construct transient solutions from the steady state ones by using a superposition technique which depends on the linearity of the governing differential equation.

Consider the integral

$$u_i(x, t) = \int_{-\infty}^{\infty} u_i(x, \omega) e^{-i\omega t} \, d\omega$$

for all the possible frequencies ω. This equation is a solution of the elastodynamic equation for a forcing function

$$\psi_i(x, t) = \int_{-\infty}^{\infty} \psi_i(x, \omega) \, e^{-i\omega t} \, d\omega$$

for a similarly specified boundary value problem. It is easy to see that $2\pi\psi_i(x, \omega)$ is the Fourier transform of a body force $\psi_i(x, t)$ with respect to time and hence that $2\pi u_i(x, \omega)$ is the solution of the Fourier-transformed elastodynamic equations. We can therefore construct a general solution of the transient elastodynamic equations via $u_i(x, \omega)$ obtained over the complete spectrum of frequencies.

It is also possible to arrive at the solution $u_i(x, t)$ via the Laplace transform of the field equation. As before (see Chapter 8), the Laplace transform of a function $\psi_j(x, t)$ is defined as

$$\psi_j(x, s) = \mathcal{L}[\psi_j(x, t)] = \int_0^{\infty} \psi_j(x, t) \, e^{-st} \, dt$$

We can thus transform the governing differential equation (10.2) to a form that is similar to Eq. (10.17) with ω replaced by $-is$ and the body force replaced by $[\psi_j(x, s) + V'(x) + su_j'(x)]$. The problem is then reduced to the solution of a static problem for any particular value of the Laplace transform parameter s. The major difficulty is then, of course, that of inverting back efficiently to the real space by some numerical means. Several examples of such BEM analyses can be found in the published literature.[15,16,21,27-29]

10.2.4 Reciprocal Work Theorem

Just as in the elastostatic case, discussed in Chapter 4, the reciprocal theorem provides a basis from which the boundary integral equations for elastodynamics can be constructed. We can, of course, derive the required reciprocal theorem using the integration by parts approach described in Chapters 2 and 8. The dynamic reciprocal theorem is in fact a direct extension of Betti's classical theorem in elastostatics and can be simply stated as follows. If there exist two unrelated elastodynamic states of body forces, boundary tractions, boundary displacements, initial displacements and initial velocities such as $(\psi_i, t_i, u_i, u_i', V_i')$ and $(\psi_i^*, t_i^*, u_i^*, u_i^{*\prime}, V_i^{*\prime})$ defined in the same region V bounded by a surface S then, for every $t \geq 0$,

$$\int_S (t_i * u_i^*)(x, t) \, dS + \int_V \rho \left[(\psi_i * u_i^*)(x, t) + V_i'(x) u_i^*(x, t) + u_i'(x) \frac{\partial u_i^*(x, t)}{\partial t} \right] dV$$

$$= \int_S (t_i^* * u_i)(x, t) \, dS + \int_V \rho \left[(\psi_i^* * u_i)(x, t) + V_i^{*\prime}(x) u_i(x, t) + u_i^{*\prime}(x) \frac{\partial u_i(x, t)}{\partial t} \right] dV$$

$$(10.19)$$

where $*$ indicates once again a Reimann convolution.

Similarly, for the steady state elastodynamic case, Eq. (10.17), which is known as the Helmholtz equation, can be used to derive the reciprocal identity for steady state elastodynamics. If $u_i(x, \omega)$ and $u_i^*(x, \omega)$ are two solutions of the reduced elastodynamic equations then

$$\int_S t_i(x, \omega) u_i^*(x, \omega) \, dS + \int_V \rho \psi_i(x, \omega) u_i^*(x, \omega) \, dV$$

$$= \int_S t_i^*(x, \omega) u_i(x, \omega) \, dS + \int_V \rho \psi_i^*(x, \omega) u_i(x, \omega) \, dV \quad (10.20)$$

Equations (10.19) and (10.20) can now be used to construct BEM formulations.

10.3 GOVERNING EQUATIONS AND FUNDAMENTAL SOLUTIONS FOR SCALAR WAVE EQUATIONS

10.3.1 Governing Equations

In a great many problems of acoustics, electromagnetic field theory and hydro-dynamics, the governing differential equations for the wave propagation problem are very similar to those of the elastodynamic case discussed above. However, because of the reduced dimensionality of the parameters involved in such problems the analytical complexities are not so severe.

Let us consider the propagation of small-amplitude acoustic waves in a gas of negligible viscosity. If the particle velocity and the excess pressure are u_i and p respectively then, by defining a potential ϕ such that $u_i = \partial\phi/\partial x_i$ and $p = -\rho_0 \, \partial\phi/\partial t$ (ρ_0 being the density of the gas at rest), the governing equations reduce to

$$\frac{\partial^2 \phi(x, t)}{\partial x_i \partial x_i} - \frac{1}{c^2} \frac{\partial^2 \phi}{\partial t^2} = 0 \quad (10.21)$$

where c^2 is a constant.

In electromagnetic field theory Maxwell's equation for a linear, homogeneous, isotropic medium of electric conductivity ε and magnetic permeability μ are

$$\nabla \times \mathbf{H} = \varepsilon \frac{\partial \mathbf{H}}{\partial t} \quad (10.22a)$$

$$\nabla \times \mathbf{E} = -\mu \frac{\partial \mathbf{H}}{\partial t} \quad (10.22b)$$

and
$$\nabla \cdot \mathbf{H} = \nabla \cdot \mathbf{H} = 0 \quad (10.22c)$$

where $\mathbf{E}$ and $\mathbf{H}$ are electric and magnetic vectors at time t.

By taking the curl of Eq. (10.22a) and using Eq. (10.22b) we have

$$\nabla \times \nabla \times \mathbf{H} = -\mu\varepsilon \frac{\partial^2 \mathbf{H}}{\partial t^2}$$

which can be reduced, by using the identity

$$\nabla \times \nabla \times \mathbf{H} = \nabla(\nabla \cdot \mathbf{H}) - \nabla^2 \mathbf{H}$$

and Eq. (10.22c), to the form

$$\nabla^2 \mathbf{H} - \mu\varepsilon \frac{\partial^2 \mathbf{H}}{\partial t^2} = 0 \tag{10.23}$$

which is similar in character to Eq. (10.21).

The governing differential equation for small-amplitude wave propagation in inviscid hydrodynamic problems is also similar to Eq. (10.21) and the differences between the various problems are seen to be physical rather than mathematical. Therefore in what follows we shall describe the systematic solution of the equation

$$\frac{\partial^2 p(x,t)}{\partial t^2} = c^2 \frac{\partial^2 p(x,t)}{\partial x_j \partial x_j} + \psi(x,t) \tag{10.24}$$

and its steady state counterpart.

If we assume a harmonic time dependence for the functions

$$\psi(x,t) = \psi(x,\omega)\,e^{-i\omega t}$$

$$p(x,t) = p(x,\omega)\,e^{-i\omega t}$$

as before, we can express Eq. (10.24) as

$$\frac{\partial^2 p(x,\omega)}{\partial x_j \partial x_j} + k^2 p(x,\omega) = -\frac{1}{c^2}\psi(x,\omega) \tag{10.25}$$

where $k = \omega/c = 2\pi/\lambda c$, with λ the frequency.

10.3.2 Fundamental Solutions

Fundamental solution of Eq. (10.24) for an impulsive source in a three-dimensional problem is given by [i.e. due to $\psi(x,t) = \delta(t - \tau)\,\delta(x,\xi)$][10]

$$p(x,t) = g(x,t;\xi,\tau) = \frac{1}{4\pi c^2 r}\delta\left(t - \frac{r}{c} - \tau\right) \tag{10.26a}$$

$$q(x,t) = \frac{\partial g}{\partial x_k}n_k = f(x,t;\xi,\tau) \tag{10.26b}$$

The corresponding two-dimensional solutions are obtained by integrating the above solutions in the third dimension.

For the periodic case governed by Eq. (10.25) the required fundamental solution is given by

$$G(x,\xi,\omega) = \frac{1}{4\pi}\left(\frac{e^{ikr}}{r}\right) \quad \text{for three-dimensional problems} \tag{10.27a}$$

$$= \frac{i}{4}H_0(kr) \quad \text{for two-dimensional problems} \tag{10.27b}$$

where $H_0(kr)$ is the Hankel function of the first kind of order zero with argument (kr), and

$$F(x, \xi, \omega) = \frac{\partial G}{\partial n} \tag{10.28}$$

Once again the two-dimensional fundamental solution can be expressed in terms of modified Bessel functions, as shown later in Eq. (10.30a).

10.4 INTEGRAL FORMULATIONS FOR STEADY STATE ELASTODYNAMICS

By using the dynamic reciprocal theorem [Eq. (10.20)], the two- and three-dimensional boundary integral representations for Eq. (10.17) can be written in the frequency domain as [with $\psi_i(x, \omega) = 0$][7,27-31]

$$\alpha_{ij}(\xi) u_i(\xi, \omega) = \int_S [G_{ij}(x, \xi, \omega) t_i(x, \omega) - F_{ij}(x, \xi, \omega) u_i(x, \omega)] \, dS(x) \tag{10.29}$$

where S is the surface enveloping the body and the fundamental solution G_{ij} and F_{ij} are defined below. It should be noted here that although the functions G_{ij} and F_{ij} become identical to their static counterpart as ω tends to zero, it is important to evaluate this limit carefully because of the presence of ω in the denominator.

The tensors G_{ij} and F_{ij} in the frequency domain for a two-dimensional problem are of the form[27-31]

$$G_{ij}(x, \xi, \omega) = \frac{1}{2\pi\mu} (A\delta_{ij} - B r_{,i} r_{,j}) \tag{10.30a}$$

$$F_{ji}(x, \xi, \omega) = \frac{1}{2\pi} \left[P\left(\delta_{ij} \frac{\partial r}{\partial n} + r_{,j} n_i \right) + Q\left(r_{,i} n_j - 2 r_{,i} r_{,j} \frac{\partial r}{\partial n} \right) + R r_{,i} r_{,j} \frac{\partial r}{\partial n} + S r_{,i} n_j \right] \tag{10.30b}$$

where

$$A = K_0(z_2) + \left(\frac{1}{z_2} \right) \left[K_1(z_2) - \left(\frac{c_2}{c_1} \right) K_1(z_1) \right]$$

$$B = K_2(z_2) - \left(\frac{c_2}{c_1} \right)^2 K_2(z_1)$$

$$P = \frac{\partial A}{\partial r} - \frac{B}{r}$$

$$Q = -2 \frac{B}{r}$$

$$R = -2\frac{\partial B}{\partial r}$$

$$S = \left[\left(\frac{c_1}{c_2}\right)^2 - 2\right]\left(\frac{\partial A}{\partial r} - \frac{\partial B}{\partial r} - \frac{B}{r}\right)$$

where $z_1 = -iwr/c_1$, $z_2 = -iwr/c_2$ and K_0, K_1 and K_2 are the modified Bessel functions of the second kind, having the following recursive properties:

$$K_2(z) = K_0(z) + \left(\frac{2}{z}\right)K_1(z)$$

$$K_0'(z) = -K_1(z)$$

$$K_1'(z) = -K_0(z) - \frac{K_1(z)}{z}$$

The tensors G_{ij} and F_{ij} for a three-dimensional problem are of the form[27-31]

$$G_{ij}(x, \xi, s) = \frac{1}{4\pi\mu}(A\delta_{ij} - Br_{,i}r_{,j}) \tag{10.31a}$$

and

$$F_{ij}(x, \xi, s) = \frac{1}{4\pi}\left[P\left(\delta_{ij}\frac{\partial r}{\partial n} + r_{,i}n_j\right) + Q\left(n_i r_{,j} - 2r_{,i}r_{,j}\frac{\partial r}{\partial n}\right) + Rr_{,i}r_{,j}\frac{\partial r}{\partial n} + Sr_{,j}n_i\right] \tag{10.31b}$$

where s is the Laplace transform parameter and is equal to $-iw$. In addition,

$$B = \left(\frac{3c_2^2}{s^2r^2} + \frac{3c_2}{sr} + 1\right)\frac{e^{-sr/c_2}}{r} - \frac{c_2^2}{c_1^2}\left(\frac{3c_1^2}{s^2r^2} + \frac{3c_1}{sr} + 1\right)\frac{e^{-sr/c_1}}{r}$$

and

$$A = \left(\frac{c_2^2}{s^2r^2} + \frac{c_2}{sr} + 1\right)\frac{e^{-sr/c_2}}{r} - \frac{c_2^2}{c_1^2}\left(\frac{c_1^2}{s^2r^2} + \frac{c_1}{sr}\right)\frac{e^{-sr/c_1}}{r}$$

where e is the exponential function and P, Q, R and S are as defined above.

Once the boundary solution is obtained, Eq. (10.29) with $\alpha_{ij} = \delta_{ij}$ can also be used to find the interior displacements, and the interior stresses can then be obtained by using the strain–displacement and stress–strain relations as[29,31,32]

$$\sigma_{jk}(\xi, \omega) = \int_S [G_{ijk}^\sigma(x, \xi, \omega)t_i(x, s) - F_{ijk}^\sigma(x, \xi, \omega)u_l(x, \omega)]\,dS(x) \tag{10.32}$$

The stresses at the surface can be calculated by combining the constitutive equations, the directional derivatives of the displacement vector and the values of field variables in an accurate matrix formulation described in earlier chapters. It should also be noted that the above integral representations can account for a certain amount of viscous damping by using complex-valued elastic constants in the usual manner.

The integral representations described above and their scalar counterparts arising from Eq. (10.25) are essentially identical to those of elastostatics and potential flow respectively for a specific value of the frequency parameter ω, except that all coefficients are now complex. The numerical treatment is therefore almost identical to that used in potential flow and elastostatics. The only differences are: the calculation of the diagonal terms of the strongly singular kernel $\mathbf{F}$, which is handled much like the transient problem described later, and during integration a greater level of element subdivisions is required, particularly at higher frequencies.[30-32] It is also possible to derive axisymmetric formulations both for the symmetric and asymmetric boundary conditions in a manner illustrated for elastostatics. This in fact was recently done by Wang and Banerjee.[33,34]

The interior, time-harmonic wave-propagation problem has a unique solution provided ω^2 is not one of the eigenvalues of the system. There are, however, also related difficulties in the corresponding exterior boundary value problem as expressed by Eq. (10.29), although it does of course satisfy the usual regularity conditions, as well as the radiation condition at infinity. There are an infinite set of values of ω for which the equation has a multiplicity of solutions that coincide with the 'resonant' wave numbers (or eigenvalues) for a related interior problem. Thus the solution of the exterior Dirichlet and Neumann problems will fail at wave numbers corresponding to eigenvalues of the Neumann and Dirichlet problems respectively. This is not a physical difficulty inherent in the exterior problem since there are no exterior eigenvalues. The difficulty of non-uniqueness stems entirely from the formulation of the problem in terms of a boundary integral. Detailed discussions of these difficulties can be found in a number of articles[11-14,22,23] where the authors have proposed modifications to both the direct and indirect formulations to overcome them, although a completely satisfactory method for the general solution of large-scale practical problems has not yet emerged. The best BEM formulation to date has been that proposed by Burton and Miller.[11]

10.5 INTEGRAL FORMULATIONS FOR TRANSIENT SCALAR WAVE PROPAGATION

Since the algebra involved in the scalar wave equation is not complex we attempt in this section a very detailed treatment of the development of the BEM.

By multiplying Eq. (10.24) by the fundamental solution g and integrating by parts we get

$$\int_0^t \int_S \left(g \frac{\partial p}{\partial n} - p \frac{\partial g}{\partial n} \right) \mathrm{d}S \, \mathrm{d}\tau - \frac{1}{c^2} \int_V \left[g(\mathbf{x}, t; \xi, 0) \frac{\partial p_0}{\partial \tau} - p_0 \frac{\partial g(\mathbf{x}, t; \xi, 0)}{\partial \tau} \right] \mathrm{d}V$$

$$+ \int_0^t \int_V p \left[\nabla^2 g - \frac{1}{c^2} \frac{\partial^2 g}{\partial \tau^2} \right] \mathrm{d}V \, \mathrm{d}\tau + \int_0^t \int_V g(\mathbf{x}, t; \xi, \tau) \, \psi(\mathbf{x}, \tau) \, \mathrm{d}V \, \mathrm{d}\tau = 0 \quad (10.33)$$

As usual, using the sifting property of the δ function, the second volume integral appearing in Eq. (10.33) becomes

$$\int_0^t \int_V p \left[\nabla^2 g - \frac{1}{c^2} \frac{\partial^2 g}{\partial \tau^2} \right] dV d\tau = -\int_0^t \int_V p(\mathbf{x}, \tau) \delta(\mathbf{x}, t; \xi, \tau) \, dV d\tau$$

$$= -p(\xi, t)$$

Equation (10.33) then reduces to (ξ being anywhere)

$$\alpha(\xi) p(\xi, t) = \int_0^t \int_S \left(g \frac{\partial p}{\partial n} - p \frac{\partial g}{\partial n} \right) dS d\tau$$

$$- \frac{1}{c^2} \int_V \left[g(\mathbf{x}, t; \xi, 0) \frac{\partial p_0}{\partial \tau} - p_0 \frac{\partial g(\mathbf{x}, t; \xi, 0)}{\partial \tau} \right] dV$$

$$+ \int_0^t \int_V g(\mathbf{x}, t; \xi, \tau) \psi(\mathbf{x}, \tau) \, dV d\tau \qquad (10.34)$$

Assuming zero initial conditions and no body source, the above equation finally reduces to

$$\alpha(\xi) p(\xi, t) = \int_S \left[g(\mathbf{x}, t; \xi, \tau) * q(\mathbf{x}, \tau) - f(\mathbf{x}, t; \xi, \tau) * p(\mathbf{x}, \tau) \right] dS(\mathbf{x}) \qquad (10.35)$$

where $*$ stands for the Reimann convolution integral and $q(\mathbf{x}, \tau) = \partial p(\mathbf{x}, \tau)/\partial n$. Equation (10.35) is the boundary integral equation for transient scalar wave propagation.

For an axisymmetric analysis, the field variables are independent of the circumferential angle, so the above equation is expressed in cylindrical coordinates (r, θ, z) as

$$\alpha(\xi) p(\xi, t) = 2 \int_L \int_0^\pi \left[g'(\mathbf{x}, t; \xi, \tau) * q'(\mathbf{x}, \tau) - f'(\mathbf{x}, t; \xi, \tau) * p'(\mathbf{x}, \tau) \right] d\theta_x r_x \, dL(\mathbf{x})$$

$$(10.36)$$

where L is the generator of the axisymmetric body. Unfortunately, the integration in θ_x indicated in Eq. (10.36) must be done numerically for the transient parts of g' and f'.

The three-dimensional problem g required in Eq. (10.35) is given by

$$g = \frac{1}{4\pi r} \delta \left[(t - \tau) - \frac{r}{c} \right] \qquad (10.37)$$

The two-dimensional g kernel is obtained from the corresponding three-dimensional one by integrating along the third spatial direction. Thus,

$$g = \begin{cases} \dfrac{c}{2\pi} \dfrac{1}{\sqrt{c^2(t-\tau)^2 - r^2}} & \text{when } (t - \tau) > r/c \\ 0 & \text{when } (t - \tau) \le r/c \end{cases} \qquad (10.38)$$

Thus, all operations on the g kernel (including the derivation of the f kernel) will be carried out only when $(t - \tau) > r/c$.

The explicit forms of the f kernels are obtained using

$$f = \frac{\partial g}{\partial n} = \frac{\partial g}{\partial r} \frac{\partial r}{\partial n}$$

Thus

$$f = \frac{1}{4\pi r} \left\{ -\frac{1}{r} \delta \left[(t - \tau) - \frac{r}{c} \right] + \frac{1}{c} \dot{\delta} \left[(t - \tau) - \frac{r}{c} \right] \right\} \frac{\partial r}{\partial n} \qquad (10.39a)$$

in three dimensions and for two-dimensional problems

$$f = \frac{c}{2\pi} \frac{r}{[c^2 (t - \tau)^2 - r^2]^{3/2}} H[c(t - \tau) - r] \frac{\partial r}{\partial n} \qquad (10.39b)$$

The axisymmetric kernels are not explicit and are obtained by integrating the three-dimensional kernels numerically along the circumferential direction.

10.5.1 Numerical Implementation[35-38]

Constant time variation   Since the field variables remain constant during a time step, they can be written in the following form:

$$f(\mathbf{x}, \tau) = \sum_{n=1}^{N} f^n(\mathbf{x}) \phi_n(\tau)$$

where $f(\mathbf{x}, \tau)$ stands for pressure $p(\mathbf{x}, \tau)$ or pressure gradient $q(\mathbf{x}, \tau)$ and

$$\phi_n(\tau) = \begin{cases} 1 & \text{for } (n-1)\Delta t \leq \tau \leq n\Delta t \\ 0 & \text{otherwise} \end{cases}$$

or

$$\phi_n(\tau) = H[\tau - (n-1)\Delta t] - H(\tau - n\Delta t)$$

where H is the Heaviside function.

Then the convolution terms appearing in Eq. (10.35) can be analytically integrated. Thus, for example,

$$g * q = \sum_{n=1}^{N} q^n(\mathbf{x}) \int_{(n-1)\Delta t}^{n\Delta t} g(\mathbf{x}, N\Delta t; \xi, \tau) \, d\tau$$

$$= \sum_{n=1}^{N} q^n(\mathbf{x}) G^{N-n+1} \qquad (10.40)$$

Thus Eq. (10.35) can now be written as

$$\alpha(\xi) p^N(\xi) = \sum_{n=1}^{N} \int_S \left[G^{N-n+1} q^n(\mathbf{x}) - F^{N-n+1} p^n(\mathbf{x}) \right] dS(\mathbf{x}) \qquad (10.41)$$

where $p^N(\xi)$ stands for the potential at time $t = N\Delta t$.

For the two-dimensional case, the convoluted kernels are

$$G^{N-n+1} = \frac{1}{2\pi} \left[H\left(\frac{cz}{r}\right) \cosh^{-1}\left(\frac{cz}{r}\right) \right]_{z=(N-n)\Delta t}^{z=(N-n+1)\Delta t} \tag{10.42a}$$

and

$$F^{N-n+1} = -\frac{1}{2\pi r}\frac{\partial r}{\partial n} \left[H\left(\frac{cz}{r}\right) \frac{cz/r}{\sqrt{(cz/r)^2 - 1}} \right]_{z=(N-n)\Delta t}^{z=(N-n+1)\Delta t} \tag{10.42b}$$

The time integrals of the three-dimensional kernels are evaluated in a special way because of the presence of the δ function in the forms: (1) $\delta(t - \tau - r/c)$ and (2) $\dot{\delta}(t - \tau - r/c)$.

The temporal integration yields:

$$\int_{(n-1)\Delta t}^{n\Delta t} \delta\left(t - \tau - \frac{r}{c}\right) d\tau = H\left[(N-n+1)\Delta t - \frac{r}{c}\right] - H\left[(N-n)\Delta t - \frac{r}{c}\right] \tag{10.43a}$$

$$\int_{(n-1)\Delta t}^{n\Delta t} \dot{\delta}\left(t - \tau - \frac{r}{c}\right) f(\mathbf{x}, \tau) d\tau = \frac{f(\mathbf{x}, n\Delta t) - f(\mathbf{x}, (n-1)\Delta t)}{\Delta t}$$

$$\times \left\{ H\left[(N-n+1)\Delta t - \frac{r}{c}\right] - H\left[(N-n)\Delta t - \frac{r}{c}\right] \right\} \tag{10.43b}$$

Linear time variation The field variables are approximated as

$$f(\mathbf{x}, \tau) = \sum_{n=1}^{N} M_1(\tau) f^n(\mathbf{x}) + M_2(\tau) f^{n-1}(\mathbf{x}) \tag{10.44}$$

where $M_1(\tau)$ and $M_2(\tau)$ are the temporal interpolation functions corresponding to the local time nodes 1 and 2 (Fig. 10.1), and are expressed by

$$M_1(\tau) = \frac{\tau - (n-1)\Delta t}{\Delta t} \phi_n(\tau) \qquad M_2(\tau) = \frac{n\Delta t - \tau}{\Delta t} \phi_n(\tau)$$

where $\phi_n(\tau)$ has been defined above. Here, the time integration involves the kernel and the temporal shape function products and yields two sets of convoluted kernels, one corresponding to each time node.

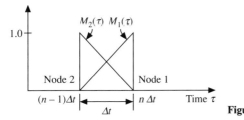

Figure 10.1 Linear temporal shape functions.

After temporal integration, the time-convoluted BEM equation takes the following form:

$$
\alpha(\xi) p^N(\xi) = \sum_{n=1}^{N} \int_S \left(\left[G_1^{N-n+1} q^n(\mathbf{x}) + G_2^{N-n+1} q^{n-1}(\mathbf{x}) \right] \right.
$$
$$
\left. - \left[F_1^{N-n+1} p^n(\mathbf{x}) + F_2^{N-n+1} p^{n-1}(\mathbf{x}) \right] \right) \, dS(\mathbf{x})
\tag{10.45}
$$

where for two-dimensional problems, the convoluted kernels are

$$
G_1^{N-n+1} = \int_{(n-1)\Delta t}^{n\Delta t} G(\mathbf{x}, N\Delta t; \xi, \tau) M_1(\tau) \, d\tau
$$
$$
= \frac{1}{2\pi} \left[H\!\left(\frac{cz}{r}\right) \left\{ N \cosh^{-1}\!\left(\frac{cz}{r}\right) - \frac{r}{c\,\Delta t} \sqrt{\left(\frac{cz}{r}\right)^2 - 1} \right\} \right]_{z=(N-n)\Delta t}^{z=(N-n+1)\Delta t}
\tag{10.46a}
$$

$$
G_2^{N-n+1} = \int_{(n-1)\Delta t}^{n\Delta t} G(\mathbf{x}, N\Delta t; \xi, \tau) M_2(\tau) \, d\tau
$$
$$
= \frac{1}{2\pi} \left[H\!\left(\frac{cz}{r}\right) \left\{ (1-N) \cosh^{-1}\!\left(\frac{cz}{r}\right) + \frac{r}{c\,\Delta t} \sqrt{\left(\frac{cz}{r}\right)^2 - 1} \right\} \right]_{z=(N-n)\Delta t}^{z=(N-n+1)\Delta t}
\tag{10.46b}
$$

$$
F_1^{N-n+1} = \frac{1}{2\pi r} \frac{\partial r}{\partial n} \left[H\!\left(\frac{cz}{r}\right) \left\{ -N \frac{cz/r}{\sqrt{(cz/r)^2 - 1}} + \frac{r}{c\,\Delta t} \frac{1}{\sqrt{(cz/r)^2 - 1}} \right\} \right]_{z=(N-n)\Delta t}^{z=(N-n+1)\Delta t}
\tag{10.46c}
$$

and

$$
F_2^{N-n+1} = \frac{1}{2\pi r} \frac{\partial r}{\partial n} \left[H\!\left(\frac{cz}{r}\right) \left\{ (-1+N) \frac{cz/r}{\sqrt{(cz/r)^2 - 1}} - \frac{r}{c\,\Delta t} \frac{1}{\sqrt{(cz/r)^2 - 1}} \right\} \right]_{z=(N-n)\Delta t}^{z=(N-n+1)\Delta t}
\tag{10.46d}
$$

The temporal integration of the time-related terms in the three-dimensional kernel are evaluated in the following way:

$$\int_{(n-1)\Delta t}^{n\Delta t} \delta\left(t - \tau - \frac{r}{c}\right) f(\mathbf{x}, \tau)\, d\tau = f^{n-1}(\mathbf{x}) \int_{(n-1)\Delta t}^{n\Delta t} M_2(\tau)\delta\left(t - \tau - \frac{r}{c}\right) d\tau$$

$$+ f^{n}(\mathbf{x}) \int_{(n-1)\Delta t}^{n\Delta t} M_1(\tau)\delta\left(t - \tau - \frac{r}{c}\right) d\tau$$

$$= f^{n-1}(\mathbf{x}) \int_{(n-1)\Delta t}^{n\Delta t} \left(n - \frac{\tau}{\Delta t}\right) \phi_n(\tau)\, \delta\left(t - \tau - \frac{r}{c}\right) d\tau$$

$$+ f^{n}(\mathbf{x}) \int_{(n-1)\Delta t}^{n\Delta t} \left[\frac{\tau}{\Delta t} - (n-1)\right] \phi_n(\tau)\, \delta\left(t - \tau - \frac{r}{c}\right) d\tau$$

The time integrals on the right-hand side of the above equation can again be isolated into the following two types:

$$\int_{(n-1)\Delta t}^{n\Delta t} \phi_n(\tau)\delta\left(t - \tau - \frac{r}{c}\right) d\tau = H\left[(N - n + 1)\Delta t - \frac{r}{c}\right] - H\left[(N - n)\Delta t - \frac{r}{c}\right]$$

$$(10.47a)$$

$$\int_{(n-1)\Delta t}^{n\Delta t} \tau\, \phi_n(\tau)\delta\left(t - \tau - \frac{r}{c}\right) d\tau = \left[(N - n + 1)\Delta t - \frac{r}{c}\right] H\left[(N - n + 1)\Delta t - \frac{r}{c}\right]$$

$$- \left[(N - n)\Delta t - \frac{r}{c}\right] H\left[(N - n)\Delta t - \frac{r}{c}\right] \quad (10.47b)$$

The time integral corresponding to $\dot{\delta}(t - \tau - r/c)$ is evaluated similarly to the procedures given by expression (10.43b). Details of these integrations can also be found in Ahmad and Banerjee.[29,39]

The convoluted two-dimensional F kernels given by expressions (10.46c,d) contain terms with $r^{-1/2}$ apparent singularity at the wave front that sometimes cause difficulty in accurate numerical integration, especially when elements with widely varying lengths are used. This difficulty is avoided by writing Eq. (10.45) in the following form:[37]

$$\alpha(\xi) p^N(\xi) = \sum_{n=1}^{N} \int_S \left[(G_1^{N-n+1} + G_2^{N-n}) q^n(\mathbf{x}) - (F_1^{N-n+1} + F_2^{N-n}) p^n(\mathbf{x}) \right] dS(\mathbf{x})$$

$$(10.48)$$

When $(F_1^{N-n+1} + F_2^{N-n})$ is evaluated these terms condense to a form that is easy to integrate numerically. The condensed form of the convoluted kernels appearing in Eq. (10.48) are

$$
G_1^{N-n+1} + G_2^{N-n} = \frac{1}{2\pi} \left\{ (N-n+1) \cosh^{-1} \left[\frac{c(N-n+1)\Delta t}{r} \right] \right.
$$

$$
- 2(N-n) \cosh^{-1} \left[\frac{c(N-n)\Delta t}{r} \right]
$$

$$
+ (N-n-1) \cosh^{-1} \left[\frac{c(N-n-1)\Delta t}{r} \right]
$$

$$
- \sqrt{(N-n+1)^2 - \left(\frac{r}{c\,\Delta t} \right)^2} + 2\sqrt{(N-n)^2 - \left(\frac{r}{c\,\Delta t} \right)^2}
$$

$$
\left. - \sqrt{(N-n-1)^2 - \left(\frac{r}{c\,\Delta t} \right)^2} \right\} \tag{10.49a}
$$

$$
F_1^{N-n+1} + F_2^{N-n} = -\frac{1}{2\pi r} \frac{\partial r}{\partial n} \left[\sqrt{(N-n+1)^2 - \left(\frac{r}{c\,\Delta t} \right)^2} \right.
$$

$$
- 2\sqrt{(N-n)^2 - \left(\frac{r}{c\,\Delta t} \right)^2}
$$

$$
\left. + \sqrt{(N-n-1)^2 - \left(\frac{r}{c\,\Delta t} \right)^2} \right] \tag{10.49b}
$$

In the above expressions the time-related terms are always non-negative because of the causality property of the wave. Also, because of the time-translation property, the convoluted kernels need to be evaluated only for $n = 1$.

For problems where the flux is specified as a sudden jump, a mixed formulation is found to produce better results. Here, the temporal variation of the flux is assumed to remain constant while the potential is varied linearly. Thus, the F kernel remains the same as that for the linear variation case. However, the G kernel is like that of the constant variation in time.

Spatial approximations   As in the previous chapters, geometry and field variables at any point over an element are given in terms of the nodal quantities as

$$
x_i = N_\alpha(\eta)\, X_{i\alpha}
$$

$$
p = N_\alpha(\eta)\, P_\alpha \tag{10.50}
$$

$$
q = N_\alpha(\eta)\, Q_\alpha
$$

where $i = 1, 2$ for two-dimensional and axisymmetric problems and $1, 2, 3$ for three-dimensional problems and $\alpha = 1, 2, \ldots, A$, A being the total number of nodes in an element.

With the spatial discretization, explained above, Eq. (10.48) takes the following form for two-dimensional and axisymmetric (after circumferential integration) problems:

$$
\alpha(\xi)\, p^N(\xi) = \sum_{n=1}^{N} \sum_{m=1}^{M} \left\{ \sum_{\alpha=1}^{A} \left[Q_\alpha^n \int_0^1 (G_1^{N-n+1} + G_2^{N-n})\, N_\alpha(\eta)\, |J|\, \mathrm{d}\eta \right. \right.
$$

$$
\left. \left. - P_\alpha^n \int_0^1 (F_1^{N-n+1} + F_2^{N-n})\, N_\alpha(\eta)\, |J|\, \mathrm{d}\eta \right] \right\} \tag{10.51a}
$$

while the three-dimensional equivalent is

$$
\alpha(\xi)\, p^N(\xi) = \sum_{n=1}^{N} \sum_{m=1}^{M} \left\{ \sum_{\alpha=1}^{A} \left[Q_\alpha^n \int_0^1 \int_0^1 (G_1^{N-n+1} + G_2^{N-n})\, N_\alpha(\eta_1,\eta_2)\, |J|\, \mathrm{d}\eta_1\, \mathrm{d}\eta_2 \right. \right.
$$

$$
\left. \left. - P_\alpha^n \int_0^1 \int_0^1 (F_1^{N-n+1} + F_2^{N-n})\, N_\alpha(\eta_1,\eta_2)\, |J|\, \mathrm{d}\eta_1\, \mathrm{d}\eta_2 \right] \right\} \tag{10.51b}
$$

where M represents the total number of boundary elements.

The indicated spatial integrations in Eqs (10.51a,b) are once again carried out using the scheme outlined in Chapter 3. However, it must be recognized that elements where the wave front has not yet been reached contribute nothing to the system. A substantial savings in the cost of integration can thus be achieved. In addition, elements must be intelligently subdivided to account for the discontinuity of the integrands over the remaining elements. The singular integrals involving the F kernel [order of singularity, $O(1/r^{n-1})$ where n = number of dimension] are evaluated indirectly using a 'rigid body' type concept, as is discussed in Banerjee *et al.*[30,40] for periodic and transient dynamic analyses, i.e.

$$
\int_S F^{\text{trans}}\, \mathrm{d}S = \int_S F^{\text{st}}\, \mathrm{d}S + \int_S \left(F^{\text{trans}} - F^{\text{st}} \right) \mathrm{d}S \tag{10.52}
$$

where F^{trans} and F^{st} are the transient and steady state F kernels respectively.

The first integral on the right-hand side of Eq. (10.52) involving the static kernel is singular and its evaluation using the uniform potential and the resulting zero flux solution is described in Chapter 3. However, for this process the body must have a closed boundary. Thus, for problems with an open boundary, the region of interest must be enclosed with a special kind of element known as an 'enclosing element'. This static kernel is integrated only once during the first time step. Moreover, these elements do not have to be as small as those used for modelling the problem. Accordingly, additional computational effort warranted by these elements is therefore not significant. Further details can be found in Ahmad and Banerjee.[32] The second integral is non-singular and its numerical treatment does not pose any special difficulty.

For axisymmetric problems, there is an additional numerical integration along the circumferential direction. For this purpose the strongly singular F kernel is divided into two parts as above. The first term on the right-hand side is the static three-dimensional kernel and its circumferential integration can be performed analytically as described in Chapter 5. The second part is non-singular and its numerical integration along the circumferential direction is not difficult. The process is similar to but algebraically simpler than that for elastodynamics, which can be found in Wang and Banerjee.[41]

10.5.2 Solution Procedure

Equation (10.51) is written for every boundary node and, after the appropriate integration, can be cast into the following set of equations:

$$\sum_{n=1}^{N} ([\mathbf{G}_1^{N-n+1} + \mathbf{G}_2^{N-n}]\{\mathbf{q}^n\} - [\mathbf{F}_1^{N-n+1} + \mathbf{F}_2^{N-n}]\{\mathbf{p}^n\}) = \{\mathbf{0}\} \tag{10.53}$$

where $\{\mathbf{p}^n\}$ and $\{\mathbf{q}^n\}$ are vectors of nodal quantities with the superscript referring to the time step index.

For a well-posed problem, at time t, there are as many unknowns as the number of equations, with the field variables at all previous times being known. Thus, putting the unknowns on the left, Eq. (10.53) can be rearranged as

$$[\mathbf{A}_1^1]\{\mathbf{X}^N\} = [\mathbf{B}_1^1]\{\mathbf{Y}^N\} - \sum_{n=1}^{N-1} ([\mathbf{G}_1^{N-n+1} + \mathbf{G}_2^{N-n}]\{\mathbf{q}^n\} - [\mathbf{F}_1^{N-n+1} + \mathbf{F}_2^{N-n}]\{\mathbf{p}^n\})$$

$$\tag{10.54}$$

which is put into the following simplified form:[27,40]

$$[\mathbf{A}_1^1]\{\mathbf{X}^N\} = [\mathbf{B}_1^1]\{\mathbf{Y}^N\} + \{\mathbf{R}^N\} \tag{10.55}$$

in which $\{\mathbf{X}^N\}$ and $\{\mathbf{Y}^N\}$ respectively are the vectors of the unknown and known variables and $\{\mathbf{R}^N\}$ represent the effects of past dynamic history on the current time node N. The superscript 1 indicates the terms at the leading time and the subscript 1 refers to the forward node at the leading time node. These terms need to be evaluated only during the first time step.

10.6 TRANSIENT ELASTODYNAMIC FORMULATION

At time t, considering a domain V bounded by surface S, the displacement at point ξ can be obtained via the dynamic reciprocal work theorem [Eq. (10.19)]:

$$\alpha_{ij}(\xi)u_i(\xi, t) = \int_S \int_0^t [g_{ij}(x, t; \xi, \tau)t_i(x, \tau) - f_{ij}(x, t; \xi, \tau)u_i(x, \tau)]\, d\tau\, dS(x)$$

$$+ \rho \int_V \int_0^t g_{ij}(x, t; \xi, \tau)\psi_i(x)\, d\tau\, dV(x)$$

$$+ \rho \int_V \left[\frac{\partial u_i(x, 0)}{\partial t} g_{ij}(x, t; \xi, 0) + u_i(x, 0) \frac{\partial g_{ij}(x, t; \xi, 0)}{\partial t} \right] dV(x) \tag{10.56}$$

The terms g_{ij} and f_{ij} are the fundamental solutions and represent respectively the displacements and tractions at the field point x at time t due to a unit force applied at source point ξ at a preceding time τ.

If the body force is absent and the domain is initially at rest, then only the surface integrals remain, i.e.[27]

$$\alpha_{ij}(\xi) u_i(\xi, t) = \int_S \int_0^t [g_{ij}(x, t; \xi, \tau) t_i(x, \tau) - f_{ij}(x, t; \xi, \tau) u_i(x, \tau)] \, d\tau \, dS(x) \quad (10.57)$$

For purely axisymmetric problems, the field variables are independent of the circumferential angle, so the above equation is expressed in cylindrical coordinates (r, θ, z) as[41]

$$\alpha_{ij}(\xi) u_i'(\xi, t) = 2 \int_L \int_0^\pi \int_0^t [g_{ij}'(x, t; \xi, \tau) t_i'(x, \tau) - f_{ij}'(x, t; \xi, \tau) u_i'(x, \tau)] \, d\tau \, d\theta_x r_x \, dL(x)$$

$$(10.58)$$

where L is the generator of the axisymmetric body and

$$u_i'(x, t) = \left\{ \begin{matrix} u_r(x, t) \\ u_z(x, t) \end{matrix} \right\}$$

$$t_i'(x, t) = \left\{ \begin{matrix} t_r(x, t) \\ t_z(x, t) \end{matrix} \right\}$$

$$g_{ij}'(x, t; \xi, \tau) = \begin{bmatrix} g_{rr}' & g_{rz}' \\ g_{zr}' & g_{zz}' \end{bmatrix} = \begin{bmatrix} g_{11}' \cos \theta_x + g_{21}' \sin \theta_x & g_{13}' \cos \theta_x + g_{23}' \sin \theta_x \\ g_{31}' & g_{33}' \end{bmatrix}$$

$$f_{ij}'(x, t; \xi, \tau) = \begin{bmatrix} f_{rr}' & f_{rz}' \\ f_{zr}' & f_{zz}' \end{bmatrix} = \begin{bmatrix} f_{11}' \cos \theta_x + f_{21}' \sin \theta_x & f_{13}' \cos \theta_x + f_{23}' \sin \theta_x \\ f_{31}' & f_{33}' \end{bmatrix}$$

Note that the factor 2 in Eq. (10.58) is due to the symmetry of kernels g_{ij}' and f_{ij}' with respect to the axis $\theta = 0$. The θ component displacement is not seen here because it is independent of other components for axisymmetric problems.

The displacement kernel for the three-dimensional problem is the well-known Stokes' solution described earlier, now recast into the form:[39,42]

$$g_{ij}(x, t; \xi, \tau) = \frac{1}{4\pi\rho} \left\{ (3a_{ij} - b_{ij}) \int_{1/c_1}^{1/c_2} \lambda \delta(t - \tau - \lambda r) \, d\lambda \right.$$

$$\left. + a_{ij} \left[\frac{\delta(t - \tau - r/c_1)}{c_1^2} - \frac{\delta(t - \tau - r/c_2)}{c_2^2} \right] + b_{ij} \frac{\delta(t - \tau - r/c_2)}{c_2^2} \right\}$$

$$(10.59)$$

where

$$a_{ij} = \frac{y_i y_j}{r^3} \qquad b_{ij} = \frac{\delta_{ij}}{r} \qquad y_i = x_i - \xi_i$$

The traction kernel obtained from the above function via the strain–displacement relation and the elastic constitutive law assumes the form,[39,42,43]

$$
\begin{aligned}
f_{ij}(x, t; \xi, \tau) = \frac{1}{4\pi} \Bigg\{ &- 6c_2^2(5a_{ij} - b_{ij}) \int_{1/c_1}^{1/c_2} \lambda \delta(t - \tau - \lambda r) \, d\lambda \\
&+ (12a_{ij} - 2b_{ij}) \left[\delta\left(t - \tau - \frac{r}{c_2}\right) - \left(\frac{c_2}{c_1}\right)^2 \delta\left(t - \tau - \frac{r}{c_1}\right) \right] \\
&+ \frac{2ra_{ij}}{c_2} \left[\delta'\left(t - \tau - \frac{r}{c_2}\right) - \left(\frac{c_2}{c_1}\right)^3 \delta'\left(t - \tau - \frac{r}{c_1}\right) \right] \\
&- c_{ij}\left(1 - \frac{2c_2^2}{c_1^2}\right) \left[\delta\left(t - \tau - \frac{r}{c_1}\right) + \left(\frac{r}{c_1}\right) \delta'\left(t - \tau - \frac{r}{c_1}\right) \right] \\
&- d_{ij} \left[\delta\left(t - \tau - \frac{r}{c_2}\right) + \left(\frac{r}{c_2}\right) \delta'\left(t - \tau - \frac{r}{c_2}\right) \right] \Bigg\}
\end{aligned}
\tag{10.60}
$$

where

$$
a_{ij} = \frac{y_i y_j y_m n_m}{r^5} \qquad c_{ij} = \frac{y_j n_i}{r^3} \qquad d_{ij} = \frac{y_i n_j + \delta_{ij} y_m n_m}{r^3} \qquad b_{ij} = c_{ij} + d_{ij}
$$

The two-dimensional displacement kernel is obtained from the three-dimensional one by integration along the third spatial axis.[3,44–46] The derivation of the two-dimensional traction kernel needs special attention because of the presence of the Heaviside function.[35–38,47–49] These procedures are similar to those of two-dimensional transient scalar kernels described earlier, except that here the functions are more complex. The final form of the two-dimensional kernels are[47–49]

$$
\begin{aligned}
g_{ij} = \frac{1}{2\pi\rho} \Bigg\{ \frac{1}{c_1} H(c_1 t' - r) \left[\frac{2c_1^2 t'^2 - r^2}{\sqrt{c_1^2 t'^2 - r^2}} \left(\frac{y_i y_j}{r^4}\right) - \frac{\delta_{ij}}{r^2} \sqrt{c_1^2 t'^2 - r^2} \right] \\
+ \frac{1}{c_2} H(c_2 t' - r) \left[-\frac{2c_2^2 t'^2 - r^2}{\sqrt{c_2^2 t'^2 - r^2}} \left(\frac{y_i y_j}{r^4}\right) + \frac{\delta_{ij}}{r^2} \sqrt{c_2^2 t'^2 - r^2} + \frac{\delta_{ij}}{\sqrt{c_2^2 t'^2 - r^2}} \right] \Bigg\}
\end{aligned}
\tag{10.61}
$$

$$f_{ij} = \frac{\mu}{2\pi\rho r} \left(\frac{1}{c_1} H\left(\frac{c_1 t'}{r} - 1 \right) \left\{ \frac{1}{[(c_1 t'/r)^2 - 1]^{3/2}} \left(\frac{A_1}{r} \right) + \frac{2(c_1 t'/r) - 1}{\sqrt{(c_1 t'/r)^2 - 1}} \left(\frac{2A_2}{r} \right) \right\} \right.$$

$$\left. - \frac{1}{c_2} H\left(\frac{c_2 t'}{r} - 1 \right) \left\{ \frac{1}{[(c_2 t'/r)^2 - 1]^{3/2}} \left(\frac{A_3}{r} \right) + \frac{2(c_2 t'/r)^2 - 1}{\sqrt{(c_2 t'/r)^2 - 1}} \left(\frac{2A_2}{r} \right) \right\} \right)$$

$$(10.62)$$

where

$$A_1 = \left(\frac{\lambda}{\mu} \right) n_i r_{,j} + 2 r_{,i} r_{,j} \frac{\partial r}{\partial n}$$

$$A_2 = n_i r_{,j} + n_j r_{,i} + \frac{\partial r}{\partial n} (\delta_{ij} - 4 r_{,i} r_{,j})$$

$$A_3 = \frac{\partial r}{\partial n} (2 r_{,i} r_{,j} - \delta_{ij}) - n_j r_{,i}$$

The time functions involved in the kernels are simple enough to carry out the time integration analytically.[39,40,47] Spatial integration still has to be done numerically.[40] For axisymmetric problems there is an additional integration along the circumferential direction which has to be carried out by splitting the kernels into a steady state (static) part which can be integrated analytically and a transient part which is integrated numerically.[41]

With the spatial discretization process the boundary integral equation for two-dimensional and axisymmetric (after circumferential integration of the kernels) analyses takes the form (see Sec. 10.5):

$$\alpha_{ij}(\xi) u_i^N(\xi) = \sum_{n=1}^{N} \sum_{m=1}^{M} \left[\sum_{\alpha=1}^{A} T_{i\alpha}^n \int_0^1 (G_{ij_1}^{N-n+1} + G_{ij_2}^{N-n}) N_\alpha(\eta) |J| \, \mathrm{d}\eta \right.$$

$$\left. - \sum_{\alpha=1}^{A} U_{i\alpha}^n \int_0^1 (F_{ij_1}^{N-n+1} + F_{ij_2}^{N-n}) N_\alpha(\eta) |J| \, \mathrm{d}\eta \right] \qquad (10.63)$$

where M is the total number of boundary elements, $|J|$ is the Jacobian of transformation and $i, j = 1, 2$.

The corresponding equation for a three-dimensional analysis involves integration over the intrinsic coordinate space η_1, η_2. The remainder of the numerical treatment is somewhat similar to that described in Sec. 10.5 for the scalar wave propagation.

10.7 EXAMPLES OF APPLICATION

10.7.1 Periodic Analysis

Vibration of a hollow and refilled caisson foundation[34] This problem was recently investigated in an excellent paper by Chen and Penzien[50] in which they combined

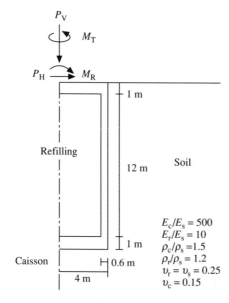

$E_c/E_s = 500$
$E_r/E_s = 10$
$\rho_c/\rho_s = 1.5$
$\rho_r/\rho_s = 1.2$
$v_r = v_s = 0.25$
$v_c = 0.15$

Figure 10.2 A typical axisymmetric caisson foundation.

the finite element method (near field) and global Ritz method (far field). A small material damping was introduced by them to obtain a better behaviour of some of the Ritz functions used. In the present analysis, this artificial material damping is not needed. Of course, it can also be included very easily if we replace the Lame constants λ and μ with their complex counterparts λ^* and μ^*. The geometry and all material properties needed for the present analysis are given in Fig. 10.2. The discretization used in the boundary element analysis is shown in Fig. 10.3. Note that

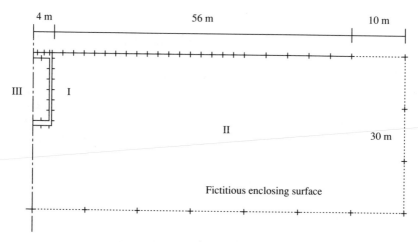

Figure 10.3 Axisymmetric boundary element model of a caisson (two-region mesh for the hollow caisson and three-region mesh for the refilled caisson).

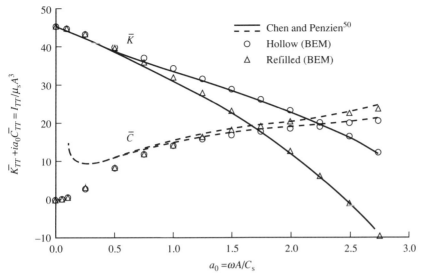

Figure 10.4 Torsional impedances of hollow and refilled caissons.

regions II and III are finite sized bodies and region III is ignored for the hollow caisson. All interfaces among caisson and other materials are assumed bonded, i.e. no-slip interfaces. The results by the BEM for both the hollow and the refilled caissons are plotted and compared with those given by Chen and Penzien in Figs 10.4 and 10.5. Very good agreement between them can be seen except that, at very

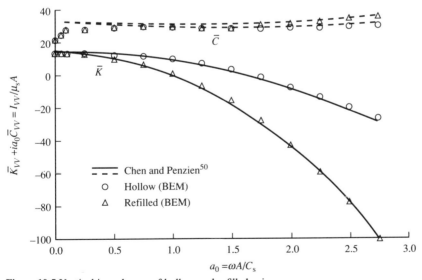

Figure 10.5 Vertical impedances of hollow and refilled caissons.

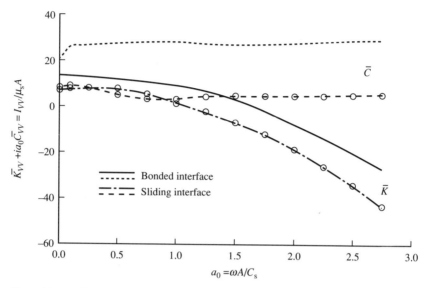

Figure 10.6 Vertical impedances of hollow caissons with bonded and sliding interfaces between caisson and soil.

low frequencies, the imaginary parts of impedance functions by the BEM approach zero instead of shooting to infinity. This is almost entirely due to the artificial damping used in the FEM analysis.

The real connection between the caisson and the soil is very hard to determine. It should be somewhere between a completely bonded connection and a frictionless connection. In Figs 10.6 and 10.7, the vertical impedances by the BEM for both cases are plotted. The assumption of a frictionless interface between the caisson and the soil and that between the caisson and the refilled materials results in a reduction in the imaginary part of vertical impedance functions, but the real part is not influenced as much.

Application to the vibration isolation problem[30,31,51]   The present analysis has been applied to study a number of vibration isolation cases, and the results were compared with the actual field observation of Woods,[52] who carried out a series of field tests in his attempt to define the screened zone and degree of amplitude reduction within the screened zone for trenches of a few specific shapes and sizes in a two-layer soil medium.

Figure 10.8 on page 286 shows the overview of contour diagram of amplitude reduction factors for a 270° circular trench with a depth of 1.0 ft under an operating frequency of 250 Hz. Both approaches (BEM and experimental) predict similar screened zones which are symmetrical about a radius from the source of excitation through the centre of the trench and bounded laterally by two radial lines extending from the center of the source of excitation through points 45° from each end of the trench.

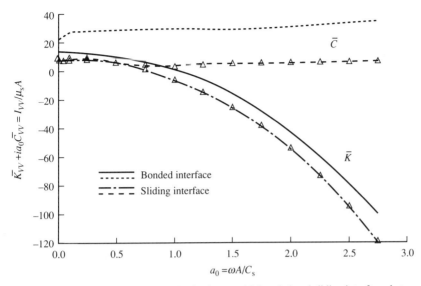

Figure 10.7 Vertical impedances of refilled caissons with bonded and sliding interfaces between caisson and soil.

10.7.2 Transient Analysis

The choice of an appropriate time step is very important for transient dynamic problems. A number of studies[35-42] have been carried out to determine the appropriate time step. It has been observed that the best results are obtained when the time step is such that the wave travels between 50 and 75 per cent of the element length during that time. Results are fairly good, even when the wave travels as low as 20 per cent of the element or covers the entire element during one time step. For a mesh with non-uniform element lengths usually the smallest element will govern, although good results were obtained even with a time step larger than the one dictated by the smallest element.

Bar under Heaviside type forcing function[36,38]   The transient scalar wave algorithm can be used to solve problems involving one-dimensional wave propagation. The pressure corresponds to axial displacement while the flux represents traction. This example has been chosen because of the availability of its analytical solution.

The rectangular bar ($a/b = 2$) shown in Fig. 10.9a,b on page 287, is discretized with 12 quadratic elements. The right end ($x = a$) of the bar is subjected to a Heaviside type loading, $q = q_0 H(t - 0)$, while the left end ($x = 0$) is fixed ($p = 0$). The sides ($y = 0, b$) are assumed to have $q = 0$. Due to the geometry and boundary conditions, the problem is actually one-dimensional and its analytical solution is simple and available in Miles.[53] The analysis was carried out for two different time steps, $\Delta t = 0.5 \, l/c$ and $\Delta t = l/c$, where l is the length of an element. Both the mixed and linear formulations were used. Results for the two time steps were found to be within 2 per cent of each other. Figure 10.10a,b on page 288 shows the nodal displacement (p) of points A($a/4, 0$), B($a/2, 0$) and C($a, 0$). For the mixed formulation, results are very accurate for the smaller time step since it captures the sudden jump at time $t = 0$.

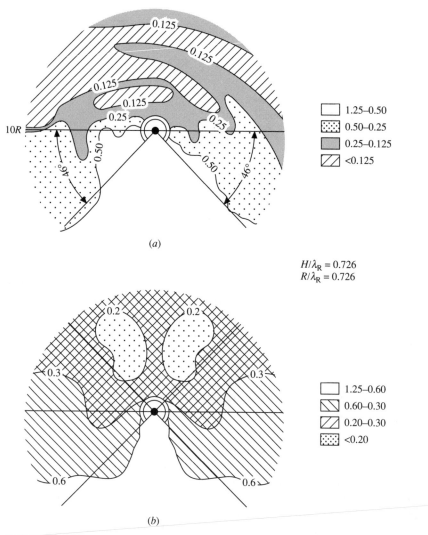

$H/\lambda_R = 0.726$
$R/\lambda_R = 0.726$

Figure 10.8 Amplitude reduction factor contour diagram: (*a*) field study (Woods[52]), (*b*) BEM results.

Reduction of noise near highway[36–38] The problem of vehicular noise near a motorway is of growing concern. The usefulness of acoustic barriers in reducing such noise is investigated here. The problem consists of the cross-section of a motorway with the following:

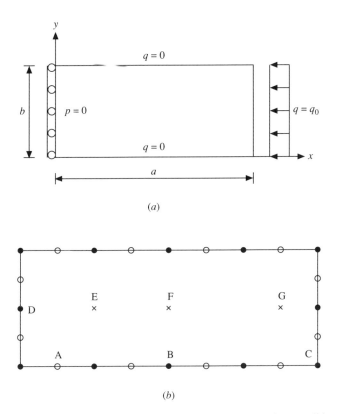

Figure 10.9 (*a*) Rectangular bar ($a/b = 2$) with the boundary conditions.

(*a*) No house, no barrier
(*b*) House, but no barrier
(*c*) Both house and barrier present

The discretization pattern is depicted in Fig. 10.11 on page 289, where L_1 and L_2 indicate the lanes of vehicular traffic and H_1 and H_2 the houses. Because of symmetry, only half of the problem is modelled. Moreover, since this problem involves an open domain, fictitious boundaries with 'enclosing elements' are used to facilitate the evaluation of the singular integral involving the F kernel. The surface of the houses H_1 and H_2 are acoustically rigid ($q = 0$), while the soil is modelled by acoustic springs with $q = 0.1p$. The barrier is modelled as a separate region and the speed of wave propagation is one-tenth of the surrounding medium. The acoustic pressure obtained at points 1 ($11b, 2.8b$) and 2 ($17b, 2.8b$) are shown in Fig. 10.12 on page 290. The results presented are for a time step of $\Delta t = b/c$ and are obtained by using a linear temporal formulation. An analysis was also done for smaller time steps which produced essentially the same answers. When there is no house and barrier, two peaks are noticed and after

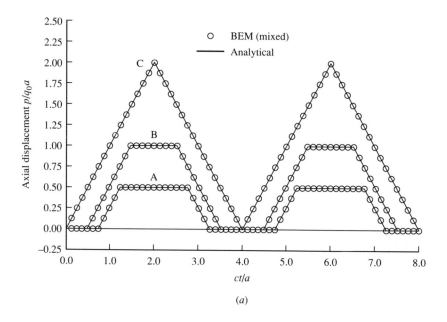

(a)

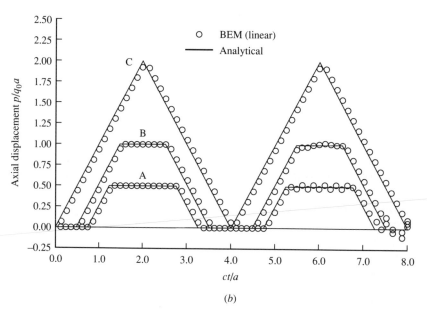

(b)

Figure 10.10 Axial displacements at boundary points A ($a/4$, 0), B ($a/2$, 0) and C (a, 0). Time step size, $t = 0.5l/c$. (a) BEM (mixed variation), (b) BEM (linear variation).

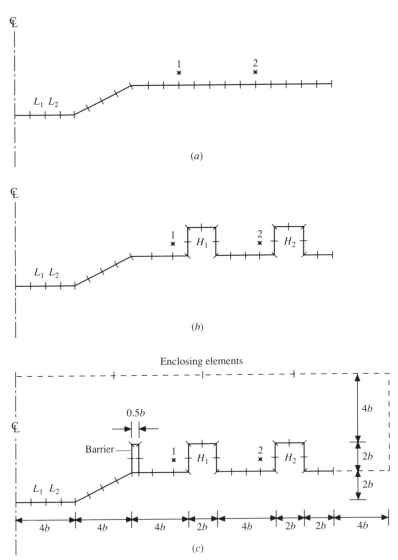

Figure 10.11 Discretization pattern for the problem of noise in the vicinity of highway. L_1, L_2 are the travelled lanes and H_1, H_2 the houses. Points 1, 2 are the locations where acoustic pressure is determined. (*a*) No house, no barrier. (*b*) House, no barrier. (*c*) House and barrier.

that the pressure decreases and smoothly decays with time. The peaks correspond to the two sources of flux on either side of the line of symmetry. The presence of the buildings produces waviness in the response and even elevates the magnitude slightly. This is probably due to the secondary waves reflected by the hard building surfaces. The use of a barrier reduces the pressure significantly for point 1, but, as expected, not as much for point 2.

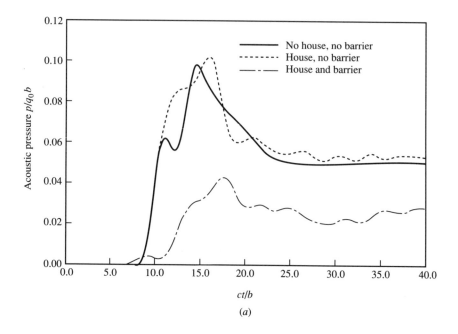

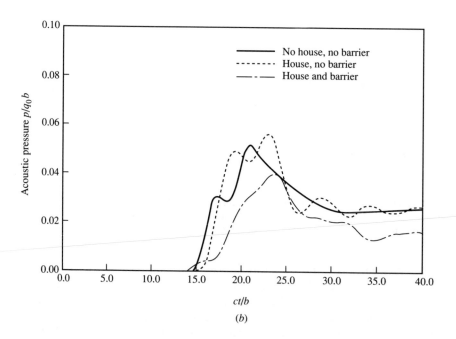

Figure 10.12 Acoustic pressure at (*a*) point 1 (11*b*, 2.8*b*), (*b*) point 2 (17*b*, 2.8*b*).

Wave propagation through layered soil[35,42,48]   In the real world, soil profile is almost never a homogeneous half-space. In many instances, it is stratified with a softer layer overlaying the stiffer ones. The presence of such a layer affects the wave propagation significantly. Waves emanating from the loaded surface undergo reflection and refraction at the interface between the layers and give rise to new types of waves, such as the Love wave, etc.

The present study involves a loaded two-layered soil profile shown in Fig. 10.13a,b with different shear modulii μ_1 and μ_2, all other properties remaining the same in the two layers. The depth of the upper layer is $H = 3b$. The portion of the top surface of width $2b$ is loaded with a triangular pulse. The rise time of

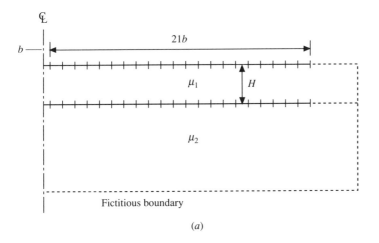

(a)

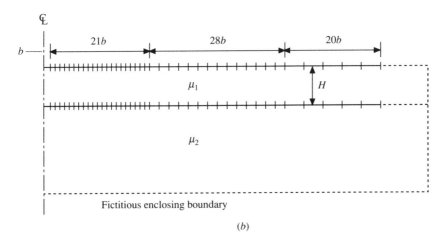

(b)

Figure 10.13 Discretization of layered half-space: (a) truncated discretization (mesh 1), (b) extended discretization (mesh 2).

the pulse is $t_r = 0.5437b/c_2$, where c_2 is the shear wave velocity in the upper layer. Because of symmetry, only one-half of the problem is modelled.

The effect of a truncated mesh on the surface displacements was found using two types of mesh pattern: one extended to a distance of $22b$ (mesh 1) and the other to $70b$ (mesh 2), as shown in Fig. 10.13a,b. The study was conducted for a layered profile with $\mu_1/\mu_2 = 0.5$ and a time step of $\Delta t = 0.2719b/c_2$. The results from the two meshes were found to be identical except for some insignificant differences at later times.

Three different layered profiles were studied with the ratio of shear modulii $\mu_1/\mu_2 = 1.0$, 0.5 and 0.25. The vertical displacements at the locations P and Q are presented in Fig. 10.14a,b. As the bottom layer becomes stiffer, more waves are reflected from the interface and a Love wave is generated in the top layer; thus surface displacement becomes oscillatory. Moreover, the wave travels faster through the lower stiffer medium and arrives early at a point as compared to when the medium is softer. This phenomena is pronounced for distant points, as can be seen in Fig. 10.14b.

Sudden expansion of spherical cavity[32,42]   The problem of radial expansion of a spherical cavity in an infinitely extending medium subjected to rectangular and triangular pressure pulses is studied. The material properties are

$$E = 8.9993 \times 10^6 \text{ lb/ft}^2$$

$$\nu = 0.25$$

$$\rho = 2.5 \times 10^{-9} \text{ lb} \cdot \text{s}^2/\text{ft}^4$$

The radius of the cavity is taken as $R = 212$ ft and using the symmetry capability only one octant was modelled with three six-noded triangles.

The time history of the radial displacements is plotted in Fig. 10.15a,b on page 294 for the triangular and rectangular pulses respectively. The numerical results from using two time steps agree fairly well. However, by comparing the results corresponding to the triangular and rectangular pulses, it can be seen that, in general, displacements at any time interval due to the rectangular pulse are twice that due to the triangular pulse. This is because the response depends upon the total impulses and the total impulse due to the rectangular pulse is double that due to the triangular one.

10.8 CONCLUDING REMARKS

The boundary element method has been presented both for periodic and time-domain transient analyses of two-dimensional, axisymmetric and three-dimensional problems. The kernel functions are explicitly defined. The algorithm incorporates curvilinear boundary elements as well as constant and linear

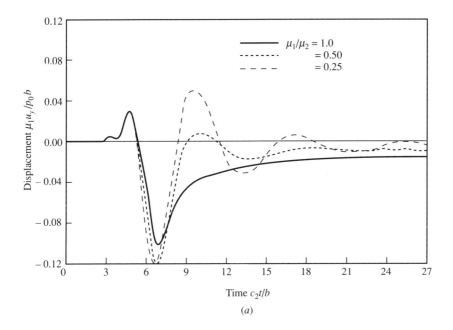

(a)

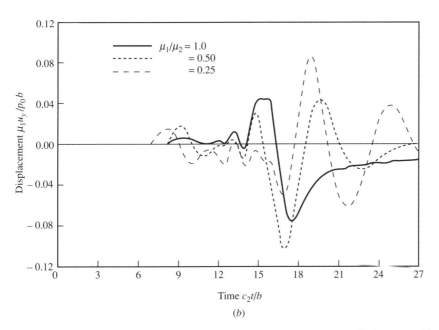

(b)

Figure 10.14 Effect of shear modulii of the layers (μ_1/μ_2) on the vertical displacement: (a) point P ($x/b = 5$), (b) point Q ($x/b = 15$).

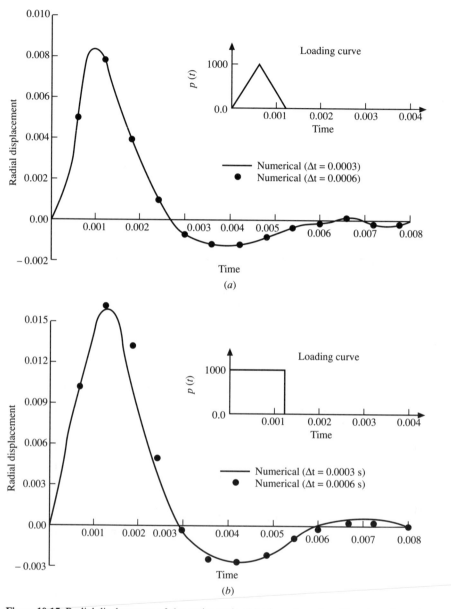

Figure 10.15 Radial displacement of the cavity under (*a*) triangular pulse, (*b*) rectangular pulse.

variations in time. The spatial integrations for the transient problems are carried out only for those parts of the problem where the wave has reached, thus saving computational time. This developed methodology has been applied to a number of practical problems here and elsewhere[54–76] to demonstrate its applicability to practical engineering problems.

REFERENCES

1 Morse, P.M. and Feshbach, H. (1953) *Methods of Theoretical Physics*, Part 2, McGraw-Hill, New York.

2. Graff, K.F. (1975) *Wave Motion in Elastic Solids*, Clarendon Press, Oxford.

3. Eringen, A.C. and Suhubi, E.S. (1975) *Elasto-dynamics*, Vol. 2, *Linear Theory*, Academic Press, New York.

4. Banerjee, P.K. and Butterfield, R. (1981) *Boundary Element Methods in Engineering Science*, McGraw-Hill, London.

5. Baker, B.B. and Copson, E.T. (1939) *The Mathematical Theory of Huygen's Principle*, Oxford University Press.

6. Kupradze, V.D. (1963) Dynamic problems in elasticity, in *Progress in Solid Mechanics*, Vol. 3, Eds I.N. Sneddon and R. Hill, North Holland, Amsterdam.

7. Friedman, M.B. and Shaw, R.P. (1962) Diffraction of a plane shock wave by an arbitrary rigid cylindrical obstacle, *Journal of Applied Mechanics*, Vol. 29, No. 1, pp. 40–46.

8. Banaugh, R.P. and Goldsmith, W. (1963) Diffraction of steady acoustic waves by surfaces of arbitrary shape, *Journal of Acoustics Society of America*, Vol. 35, No. 10, pp. 1590–1601.

9. Chen, L.H. and Schweikert, J. (1963) Sound radiation from an arbitrary body, *Journal of Acoustics Society of America*, Vol. 35, pp. 1626–1632.

10. Mitzner, K.M. (1967) Numerical solution for transient scattering from a hard surface of arbitrary shape—retarded potential technique, *Journal of Acoustics Society of America*, Vol. 42, No. 2, pp. 391–397.

11. Burton, A.J. and Miller G.F. (1971) The application of integral equation methods to the numerical solution of some exterior boundary value problems, *Proceedings of Royal Society, London*, Vol. 323(A), pp. 201–210.

12. Greenspan, D. and Werner, P. (1966) A numerical method for the exterior Dirichlet problem for the reduced wave equation, *Archives of Rational Mechanical Analysis*, Vol. 23, pp. 288–316.

13. Jones, D.S. (1974) Integral equations for the exterior acoustic problems, *Quarterly Journal of Mechanical Applied Mathematics*, Vol. 27, No. 1, pp. 129–142.

14. Schenck, H.A. (1966) Improved integral formulation for acoustic radiation problems, *Journal of Acoustics Society of America*, Vol. 44, No. 1, pp. 41–58.

15. Cruse, T.A. and Rizzo, F.J. (1968) A direct formulation and numerical solution of the general transient elastodynamic problem—I, *Journal of Mathematical Analysis Applications*, Vol. 22, No. 1, pp. 244–259.

16. Cruse, T.A. (1968) A direct formulation and numerical solution of the general transient elastodynamic problem—II, *Journal of Mathematical Analysis Applications*, Vol. 22, No. 2, pp. 341–355.

17. Hwang, L.S. and Tuck, E.O. (1970) On the oscillations of harbours of arbitrary shape, *Journal of Fluid Mechanics*, Vol. 42, pp. 447–464.

18. Eatock Taylor, R. and Waite, J.B. (1978) The dynamics of offshore structures evaluated by boundary integral techniques, *International Journal of Numerical Methods in Engineering*, Vol. 13, pp. 73–92.

19. Kobayashi, S., Fukui, T. and Azuma, N. (1975) An analysis of transient stresses produced around a tunnel by the integral equation method, *Proceedings of Symposium on Earthquake Engineering*, Japan, pp. 631–638.

20. Niwa, Y., Kobayashi, S. and Azuma, N. (1975) An analysis of transient stresses produced around cavities of arbitrary shape during the passage of travelling waves, Memo, Vol. 36, pp. 1–2, 28–46, Faculty of Engineering, Kyoto University, Japan.

21. Doyle, J.M. (1966) Integration of the Laplace transformed equations of classical elasto-kinetics, *Journal of Mathematical Analysis Applications*, Vol. 13, pp. 118–131.

22. Ursell, F. (1973) On exterior problems of acoustics, *Proceedings of Cambridge Philosophical Society*, Vol. 74, pp. 117–125.

23. Kleinmann, R.E. and Roach, G.F. (1974) Boundary integral equations for the three-dimensional Helmholtz equations, *SIAM Review*, Vol. 16, pp. 214–236.

24. Garnet, H. and Pascal, J.C. (1966) Transient response of circular cylinder of arbitrary

thickness in an elastic medium to a plane dilational wave, *Journal of Applied Mechanics*, Vol. 33, pp. 521–531.

25. Chen, L.H. (1967) Projections for the near terms—application of approaches to real submersibles, *Proceedings of Acoustics of Submerged Structs*, Vol. II, 15–17 February 1967, Office of Naval Research, Washington, D.C.

26. Banaugh, R.P. and Goldsmith, W. (1963) Diffraction of steady elastic waves by surfaces of arbitrary shape, *Journal of Applied Mechanics*, Vol. 30, pp. 589–597.

27. Banerjee, P.K. and Ahmad, S. (1985) Advanced three-dimensional dynamic analysis of boundary element method, in *Advanced Topics in Boundary Element Analysis*, Eds T.A. Cruse, A.B. Pifko and H. Armen, AMD Vol. 72, pp. 65–81, ASME, New York.

28. Ahmad, S. and Manolis, G.D. (1987) Dynamic analysis of 3D structures by a transformed boundary element method, *Computational Mechanics*, Vol. 2, pp. 185–196.

29. Ahmad, S. (1986) Linear and non-linear dynamic analysis by boundary element method, PhD Dissertation, State University of New York at Buffalo, New York.

30. Banerjee, P.K., Ahmad, S. and Chen, K. (1988) Advanced application of BEM in wave barriers in multilayer soil media, *Earthquake Engineering and Structural Dynamics*, Vol. 16, pp. 1041–1060.

31. Banerjee, P.K., Wang, H.C. and Ahmad, S. (1992) Multiregion periodic dynamic analysis by BEM of 2D, axisymmetric and 3D problems, Chapter II in *Advanced Dynamic Analysis by BEM*, Eds P.K. Banerjee and S. Kobayashi, pp. 27–74, Elsevier Applied Science, London.

32. Ahmad, S. and Banerjee, P.K. (1988) Multi-domain BEM for two-dimensional problems of elastodynamics, *International Journal of Numerical Methods in Engineeering*, Vol. 26, pp. 891–911.

33. Wang, H.C. (1989) A general purpose development of BEM for axisymmetric solids, PhD Dissertation, State University of New York at Buffalo, New York.

34. Wang, H.C. and Banerjee, P.K. (1990) Advanced periodic dynamic analysis of three-dimensional solids by BEM, *International Journal of Numerical Methods in Engineering*, Vol. 30, pp. 115–131.

35. Israil, A.S.M. (1990) Time-domain elastic and inelastic transient dynamic analysis of 2D solids by boundary element method, PhD Dissertation, State University of New York at Buffalo, New York.

36. Israil, A.S.M. and Banerjee, P.K. (1990) Advanced development of time-domain BEM for two-dimensional scalar wave propagation, *International Journal of Numerical Methods in Engineering*, Vol. 29, pp. 1003–1020.

37. Israil, A.S.M. and Banerjee, P.K. (1991) Calculations of interior flux by time-domain BEM, *Communications in Applied Numerical Methods*, Vol. 67, pp. 411–420.

38. Israil, A.S.M., Banerjee, P.K. and Wang, H.C. (1992) Time domain BEM formulations of BEM for two-dimensional, axisymmetric and three-dimensional scalar wave propagation, Chapter III in *Advanced Dynamic Analysis by BEM*, Eds P.K. Banerjee and S. Kobayashi, pp. 75–114, Elsevier Applied Science, London.

39. Ahmad, S. and Banerjee, P.K. (1988) Time-domain transient elastodynamic analysis of 3D solids by BEM, *International Journal of Numerical Methods in Engineering*, Vol. 26, pp. 1709–1728.

40. Banerjee, P.K., Ahmad, S. and Manolis, G.D. (1986) Transient elastodynamic analysis of three-dimensional problems by BEM, *International Journal of Earthquake Engineering and Structural Dynamics*, Vol. 14, pp. 933–949.

41. Wang, H.C. and Banerjee, P.K. (1990) Axisymmetric transient elastodynamic analysis by boundary element method, *International Journal of Solids and Structures*, Vol. 26, pp. 401–415.

42. Banerjee, P.K., Israil, A.S.M. and Wang, H.C. (1992) Time domain BEM for two-dimensional axisymmetric and three-dimensional transient elastodynamics, Chapter IV in *Advanced Dynamic Analysis by BEM*, Eds P.K. Banerjee and S. Kobayashi, pp. 115–154, Elsevier Applied Science, London.

43. Ahmad, S. and Banerjee, P.K. (1990) Inelastic transient dynamic analysis of three-dimensional problems by BEM, *International Journal of Numerical Methods in Engineering*, Vol. 29, pp. 371–390.

44. Cole, D.M., Kosloff, D.D. and Minster, J.B. (1978) A numerical boundary integral equation method for elastodynamics I, *Bulletin of Seismic Society America*, Vol. 68, pp. 1331–1357.

45. Mansur, W.J. (1983) A time-stepping technique to solve wave propagation problems using the boundary element method, PhD Dissertation, University of Southampton.

46. Antes, H. (1985) A boundary element procedure for transient wave propagation in two-dimensional isotropic elastic media, *Finite Element Analysis and Design*, Vol. 1, pp. 313–322.

47. Israil, A.S.M. and Banerjee, P.K. (1990) Advanced time-domain formulation of BEM for two-dimensional transient elastodynamics, *International Journal of Numerical Methods in Engineering*, Vol. 29, pp. 1421–1440.

48. Israil, A.S.M. and Banerjee, P K (1990) Two- dimensional transient wave-propagation problems by time-domain BEM, *International Journal of Solids and Structures*, Vol. 26, pp. 851–864.

49. Israil, A.S.M. and Banerjee, P.K. (1991) Interior stress calculations in 2-D time-domain transient BEM analysis, *International Journal of Solids and Structures*, Vol. 27, No. 7, pp. 915–927.

50. Chen, C.H. and Penzien, J. (1986) Dynamic modelling of axisymmetric foundations, *Earthquake Engineering and Structural Dynamics*, Vol. 14, pp. 823–840.

51. Banerjee, P.K., Ahmad, S. and Wang, H.C. (1989) Advanced development of BEM for elastic and inelastic dynamic analysis of solids, Chapter 3 in *Industrial Applications of BEM, Volume 5, Developments in BEM*, Eds P.K. Banerjee and R.B. Wilson, pp. 77–117, Elsevier Applied Science, London.

52. Woods, R.D. (1968) Screening of surface waves in soils, *Proceedings of ASCE*, Vol. 94 (SM4), pp. 951–979.

53. Miles, J.W. (1961) *Modern Mathematics for Engineers*, Ed. E.F. Beckenback, McGraw-Hill, London.

54. Abscal, R. and Dominguez, J. (1984) Dynamic behavior of strip footings on non-homogeneous viscoelastic soils, *Proceedings of the International Symposium on Dynamic Soil–Structure Interaction*, Ed. D.E. Beskos, Minneapolis, pp. 25–35, A.A. Balkema Publishers.

55. Estorff, O.V. and Schmid, G. (1984) Application of the boundary element method to the analysis of the vibration behavior of strip foundations on a soil layer, *Proceedings of the International Symposium on Dynamic Soil–Structure Interaction*, Ed. D.E. Beskos, Minneapolis, A.A. Balkema Publishers.

56. Israil, A.S.M. and Ahmad, S. (1989) Dynamic vertical compliance of strip foundations in layered soils, *Earthquake Engineering and Structural Dynamics*, Vol. 18, pp. 933–950.

57. Israil, A.S.M. and Banerjee, P.K. (1990) Dynamic analysis of 3D buried foundations by BEM, *International Journal of Numerical and Analytical Methods in Geomechanics*, Vol. 14, No. 1, pp. 49–70.

58. Karabalis, D.L. and Beskos, D.E. (1984) Dynamic response of 3D rigid surface foundations by time-domain boundary element method, *Earthquake Engineering and Structural Dynamics*, Vol. 12, pp. 73–93.

59. Nishimura, N. and Kobayashi, S. (1989) A regularized boundary integral equation method for elastodynamic crack problems, *Computational Mechanics*, Vol. 4, pp. 319–328.

60. Zhang, C.L. and Achenbach, J.D. (1988) Scattering by multiple crack configurations, *Journal of Applied Mechanics*, Vol. 55, pp. 104–110.

61. Hirose, S. and Achenbach, J.D. (1989) Time domain BEM of elastic wave interaction with a crack, *International Journal of Numerical Methods in Engineering*, Vol. 28, pp. 629–644.

62. Hirose, S. and Achenbach, J.D. (1991) Acoustic emission and near-tip elastodynamic fields of a growing penny-shape crack, *Engineering Fracture Mechanics*, Vol. 39, pp. 21–36.

63. Budreck, D.E. and Achenbach, J.D. (1988) Scattering from three-dimensional planar crack by boundary integral equation method, *Journal of Applied Mechanics*, Vol. 55, pp. 405–412.

64. Martin, P.A. and Rizzo, F.J. (1989) On boundary integral equations for crack problems, *Proceedings of Royal Society*, Vol. 421(A), pp. 341–355.

65. Jia, Z.H., Shippy, A.J. and Rizzo F.J. (1990) Boundary element analysis of wave scattering from cracks, *Communications in Applied Numerical Methods*, Vol. 6, pp. 591–601.

66. Krishnaswamy, G., Schmerr, L., Rudolphi, T.J. and Rizzo, F. (1990) Hypersingular boundary integral equations: some applications in acoustic and elastic wave scattering, *Journal of Applied Mechanics*, Vol. 57, pp. 404–414.

67. Rizzo, F.J., Shippy, D.J. and Rezayat, M. (1985) A boundary integral equation method for radiation and scattering of elastic waves in three-dimensions, *International Journal of Numerical Methods in Engineering*, Vol. 21, pp. 115–129.

68. Sybert, A.F., Soenarko, B., Rizzo, F. and Shippy, D.J. (1988) An advanced computational method for radiation and scattering of acoustic waves in three-dimensions, *Journal of Acoustical Society of America*, Vol. 77, No. 2, pp. 362–368.

69. Nishimura, N. and Kobayashi, S. (1992) Advanced dynamic fracture mechanics analysis, Chapter 1 in *Developments in Boundary Element Methods*, Vol. 7, Eds P.K. Banerjee and S. Kobayashi, pp. 1–27, Elsevier Applied Science, London.

70. Tosaka, N. and Ohmi, M. (1992) Transient elastic wave scattering problems by BEM, Chapter 5 in *Developments in Boundary Element Methods*, Vol. 7, Eds P.K. Banerjee and S. Kobayashi, pp. 155–206, Elsevier Applied Science, London.

71. Kitahara, M., Hiroshi, S. and Achenbach, J.D. (1992) Transient elastodynamic analysis for three-dimensional configurations, Chapter 7 in *Developments in Boundary Element Methods*, Vol. 7, Eds P.K. Banerjee and S. Kobayashi, pp. 253–282, Elsevier Applied Science, London.

72. Cao, Z.Y., Zhu, J.X. and Cheung, Y.K. (1990) A semi-analytical BEM for scattering of waves in a half space, *Earthquake Engineering and Structural Dynamics*, Vol. 19, pp. 1073–1082.

73. Zhu, J.X., Cao, Z.Y. and Cheung, Y.K. (1991) Transient stress analysis of grooved and stepped shafts by time domain semi-analytical BEM, *Computers and Structures*, Vol. 39, pp. 243–247.

74. Israil, A.S.M. and Banerjee, P.K. (1992) Advanced development of BEM for two-dimensional dynamic elastoplasticity, *International Journal of Solids and Structures*, Vol. 29, No. 11, pp. 1433–1451.

75. Antes, H. and Panagiotopoulos, P.D. (1992) *The Boundary Integral Approach to Static and Dynamic Contact Problems*, Birkhäuser-Verlag, Basle, Boston and Berlin.

76. Banerjee, P.K., and Ed. Kobayashi, S. (1992) *Advanced Dynamic Analysis by Boundary Element Methods*, pp. 410, Elsevier Applied Science, London.

ELEVEN

FRACTURE MECHANICS

11.1 INTRODUCTION

The residual fatigue life of a cracked structural element can be characterized by crack tip stress intensity factors. The stress intensity factors relate the crack size and shape, and the structural geometry in a systematic manner.[1-10] This chapter illustrates some of the basic modelling problems associated with fracture mechanics analysis and provides numerical results for complex cracked bodies. Basic aspects of boundary element modelling of linear elastic fracture mechanics are first reviewed here.

11.2 AN OVERVIEW OF LEFM

11.2.1 Linear Elastic Fracture Mechanics (LEFM) Analysis

The presence of a crack in a structure generally induces high stress concentration at the crack tip. Fracture mechanics provides the means by which local crack tip stress fields as well as the elastic deformations can be characterized. In LEFM, the inelastic deformation in the vicinity of the crack tip due to stress concentration is deemed to be small compared to the size of the crack and other characteristic lengths.

Elastic modelling of crack tip behaviour makes use of deformation due to three primary modes of loading, as illustrated in Fig. 11.1. The three modes are: the opening mode (mode I) due to normal stress, the sliding mode (mode II) due to in-plane shear stress and the tearing mode (mode III), due to out-of-plane shear.

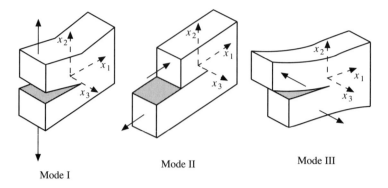

Mode I

Mode II

Mode III

Figure 11.1 Three primary loading modes of a cracked body.

Consider the crack problem defined in Fig. 11.2. The stresses and displacements for traction free cracks may be given as an infinite series in r, where r is the distance from the crack tip. For the anti-plane problem the near crack tip field is given by [1,9]

$$\begin{Bmatrix} \sigma_{31} \\ \sigma_{32} \end{Bmatrix} = \frac{K_{\mathrm{III}}}{(2\pi r)^{1/2}} \begin{Bmatrix} \sin(\theta/2) \\ \cos(\theta/2) \end{Bmatrix} \tag{11.1}$$

and

$$u_3 = \frac{2K_{\mathrm{III}}}{\mu} \left(\frac{r}{2\pi}\right)^{1/2} \sin(\theta/2) \tag{11.2}$$

where u_i is the displacement, σ_{ij} is the stress and μ is the shear modulus. The mode III stress intensity factor, K_{III}, is established by the far-field boundary conditions and is a function of applied loading and the geometry of the cracked body. The

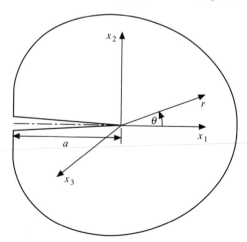

Figure 11.2 Infinite plate under triaxial load.

mode III stress intensity factor is defined by

$$K_{III} = \lim_{r \to 0} \left\{ (2\pi r)^{1/2} \sigma_{32} \big|_{\theta=0} \right\} \tag{11.3}$$

For the plane problem, assuming plane strain conditions,[1,9]

$$\begin{Bmatrix} \sigma_{11} \\ \sigma_{12} \\ \sigma_{22} \end{Bmatrix} = \frac{K_I}{(2\pi r)^{1/2}} \cos(\theta/2) \begin{Bmatrix} 1 - \sin(\theta/2)\sin(3\theta/2) \\ \sin(\theta/2)\cos(3\theta/2) \\ 1 + \sin(\theta/2)\sin(3\theta/2) \end{Bmatrix} \tag{11.4}$$

and

$$\begin{Bmatrix} u_1 \\ u_2 \end{Bmatrix} = \frac{K_I}{2\mu} \left(\frac{r}{2\pi} \right)^{1/2} \begin{Bmatrix} \cos(\theta/2)[\alpha - 1 + 2\sin^2(\theta/2)] \\ \sin(\theta/2)[\alpha + 1 - 2\cos^2(\theta/2)] \end{Bmatrix} \tag{11.5}$$

where K_I is the mode I stress intensity factor, defined by

$$K_I = \lim_{r \to 0} \left\{ (2\pi r)^{1/2} \sigma_{22} \big|_{\theta=0} \right\} \tag{11.6}$$

and

$$\alpha = 3 - 4\nu$$

The corresponding relationships for mode II fields are[1,9]

$$\begin{Bmatrix} \sigma_{11} \\ \sigma_{12} \\ \sigma_{22} \end{Bmatrix} = \frac{K_{II}}{(2\pi r)^{1/2}} \begin{Bmatrix} -\sin(\theta/2)[2 + \cos(\theta/2)\cos(3\theta/2)] \\ \cos(\theta/2)[1 - \sin(\theta/2)\sin(3\theta/2)] \\ \sin(\theta/2)\cos(\theta/2)\cos(3\theta/2) \end{Bmatrix} \tag{11.7}$$

$$\begin{Bmatrix} u_1 \\ u_2 \end{Bmatrix} = \frac{K_{II}}{2\mu} \left(\frac{r}{2\pi} \right)^{1/2} \begin{Bmatrix} \sin(\theta/2)[\alpha + 1 + 2\cos^2(\theta/2)] \\ -\cos(\theta/2)[\alpha - 1 - 2\sin^2(\theta/2)] \end{Bmatrix} \tag{11.8}$$

where the mode II stress intensity factor, K_{II}, is defined by

$$K_{II} = \lim_{r \to 0} \left\{ (2\pi r)^{1/2} \sigma_{12} \big|_{\theta=0} \right\} \tag{11.9}$$

11.2.2 BEM Modelling of Cracked Bodies

The presence of two coplanar surfaces precludes the use of the BEM for the general solution of cracked bodies. Several different modelling strategies have been employed for three-dimensional cracked bodies, as illustrated in Fig. 11.3. The first approach is to model the crack as an open notch. The major deficiency of this modelling approach is that the system equation becomes badly conditioned when the surfaces modelled are too close together and the results become inaccurate. One form of avoiding this difficulty is the dislocation or traction BEM modelling approach.[2–5] However, the highly singular nature of the integrals in this method presents difficulty in the numerical implementation although recent improvements have been reported by Polch et al.[6] and others. Further research is required before the full potential of this hyper-singular method could be realized.

The modelling of symmetric cracks is rather straightforward using the usual

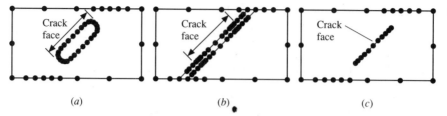

Figure 11.3 BEM modelling of cracks.

symmetry condition during the discretization. Improved results have been obtained subsequently using isoparametric interpolation functions.[7] The accuracy of the results has been further improved by the use of special crack tip elements that accurately model the crack tip field as discussed below.

The best strategy applicable for any general two- and three-dimensional modelling of a non-symmetric crack is the sub-region model.[8,9] In this approach, the body is sub-structured into separate regions through the crack plane such that each crack surface is in a different region. By satisfying compatibility and continuity conditions at the interface of the regions except along the crack surface the solutions are obtained. Numerical results using this strategy for elasticity involving two-dimensional structures have been reported, among others, by Blandford et al.[8] and by Raveendra and Cruse[9] for three-dimensional problems.

11.2.3 Special Elements

Quarter-point (QP) elements   It is well known that the behaviour of displacement and stress near the crack tip are $O(\sqrt{r})$ and $O(1/\sqrt{r})$ respectively.[1] By shifting the mid-node to the quarter-point from the crack tip, the correct variation of displacement can be obtained, as explained below.[10] The isoparametric representation for the field variable (f_i) is

$$f_i = N_1 f_i^1 + N_2 f_i^2 + N_3 f_i^3 \tag{11.10}$$

where f_i^m is the ith component of the mth nodal value of the field variable, and

$$N_1 = 2(\eta - 1)(\eta - \tfrac{1}{2})$$

$$N_2 = -4\eta(\eta - 1)$$

$$N_3 = 2\eta(\eta - \tfrac{1}{2})$$

Similarly, the coordinate ($\bar{r}$) along the length of an element is given by (see Fig. 11.4)

$$\bar{r} = N_1 \bar{r}^{(1)} + N_2 \bar{r}^{(2)} + N_3 \bar{r}^{(3)}$$

$$= 2(\eta - 1)(\eta - \tfrac{1}{2})\bar{r}^{(1)} - 4\eta(\eta - 1)\bar{r}^{(2)} + 2\eta(\eta - \tfrac{1}{2})\bar{r}^{(3)} \tag{11.11}$$

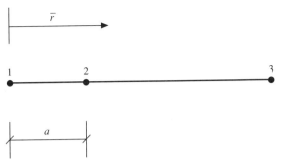

Figure 11.4 The quarter-point element.

and the Jacobian is

$$\frac{\partial \bar{r}}{\partial \eta} = 2\left(2\eta - \frac{3}{2}\right)\bar{r}^{(1)} - 4(2\eta - 1)\bar{r}^{(2)} + 2\left(2\eta - \frac{1}{2}\right)\bar{r}^{(3)} \tag{11.12}$$

By letting $\bar{r}^{(1)} = 0, \bar{r}^{(2)} = a$ and $\bar{r}^{(3)} = 1$,

$$\frac{\partial \bar{r}}{\partial \eta} = (4a - 1) + 4\eta(1 - 2a) \tag{11.13}$$

and

$$\bar{r} = -4(\eta^2 - n)a + 2\left(\eta^2 - \frac{n}{2}\right) \tag{11.14}$$

from which

$$\eta = \frac{-(4a - 1) + \sqrt{(4a - 1)^2 - 4(1 - 2a)\bar{r}}}{4(1 - 2a)} \tag{11.15}$$

Using Eq. (11.15) in Eq. (11.13)

$$\frac{\partial \bar{r}}{\partial \eta} = (4a - 1) - \frac{1}{4}(4a - 1) + \frac{1}{4}\sqrt{(4a - 1)^2 - 8(1 - 2a)\bar{r}^1}$$

The Jacobian at node 1 is zero, if $\bar{r} = 0$; thus

$$0 = (4a - 1) - \tfrac{1}{4}(4a - 1) + \tfrac{1}{4}(4a - 1)$$

which yields $a = \tfrac{1}{4}$; i.e. the mid-node should be at the quarter-point. From Eq. (11.14), we have

$$\bar{r} = \eta^2$$

i.e.

$$\eta = \sqrt{\bar{r}}$$

Using $\eta = \sqrt{\bar{r}}$ in Eq. (11.10),

$$f_i = f_i^{(1)} + \sqrt{\bar{r}}\left(-3f_i^{(1)} + 4f_i^{(2)} - f_i^{(3)}\right) + \bar{r}\left(2f_i^{(1)} - 4f_i^{(2)} + 2f_i^{(3)}\right) \tag{11.16}$$

The above relation shows the desired $\sqrt{r}$ variation of the field variable near the crack tip.

If the same shape functions are used for describing traction variations, it also shows $\sqrt{r}$ behaviour which is incorrect. To obtain $1/\sqrt{r}$ behaviour of traction, modified shape functions are needed, as described in the next section.

Traction singular (TS) elements   For traction, expression (11.10) can be modified through division by $1/\sqrt{\bar{r}}$ and written as

$$\bar{t}_i = \bar{t}_i^1 N_1 \frac{1}{\sqrt{\bar{r}}} + \bar{t}_i^{(2)} N_2 \frac{1}{\sqrt{\bar{r}}} + t_i^{(3)} N_3 \frac{1}{\sqrt{\bar{r}}}$$

$$= b_i^{(1)} \frac{1}{\sqrt{\bar{r}}} + b_i^{(2)} + b_i^{(3)} \sqrt{\bar{r}}$$

where

$$b_i^{(1)} = \bar{t}_i^{(1)}$$

$$b_i^{(2)} = -3\bar{t}_i^{(1)} + 4\bar{t}_i^{(2)} - \bar{t}_i^{(3)}$$

$$b_i^{(3)} = 2\bar{t}_i^{(1)} - 4\bar{t}_i^{(2)} + 2\bar{t}_i^{(3)}$$

The above equation for traction can alternately be written as

$$\bar{t}_i = \bar{t}_i^{(1)} M_1 + \bar{t}_i^{(2)} M_2 + \bar{t}_i^{(3)} M_3$$

where M_1, M_2, M_3 are traction singular shape functions given by[8]

$$M_1 = -3 + \frac{1}{\rho} + 2\rho$$

$$M_2 = 4(1 - \rho)$$

$$M_3 = -1 + 2\rho$$

The relation between ρ and η is $\rho = \eta$ when node 1 is the singular node. When node 3 is the singular node, $\rho = 1 - \eta$ and M_1 and M_3 interchange expressions. Traction singular elements were first introduced in the boundary element literature by Blandford et al.[8]

11.2.4 Evaluation of Stress Intensity Factors (SIFs)

The stress intensity factor can be determined either from nodal tractions or from the crack opening displacements (COD).

From nodal traction

$$K_I = t_A^2 \sqrt{2\pi \ell}$$

$$K_{II} = t_A^1 \sqrt{2\pi \ell}$$

where K_I and K_{II} are the mode I (opening mode) and mode II (shearing mode) stress intensity factors respectively; t_A refers to the traction at node A (the crack

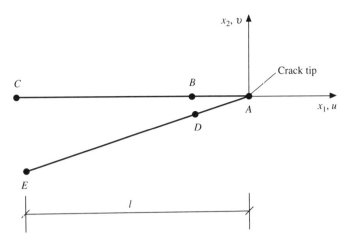

Figure 11.5 The crack tip modelling.

tip), while superscripts 2 and 1 indicate, respectively, the opening and shearing mode directions. The length of the element in the direction perpendicular to the crack front is represented by ℓ.

From COD Referring to Fig. 11.5, the stress intensity factors are given by (Blandford *et al.*[8])

$$K_{\mathrm{I}} = \frac{\mu}{\alpha + 1} \sqrt{\frac{2\pi}{\ell}} [4(v_{\mathrm{B}} - v_{\mathrm{D}}) + v_{\mathrm{E}} - v_{\mathrm{C}}]$$

$$K_{\mathrm{II}} = \frac{\mu}{\alpha + 1} \sqrt{\frac{2\pi}{\ell}} [4(u_{\mathrm{B}} - u_{\mathrm{D}}) + u_{\mathrm{E}} - u_{\mathrm{C}}]$$

where μ is the shear modulus, $\alpha = 3 - 4\nu$ for plane strain and $(3 - \nu)/(1 + \nu)$ for plane stress; u and v indicate the displacements along the shearing and opening mode directions respectively. The points B, C, D and E are as defined in Fig. 11.5.

For a symmetric crack $v_{\mathrm{B}} = -v_{\mathrm{D}}$ and $v_{\mathrm{C}} = -v_{\mathrm{E}}$; then the expression for K_{I} simplifies to

$$K_{\mathrm{I}} = \frac{2\mu}{\alpha + 1} \sqrt{\frac{2\pi}{\ell}} (4v_{\mathrm{B}} - v_{\mathrm{C}})$$

and of course $K_{\mathrm{II}} = 0$.

Alternatively, the SIFs can also be determined from crack opening displacements Δu_1 and Δu_2 of the quarter-point nodes as

$$\left\{ \begin{array}{c} K_{\mathrm{I}} \\ K_{\mathrm{II}} \end{array} \right\} = \frac{H}{8} \sqrt{\frac{2\pi}{\rho}} \left\{ \begin{array}{c} \Delta u_2 \\ \Delta u_1 \end{array} \right\}$$

where $\Delta u_1 = u_{\mathrm{B}} - u_{\mathrm{D}}$, $\Delta u_2 = v_{\mathrm{B}} - v_{\mathrm{D}}$, $H = E$ for the plane stress case and

$H = E/(1 - \nu^2)$ for the plane strain case. The three-dimensional solutions are obtained from the plane strain value of H. Similar expressions can also be derived for K_{III}.

11.3 NUMERICAL EXAMPLES

11.3.1 Cracks in Bodies Subjected to Mechanical Forces

Figure 11.6 shows the BEM mesh used for the analysis of a buried circular crack problem.[9] Due to symmetry, only one-eighth of the body was modelled. The crack opening displacements using standard quadratic element and modified crack tip elements are compared to analytical results in Fig. 11.7 which indicates that the results are improved by the use of modified crack tip elements. The analytical solution plotted in the figure is for an infinite body. To further assess the accuracy, the mode I stress intensity factors were evaluated for standard and modified crack tip element models. A comparison with the empirical value that takes the finite dimension of the body into account demonstrated that the stress intensity factors using quarter-point (QP) elements are approximately 5 per cent in error, compared to standard element results which are approximately 10 per cent in error. Further enhancement was obtained by using traction singular (TS) modification to the quarter-point element, which improved the stress intensity factor value to within 1 per cent of the predicted value.

11.3.2 Cracks at the Interface of Different Materials

A study of interfacial cracks has been conducted by many investigators.[11-14] A full form of the near tip field solution, in the sense of a complete Williams expansion,[15]

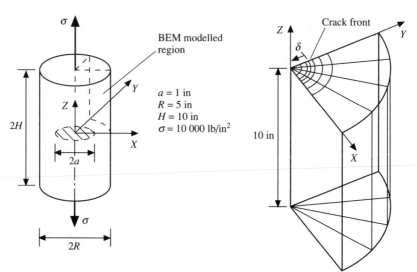

Figure 11.6 Circular crack—BEM map.

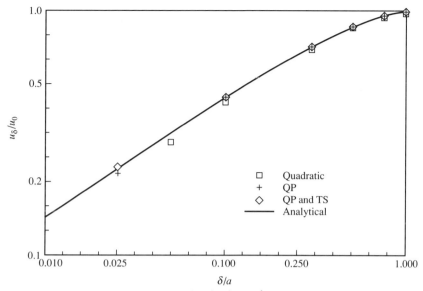

Figure 11.7 Crack opening displacement for circular track.

however, has been presented only recently by Rice,[16] who solved the problem of a plane crack at a bimaterial interface in an infinite body by using analytic functions.

BEM-based interface crack analysis has been performed by Lee and Choi[17] who erroneously referred to the stress intensity factors as classical mode I and mode II types. As discussed by Rice,[16] the stress intensity factors for the bimaterial problem are coupled and the crack tip behaviour is characterized by a complex stress intensity factor and therefore this separation of stress intensity factors is pertinent only in limited circumstances where materials are sufficiently similar. Recently Raveendra and Banerjee[18] used Rice's definition to compute stress intensity factors in interfacial cracks.

For interfacial cracks, the stress field ahead of the crack along the interface is determined by Rice[16] as

$$(\sigma_{yy} + i\sigma_{xy})|_{\theta=0} = \frac{K}{\sqrt{2\pi r}} \exp[i\epsilon \ln(\rho)]$$

where

ρ = distance from the crack tip
$K = k_1 + ik_2$, where k_1 and k_2 are real numbers, is a complex stress intensity factor
that characterizes near tip singular field

$$\epsilon = \frac{1}{2\pi} \ln(\gamma)$$

$$\gamma = \frac{\alpha_1 \mu_2 + \mu_1}{\alpha_2 \mu_1 + \mu_2}$$

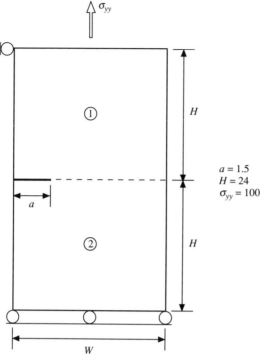

Figure 11.8 Edge crack model.

$$
\alpha_i = \begin{cases} 3 - 4\nu_i & \text{for plane strain} \\ (3 - \nu_i)/(1 + \nu_i) & \text{for plane stress, } i = 1, 2 \end{cases}
$$

$$
\nu = \text{Poisson's ratio}
$$

$$
\mu = \text{shear modulus}
$$

and subscripts 1 and 2 refer to the material in $y > 0$ and $y < 0$ respectively.

The crack length effect on different material combinations was investigated[18] for an edge crack subjected to uniaxial loading using the model shown in Fig. 11.8. The k_1 plot shown in Fig. 11.9 indicates that the increase of k_1 values with crack length for the edge crack is rather rapid when compared to the similar values for a centre cracked plate given in Ref. 18. The values also show that, at small crack lengths, k_1 increases with increasing γ but remains the same at large crack lengths. The k_2 plot shown in Fig. 11.10 indicates that k_2 values increase rapidly with crack lengths and also with an increase of γ.

11.3.3 Cracks in Rotating Solids

Investigation of rotating bodies with cracks has been performed by many investigators. For example, internal as well as edge cracks in finite rotating discs were solved by Rooke and Tweed[19] in terms of the solution of a Fredholm integral

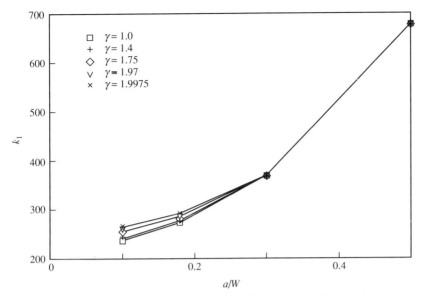

Figure 11.9 Variation of k_1 with a/W due to uniaxial loading under plane strain.

equation; problems of cracks emanating from a central hole were solved by Owen and Griffiths[20] and Riccardella and Bamford[21] using the finite element method (FEM); the FEM was also used by Chen and Lin[22] to solve arbitrary cracks in a rotating disc; Isida[23] has used the eigenfunction expansion of the complex potentials together with the boundary collocation technique to solve internal cracks

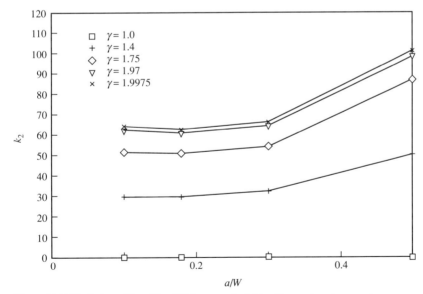

Figure 11.10 Variation of k_2 with a/W due to uniaxial loading under plane strain.

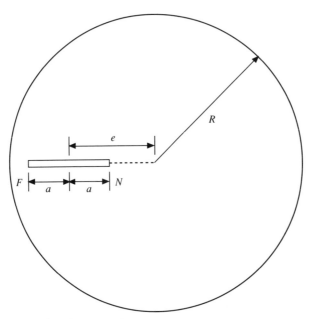

Figure 11.11 Geometry of a radial crack in a rotating disc ($R = 200$).

located at arbitrary positions in a disc; the method of caustics was used by Sukere[24] to compute the stress intensity factors of internal radial cracks in rotating discs. The stress intensity factors of rotating cracks have also been calculated using the weight function approach.[25] While the boundary element method is known to give very accurate solutions for stationary elastic bodies with cracks, it appears that the only extension of BEM to the solution of cracked rotating bodies is the application of Smith,[26] who computed stress intensity factors for arc cracks in a rotating disc. Recently Raveendra and Banerjee[27] used a surface-only formulation based on particular integrals to solve two- and three-dimensional crack problems in rotating bodies.

The disc shown in Fig. 11.11 is analysed by changing the location and size of the crack.[27] The normalized SIFs computed by the BEM at both crack tips for an arbitrary crack length parameter $\lambda (= a/(R - e))$ at eccentricities $\epsilon (= e/R)$ are compared in Figs 11.12 and 11.13 to the results of Isida,[23] who obtained the results by using eigenfunction expansion of complex potentials together with the boundary collocation technique. The results show excellent agreement between the BEM and eigenfunction solutions.

11.3.4 Thermal Fracture Mechanics

The stress intensity factors in thermally stressed structures have been generally obtained analytically for simple structures under steady state conditions and experimentally and numerically for arbitrary structures under steady state and time-dependent thermal loading. For example, Emery *et al.*[28] have used the

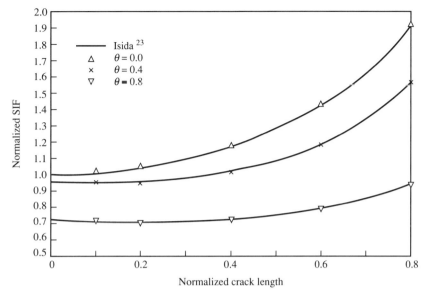

Figure 11.12 Comparison of mode I SIFs at the crack tip N ($\nu = 0.3$).

finite element method (FEM) with singularity and quarter-point crack elements to compute stress intensity factors for edge cracks. Stress intensity factors for centre cracks under a steady state temperature field were calculated by Sumi and

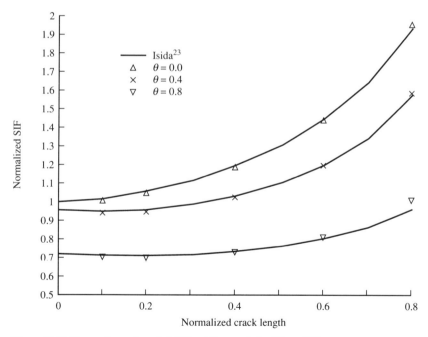

Figure 11.13 Comparison of mode I SIFs at the crack tip F ($\nu = 0.3$).

Katayama[29] using the complex variable method together with analytic continuation and modified mapping collocation techniques. The stress intensity factors for cracks in planar structures under time-dependent thermal loading have been computed by Hellen *et al.*[30] using the FEM together with quarter-point crack elements, *J* integral and virtual crack extension techniques and by Emmel and Stamm[31] using the FEM with quarter-point elements.

Tanaka *et al.*[32] have used extended boundary integral equations involving the volume integral of the temperature field for the solution of steady and non-steady state problems including fracture mechanics problems. A BEM formulation for steady state thermoelasticity based on converting the volume integral involving the temperature field to equivalent surface integrals was used by Lee and Choi[33] to compute thermal stress intensity factors of cusp cracks. As indicated, the above work is confined to steady state analysis and also only symmetric cracks were considered.

A complete boundary element solution procedure for time-dependent thermoelastic problems, based on integral equations expressed only over the surface of the body, was used recently for the solution of thermoelastic fracture mechanics problems by Raveendra, Banerjee and Dargush.[34,35]

11.3.5 Dynamic Fracture Mechanics

The finite difference method and the finite element method have been used with some success in dynamic crack problems. The former method has difficulty in modelling dynamic crack initiation while the latter suffers from being unable to predict the time-dependent nature of crack propagation. However, the finite element method can be used by applying a series of incremental static loads to approximate time variations. For a detailed discussion, interested readers are referred to the article by Kanninen.[36]

Transient dynamic fracture mechanics studies were conducted by Sladek and Sladek[37] using a Laplace domain boundary element formulation, while Dominguez and Gallego[38] used a time-domain formulation. Recently Israil and Dargush[39] used a time-domain boundary element method for the study of dynamic fracture mechanics problems.

A plate with an inclined crack is supported on rollers on three of its sides while it is acted upon by a suddenly applied tensile load on the fourth side (Fig. 11.14) under conditions of plane strain. This problem has been investigated by Dominguez and Gallego [38] using a time-domain boundary element method, while Aoki *et al.*[40] provided solutions via the finite element method. The plate has the following dimensions and material properties: $b = 44\,\text{mm}$, $w = 32\,\text{mm}$, $a_0 = 22.63\,\text{mm}$, $c = 6\,\text{mm}$ and $\sigma_0 = 400\,\text{MPa}$; Young's modulus $E = 7.35 \times 10^4\,\text{MPa}$, $\nu = 0.25$, $\rho = 2.45 \times 10^3\,\text{kg/m}^3$.

The plate is modelled as a two-region problem with 28 elements in each region. The time step is chosen to be $\Delta t = 0.25\,\mu\text{s}$. The present results[39] are compared with those of previously mentioned researchers in Figs 11.15 and 11.16. Results by Aoki *et al.*[40] are lower than the other solutions at later times. The kink observed near

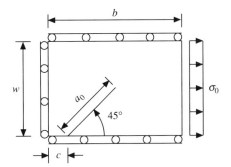

Figure 11.14 Rectangular plate with an inclined crack.

$t = 12\,\mu s$ in the present results and also in the results of Dominguez and Gallego[38] is due to the waves reflecting back from the left edge of the plate and impinging on to the crack.

11.4 CONCLUDING REMARKS

The suitability of the BEM for the solution of fracture mechanics problems is demonstrated by solving a number of problems with different types of loading. The approach utilizes the concept of the multiple region for the modelling of non-symmetric cracks. The accurate evaluation of SIFs requires proper modelling of displacement and stress fields in the vicinity of the crack tip. Special crack tip elements that provide the correct variation must be used in the analysis. The high level of accuracy obtained for various analyses make this approach very

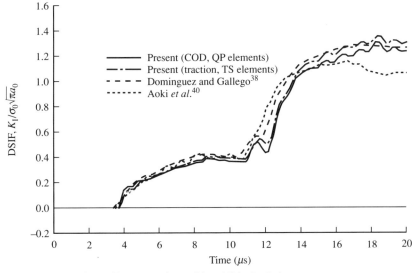

Figure 11.15 Mode I DSIF: comparison with published solutions.

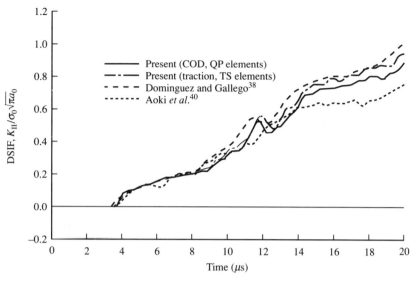

Figure 11.16 Mode II DSIF: comparison with published solutions.

useful for the analysis of fracture mechanics problems. This is evident in the many other applications of BEM to fracture mechanics problems described in Refs 8, 9, 18, 26, 27, 34, 35, 37 to 39 and 41 to 56 and Refs 61 to 71 of Chapter 10.

REFERENCES

1. Kanninen, M.F. and Popelar, C.H. (1985) *Advanced Fracture Mechanics*, Oxford Science Series 15, Oxford University Press, New York.
2. Guidera, T.J. and Lardner, R.W. (1975) Penny-shaped cracks, *Journal of Elasticity*, Vol. 5, pp. 59–73.
3. Cruse, T.A. (1975) Boundary-integral equation method for three dimensional elastic fracture mechanics analysis, AFOSR-TR-75-0813, Accession No. ADA 011660.
4. Bui, H.D. (1977) An integral equation method for solving the problem of a plane crack of arbitrary shape, *Journal of Mechanical Physics*, Vol. 25, pp. 29–37.
5. Weaver, J. (1977) Three dimensional crack analysis, *International Journal of Solids and Structures*, Vol. 13, pp. 321–330.
6. Polch, E.Z., Cruse, T.A. and Huang, C.-J. (1987) Traction BIE solutions for flat cracks, *Computational Mechanics*, Vol. 2, pp. 253–267.
7. Cruse, T.A. and Wilson, R.B. (1977) Boundary-integral equation method for elastic fracture mechanics analysis, AFOSR-TR-0355, Accession No. ADA 051992.
8. Blandford, G.E., Ingraffea, A.R. and Liggett, J.A. (1981) Two dimensional stress intensity factor computations using the boundary element method, *International Journal of Numerical Methods in Engineering*, Vol. 17, pp. 387–404.
9. Raveendra, S.T. and Cruse, T.A. (1989) BEM analysis of problems of fracture mechanics, Chapter 6 in *Industrial Applications of Boundary Element Methods*, Eds P.K. Banerjee and R.B. Wilson, pp. 187–204, Elsevier Applied Science, London.
10. Barsoum, R.S. (1976) On the use of isoparametric finite elements in linear fracture mechanics, *International Journal of Numerical Methods in Engineering*, Vol. 10, pp. 25–37.
11. Cherepanov, G.P. (1979) *Mechanics of Brittle Fracture*, McGraw-Hill, New York.
12. England, A.H. (1965) A crack between dissimilar media, *Journal of Applied Mechanics*, Vol. 32, pp. 400–402.

13. Erdogan, F. (1965) Stress distribution in bonded dissimilar materials with cracks, *Journal of Applied Mechanics*, Vol. 32, pp. 403–410.

14. Rice, J.R. and Sih, G.C. (1965) Plane problems of cracks in dissimilar media, *Journal of Applied Mechanics*, Vol. 32, pp. 418–423.

15. Williams, M.L. (1959) The stresses around a fault or crack in dissimilar media, *Bulletin of Seismological Society of America*, Vol. 49, pp. 199–204.

16. Rice, J.R. (1988) Elastic fracture mechanics concepts for interfacial cracks, *Transactions of the ASME*, Vol. 55, pp. 98–103.

17. Lee, K.Y. and Choi, H.J. (1988) Boundary element analysis of stress intensity factors for bi-material interface cracks, *Engineering Fracture Mechanics*, Vol. 29, No. 4, pp. 461–472.

18. Raveendra, S.T. and Banerjee, P.K. (1991) Computation of stress intensity factors for interfacial cracks. *Engineering Fracture Mechanics*, Vol. 40, No. 1, 89–103.

19. Rooke, D.P. and Tweed, J. (1972) The stress intensity factors of a radial crack in a finite rotating elastic disc, *International Journal of Engineering Science*, Vol. 10, pp. 709–714.

20. Owen, D.R.J. and Griffiths, J.R. (1973) Stress intensity factors of cracks in a plate containing a hole and in a spinning disc, *International Journal of Fracture*, Vol. 9, pp. 471–476.

21. Riccardella, P.C. and Bamford, W.H. (1974) Reactor coolant pump flywheel overspeed evaluation. *Journal of Pressure Vessel Technology, ASME*, Vol. 96, No. 4, pp. 279–285.

22. Chen, W.H. and Lin, T.C. (1983) A mixed-mode crack analysis of rotating disk using FEM, *Engineering Fracture Mechanics*, Vol. 18, No. 1, pp. 133–143.

23. Isida, M. (1981) Rotating disk containing an internal crack located at an arbitrary position, *Engineering Fracture Mechanics*, Vol. 14, pp. 549–555.

24. Sukere, A.A. (1987) The stress intensity factors of internal radial cracks in rotating disks by the method of caustics, *Engineering Fracture Mechanics*, Vol. 26, No. 1, pp. 65–74.

25. Schneider, G.A. and Danzer, R. (1989) Calculation of the stress intensity factor of an edge crack in a finite elastic disc using the weight function method, *Engineering Fracture Mechanics*, Vol. 34, No. 3, pp. 547–552.

26. Smith, R.N.L. (1985) Stress intensity factors for an arc crack in a rotating disc, *Engineering Fracture Mechanics*, Vol. 21, No. 3, pp. 579–587.

27. Raveendra, S.T. and Banerjee, P.K. (1991) Analysis of rotating solids with cracks by the boundary element method, *International Journal of Solids and Structures*, Vol. 28, No. 9, pp. 1155–1170.

28. Emery, A.F., Neighbors, P.K., Kobayashi, A.S. and Love, W.J. (1977) Stress intensity factors in edge-cracked plates subjected to transient thermal singularities, *Journal of Pressure Vessel Technology*, Vol. 99, pp. 100–104.

29. Sumi, N. and Katayama, T. (1980) Thermal stress singularities at tips of a Griffith crack in a finite rectangular plate. *Nuclear Engineering and Design*, Vol. 60, pp. 389–394.

30. Hellen, T.K., Cesari, F. and Maitan, A. (1982) The application of fracture mechanics in thermally stressed structure. *International Journal of Pressure Vessel and Piping*, Vol. 10, pp. 181–204.

31. Emmel, E. and Stamm, H. (1985) Calculation of stress intensity factors of thermally loaded cracks using the finite element method, *International Journal of Press Vessel and Piping*, Vol. 19, 1–17.

32. Tanaka, M., Togoh, H. and Kikuta, M. (1984) Boundary element method applied to 2-D thermoelastic problems in steady and non-steady states. *Engineering Analysis*, Vol. 1, No. 1, pp. 13–19.

33. Lee, K.Y. and Choi, Y.H. (1990) Boundary element analysis of thermal stress intensity factors for cusp crack, *Engineering Fracture Mechanics*, Vol. 37, No. 4, pp. 787–798.

34. Raveendra, S.T. and Banerjee, P.K. (1992) Boundary element analysis of cracks in thermally stresses planar structures, *International Journal of Solids and Structures*, Vol. 29, No. 18, pp. 2301–2317.

35. Raveendra, S.T., Banerjee, P.K. and Dargush, G.F. (1992) Three-dimensional analysis of thermally loaded cracks, *International Journal of Numerical Methods in Engineering*, Vol. 36, pp. 1909–1926.

36. Kanninen, M.F. (1978) A critical appraisal of solution techniques in dynamic fracture mechanics, in *Numerical Methods in Fracture Mechanics*, Eds A.R. Luxmore and D.R.J. Owen, pp. 612–634, Swansea.

37. Sladek, J. and Sladek, V. (1987) A boundary integral equation method for dynamic crack problems, *Engineering Fracture Mechanics*, Vol. 27, pp. 269–277.

38. Dominguez, J. and Gallego, R. (1989) Time-domain boundary element analysis of two dimensional crack problems, *International Seminar on BEM*, Hartford, Connecticut.

39. Israil, A.S.M. and Dargush, G.F. (1991) Dynamic fracture mechanics studies by time-domain BEM, *Engineering Fracture Mechanics*, Vol. 39, pp. 315–328.

40. Aoki, S., Kishimoto, K., Kondo, H. and Sakata, M. (1978) Elastodynamic analysis of crack by finite element method using singular element, *International Journal of Fracture*, Vol. 14, pp. 59–68.

41. Snyder, M.D. and Cruse, T.A. (1975) Boundary-integral equation analysis of cracked anisotropic plates, *International Journal of Fracture*, Vol. 11, pp. 315–328.

42. Ingraffea, A.R. and Manu, C. (1980) Stress intensity factor computation in three dimensions with quarter-point elements, *International Journal of Numerical Methods in Engineering*, Vol. 15, pp. 1427–1445.

43. Martinez, J. and Dominguez, J. (1984) On the use of quarter-point boundary elements for stress intensity factor computations, *International Journal of Numerical Methods in Engineering*, Vol. 20, pp. 1941–1950.

44. Gangming, L. and Yongyuan, Z. (1988) Improvement of stress singular element for crack problems in three dimensional boundary element method, *Engineering Fracture Mechanics*, Vol. 31, pp. 993–999.

45. Xanthis, L.S., Bernal, M.J.M. and Atkinson, C. (1981) The treatment of singularities in the calculation of stress intensity factors using the integral equation method, *Computational Methods in Applied Mechanics Engineering*, Vol. 26, pp. 285–304.

46. Aliabadi, M.H., Rooke, D.P. and Cartwright, D.J. (1987) An improved boundary element formulation for calculating stress intensity factors: application to aerospace structures, *Journal of Strain Analysis*, Vol. 22, pp. 203–207.

47. Cruse, T.A. (1978) Two dimensional BIE fracture mechanics analysis, *Applied Mathematical Modelling*, Vol. 2, pp. 287–293.

48. Stern, M. and Becker, E.B. (1978) A conforming crack tip element with quadratic variation in the singular fields, *International Journal of Numerical Methods in Engineering*, Vol. 12, pp. 279–288.

49. Jia, Z.H., Shippy, D.J. and Rizzo, F.J. (1988) On the computation of two dimensional stress intensity factors using the boundary element method, *International Journal of Numerical Methods in Engineering*, Vol. 26, pp. 2739–2753.

50. Luchi, M.L. and Rizzuti, S. (1987) Boundary elements for three dimensional elastic crack analysis, *International Journal of Numerical Methods in Engineering*, Vol. 24, pp. 2253–2271.

51. Young, A., Cartwright, D.J. and Rooke, D.P. (1988) The boundary element method for analyzing repair patches on cracked finite sheets, *Aero Journal*, Vol. 92, pp. 416–421.

52. Mews, H. (1987) Calculation of stress intensity factors for various crack problems with the boundary element method. *Proceedings of 9th International Conference on BEM*, Eds C.A. Brebbia, W.L. Wendland and G. Kuhn, Vol. 2, pp. 259–278, Springer-Verlag, Berlin.

53. Newman, J.C. (1971) An improved method of collocation for the stress analysis of cracked plates with various shaped boundaries, NASA TN D-6376.

54. Cruse, T.A. and Raveendra, S.T. (1986) A general solution procedure for fracture mechanics weight function evaluation based on the boundary element methods, *Computational Mechanics*, Vol. 3, pp. 157–166.

55. Tan, C.L. and Fenner, R.T. (1979) Elastic fracture mechanics analysis by the boundary integral equation method. *Proceedings of Royal Society, London*, Vol. 369(A), pp. 243–260.

56. Mukherjee, S. (1982) *Boundary Element Method in Creep and Fracture*, Applied Science Publishers, London.

TWELVE

PLATE-BENDING PROBLEMS

12.1 INTRODUCTION

The Kirchoff thin-plate-bending theory, in the absence of membrane forces, is a natural, two-dimensional extension of the simple Bernoulli beam-bending theory. Both are based on the assumption that 'plane sections remain plane' during bending and that the displacements are small enough for changes of geometry to be negligible and, thereby, small strain theory to be applicable.

Some of the earlier efforts in using the BEM for plate-bending problems include the work of Tottenham[1] who mentions using the reciprocal theorem and homogeneous solutions from the early sixties; Niwa, Kobayashi, and Fukui[2] as well as Altiero and Sikarski[3] published IBEM solutions. Bezine and Gamby[4,5] described a DBEM formulation while Segdin and Bricknell,[6] as well as Jaswon and Maiti,[7] discussed a semi-direct formulation for plates with and without a re-entrant corner. Recent work in this topic includes those of Stern, Tanaka, Hartman, Mukherjee and others.[8-13] The BEM formulations presented below follow the spirit of the above referenced work and includes simple extensions to plates supported on spring-mounted and elastic supports. In the later part of this chapter we consider Reissner's plate theory utilizing the recent work of Van der Weeën.[14,15]

12.2 STATEMENT OF THE PROBLEM AND THE GOVERNING DIFFERENTIAL EQUATIONS

An element of our elastic plate, of thickness h and modulus $D = Eh^3/12(1 - \nu^2)$, is shown in Fig. 12.1a,b, together with the sign convention for the edge moments

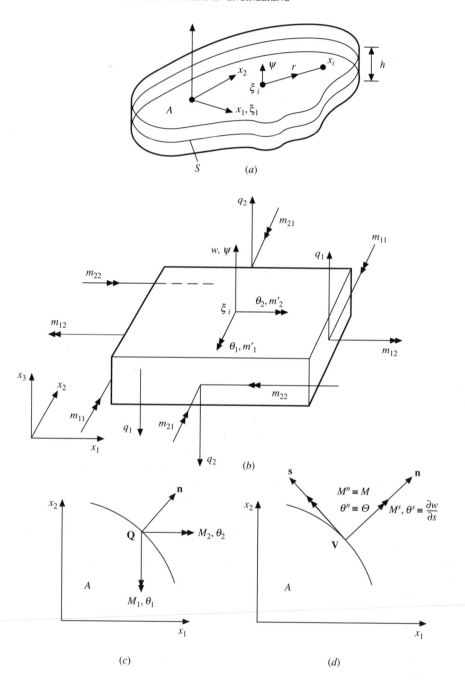

Figure 12.1 Definitions of moments, shears and loading for the plate-bending problem.

(m_{ij}) and shears (Q_i), all per unit edge length; the plate displacement (w), slopes (θ_i), all referred to the middle surface of the plate; and the applied surface loads (ψ) and external moments (m_i'). The double-arrow sign convention on the moment and slope components indicates their sense as that of a right-handed screw progressing along the arrow.

The thin-plate theory relationships between these quantities are well known to be[16]

$$\theta_i = \frac{\partial w}{\partial x_i} = w_{,i} \tag{12.1a}$$

with the comma denoting partial differentiation with respect to x_i and i ranging over $(1, 2)$ only

$$m_{ij} = m_{ji} = D[(1 - \nu)w_{,ij} + \nu\delta_{ij}w_{,kk}] \tag{12.1b}$$

(which is somewhat simplified in cases where $\nu = 0$ can be assumed) and

$$q_i = -m_{ij,j} - m_j' = -Dw_{,kki} - m_i' \tag{12.1c}$$

Finally, by noting that for vertical equilibrium $q_{i,i} + \psi = 0$ and differentiating Eq. (12.1c) again, we arrive at the governing biharmonic equation

$$\psi - m_{i,i}' = Dw_{,iijj} \tag{12.1d}$$

As before we shall be very much concerned with the components of (m, θ, q) 'resolved' along an outward boundary normal n_i at some point on the surface S bounding an arbitrary plan-form plate of area A (Fig. 12.1c), namely

$$Q = q_i n_i \quad \text{and} \quad M_i = m_{ij} n_j \tag{12.2}$$

where the sense of the M_i and corresponding θ_i components is shown in the figure. An immediate, but erroneous, conclusion is that at any point the independent boundary variables will number six $(Q, w, M_1, M_2, \theta_1, \theta_2)$. Since Eq. (12.1d) is fourth order we must be able to reduce these four. This is accomplished by noting that:

1. M_i can be considered more usefully in terms of its boundary normal and tangential components (M^n, M^s) (Fig. 12.1d) with

$$m_{ij} n_j = M_i = -M^n n_i + M^s s_i \tag{12.3}$$

   where s_i is a tangential unit boundary vector.

2. The edge twisting moment M^s can be combined with Q (Fig. 12.2a) to produce a resultant boundary shear force V as

$$V = Q - \frac{\partial M^s}{\partial s} \tag{12.4}$$

3. The slopes corresponding to M_i will be related to the normal and tangential slope components, θ^n and θ^s say (Fig. 12.1d), by an equation equivalent to

(12.3), namely

$$\theta_i = -\theta^n n_i + \theta^s s_i \tag{12.5a}$$

whence

$$\theta_i n_i = -\theta^n \quad \text{and} \quad \theta_i s_i = \theta^s = \frac{\partial w}{\partial s} \tag{12.5b}$$

We have now reduced our boundary variables to four (V, w, M, Θ) if we write $M \equiv M^n$ and $\Theta \equiv \theta^n$.

However, by incorporating M^s into V in Eq. (12.4) (Fig. 12.2a), we are then obliged to consider the so-called 'corner forces' Q_c (see Fig. 12.2b) which arise from M^s at any discontinuous change of boundary direction. From the sign convention of Fig. 12.1d we see that the M^s effects at each side of any corner (C) will be opposed and therefore if $(M^s)_{c^-}$ and $(M^s)_{c^+}$ relate to the boundary on each side of any corner, we have

$$Q_c = (M^s)_{c^-} - (M^s)_{c^+} \tag{12.6}$$

These corner forces are analysed more thoroughly in Sec. 12.5 where the BEM equations are developed.

12.3 SINGULAR SOLUTIONS

Our fundamental solution is that for the displacement $w^0(x)$ at any point (x) in a plate of infinite extent produced by a unit load acting at the point (ξ) (Fig. 12.1a), which is

$$w^0(x) = \frac{1}{8\pi D} r^2 \left(\ln \frac{r}{r_0} \right) = G^0(x, \xi) \tag{12.7}$$

where $r = y_i y_i$, $y_i = (x - \xi)_i$ and r^0 locates an arbitrary circle on which the displacement is zero (i.e. once more the displacements are relative to a datum at $r = r_0$ and therefore we shall have to introduce auxiliary terms in any IBEM development to eliminate 'reactions' from the infinite boundary as in Chapter 3). For a

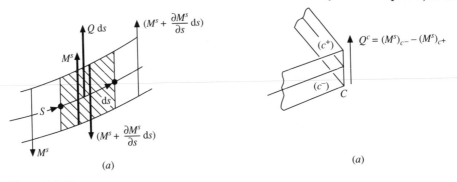

(a)   (a)

Figure 12.2 The corner force.

DBEM formulation r_0 could be once again chosen as unity. By defining $\rho = r/r_0$ and differentiating, etc., Eq. (12.7) in accordance with Eqs (12.1) we obtain the slopes, moments and shears related to Eq. (12.7) as, say,

$$\theta_i^0(x) - \frac{y_i \ln \rho}{4\pi D}$$

Therefore

$$\Theta^0(x) = -\theta_i^0(x) n_i(x) = F^0(x, \xi) \tag{12.8a}$$

$$m_{ij}^0(x) = \frac{1}{4\pi} \left\{ \frac{(1-\nu)y_i y_j}{r^2} + \delta_{ij}[(1+\nu)\ln \rho + \nu] \right\}$$

whence

$$M_i^0(x) = m_{ij}^0(x) n_j(x)$$

and, from Eq. (12.3),

$$M^{n0} = -M_i^0 n_i = -m_{ij}^0 n_i n_j = E^0(x, \xi) \tag{12.8b}$$

and

$$M^{s0} = M_i^0 s_i = C^0(x, \xi)$$

Finally,

$$q_i^0(x) = -\frac{1}{2\pi} \frac{y_i}{r^2}$$

leading to

$$V^0(x) = q_i^0(x) n_i(x) - \frac{\partial M^{so}}{\partial s}(x) = D^0(x, \xi) \tag{12.8c}$$

where, in general, s defines any curve of interest through (x).

We shall also require the corresponding results due to a unit moment acting at (ξ), which is most conveniently introduced by considering two 'equal and opposite' forces $(\psi, -\psi)$ acting at (ξ_i) and $(\xi + d\xi)_i$ (Fig. 12.3), with μ_i a unit vector colinear with $d\xi_i$. The applied forces define a couple $(\mathcal{M})$ of magnitude

$$|\mathcal{M}| = \psi|d\xi| \qquad \mathcal{M} = \psi \, d\xi \tag{12.9}$$

and if we use a superfix 1, $w^1(x)$, etc., to denote displacements, slopes, etc., due to the 'first-order' moment $(\mathcal{M})$ we have

$$w^1(x) = \psi G^0(x, \xi) - \psi G^0(x, \xi + d\xi)$$

$$= -\psi \frac{\partial G^0}{\partial \xi_i} \, d\xi_i = -(\psi|d\xi|) \frac{\partial G^0}{\partial \xi_i} \mu_i$$

i.e.

$$w^1(x) = \mathcal{M}(\xi)\mu_i(\xi) \frac{\partial G^0}{\partial x_i}(x, \xi) = \mathcal{M}\mu_i G^0_{,i} \tag{12.10a}$$

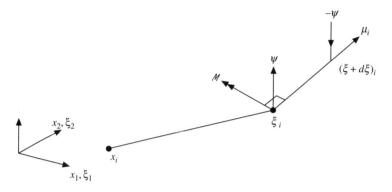

Figure 12.3 The higher-order fundamental solution.

If, in particular, $\mathcal{M}$ is a unit moment and μ_i is directed along the normal to S then the solution for a unit 'normal' boundary moment is seen to be

$$w^1(x) = \mu_i G^0_{,i} = G^1(x, \xi) \tag{12.10b}$$

and, for a direction specified by n_i at x,

$$\Theta^1(x) = -\theta_j\, n_j = -w^1_{,j}\, n_j = -\mathcal{M}_i\, n_j G^0_{,ij} = F^1(x, \xi) \qquad \text{say}$$

By carrying out the appropriate differentiations, etc., a complete set of functions C^1, D^1 and E^1 [for all functions of (x, ξ)] corresponding to C^0 to E^0 can be obtained from Eqs (12.10b) and (12.1). Tottenham[1] points out that a hierarchy of 'higher' moment unit solutions can be developed using the steps (12.9) to (12.10b). For example, by considering two 'equal and opposite' $(\mathcal{M}, -\mathcal{M})$ couples, in lieu of the $(\psi, -\psi)$ forces in Fig. 12.3, and defining a 'second-order' moment $(\mathcal{N})$ such that

$$\mathcal{N} = \mathcal{M}\, d\xi \tag{12.11}$$

we obtain, say,

$$w^2(x) = \mathcal{M}(\xi)\mu_j(\xi)\frac{\partial G^1}{\partial x_j}(x, \xi) = \mathcal{M}\mu_j G^1_{,j}$$

i.e.

$$w^2(x) = \mathcal{N}\mu_j\mu_i G^0_{,ij} \tag{12.12a}$$

and if, again, μ_i is normal to S then, due to $|\mathcal{N}| = 1$,

$$w^2(x) = \mu_i\mu_j G^0_{,ij} = G^2(x, \xi)$$

and

$$\Theta^2(x) = -\mu_i\,\mu_j\, n_k G^0_{,ijk} = F^2(x, \xi) \tag{12.12b}$$

etc., and similarly for higher-order moments generating $w^3(x) = G^3(x, \xi)$, and so on. Tottenham[1] has solved plate problems using different combinations of these (as also have Niwa, Kobayashi and Fukui[2]) depending upon the form of the boundary condition specified. One obvious difficulty of using these higher-order functions is that one needs to deal with higher-order singularities.

12.4 THE DIRECT BOUNDARY ELEMENT EQUATIONS

These can be written down immediately from the Maxwell–Betti reciprocal relationships between two distinct equilibrium states in an elastic body. It is quite clear that, considering only normal surface loading ψ initially, one such state can be defined by $(w, \theta_i, M_i, Q, \psi)$ and the second by a corresponding starred (*) set of quantities, whence

$$\int_S (Q^* w + M_i^* \theta_i)\, dS + \int_A (\psi^* w)\, dA = \int_S (Qw^* + M_i \theta_i^*)\, dS + \int_A (\psi w^*)\, dA \quad (12.13)$$

However, we really want to work in terms of the resultant boundary shear V [Eq. (12.4)] and the boundary normal moment components (M^n) in order to have numbers of unknowns consistent with our fourth-order equation. We can achieve this by noting that, from Eq. (12.3),

$$\int_S M_i \theta_i^*\, dS = \int_S (M^n n_i - M^s s_i)(\theta^{n*} n_i - \theta^{s*} s_i)\, dS$$

$$= \int_S (M^n \theta^{n*} + M^s \theta^{s*})\, dS = \int_S \left(M^n \Theta^* + M^s \frac{\partial w^*}{\partial s} \right) dS \quad (12.14)$$

If the partial derivative $\partial F / \partial s$ of any continuous function F is integrated along a smooth curve S between two points (a, b) then

$$I = \int_S \frac{\partial F}{\partial s}\, dS = \int_a^b dF = F(b) - F(a)$$

and if S is continuous and closed $a \equiv b$ and $I = 0$. However, if a corner is introduced at C (Fig. 12.4), we then have

$$I = [F]_a^c + [F]_c^b = F(c^-) - F(c^+) \quad (12.15a)$$

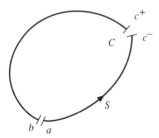

Figure 12.4 The boundary discontinuity.

Thus, along such a boundary ΔS,

$$\int_{\Delta S}\frac{\partial}{\partial s}(M^s w^*)\,\mathrm{d}S = \int_{\Delta S}\left(M^s\frac{\partial w^*}{\partial s}+\frac{\partial M^s}{\partial s}w^*\right)\mathrm{d}S = (M^s_{(c^-)} - M^s_{(c^+)})w^* \quad (12.15b)$$

If the boundary has N corners a corner force can occur at each of them and the right-hand side of Eq. (12.15b) then becomes $Q_c w_c^*$, from Eq. (12.6), with $c = 1, 2, \ldots, N$.

Therefore we can rewrite Eq. (12.14), with summation over c implied, as

$$\int_S M_i \theta_i^* \,\mathrm{d}S = \int_S \left(M^n \Theta^* - \frac{\partial M^s}{\partial s}w^*\right)\mathrm{d}S + Q_c w_c^* \quad (12.16a)$$

and, similarly,

$$\int_S M_i^* \theta_i \,\mathrm{d}S = \int_S \left(M^{n^*}\Theta - \frac{\partial M^{s^*}}{\partial s}w\right)\mathrm{d}S + Q_c^* w_c \quad (12.16b)$$

Finally, noting that $V = Q - \partial M^s/\partial s$ and $V^* = Q^* - \partial M^{s^*}/\partial s$, substitution of Eqs (12.16a,b) into Eq. (12.13) produces (with $M \equiv M^n$)[17]

$$\int_S (V^* w + M^* \Theta)\,\mathrm{d}S + \int_A (\psi^* w)\,\mathrm{d}A + Q_c^* w_c = \int_S (Vw^* + M\Theta^*)\,\mathrm{d}S$$

$$+ \int_A (\psi w^*)\,\mathrm{d}A + Q_c w_c^* \quad (12.17)$$

If we take the starred quantities to be the fundamental solution [Eqs (12.7) and (12.8)], Eq. (12.17) becomes

$$\alpha w(\xi) = \int_S (VG^0 + MF^0 - \Theta E^0 - wD^0)\,\mathrm{d}S + \int_A (\psi G^0)\,\mathrm{d}A + G_c^0 Q_c \quad (12.18)$$

where $\alpha = \frac{1}{2}$ for ξ on S and 0(1) for ξ outside (inside) S.

Equation (12.18) has to be complemented by a further one, for $\Theta(\xi) = -[\partial w(\xi)/\partial \xi_i]n_i(\xi)$, in order to build up sufficient equations so that, when (ξ) is taken to (ξ_0) on the boundary, the unspecified components of (V, M, Θ, w) can be determined.

By writing $G' = [\partial G^0(x,\xi)/\partial \xi_i]n_i$, $F' = (\partial F^0/\partial \xi_i)\,n_i$ and similarly E' and D', etc., all of which are easily calculated from Eqs (12.7) and (12.8), the additional equation which we need becomes

$$\alpha\Theta(\xi) = \int_S (VG' + MF' - \Theta E' - wD')\,\mathrm{d}S + \int_A (\psi G')\,\mathrm{d}A + G_c^1 Q_c \quad (12.19)$$

In order to clarify the role of the corner forces and matching displacements in Eqs (12.18) and (12.19) it is worth while listing the most common boundary conditions that occur in plate problems:

1. Clamped edges:

$$w = \Theta = 0 \qquad Q_c = w_c = 0$$

2. Simply supported edges:
$$w = M = 0 \qquad w_c = 0 \qquad Q_c \neq 0$$

3. Free edges:
$$V = M - 0 \qquad Q_c = 0 \qquad w_r \neq 0$$

Had we included external moments m_i' in the loading vector then Eq. (12.18) would have been augmented by a further term:

$$\int_A -m_i' \frac{\partial G^0(x, \xi)}{\partial x_i} \, \mathrm{d}A$$

and Eq. (12.19) by the term:

$$\int_A -m_i' \frac{\partial G^0(x, \xi)}{\partial x_i \partial \xi_j} n_j(\xi) \, \mathrm{d}A$$

Equations (12.18) and (12.19) can now be used to form the boundary integral representation in our familiar form:

$$\begin{Bmatrix} \alpha w \\ \alpha \Theta \end{Bmatrix} = \int_S \left(\begin{bmatrix} G^0 & F^0 \\ G' & F' \end{bmatrix} \begin{Bmatrix} V \\ M \end{Bmatrix} - \begin{bmatrix} D^0 & E^0 \\ D' & E' \end{bmatrix} \begin{Bmatrix} \omega \\ \Theta \end{Bmatrix} \right) \mathrm{d}S$$

$$+ \int_A \begin{bmatrix} G^0 \\ G' \end{bmatrix} \psi \, \mathrm{d}V + \begin{bmatrix} G_c^o \\ G_c' \end{bmatrix} Q_c \qquad (12.20)$$

In the numerical implementation of the above equation using simple constant elements we can simply follow the previously established procedures. In fact all of the integrals over boundary elements can be calculated analytically using a polar coordinate system through the field point ξ.[1] For a higher-order implementation[8,13,17,18] where the field point ξ can be at a non-smooth point of the surface, Eq. (12.19) becomes ambiguous. Furthermore, housekeeping efforts involving the treatment of the corner force become cumbersome. Hartman[8] has given a very good detailed account of the numerical process. Note, however, that Kirchoff's theory of plate does not give the correct result at a corner; it is therefore perfectly valid to round off the corners for the convenience of numerical implementation.

In the numerical implementation of Eq. (12.20) there are several singular integrals involved. Any numerical treatment of these integrals could be easily avoided by utilizing a set of admissible rigid body translation and rigid body rotations of the whole plate, which obviously will not generate any bending moment or shear. It is also possible to use other admissible solutions of displacements, rotations, moments and shears that satisfy the governing equations to obtain the values of these strongly singular integrals.

12.5 INTERNAL LOADING

For a completely general normal loading ψ or moment m_i' it is obviously necessary to evaluate the above area integrals as a series of summations over internal cells. For simple distributions of ψ and m_i' it is possible to reduce the area integrals involved into a boundary integral by using the divergence theorem. For example, if a uniformly distributed load ψ over an area B enclosed by a boundary C exists within the plate then

$$\int_B G^0(x, \xi)\psi \, dB = \frac{\psi}{8\pi D} \int_B r^2 \log r \, dB$$

$$= \frac{\psi}{64\pi D} \int_C r^3(2\log r - 0.5) n_r \, dC \qquad (12.21)$$

Kamiya and Sawaki[19] have considered several polynomial distributions of ψ and have given explicit expressions that could be utilized for this purpose. One can then treat these additional boundary integrals involving known quantities using the standard boundary element discretization.

It is, of course, possible to use the theory of particular integrals by developing solutions in terms of a complimentary solution ω^c satisfying

$$D\omega^c_{,iijj} = 0 \qquad (12.22a)$$

and particular integrals involving polynomial distributions of ψ and m_i such as those considered in Chapters 3 and 4 by satisfying

$$Dw^p_{,iijj} = \psi - m'_{i,i} \qquad (12.22b)$$

Jaswon and Maiti[7] were the first to consider such a method in plate-bending problems by BEM.

12.6 PLATES AND BEAMS ON ELASTIC SUPPORT

The simplest practicable form of elastic support to beam and plate structures is that of a series of identical, closely spaced, linear springs without any shear coupling between them—a Winkler foundation. This simplification of truly continuous elastic support is accurately realized in both floating-plate structures and plates supported on quite a large class of orthotropic elastic half-spaces in which moduli increase linearly with depth, from zero at the 'ground surface'. It is also used to approximate many other types of structure, particularly in civil engineering.

If the spring support intensity is k per unit of supported area, then the governing equation is a minor extension of Eq. (12.1d) to

$$(\psi + m'_{i,i}) = Dw_{,iijj} + kw \qquad (12.23)$$

The only modifications required to the BEM formulation are the use of a different and rather more complicated fundamental solution. An explicit treatment of the BEM to the use of such fundamental solutions in a one-dimensional beam on a Winkler foundation can be seen in Butterfield.[20]

Timoshenko and Woinowski-Krieger[16] discuss a solution, in the form of an infinite integral, for a loaded plate on quite general elastic support which reduces, for a simple elastic half-space, to

$$w^0(x) = \frac{l^3}{2\pi D} \int_0^\alpha \frac{J_0(\alpha r)\, d\alpha}{1 + (\alpha l)^3} = G^0(x, \xi) \qquad (12.24)$$

with $J_0(\alpha r)$ a zero-order Bessel function and $l^3 = 2D(1 - \nu_0^2)/E_0^2$, (E_0, ν_0) being Young's modulus and Poisson's ratio for the supporting half-space. Whereas one can, in principle, derive all of G^0 to D^0 and G^1 to D^1 from Eq. (12.24) and use these directly in Eq. (12.20), the BEM formulation, considerable algebraic complexity arises from having to deal with integrals of Bessel functions.

Alternatively, we can treat the plate and the half-space separately as components of a 'two-zone' problem with compatibility enforced at the plate-foundation interface.[21] As an example, consider a BEM formulation for a plate under vertical loading with the simplest continuity requirement across a smooth interface—that of matching vertical displacements only. (Note that interface couples would be required in addition if local rotations were to be matched and for a perfectly bonded plate, surface tractions, and therefore membrane forces, would arise.) The final discretized matrix form of the BEM equations for the plate alone will be, say,[21]

$$[W]\{w\} + [X]\{\Theta\} + [Y]\{M\} + [Z]\{V\} = [A]\{\psi\} - [B]\{p\} \qquad (12.25)$$

in which all of the coefficient matrices (W to Z, A, B), the loading vector ψ and half of the total number of components in the (w, Θ, M, V) vectors will be known. For, say, n distinct values of each boundary variable the left-hand side of Eq. (12.25) will comprise $(2n \times 2n)(2n \times 1)$ matrix and vector pairs, the $[A]\{\psi\}$ terms will be $(2n \times l)(l \times 1)$, for l loaded cells, and the final $[B]\{p\}$ terms will be $(2n \times m)(m \times 1)$, which represent the foundation support reaction vector $\{p\}$ acting over the whole plate surface divided into cells producing m distinct terms.

The half-space loading is now also $\{p\}$ over an identical pattern of surface elements and this will generate an $m \times 1$ displacement vector $\{\bar{w}\}$, say, where

$$\{p\} = [C]\{\bar{w}\} \qquad (12.26)$$

If only a simple spring support foundation is being modelled then C will be a diagonal matrix; otherwise it will be fully populated, with terms calculated using either Boussinesq's solution integrated over the surface elements for a uniform half-space or any other surface load solutions available for anisotropic or inhomogeneous half-spaces.

12.7 BEM FOR REISSNER'S PLATE THEORY

The deficiencies of the Kirchoff's plate theory are well known; e.g. the transverse shear deformability of the plate is not taken into account, only two of three physical boundary conditions are satisfied and results are inaccurate in edge zones or around holes having a small diameter in comparison to plate thickness. A direct BEM is considered here for Reissner's isotropic plate theory. This refined bending theory leads to a sixth-order boundary value problem in contrast to the biharmonic problem of classical theory. Consequently one needs to consider a system of three integral representations involving generalized displacements (two rotation components plus deflection) and associated tractions (normal flexural moment, torsional moment and transverse shear force). Because Reissner's plate theory contains some elements of elastostatics we shall revert to nomenclature similar to those of elasticity in our discussion.

Consider a homogeneous isotropic plate of uniform thickness h subjected to a transverse load ψ per unit area. Reissner's constitutive equations relating rotation components u_α and deflection u_3 to bending stress couples $\sigma_{\alpha\beta}$ and shear stress resultants $\sigma_{3\alpha}$ are given by[14,15]

$$\sigma_{\alpha\beta} = D\frac{1-\nu}{2}\left(u_{\alpha,\beta} + u_{\beta,\alpha} + \frac{2\nu}{1-\nu}u_{\gamma,\gamma}\delta_{\alpha\beta}\right) - \frac{\nu}{1-\nu}\lambda^{-2}\psi\delta_{\alpha\beta} \qquad (12.27)$$

$$\sigma_{3\alpha} = D\frac{1-\nu}{2}\lambda^2(u_\alpha + u_{3,\alpha})$$

where $\lambda = \sqrt{10}/h$, ν is Poisson's ratio, D denotes the flexural plate rigidity and α, β and γ range from 1 to 2.

Substitution of Eqs (12.27) into the equilibrium equations

$$\sigma_{\alpha\beta,\beta} - \sigma_{3\alpha} = 0 \qquad (12.28)$$

$$\sigma_{3\alpha,\alpha} + \psi = 0 \qquad (12.29)$$

yields a sixth-order system of equations requiring satisfaction of three boundary conditions. On the boundary S of the mid-plane A either the displacement component u_i ($i = 1, 2$ and 3) or the corresponding traction component $t_i = \sigma_{i\beta}n_\beta$ must be prescribed; here, n_β are the direction cosines of the outward normal on S.

12.7.1 Integral Representations

It is now possible to combine Eqs (12.27), (12.28) and (12.29) to form the governing equations of the problem which could be multiplied with the fundamental solution and integrated by parts to yield the necessary integral representation:[14,15]

$$c_{ij}(\xi)u_i(\xi) = \int_S [G_{ij}(x,\xi)t_i(x) - F_{ij}(x,\xi)u_i(x)]\,\mathrm{d}S$$

$$+ \int_A \left[G_{3j}(x,\xi) - \frac{\nu}{1-\nu}\lambda^{-2}G_{\alpha j,\alpha}(x,\xi)\right]\psi(x)\,\mathrm{d}A \qquad (12.30)$$

where $G_{ij}(x,\xi)$ represent the rotations (with $i = 1, 2$) and the transverse deflection (with $i = 3$) at point x due to a unit point couple ($j = 1, 2$) and a unit point force

($j = 3$) at the load point ξ in an infinite plate and are given by[14,15]

$$G_{\alpha\beta} = [8\pi D(1-\nu)]^{-1}\{[8B(z) - (1-\nu)(2\ln z - 1)]\delta_{\alpha\beta} - [8A(z) + 1 - \nu]r_{,\alpha}r_{,\beta}\}$$

$$(12.31a)$$

$$G_{3\alpha} = -G_{\alpha 3} = (8\pi D)^{-1}(2\ln z - 1)rr_{,\alpha} \tag{12.31b}$$

$$G_{33} = [8\pi D(1-\nu)\lambda^2]^{-1}[(1-\nu)z^2(\ln z - 1) - 8\ln z] \tag{12.31c}$$

where

$$A(z) = K_0(z) + 2z^{-1}[K_1(z) - z^{-1}]$$

$$B(z) = K_0(z) + z^{-1}[K_1(z) - z^{-1}]$$

$$z = \lambda r$$

$$r^2 = y_i y_i$$

$$y_i = (x_i - \xi_i)$$

$$r_{,\alpha} = r^{-1}y_\alpha$$

Expanding the modified Bessel functions $K_0(z)$ and $K_1(z)$ for small arguments it can be found that $A(z)$ is continuous whereas $B(z)$ has a singularity ($-\frac{1}{2}\ln z$). The corresponding components of the traction kernel $F_{ij}(x, \xi)$ are given by

$$\left\{ \begin{array}{l} F_{\alpha\gamma} = -T_{\alpha\beta\gamma}n_\beta \\ F_{3\gamma} = T_{3\beta\gamma}n_\beta \\ F_{\alpha 3} = T_{\alpha\beta 3}n_\beta \\ F_{33} = -T_{3\beta 3}n_\beta \end{array} \right\} \tag{12.32}$$

where n_β are the direction cosines of the outward normal at x over S and

$$T_{\alpha\beta\gamma} = (4\pi r)^{-1}[(4A + 2zK_1 + 1 - \nu)(\delta_{\alpha\gamma}r_{,\beta} + \delta_{\beta\gamma}r_{,\alpha})$$

$$+ (4A + 1 + \nu)\delta_{\alpha\beta}r_{,\gamma} - (16A + 4zK_1 + 2 - 2\nu)r_{,\alpha}r_{,\beta}r_{,\gamma}]$$

$$T_{\alpha\beta 3} = -(8\pi)^{-1}(1 - \nu)\left[\left(2\frac{1+\nu}{1-\nu}\ln z - 1\right)\delta_{\alpha\beta} + 2r_{,\alpha}r_{,\beta}\right]$$

$$T_{3\beta\gamma} = (2\pi)^{-1}\lambda^2(B\delta_{\beta\gamma} - Ar_{,\beta}r_{,\gamma})$$

$$T_{\beta 33} = (2\pi r)^{-1}r_{,\beta}$$

Equation (12.30) can be used to develop the numerical solution for Reissner's plate-bending problem. In the numerical implementation we can once again use, for example, the admissible rigid body motion and rigid body rotations to evaluate the strongly singular integrals. Thus with $t_i = 0$ and $\psi = 0$, u_1, u_2 and u_3 are three rigid body components, $(1, 0, \xi_1 - x_1)$, $(0, 1, \xi_2 - x_2)$ and $(0, 0, 1)$, and therefore the jump terms c_{ij} in Eq. (12.30) are given by

$$c_{\beta i}(\xi) = -\int_S [F_{\beta i}(x, \xi) + (\xi_\beta - x_\beta)F_{3i}(x, \xi)] \, dS(x) \tag{12.33a}$$

$$c_{3i}(\xi) = -\int_S F_{3i}(x, \xi) \, dS(x) \tag{12.33b}$$

Finally, if the applied loading is uniform over an area B bounded by a contour C it is possible to convert the area integral in Eq. (12.30) into one around C. Moreover, it is also possible to calculate particular solutions of the governing equations for simpler forms of applied loading, as was shown in Chapters 3 and 4.

12.8 EXAMPLES

12.8.1 A Uniformly Loaded Square Plate with a Central Square Opening

Tottenham[1] analysed this problem by both direct and indirect BEM. The outer edges of the plate were simply supported and the inner edges free. In the IBEM solution, corner forces were ignored and the fictitious moments at the corners applied to one-quarter of the symmetrical system (Fig. 12.5) with nodes at the quarter-points along each side. For the direct method solution, using a 'boundary' exterior to S, a linear variation of parameters was used. Table 12.1 shows a comparison between the displacements calculated at typical points by BEM, finite element and finite difference analyses, the latter two using mesh sizes of $a/16(\nu = 0.25)$. Tottenham also provides other examples, in one of which the

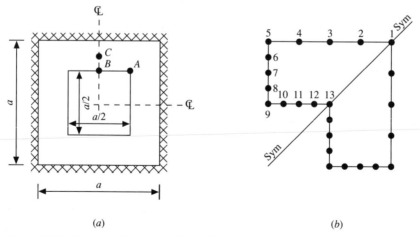

(a)   (b)

Figure 12.5 Uniformly loaded square plate with central square opening.

Table 12.1

| Point | Displacement $\times qa^4/100D$ | | | |
| | Boundary element | | Finite element | Finite difference |
	Indirect	Direct		
A	0.2188	0.2188	0.2185	0.2174
B	0.3107	0.3141	0.3156	0.3006
C	0.1558	0.1565	—	0.1541

BEM equations had been made symmetrical by using a least square error minimization technique.

12.8.2 Bending of a Circular Plate

A clamped, circular plate subjected to a uniformly distributed, circular, patch load is shown in Fig. 12.6a. The plate is modelled using the theory of elasticity with a five-region, axisymmetric mesh shown in Fig. 12.6b. The modulus of elasticity is $E = 210\,000$, Poisson's ratio is $\nu = 0.3$ and $a = 1000$, $b = h = 100$.

In Figs 12.7 and 12.8, the radial, tangential and shear stresses obtained from an axisymmetric solid BEM analysis are compared with a boundary element solution based on Reissner's plate theory.[14,15] General agreement is exhibited by the various analyses. The BEM solution of the axisymmetric analysis, as expected,

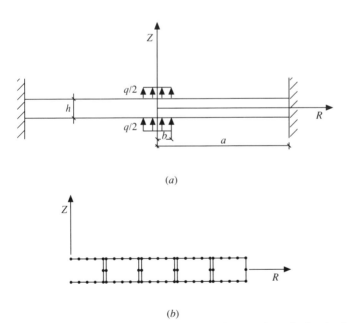

(a)

(b)

Figure 12.6 (a) Clamped circular plate. (b) Axisymmetric mesh of a circular plate.

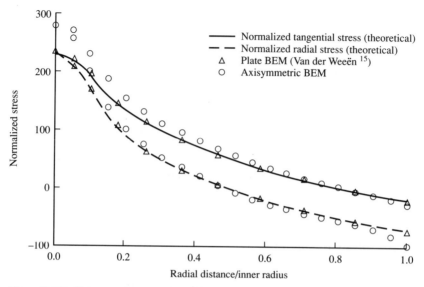

Figure 12.7 Radial and tangential stress through a circular plate.

deviates slightly from the plate theory results under the load and near the support. The same problem was also analysed with a single-region axisymmetric solid mesh. Results were almost identical to those of the multi-region model shown here.

In recent years, Wang et al.[22] and Lei et al.[23] have published extensions of the BEM formulations based on Reissner's plate theory, to include foundation support and finite elastic deformation.

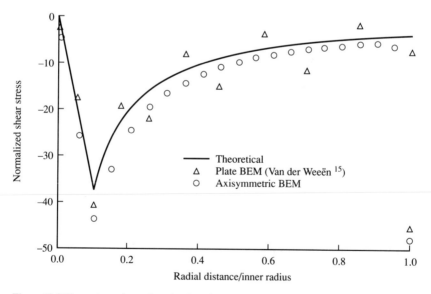

Figure 12.8 Shear stress through a circular plate.

12.9 CONCLUDING REMARKS

The plate analyses, based on Kirchoff and Reissner's plate theories, presented in this chapter follow the established pattern of BEM formulations. In addition, the extension of the analysis to include elastically supported plates provides examples of how new unit solutions of ever-increasing complexity can become useful. The plate and its elastic support can also be treated as a special form of 'two-zone' problem, where the interface links the boundary of one zone with the other.[20,21,24]

The flat, thin, spring-supported plate theory is closely related to the thin shallow shell theory. Tottenham,[1] Newton and Tottenham[25] and Lu and Huang[26] have published analyses of such problems.

REFERENCES

1. Tottenham, H. (1979) The boundary element method for plates and shells, in *Developments in Boundary Element Methods*, Eds P.K. Banerjee and R. Butterfield, Elsevier Applied Science Publishers, London.
2. Niwa, Y., Kobayashi, S. and Fukui, T. (1974) An application of the integral equation method to plate bending, Memo, Vol. 36, Part 2, pp. 140–158, Faculty of Engineering, Kyoto University, Japan.
3. Altiero, N.J. and Sikarski, D.L. (1978), A boundary integral method applied to plates of arbitrary plan form, *Computers and Structures*, Vol. 9, pp. 163–168.
4. Bezine, G.P. and Gamby, D.A. (1978) A new integral equation formulation for plate bending problems, in *Recent Advances in Boundary Element Methods*, Ed. C.A. Brebbia, Pentech Press, London.
5. Bezine, G.P. (1979) Boundary integral formulation for plate flexure with arbitrary boundary conditions, *Mechanics Research Communications*, Vol. 5, No. 4.
6. Segdin, C.M. and Bricknell, G.A. (1968) Integral equation method for a corner plate, *Journal of Structures Division, ASCE*, Vol. ST-1, pp. 43–51.
7. Jaswon, M.A. and Maiti, M. (1968) An integral formulation of plate bending problems, *Journal of Engineering Mathematics*, Vol. 2, No. 1, pp. 83–93.
8. Hartman, F. (1989) *Introduction to Boundary Elements*, Springer-Verlag, Berlin.
9. Mukherjee, S. (1982) *Boundary Element Methods in Creep and Fracture*, Elsevier Applied Science, London.
10. Stern, M. (1979) A general boundary integral formulation for the numerical solution of plate bending problems, *International Journal of Solids and Structures*, Vol. 15, pp. 769–782.
11. Morjaria, M. and Mukherjee, S. (1980) Inelastic analysis of transverse deflection of plates by BEM, *Journal of Applied Mechanics*, Vol. 47, pp. 291–296.
12. Barone, M.R. and Robinson, A.R. (1972) Integral formulations for plates and shells, *International Journal of Solids and Structures*, Vol. 8, pp. 1319–1338.
13. Tanaka, M. (1984) Large deflection analysis of thin elastic plates by BEM, Chapter 5 in *Developments in Boundary Element Methods*, Vol. 3, Eds P.K. Banerjee and S. Mukherjee, Elsevier Applied Science, London.
14. Van der Weeën, F. (1981) Reissner's plate theory by BEM, PhD Dissertation, State University of Ghent, Belgium.
15. Van der Weeën, F. (1982) Application of BEM to Reissner plate model, *International Journal of Numerical Methods in Engineering*, Vol. 18, pp. 1–10.
16. Timoshenko, S. and Woinowski-Krieger, S. (1959) *Theory of Plates and Shells*, McGraw-Hill, New York.

17. Bergman, S. and Schiffer, M. (1953) *Kernel Functions and Elliptic Differential Equations in Mathematical Physics*, Academic Press, New York.
18. Stern, M. and Lin, T.L. (1986) Thin elastic plate in bending, in *Development in Boundary Element Methods*, Vol. 4, Eds P.K. Banerjee and J.O. Watson, pp. 91–119, Elsevier Applied Science, London.
19. Kamiya, N. and Sawaki, Y. (1985) An efficient BEM for some inhomogeneous and nonlinear problem, *Proceedings of 7th International Conference on BEM*, pp. 13/59–13/68, Lake Como, Italy.
20. Butterfield, R. (1979) New concepts illustrated by old problems, Chapter 1 in *Developments in Boundary Element Methods*, Vol. 1, Eds P.K. Banerjee and R. Butterfield, Elsevier Applied Science, London.
21. Banerjee, P.K. and Butterfield, R. (1981) *Boundary Element Methods in Engineering Science*, McGraw-Hill, London.
22. Wang, J., Wang, X. and Huang, M. (1991) Fundamental solutions and boundary integral equations for Reissner's plates on two-parameter foundations, *International Journal of Solids and Structures*, Vol. 29, No. 10, pp. 1233–1239.
23. Lei, X.Y., Huang, M. and Wang, X. (1990) Geometrically nonlinear analysis of a Reissner type plate by BEM, *Computers and Structures*, Vol. 37, No. 6, pp. 991–996.
24. Ng, S.F., Cheung, M.S. and Xu, T. (1990) A combined boundary element and finite element solution of slab and slab on girder bridges, *Computers and Structures*, Vol. 37, No. 6, pp. 1069–1075.
25. Newton, D.A. and Tottenham, H. (1968) Boundary value problems in thin shallow shells of arbitrary plan form, *Journal of Engineering Mathematics*, Vol. 2, No. 3, pp. 211–224.
26. Lu, P. and Huang, M. (1992) Boundary element analysis of shallow shells involving shear deformation, *International Journal of Solids and Structures*, Vol. 29, No. 10, pp. 1273–1282.

NON-LINEAR ANALYSIS OF SOLIDS

13.1 INTRODUCTION

In previous chapters we have dealt with problems governed by linear differential equations and explored the essential features of BEM algorithms for them. However, in order to qualify as a completely general problem-solving tool, it is essential that BEM must be demonstrably applicable to non-linear systems. Non-linearities do occur in almost every realistic idealization of practical problems.

Like any other general numerical method BEMs are fully capable of solving non-linear differential equations by an incremental or iterative procedure—in this case via a volume integral defined over the region within which the non-linearities occur. For the vast majority of such problems the non-linear regions are mainly confined to small sub-regions of the system and it will be found that, particularly for non-linear three-dimensional systems, the BEM provides a relatively attractive tool for dealing with non-linearities.

13.2 CONSTITUTIVE RELATIONSHIPS FOR SOLIDS

In order to appreciate the essential features of the solution process it is important to explore first the behaviour of an infinitesimal element of the region. This is described by the constitutive relations for the materials concerned. Many such models have been proposed to cover the wide range of materials that engineers have to deal with and a detailed discussion of these would be out of place in this book. We shall therefore outline below a few, well-established theories that model the major non-linear features of certain classes of solids.

335

13.2.1 Incremental Theory of Plasticity

The necessary ingredients of an incremental stress–strain relationship based on the 'path independence in the small' theory of plasticity (the incremental theory of plasticity) are[1-3]

1. A yield function defining the limits of the elastic behaviour.
2. A flow rule relating the irrecoverable plastic part of the strain rate to the stress state in the material.
3. A hardening rule defining the subsequent yield surfaces resulting from continuous plastic deformation.

The nature of the yield surface and the hardening parameters are obviously dependent on the type of material. Therefore, we need to derive stress–strain relationships for quite general cases of isotropic and kinematic hardening.

Isotropic hardening In this theory it is assumed that, during plastic flow, the yield surface expands uniformly about the origin in stress space, maintaining its shape, centre and orientation as shown in Fig. 13.1. For example, the path OA is elastic; at A the material is in a state of incipient yielding and, upon further loading, the yield surface expands to B. The path AB represents isotropic strain-hardening behaviour. If we unload along BC the behaviour will be elastic until point C is reached, when the material will start to yield again. If we assume that the continuously changing yield surfaces can be represented by a loading function of type

$$F(\sigma_{ij}, \varepsilon_{ij}^{P}, h) = 0 \tag{13.1}$$

where h is a hardening parameter, ε_{ij}^{P} the current total plastic strain in the material element and for plastic states $F = 0$, then for elastic states $F < 0$ and $F > 0$ is meaningless.

Since during plastic deformation the new stress state must lie on a newly developed yield surface characterized by a new strain-hardening parameter $(h + \dot{h})$ (where a superior dot will be used to denote an increment, $\dot{h} = dh$, etc.),

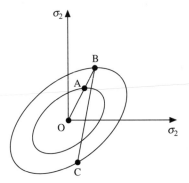

Figure 13.1 Isotropic hardening.

we have

$$dF = \frac{\partial F}{\partial \sigma_{ij}} \dot{\sigma}_{ij} + \frac{\partial F}{\partial \varepsilon_{ij}^{p}} \dot{\varepsilon}_{ij}^{p} + \frac{\partial F}{\partial h} \dot{h} = 0 \qquad (13.2)$$

Using Drucker's postulate for associated plastic flow we obtain[1]

$$\dot{\varepsilon}_{ij}^{p} = \dot{\lambda} \frac{\partial F}{\partial \sigma_{ij}} \qquad (13.3)$$

where $\dot{\lambda}$ is a non-negative scalar variable.

Since the hardening parameter h is a function of plastic strain we can write Eq. (13.2) as

$$\frac{\partial F}{\partial \sigma_{ij}} \dot{\sigma}_{ij} + \frac{\partial F}{\partial \varepsilon_{ij}^{p}} \dot{\varepsilon}_{ij}^{p} + \frac{\partial F}{\partial h} \frac{\partial h}{\partial \varepsilon_{ij}^{p}} \dot{\varepsilon}_{ij}^{p} = 0 \qquad (13.4)$$

Substituting Eq. (13.3) in Eq. (13.4) and solving for $\dot{\lambda}$ we get

$$\dot{\lambda} = L_{ij}^{\sigma} \dot{\sigma}_{ij} \qquad (13.5a)$$

where

$$L_{ij}^{\sigma} = \frac{1}{H} \frac{\partial F}{\partial \sigma_{ij}} \qquad (13.5b)$$

$$H = -\left(\frac{\partial F}{\partial \varepsilon_{mn}^{p}} + \frac{\partial F}{\partial h} \frac{\partial h}{\partial \varepsilon_{mn}^{p}} \right) \frac{\partial F}{\partial \sigma_{mn}} \qquad (13.5c)$$

It should be noted that L_{ij}^{σ} depends upon the current state variables, not on the incremental quantities, and this tensor has a very important role in the development of BEM formulation for elastoplasticity.

However, the relationship given by Eq. (13.5a) does not exist for ideal plasticity as H vanishes for zero hardening. This can be avoided by reformulating the preceding expression in terms of strain increments:

$$\dot{\lambda} = L_{ij}^{\epsilon} \dot{\epsilon}_{ij} \qquad (13.6a)$$

where

$$L_{ij}^{\epsilon} = \frac{1}{H'} \frac{\partial F}{\partial \sigma_{ij}} \qquad (13.6b)$$

$$H' = \frac{\partial F}{\partial \sigma_{kl}} D_{klmn}^{e} \frac{\partial F}{\partial \sigma_{mn}} - \left(\frac{\partial F}{\partial \varepsilon_{kl}^{p}} + \frac{\partial F}{\partial h} \frac{\partial h}{\partial \varepsilon_{kl}^{p}} \right) \frac{\partial F}{\partial \sigma_{kl}} \qquad (13.6c)$$

where D_{ijkl}^{e} = elastic constitutive tensor. It is evident that H' does not vanish for zero hardening (ideal plasticity).

Next, the plastic strain increment can be obtained from

$$\dot{\varepsilon}_{ij}^{p} = G \frac{\partial F}{\partial \sigma_{ij}} \frac{\partial F}{\partial \sigma_{kl}} \dot{\sigma}_{kl} \qquad (13.7a)$$

where

$$G = - \frac{1}{[\partial F/\partial \varepsilon_{mn}^{p} + (\partial F/\partial h)(\partial h/\partial \varepsilon_{mn}^{p})]\partial F/\partial \sigma_{mn}} \tag{13.7b}$$

Equation (13.7a) together with the elastic component of the strain increment ($\dot{\varepsilon}_{ij}^{e}$) may be written as

$$\dot{\varepsilon}_{ij} = \dot{\varepsilon}_{ij}^{e} + \dot{\varepsilon}_{ij}^{p} = C_{ijkl}^{ep} \dot{\sigma}_{kl} \tag{13.8a}$$

or

$$\dot{\sigma}_{ij} = D_{ijkl}^{ep} \dot{\varepsilon}_{kl} \tag{13.8b}$$

with D_{ijkl}^{ep} a fourth-rank tensor quantity, the incremental elastoplastic material modulus. Equations (13.8a,b) provide the necessary incremental stress–strain relations for a general case of isotropic hardening. Specifically, for an isotropic strain-hardening von Mises material we can write Eq. (13.8b) explicitly as

$$\dot{\sigma}_{ij} = 2\mu \left[\dot{\varepsilon}_{ij} + \frac{\nu}{1 - 2\nu} \delta_{ij} \dot{\varepsilon}_{kk} - \frac{3 S_{ij} S_{kl}}{2\sigma_0^2 (1 + H/3\mu)} \dot{\varepsilon}_{kl} \right] \tag{13.9a}$$

and

$$L_{ij}^{\sigma} = \frac{1}{H} \left(\frac{3}{2\sigma_0} \right)^2 S_{ij} \tag{13.9b}$$

where

μ = elastic shear modulus
σ_0 = uniaxial yield stress = $\sqrt{\frac{3}{2} S_{ij} S_{ij}}$
S_{ij} = deviatoric stress = $\sigma_{ij} - \frac{1}{3} \delta_{ij} \sigma_{kk}$

H = plastic-hardening modulus, the current slope of the uniaxial plastic stress–strain curve

With a proper definition of the yield function F and the hardening parameter h in Eq. (13.2) we can derive equations similar to Eq. (13.9) for isotropic yielding and hardening of any material.[1-4]

Kinematic hardening The kinematic hardening theory of plasticity, originally developed by Prager,[5] assumes that during plastic deformation the loading surface (or the yield surface) translates as a rigid body in stress space. The basic aim of this theory was to model the Bauschinger effect which is apparent during the cyclic loading of metals.

Figure 13.2 shows typical kinematic hardening behaviour. The loading path OA is elastic. At A yielding starts and the loading path AB initiates elastoplastic kinematic hardening. The translation of the yield surface during this loading causes its centre to move from O to O'. Any unloading from B along BC results in purely elastic behaviour until the loading path reaches C, when the material yields once again—now at a lower yield stress.

If we assume that the translation of the centre of the yield surface can be represented by a tensor $\dot{\alpha}_{ij}$, then the loading function may be written as

$$F(\sigma_{ij}, \alpha_{ij}) = 0 \tag{13.10}$$

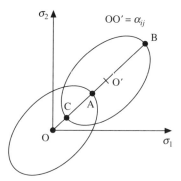

Figure 13.2 Kinematic hardening.

and the consistency relation during the loading becomes

$$dF = \frac{\partial F}{\partial \sigma_{ij}} \dot{\sigma}_{ij} + \frac{\partial F}{\partial \alpha_{mn}} \frac{\partial \alpha_{mn}}{\partial \varepsilon_{ij}^{\mathrm{p}}} \dot{\varepsilon}_{ij}^{\mathrm{p}} = 0 \qquad (13.11)$$

if the translation α_{mn} is assumed to be a function of the plastic strain.

Equation (13.11) represents the kinematic hardening theory of Prager which postulates that the translation increments $(\dot{\alpha}_{ij})$ of the loading surface, in nine-dimensional stress space, occur in the direction of the outward normal to the surface at the current stress state and therefore

$$\dot{\alpha}_{ij} = C \dot{\varepsilon}_{ij}^{\mathrm{p}} \qquad (13.12)$$

where C now is like the hardening parameter H in Eq. (13.9). By using Eqs (13.3) and (13.12) in Eq. (13.11) we can obtain the plastic flow factor λ and thus derive an incremental stress–strain relationship in the manner discussed for isotropic hardening.

This was the theory originally proposed by Prager, but when its consequences are investigated in certain sub-spaces of the stress space several inconsistencies arise, as discussed by Shields and Ziegler.[6,7] To avoid these difficulties Ziegler proposed a modification to Prager's hardening rule. Instead of Eq. (13.12) he proposed that

$$\dot{\alpha}_{ij} = \dot{\mu}_0(\sigma_{ij} - \alpha_{ij}) \qquad (13.13)$$

where $\dot{\mu}_0$ is a positive scalar parameter.

Figure 13.3 illustrates the difference between the Prager and the Ziegler postulates. The increment of translation $\dot{\alpha}_{ij}$ in Ziegler's theory is colinear with the radius vector from the centre of the yield surface O' to the stress point. The scalar function $\dot{\mu}_0$ can be determined by using the 'consistency condition' that

$$dF(\bar{\sigma}_{ij}) = 0 \qquad \text{where } \bar{\sigma}_{ij} = \sigma_{ij} - \alpha_{ij}$$

i.e.

$$\frac{\partial F}{\partial \bar{\sigma}_{ij}} (\dot{\sigma}_{ij} - \dot{\alpha}_{ij}) = 0$$

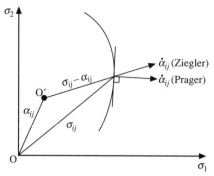

Figure 13.3 Ziegler's and Prager's hardening rules.

Substituting Eq. (13.13) in the above gives

$$\dot{\mu}_0 = \frac{(\partial F/\partial \sigma_{ij})\dot{\sigma}_{ij}}{(\sigma_{kl} - \alpha_{kl})\partial F/\partial \sigma_{kl}} \tag{13.14}$$

In order to determine the plastic strain we have to determine $\dot{\lambda}$ in the flow rate (13.3). Ziegler assumed that the vector $C\dot{\varepsilon}_{ij}^{p}$ is the projection of $\dot{\sigma}_{ij}$ on the outward normal through the instantaneous stress state, as shown in Fig. 13.4. This assumption is consistent with the 'path independent in the small' theory of plasticity and we can therefore write

$$(\dot{\sigma}_{ij} - C\dot{\varepsilon}_{ij}^{p})\frac{\partial F}{\partial \sigma_{ij}} = 0$$

Substituting Eq. (13.3) in the above we have

$$\dot{\lambda} = \frac{1}{C}\frac{(\partial F/\partial \sigma_{ij})\dot{\sigma}_{ij}}{(\partial F/\partial \sigma_{kl})(\partial F/\partial \sigma_{kl})}$$

Therefore the plastic strain increments are given by

$$\dot{\varepsilon}_{ij}^{p} = \frac{1}{C}\frac{(\partial F/\partial \sigma_{ij})(\partial F/\partial \sigma_{mn})\dot{\sigma}_{mn}}{(\partial F/\partial \sigma_{kl})(\partial F/\partial \sigma_{kl})} \tag{13.15}$$

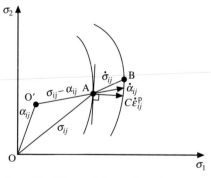

Figure 13.4 Kinematic hardening rule.

The elastic and plastic strain components may be added to provide the complete stress–strain relations [Eqs (13.8a,b)]. Specifically, if we choose a von Mises yield criterion, i.e.

$$F = \tfrac{1}{2}\overline{S}_{ij}\overline{S}_{ij} - \tfrac{1}{3}\overline{\sigma}_0^2 = 0 \tag{13.16}$$

where

$$\overline{S}_{ij} = \overline{\sigma}_{ij} - \tfrac{1}{3}\delta_{ij}\overline{\sigma}_{kk}$$

$\overline{\sigma}_{ij} = \sigma_{ij} - \alpha_{ij}$, which may be called a reduced stress state

and $\quad \overline{\sigma}_0 = \sqrt{\tfrac{3}{2}\overline{S}_{ij}\overline{S}_{ij}}$, the reduced uniaxial yield stress

we can write the explicit stress–strain relations as

$$\dot{\sigma}_{ij} = 2\mu\left[\dot{\varepsilon}_{ij} + \frac{\nu}{1-2\nu}\delta_{ij}\dot{\varepsilon}_{kk} - \frac{3\overline{S}_{ij}\overline{S}_{kl}}{2\overline{\sigma}_0^2(1+C/2\mu)}\dot{\varepsilon}_{kl}\right] \tag{13.17}$$

and $\quad \dot{\alpha}_{ij} = C\dot{\varepsilon}_{ij}^{\text{p}} = \dfrac{3\overline{S}_{ij}\overline{S}_{kl}}{2\overline{\sigma}_0^2}\dot{\sigma}_{kl}$

It is of interest to note that since the normal to the von Mises yield surface is colinear with the line joining the centre of the yield surface and the material point, the stress–strain relations derived from Prager's and Ziegler's hardening rules become identical.

By examining the uniaxial loading case, it is easy to see that the relationship between C and H is $C = 2H/3$. Equations (13.9) and (13.17) are the stress–strain relationships that we need for carrying out a non-linear stress analysis involving a von Mises material. Similar stress–strain relations can be derived for any material to describe its mechanical behaviour under monotonic and cyclic loading provided that an appropriate choice of the yield function F and the hardening parameter are made from physical test data.

The isotropic and kinematic hardening theories described above are based upon distinctly defined elastic and elastoplastic behaviour and are inadequate in representing the smooth transition from elastic to a fully plastic state as observed in experiments conducted on metals and soils under cyclic loading conditions. Consequently, a number of more advanced models have been proposed in the literature.[8-11] Many of these are based on the original work on nested yield surfaces by Mröz.[8,11] One such practical two-surface model was developed by Krieg.[10] It takes into account both kinematic and isotropic hardening behaviour and allows a smooth transition of states.

Thermally sensitive von Mises model Some materials exhibit a strong dependence upon the temperature in their yielding behaviour. A simple thermally sensitive von Mises model was developed by Boley and Weiner[12] for elastic–perfectly plastic materials which is modified and presented here. For incremental thermoplasticity, the total strain increment may be expressed as

$$\dot{\varepsilon}_{ij} = \dot{\varepsilon}_{ij}^{e} + \dot{\varepsilon}_{ij}^{p} + \dot{\varepsilon}_{ij}^{T} \tag{13.18}$$

and the mechanical strain $\dot{\varepsilon}_{kl}^{m}$ is defined as

$$\dot{\varepsilon}_{kl}^{m} = \tfrac{1}{2}(\dot{u}_{k,l} + \dot{u}_{i,k}) - \delta_{kl}\alpha\dot{T} \tag{13.19}$$

A temperature-dependent von Mises yield function can be stated as follows:

$$f(\sigma_{ij}, T, \varepsilon_{ij}^{p}) = \frac{1}{2}S_{ij}S_{ij} - \frac{\sigma_{0}^{2}}{3}\Gamma \tag{13.20}$$

where Γ is a dimensionless thermal parameter defined as

$$\Gamma = \left[1 - \left(\frac{T - T_{\text{ref}}}{T_{\text{melt}} - T_{\text{ref}}}\right)^{2}\right]^{1/2} \tag{13.21}$$

where T_{melt} represents the melting temperature while σ_{0} becomes the uniaxial yield stress defined at T_{ref} associated with the current level of plastic strain.

The yield function is shown graphically in $(J_{2})^{1/2}$–βT space in Fig. 13.5 as an elliptical curve. The material constant β is defined as $\beta = (3\lambda + 2\mu)\alpha$. As is evident from Fig. 13.5, the thermal portion of the dilatational deformation now has a marked influence on the yield strength. Thus, although the stress field results from the mechanical components of strain, the thermal components influence the inception of yielding in the material.

Making use of an associated flow rule and isotropic hardening behaviour and by following the same steps for the derivation of the constitutive relations as before, the stress increment in the thermoplastic case may be related to the strain increment and temperature change as

$$\dot{\sigma}_{ij} = D_{ijkl}^{ep}\dot{\varepsilon}_{kl}^{m} - \beta_{ij}^{ep}\dot{T} \tag{13.22}$$

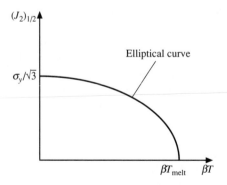

Figure 13.5 Yield function in stress–temperature space.

where

$$D^{ep}_{ijkl} = D^e_{ijkl} - 2\mu \left[\frac{3 S_{ij} S_{kl}}{2\sigma_y^2 \Gamma (1 + \Gamma H / 3\mu)} \right] \qquad (13.23a)$$

and

$$\beta^{ep}_{ij} = \frac{S_{ij} T}{T^2_{\text{melt}} \Gamma (1 + \Gamma H / 3\mu)} \qquad (13.23b)$$

In addition, for the thermally sensitive von Mises model, the characteristic tensor L^σ_{ij} is obtained explicitly as follows:

$$\dot{\lambda} = L^\sigma_{ij} \dot{\sigma}_{ij} + L^T \dot{T} \qquad (13.24a)$$

where

$$L^\sigma_{ij} = \frac{1}{H} \left(\frac{3}{2\sigma_y} \right)^2 S_{ij} \qquad (13.24b)$$

and

$$L^T = \frac{1}{H} \left(\frac{3}{2\Gamma^2} \right) \frac{T}{T^2_{\text{melt}}} \qquad (13.24c)$$

13.2.2 Viscoplasticity

A viscoplastic model is often used to describe time-dependent inelastic deformation in solids. One of the most attractive features of this model is that steady state viscoplasticity turns out to be an alternative description of the elastoplastic behaviour.

In such a model a set of viscoplastic strains ε^{vp}_{ij} are present in addition to the elastic strains ε^e_{ij}. Whenever a certain threshold value of stress (such as yield) is exceeded a non-zero value of $\dot{\varepsilon}^{vp}_{ij}$ will occur. A comprehensive account of the present state of development of viscoplasticity can be found in a review by Perzyna.[13] A viscoplastic model was used by Zienkiewicz and Cormeau[14-16] in developing a very efficient finite element algorithm for non-linear deformation analyses of solids.

In viscoplasticity the stress state can be outside the yield surface, but then creep occurs. When the creep stops the stress state returns to the current yield surface, an elastoplastic solution is established and it is therefore often possible to use viscoplasticity as a purely fictitious device to obtain elastoplastic solutions. There are, of course, in addition many problems where a viscoplastic idealization is used to represent the truly time-dependent behaviour.

Viscoplastic strains can be defined if a yield condition is specified as

$$F(\sigma_{ij}, h) = 0 \qquad (13.25)$$

where $Y(h)$ = yield stress which in turn is a function of the magnitude of the hardening (h) and $F < 0$ indicates purely elastic behaviour and $F > 0$ is admissible. If we define a plastic potential function $Q(\sigma_{ij})$ we can obtain the viscoplastic strain rate from

$$\frac{\partial \varepsilon_{ij}^{\text{vp}}}{\partial t} = \dot{\varepsilon}_{ij}^{\text{vp}} = \langle \phi(F) \rangle \frac{\partial Q}{\partial \sigma_{ij}} \tag{13.26}$$

where $\langle \phi(F) \rangle$ is a viscoplastic flow factor which may be dependent upon certain state variables such as time, strain invariants, etc.:

$$\langle \phi \rangle = 0 \qquad \text{if } F \leq 0, \text{ in the elastic region}$$

$$\langle \phi(F) \rangle = \phi(F) \qquad \text{if } F > 0, \text{ in the viscoplastic region}$$

A sufficiently general viscoplastic strain rate expression for our purposes is

$$\phi(F) = \gamma F^n \tag{13.27}$$

where γ is a fluidity parameter which again may be a function of time, strain invariants, etc., and n is a suitable material parameter.

If the plastic flow is assumed to be governed by the associated flow rule (i.e. the plastic strain increment vector is normal to the yield surface at the material point) then $Q = F$; on the other hand, if the flow is non-associated the plastic potential function Q may be chosen to be different from F.

By defining Q and/or F surfaces for a given material and using Eqs (13.26) and (13.27) we can express the viscoplastic strain rate as

$$\dot{\varepsilon}_{ij}^{\text{vp}} = C_{ijkl}^{\text{vp}} \sigma_{kl} \tag{13.28}$$

Equation (13.28) differs from that developed for incremental elastoplasticity in one, most important, respect—it specifies the viscoplastic strain rate as a function of the current stress state. In the incremental theory of plasticity, the fact that the strain rates are functions not only of the current stress level but also of the stress increments is the major source of difficulty in all solution processes [compare Eqs (13.28) and (13.8a)].

13.2.3 State Variable Theories for Inelastic Deformation of Metals

While the incremental theory of plasticity provides the most general theory for the inelastic deformation of a wide class of materials, it is essentially restricted to strain hardening and elastic–ideally plastic materials, primarily as a result of adopting Drucker's postulate for a stable yielding. Therefore, the behaviour of metals at elevated temperatures and that of some geological materials which may violate this postulate cannot be described entirely satisfactorily by such a theory.

In recent years considerable effort has been directed towards developing so-called 'state variable theories'.[17-24] Many of these state variable models have a mathematical structure similar to the viscoplastic model outlined above (i.e. the inelastic strain rate at any time is a function of stress and state variables but not of the rates of stress). Since no specific yield surfaces and special loading criteria in the sense of classical plasticity appear in such theories, inclusion of strain softening does not present any problem.

State variable theories assume that elastic and inelastic behaviour are present at every stage of loading, unloading and reloading and include within them features such as strain rate sensitivity, work hardening, Bauschinger's effect, history dependence, creep recovery, strain softening, etc., which may be present in the deformation of metals at elevated temperature, etc.

According to these models the total strain rate tensor $\dot{\varepsilon}_{ij}$, at any time t, is the sum of the elastic part $\dot{\varepsilon}_{ij}^{e}$, the anelastic part $\dot{\varepsilon}_{ij}^{n}$ and the thermal strain rate $\dot{\varepsilon}_{ij}^{T}$, i.e.

$$\dot{\varepsilon}_{ij} = \dot{\varepsilon}_{ij}^{e} + \dot{\varepsilon}_{ij}^{n} + \dot{\varepsilon}_{ij}^{T} \tag{13.29}$$

where the dot now denotes differentiation with respect to the real time variable t (as opposed to any time-like monotonically increasing parameter in elastoplasticity). The elastic strain is related to the stress tensor σ_{ij} by Hooke's law and $\dot{\varepsilon}_{ij}^{T} = \delta_{ij} \alpha \dot{T}$, where α is the coefficient of thermal expansion. The anelastic strain rate tensor $\dot{\varepsilon}_{ij}^{n}$ is time and history dependent in nature and usually consists of a recoverable part and a permanent part, and is given by

$$\dot{\varepsilon}_{ij}^{n} = M_{ij}(\sigma_{ij}, q_{ij}, T) \tag{13.30}$$

where q_{ij} denotes the state variables of the model with

$$\dot{q}_{ij} = N_{ij}(\sigma_{ij}, q_{ij}, T) \tag{13.31}$$

The state variables q_{ij} are assumed to completely characterize the present deformation state of the material. They vary along the deformation path according to certain laws, and their values at some time t depend upon the deformation history up to that time. Moreover, the history dependence of the rate of anelastic strain $\dot{\varepsilon}_{ij}^{n}$ up to time t is taken into account completely by q_{ij} at t and no further knowledge of the prior deformation history is required. The number of state variables q_{ij} varies in different models and they can be either scalar or tensor quantities. For example:

1. Miller uses a scalar 'drag' stress D and a 'back' stress tensor R_{ij}.[17,18]
2. Hart and his coworkers use a strain tensor ε_{ij}^{a} and a scalar hardness σ^{*}.[19]
3. Lagneborg and Robinson use an internal flow stress S_{ij}.[20,21]
4. Bodner and Partom use the plastic work W^{p}.[22]

A very general thermally sensitive viscoplastic model was described in Walker[23] and Cassenti and Thompson.[24] The former reference[23] discusses the basic theory while the latter[24] describes the modifications to the material parameters for Hestalloy X, which produces more accurate results at relatively low temperatures.

13.3 GOVERNING DIFFERENTIAL EQUATIONS FOR ELASTOPLASTICITY

An elastoplastic material must obey the equilibrium equations for the stress rate (increments) which are, in the absence of time-dependent body forces,

$$\frac{\partial \dot{\sigma}_{ij}}{\partial x_j} = 0 \tag{13.32}$$

where the dot indicates an increment. The stresses are related to strains via

$$\dot{\sigma}_{ij} = \lambda \delta_{ij}(\dot{\varepsilon}_{kk} - \dot{\varepsilon}_{kk}^{\mathrm{P}}) + 2\mu(\dot{\varepsilon}_{ij} - \dot{\varepsilon}_{ij}^{\mathrm{P}}) \tag{13.33}$$

By substituting Eq. (13.33) in Eq. (13.32) and using the strain–displacement relations we obtain the governing differential equations for elastoplastic flow:

$$\mu \frac{\partial^2 \dot{u}_i}{\partial x_j \partial x_j} + (\lambda + \mu) \frac{\partial^2 \dot{u}_j}{\partial x_i \partial x_j} - \left(\lambda \delta_{ij} \frac{\partial \dot{\varepsilon}_{kk}^{\mathrm{P}}}{\partial x_j} + 2\mu \frac{\partial \dot{\varepsilon}_{ij}^{\mathrm{P}}}{\partial x_j} \right) = 0 \tag{13.34}$$

In the absence of any plastic strain the second pair of terms in parentheses disappear and Eq. (13.34) becomes Navier's equation for elasticity in incremental form. An elastoplastic problem may therefore be posed as the solution of an elastic problem governed by the differential equation[25,26]

$$\mu \frac{\partial^2 \dot{u}_i}{\partial x_j \partial x_j} + (\lambda + \mu) \frac{\partial^2 \dot{u}_j}{\partial x_i \partial x_j} + \dot{f}_i = 0 \tag{13.35a}$$

over the volume and subject to the boundary condition

$$\dot{t}_i' = \dot{\sigma}_{ij} n_j - \dot{t}_i^0 \tag{13.35b}$$

over the surface, where

$$\dot{f}_i = -\left(\lambda \delta_{ij} \frac{\partial \dot{\varepsilon}_{kk}^{\mathrm{P}}}{\partial x_j} + 2\mu \frac{\partial \dot{\varepsilon}_{ij}^{\mathrm{P}}}{\partial x_j} \right)$$

$$\dot{t}_i^0 = -(\lambda \delta_{ij} \dot{\varepsilon}_{kk}^{\mathrm{P}} + 2\mu \dot{\varepsilon}_{ij}^{\mathrm{P}}) n_j$$

with n_j the outward normal to the boundary.

We see at once that Eq. (13.35a) is identical to the incremental Navier equation with suitably modified body forces $\dot{f}_i$ generated by the irrecoverable component of the deformation field. Moreover, the non-linear terms occur only in the body force term of the inhomogeneous differential equation.

We may assume an imaginary elastic body whose body forces and the traction boundary conditions have been modified according to Eqs (13.35a,b). The displacement field obtained from the solution of Eq. (13.35a) would thus be correct for the real elastoplastic body. The stresses corresponding to this displacement field would then be given by the elastic stress–strain relations in the elastic regions and elastoplastic stress–strain relations in the elastoplastic regions. This procedure for solving an elastoplastic problem by means of a modified elastic one is not new. Reisner[27] suggested an essentially identical scheme using intuitive reasoning, and called it an initial stress (eigenspannungen) approach. Zienkiewicz and his coworkers[28,29] also developed an initial stress algorithm for the solution of elastoplastic problems by the finite element method.

The governing differential equations for the viscoplastic and the state variable models are similar in character to those developed for the elastoplastic case outlined above. However, the variations of quantities such as displacements, stresses, etc., with time give rise to a stiff system of differential equations such as[14–16,30–33]

$$\frac{d\sigma_{ij}}{dt} = f(\sigma_{ij}, t) \tag{13.36}$$

It is therefore important to choose the correct automatic time step[16] in a time-integration scheme.

13.4 DIRECT BEM FOR MATERIAL NON-LINEARITIES

It is evident, from the discussions in the preceding section, that the BEM formulations developed in Chapter 7 for dealing with an interior distribution of body forces generated by initial stress/strains, converted into incremental form, will be directly applicable to the present case. Thus, depending on the type of formulation used, the BEM algorithm for dealing with material non-linearities may be classified as initial stress and initial strain.

13.4.1 Initial Stress Approach[34–57]

In initial stress formulation the initial stress rates are defined as

$$\dot{\sigma}_{ij}^0 = \dot{\sigma}_{ij}^e - \dot{\sigma}_{ij}^{ep} \tag{13.37}$$

where

$$\dot{\sigma}_{ij}^e = D_{ijkl}^e \dot{\varepsilon}_{kl} \qquad \dot{\sigma}_{ij}^{ep} = D_{ijkl}^{ep} \dot{\varepsilon}_{kl}$$

and D_{ijkl}^e, D_{ijkl}^{ep} are the elastic and elastoplastic constitutive tensors respectively.

The conventional boundary integral equation for displacement satisfying the governing equation for elastoplasticity can be expressed as (see Chapter 7)

$$c_{ij}(\xi)\dot{u}_i = \int_S [G_{ij}(x, \xi)\dot{t}_i(x) - F_{ij}(x, \xi)\dot{u}_i(x)] \, dS(x)$$

$$+ \int_V B_{ikj}(x, \xi)\dot{\sigma}_{ik}^0(x) \, dV(x) \tag{13.38}$$

The two- and three-dimensional kernel functions in Eq. (13.38) are also given in Banerjee and Raveendra,[41] and the axisymmetric kernels can be found in Henry and Banerjee.[46,47]

The initial stress rate $\dot{\sigma}_{ij}^0$ can be defined in terms of the plastic strain rate as

$$\dot{\sigma}_{ij}^0 = D_{ijkl}^e \dot{\varepsilon}_{kl}^p \tag{13.39}$$

The equation for the strain at an interior point is found analytically by substituting

Eq. (13.38) (with $c_{ij} = \delta_{ij}$) into the strain–displacement relations (see Chapter 7), where differentiation is with respect to the field point ξ:[41,46,47]

$$\dot{\varepsilon}_{ij}(\xi) = \int_S [G^{\varepsilon}_{kij}(x,\xi)\dot{t}_k(x) - F^{\varepsilon}_{kij}(x,\xi)\dot{u}_x(x)]\,\mathrm{d}S(x)$$

$$+ \int_V B^{\varepsilon}_{klij}(x,\xi)\dot{\sigma}^0_{kl}(x)\,\mathrm{d}V(x) + J^{\varepsilon}_{klij}\dot{\sigma}^0_{kl}(\xi) \tag{13.40}$$

By introducing this result into the inelastic stress–strain equation $(\sigma_{ij} = D_{ijkl}\dot{\varepsilon}_{kl} - \dot{\sigma}^0_{ij})$, the stress integral equation is derived:[41,46,47]

$$\dot{\sigma}_{ij}(\xi) = \int_S [G^{\sigma}_{kij}(x,\xi)\dot{t}_k(x) - F^{\sigma}_{kij}(x,\xi)\dot{u}_k]\,\mathrm{d}S(x)$$

$$+ \int_V B^{\sigma}_{klij}(x,\xi)\dot{\sigma}^0_{kl}(x)\,\mathrm{d}V(x) + J^{\sigma}_{klij}\dot{\sigma}^0_{kl}(\xi) \tag{13.41}$$

The last integrals in Eqs (13.40) and (13.41) are strongly singular and must be treated in the Cauchy principle sense with additional jump terms J^{σ} and J^{ε} (see Chapter 7).

For a point on the boundary, all kernel functions of the stress rate equation are all strongly singular and difficult to integrate numerically. However, the stress rate on the boundary can be obtained from the boundary traction and displacement rates without any integration.[41] For numerical purposes, this equation can be arranged in a form consistent with the discretized form of Eq. (13.40) or Eq. (13.41), as shown in Chapter 7. The above integral formulations have also been extended to thermoplasticity,[43,53,55] non-linear consolidation[55] and dynamic problems of elastoplasticity.[40,50,56,57]

13.4.2 Variable Stiffness Plasticity Formulation Based on Initial Stress

A non-iterative, direct solution method to determine the unknown initial stress is made feasible by reducing the number of unknowns in the governing equations by utilizing certain features of the incremental theory of plasticity expressed by Eq. (13.5) and of course using small increments.

The initial stress rates $\dot{\sigma}^0_{ij}$ appearing in the incremental form of integral equations above can be expressed in the context of an elastoplastic deformation as[39,41,46,47]

$$\dot{\sigma}^0_{ij} = K_{ij}\dot{\lambda} \tag{13.42a}$$

and

$$\dot{\lambda} = L^{\sigma}_{ij}\dot{\sigma}_{ij} \tag{13.42b}$$

where

$$K_{ij} = D^{e}_{ijkl}\left(\frac{\partial F}{\partial \sigma_{kl}}\right) \quad \text{and} \quad L^{\sigma}_{ij} = \frac{1}{H}\frac{\partial F}{\partial \sigma_{ij}}$$

Substituting the above into Eqs (13.38) and (13.41) yields[39,41,46,47]

$$c_{ij}(\xi)\dot{u}_i = \int_S [G_{ij}(x,\xi)\dot{t}_i(x) - F_{ij}(x,\xi)\dot{u}_i(x)]\,dS(x)$$

$$+ \int_V B_{ipj}(x,\xi)K_{ip}(x)\dot{\lambda}(x)\,dV(x) \tag{13.43}$$

and $\dot{\lambda}(\xi) = L_{jk}^\sigma(\xi) \int_S [G_{ijk}^\sigma(x,\xi)\dot{t}_i(x) - F_{ijk}^\sigma(x,\xi)\dot{u}_i(x)]\,dS(x)$

$$+ L_{jk}^\sigma(\xi)\left[\int_V B_{ipjk}^\sigma(x,\xi)K_{ip}(x)\dot{\lambda}(x)\,dV(x) + J_{ipjk}^\sigma K_{ip}(\xi)\dot{\lambda}(\xi)\right] \tag{13.44}$$

Equations (13.43) and (13.44) can now be solved simultaneously to evaluate the unknown values of displacement rates, traction rates and the scalar variable $\dot{\lambda}$.

Although Eq. (13.44) can be applied to any elastoplastic strain-hardening problem, L_{ij}^σ becomes indeterminate for the case of ideal plasticity (zero strain hardening), and it cannot therefore be used in this circumstance. This minor problem, however, can be circumvented either by specifying a small amount of strain hardening or, more appropriately, by replacing Eq. (13.44) by the strain rate equations (13.40) and (13.6a). Thus, using Eq. (13.6a) in Eq. (13.40), we obtain[39,41,46,47]

$$\dot{\lambda}(\xi) = L_{jk}^\varepsilon(\xi) \int_S [G_{ijk}^\varepsilon(x,\xi)\dot{t}_i(x) - F_{ijk}^\varepsilon(x,\xi)\dot{u}_i(x)]\,dS(x)$$

$$+ L_{jk}^\varepsilon(\xi)\left[\int_V B_{ipjk}^\varepsilon(x,\xi)K_{ip}(x)\dot{\lambda}(x)\,dV(x) + J_{ipjk}^\varepsilon K_{ip}(\xi)\dot{\lambda}(\xi)\right] \tag{13.45}$$

Equation (13.45) can now be applied to any problems of elastoplasticity (both strain hardening and ideal plasticity).

13.4.3 Initial Strain Approach[25,26,30–33,58–62]

In an initial strain formulation the initial strain rates are defined as

$$\dot{\varepsilon}_{ij}^0 = \dot{\varepsilon}_{ij} - \dot{\varepsilon}_{ij}^e \tag{13.46}$$

and related to the initial stress rate $\dot{\sigma}_{ij}^0$ via Eq. (13.39).

For elastoplasticity the initial strain is simply the plastic strain increment, but for viscoplasticity and state variable theories the initial strain rate takes on the role of the viscoplastic strain rate or the anelastic strain rate respectively. The necessary integral representation can be expressed for a boundary point ξ (see Chapter 7) as[25,26]

$$c_{ij}\dot{u}_i(\xi) = \int_S [\dot{t}_i(x)G_{ij}(x,\xi) - F_{ij}(x,\xi)\dot{u}_i(x)]\,dS$$

$$+ \int_V T_{ikj}(x,\xi)\dot{\varepsilon}_{ik}^0(x)\,dV \tag{13.47}$$

Although Eqs (13.38) and (13.47) look very similar, their implementation in an incremental solution process involves the development of quite distinct algorithms. Moreover, Eq. (13.47) has two distinctly different forms for plane stress and plane strain problems. On the other hand, Eqs (13.38) to (13.41) have the same forms for plane stress and plane strain problems. Riccardella, Mendelson, Mukherjee, Telles and their coworkers have developed numerical algorithms based on this formulation.[31-33,58-62]

By using L_{ij}^{σ}, L_{ij}^{ε} and K_{ij}, Eq. (13.47) and its interior stress counterpart can be converted to forms presented in Sec. 13.4.2 to develop a variable stiffness formulation.

13.5 INCREMENTAL COMPUTATIONS FOR ELASTOPLASTICITY

If the initial stresses and strains are known, equations developed above or similar ones via particular integrals (see Chapter 7) provide the necessary formulation of the inelastic problem for any general, well-posed set of boundary conditions. For thermal problems, similar integral formulations could be developed by adding the initial stress or strain volume integrals to the elastic case described in Chapter 9.

For our present purpose, in addition to the usual boundary discretization, the interior region, which is expected to yield as a result of the loading, is divided into a suitable number of triangular or quadrilateral cells for two-dimensional problems and tetrahedra or brick-like cells in three dimensions. Although such a discretization has the appearance of that used in the finite element method, the cells are used here simply to evaluate the various volume integrals as a piece-wise summation, as discussed in Chapter 7. The formation of the discretized system equations is therefore essentially identical to that described in Chapters 3 to 8. Thus, for example, we can write Eq. (13.38) as

$$\dot{\mathbf{G}}\mathbf{t} - \mathbf{F}\dot{\mathbf{u}} = \mathbf{B}\dot{\sigma}^0 \tag{13.48}$$

which can be solved to obtain the unknown boundary data, provided that the initial stress increments are known. The determination of the initial stresses is discussed in the next section.

Having obtained the unknown boundary displacement components and the tractions, the displacements and stresses at interior points can be calculated by using the interior point versions of the displacement and stress integral equations [e.g. Eq. (13.41)] respectively. It is also of considerable interest to note that in recent years alternative symmetric forms of Eq. (13.48) have been proposed by Maier and his colleagues.[63,64]

13.5.1 Iterative Algorithm[34-38]

The above analysis requires complete knowledge of the initial stress distribution $\dot{\sigma}_{ik}^0$ within the yielded region, which is not known a priori for a particular loading increment. This must be obtained from an iterative process as described below. For this purpose, it is convenient to write Eq. (13.48) as

$$\mathbf{A}\dot{\mathbf{x}} = \dot{\mathbf{b}} + \mathbf{B}\dot{\sigma}^0 \tag{13.49}$$

where

$\dot{\mathbf{x}}$ = vector of unknown boundary tractions and boundary displacements

$\dot{\mathbf{b}}$ = matrix formed by multiplying the appropriate columns of $\mathbf{G}$ and $\mathbf{F}$ and the given increments of the boundary tractions and boundary displacements

The steps of the incremental solution may then be described as follows:

1. Apply a load increment $\dot{\mathbf{b}}$, assuming $\dot{\boldsymbol{\sigma}}^0$ to be zero. Calculate $\mathbf{x}$, the strains $\dot{\boldsymbol{\varepsilon}}$ and the elastic stress increments $\dot{\boldsymbol{\sigma}}$. Scale the solution so that the most highly stressed cell is at the point of yielding. Store the current values of stresses $\boldsymbol{\sigma}_1$, say.

2. Apply a small-load increment $\dot{\mathbf{b}}$. Calculate the stress increments in all the cells by using the elastic stress–strain relations, that is $\dot{\boldsymbol{\sigma}}^e = \mathbf{D}^e \dot{\boldsymbol{\varepsilon}}$. Calculate the value of $\boldsymbol{\sigma}_0$, the equivalent stresses, by using $\boldsymbol{\sigma} = \boldsymbol{\sigma}_1 + \dot{\boldsymbol{\sigma}}^e$ as the stress history and compile a list of yielded cells. Calculate the correct stresses in the elastoplastic cells by using the elastoplastic stress–strain relations $\dot{\boldsymbol{\sigma}}^{ep} = \mathbf{D}^{ep} \dot{\boldsymbol{\varepsilon}}$, using the elastic strain increments as a first approximation. The initial stresses generated are $\dot{\boldsymbol{\sigma}}^0 = \dot{\boldsymbol{\sigma}}^e - \dot{\boldsymbol{\sigma}}^{ep}$ as a first approximation. Modify the stress history for the yielded cells to $\boldsymbol{\sigma}_2 = \boldsymbol{\sigma}_1 + \dot{\boldsymbol{\sigma}}^{ep}$ and make $\boldsymbol{\sigma}_1 = \boldsymbol{\sigma}_2$.

3. Assume $\dot{\mathbf{b}} = 0$ and, with the generated initial stresses $\dot{\boldsymbol{\sigma}}^0$, calculate a new $\dot{\mathbf{x}}$ vector by using Eq. (13.61) and also the nodal displacement increments $\dot{\mathbf{u}}_n$, the strain increment $\dot{\boldsymbol{\varepsilon}}$ and the stress increments by using the elastic stress–strain relations $\dot{\boldsymbol{\sigma}}^e = \mathbf{D}^e \dot{\boldsymbol{\varepsilon}}$. Calculate the equivalent stresses by using the history $\boldsymbol{\sigma}_2 = \boldsymbol{\sigma}_1 + \dot{\boldsymbol{\sigma}}^e$ and compile a list of yielded cells. For the elastoplastic cells calculate the correct stresses $\dot{\boldsymbol{\sigma}}^{ep} = \mathbf{D}^{ep} \dot{\boldsymbol{\varepsilon}}$. The initial stresses generated are $\dot{\boldsymbol{\sigma}}^0 = \dot{\boldsymbol{\sigma}}^e - \dot{\boldsymbol{\sigma}}^{ep}$. Modify the stress history for the yielded cells to $\boldsymbol{\sigma}_2 = \boldsymbol{\sigma}_1 + \dot{\boldsymbol{\sigma}}^{ep}$ and make $\boldsymbol{\sigma}_1 = \boldsymbol{\sigma}_2$.

4. Check whether the initial stresses $\dot{\boldsymbol{\sigma}}^0$ are less than an acceptable norm: if so, go to step 2; if not, go back to step 3. If the number of iterations exceeds, say, 50 then it is reasonable to assume that 'collapse' has occurred or the solution cannot be converged by this simple algorithm.

Within the framework of this algorithm several improvements are possible. For example, a certain degree of forward extrapolation may result in quicker convergence.[41,46] The accuracy can be improved if a check is carried out on the sign of the dissipated plastic work rate to ensure that all the plastic cells are, in fact, plastic. The incremental algorithm outlined above is essentially similar to that described by Zienkiewicz.[29]

An algorithm based on an initial strain formulation would be somewhat different, and for explicit details the reader is referred to Riccardella, Mendelson and Telles, Chandra, and Mukherjee.[58–62] For conventional elastoplasticity, the initial stress method which we have described appears to provide the most generally applicable procedure for a wide class of materials.

13.5.2 Variable Stiffness Algorithm for Elastoplasticity[39,42,47,51,52]

The displacement and stress rate equations are assembled by collecting the known and unknown values of traction and displacement rates and their coefficients together. The assembled equations are modified appropriately for cases where the functions (boundary conditions) are referred to local coordinates. The equilibrium and compatibility conditions are invoked at common interfaces between regions in a manner described in Chapter 3, and the final system equations can be cast as

$$\mathbf{A}^b\dot{\mathbf{x}} = \mathbf{B}^b\dot{\mathbf{y}} + \mathbf{C}^b\mathbf{K}\dot{\lambda} \tag{13.50a}$$

and
$$\dot{\lambda} = \mathbf{L}\mathbf{A}^\sigma\dot{\mathbf{x}} + \mathbf{L}\mathbf{B}^\sigma\dot{\mathbf{y}} + \mathbf{L}\mathbf{C}^\sigma\mathbf{K}\dot{\lambda}, \tag{13.50b}$$

where $\dot{\mathbf{y}}$ are the known incremental boundary conditions, $\dot{\mathbf{x}}$ are the unknown boundary conditions, $\mathbf{A}^b$, $\mathbf{B}^b$, $\mathbf{C}^b$ are the coefficient matrices of the boundary (displacement) system and $\mathbf{A}^\sigma$, $\mathbf{B}^\sigma$, $\mathbf{C}^\sigma$ are the coefficient matrices of the stress equations.

Matrices $\mathbf{A}^b$, $\mathbf{B}^b$, $\mathbf{C}^b$ and $\mathbf{A}^\sigma$, $\mathbf{B}^\sigma$, $\mathbf{C}^\sigma$ are constant for a given problem and are identical to those used in the iterative algorithm; on the other hand, matrices $\mathbf{K}$ and $\mathbf{L}$ are dependent upon state variables and are assumed to be constant during each small load step. Further, in a sub-structured system the matrices $\mathbf{A}^b$ and $\mathbf{B}^b$ are block-banded, and since each sub-structured region can only communicate through the common interfaces, matrices $\mathbf{C}^b$, $\mathbf{A}^\sigma$, $\mathbf{B}^\sigma$ and $\mathbf{C}^\sigma$ are block diagonal.

Note that only equations for the cell nodes that are expected to yield during the current load step are used in the system equation (13.50b). At elastic nodes, the values of λ are assumed to be zero. This renders a small set of equations (13.50b) at the start of the analysis which eventually grows larger as more nodes yield.

These equations can be rewritten as

$$\mathbf{A}^b\dot{\mathbf{x}} = \dot{\mathbf{b}}^b + \mathbf{C}^{\lambda b}\dot{\lambda} \tag{13.51a}$$

and
$$\dot{\lambda} = \mathbf{A}^\lambda\dot{\mathbf{x}} + \dot{\mathbf{b}}^\lambda + \mathbf{C}^\lambda\dot{\lambda} \tag{13.51b}$$

or, upon rearranging Eq. (13.51b), we have

$$\mathbf{H}\dot{\lambda} = \mathbf{A}^\lambda\dot{\mathbf{x}} + \dot{\mathbf{b}}^\lambda \tag{13.52}$$

where $\dot{\mathbf{b}}^b = \mathbf{B}^b\dot{\mathbf{y}}$, $\mathbf{C}^{\lambda b} = \mathbf{C}^b\mathbf{K}$, $\mathbf{A}^\lambda = \mathbf{L}\mathbf{A}^\sigma$, $\dot{\mathbf{b}}^\lambda = \mathbf{L}\mathbf{B}^\sigma\dot{\mathbf{y}}$, $\mathbf{C}^\lambda = \mathbf{L}\mathbf{C}^\sigma\mathbf{K}$, $\mathbf{H} = \mathbf{I} - \mathbf{C}^\lambda$ and $\mathbf{I}$ is the identity matrix. Equation (13.52) can then be recast as

$$\dot{\lambda} = \mathbf{A}^0\dot{\mathbf{x}} + \mathbf{b}^0 \tag{13.53}$$

where $\mathbf{A}^0 = \mathbf{H}^{-1}\mathbf{A}^\lambda$ and $\dot{\mathbf{b}}^0 = \mathbf{H}^{-1}\dot{\mathbf{b}}^\lambda$. Substituting the above equation into Eq. (13.51a) results in the final system equations:

$$\mathbf{A}^r\dot{\mathbf{x}} = \dot{\mathbf{b}}^r \tag{13.54}$$

where $\mathbf{A}^r = \mathbf{A}^b - \mathbf{C}^{\lambda b}\mathbf{A}^0$ and $\dot{\mathbf{b}}^r = \dot{\mathbf{b}}^b + \mathbf{C}^{\lambda b}\dot{\mathbf{b}}^0$.

Equation (13.54) is constructed and solved to evaluate the unknown vector $\dot{\mathbf{x}}$ at boundary nodes for every increment of loading. This formulation is similar to the variable stiffness approach used in the finite element method, since the system matrix on the boundary, as well as the right-hand-side vector, is modified for each increment of loading. For a multiregion system the inversion of **H** can be carried out for each region separately.

13.5.3 Solution Process for the Variable Stiffness Approach

As previously mentioned, this solution process using very small load increment need not involve any extensive iterative procedure; instead the substantial part of the solution effort is spent on assembly of the system equations for each load step. These operations using the simplest algorithm can be described as follows:

1. Impose an arbitrary boundary loading and solve the elastic problem in the usual manner.
2. Scale the elastic solution such that the most highly stress node is at yield.
3. Apply a small load increment (usually 5 per cent of the yield load) and compute **K** and **L** matrices using the past stress history.
4. Form the system equation (13.54) and solve for $\dot{\mathbf{x}}$.
5. Evaluate the initial stress rates $\dot{\sigma}^0$ using

$$\dot{\sigma}^0 = \mathbf{K}\dot{\lambda} = K[\mathbf{A}^0\dot{\mathbf{x}} + \dot{\mathbf{b}}^0]$$

6. Evaluate interior quantities displacement and stress rates using the discretized form of Eqs (13.38) and (13.41).
7. Return to step 3 if the strains are less than a specified norm. Otherwise, failure is assumed to have occurred.

It is of importance to note that the matrices **K** and **L** do not exist in the elastic region; the equations involving these matrices are therefore formed (and σ^0 is determined) only for the nodes that are at yield. Furthermore, any small deviation from the yield surface can be corrected by applying the stress rate difference (i.e. the initial stress rate) during the next load step. A simple radial return algorithm can be used.

13.5.4 Newton-Raphson Algorithm for Elastoplasticity[55]

Starting with the same set of system equations as in the variable stiffness algorithm, it was observed that improved results were obtained through the use of a Newton–Raphson algorithm, particularly for thermoplasticity involving a high degree of non-linearities.

Rewriting the system equations (13.50a,b) in a slightly different form of residue (or error) terms for the Nth time step (increment) yields

$$\mathbf{g}^b(\dot{\mathbf{x}}, \dot{\lambda}) = \mathbf{A}^b\dot{\mathbf{x}} - \dot{\mathbf{b}} - \mathbf{C}^b\mathbf{K}\dot{\lambda} = 0 \tag{13.55a}$$

$$\mathbf{g}^\lambda(\dot{\mathbf{x}}, \dot{\lambda}) = \mathbf{L}^\sigma\mathbf{A}^\sigma\dot{\mathbf{x}} - \mathbf{L}^\sigma\dot{\mathbf{b}}^\sigma - \mathbf{L}^\sigma\mathbf{C}^\sigma\mathbf{K}\dot{\lambda} + \mathbf{I}\dot{\lambda} = 0 \tag{13.55b}$$

To obtain the correct solution for the unknowns $\dot{\mathbf{x}}$ from the equations above, it is necessary to evaluate the additional unknown scalar parameter $\dot{\lambda}$ (and thereby the initial stress rate $\dot{\sigma}^0$). This is done with the help of an iterative scheme wherein the error functions $\mathbf{g}^b$ and $\mathbf{g}^\lambda$ are minimized.

Writing a Taylor series expansion for $\mathbf{g}^b(\dot{\mathbf{x}}, \dot{\lambda})$ with respect to $\dot{\mathbf{x}}$ and $\dot{\lambda}$ for the mth iteration produces

$$\mathbf{g}^b(\dot{\mathbf{x}}, \dot{\lambda}) = \mathbf{g}^b(\dot{\mathbf{x}}^m, \dot{\lambda}^m) + \frac{\partial \mathbf{g}^b}{\partial \dot{\mathbf{x}}}(\Delta \dot{\mathbf{x}}^m) + \frac{\partial \mathbf{g}^b}{\partial \dot{\lambda}}(\Delta \dot{\lambda}^m) + \text{higher-order terms}\ldots = 0$$

$$(13.56a)$$

$$\mathbf{g}^\lambda(\dot{\mathbf{x}}, \dot{\lambda}) = \mathbf{g}^\lambda(\dot{\mathbf{x}}^m, \dot{\lambda}^m) + \frac{\partial \mathbf{g}^\lambda}{\partial \dot{\mathbf{x}}}(\Delta \dot{\mathbf{x}}^m) + \frac{\partial \mathbf{g}^\lambda}{\partial \dot{\lambda}}(\Delta \dot{\lambda}^m) + \text{higher-order terms}\ldots = 0$$

$$(13.56b)$$

where $(\Delta \dot{\mathbf{x}}^m)$ and $(\Delta \dot{\lambda}^m)$ are the changes in the incremental quantities during the mth iteration of the current load increment. The quantities $\dot{\mathbf{x}}^m$ and $\dot{\lambda}^m$ are the values of $\dot{\mathbf{x}}$ and $\dot{\lambda}$ computed at the end of the $(m-1)$th increment. Therefore, $\dot{\mathbf{x}}^m$ and $\dot{\lambda}^m$ are known quantities for the mth iteration. Neglecting the higher-order terms and rearranging, Eqs (13.56a,b) yield

$$\frac{\partial \mathbf{g}^b}{\partial \dot{\mathbf{x}}}(\Delta \dot{\mathbf{x}}^m) + \frac{\partial \mathbf{g}^b}{\partial \dot{\lambda}}(\Delta \dot{\lambda}^m) = -\mathbf{g}^b(\dot{\mathbf{x}}^m, \dot{\lambda}^m) \qquad (13.57a)$$

$$\frac{\partial \mathbf{g}^\lambda}{\partial \dot{\mathbf{x}}}(\Delta \dot{\mathbf{x}}^m) + \partial \frac{\mathbf{g}^\lambda}{\partial \dot{\lambda}}(\Delta \dot{\lambda}^m) = -\mathbf{g}^\lambda(\dot{\mathbf{x}}^m, \dot{\lambda}^m) \qquad (13.57b)$$

Rewriting the set of equations (13.57a,b) in a global matrix form, one obtains the following system equation:

$$\begin{bmatrix} \dfrac{\partial \mathbf{g}^b}{\partial \dot{\mathbf{x}}} & \dfrac{\partial \mathbf{g}^b}{\partial \dot{\lambda}} \\ \dfrac{\partial \mathbf{g}^\lambda}{\partial \dot{\mathbf{x}}} & \dfrac{\partial \mathbf{g}^\lambda}{\partial \dot{\lambda}} \end{bmatrix} \left\{ \begin{array}{c} (\Delta \dot{\mathbf{x}}^m) \\ (\Delta \dot{\lambda}^m) \end{array} \right\} = \left\{ \begin{array}{c} -\mathbf{g}^b(\dot{\mathbf{x}}^m, \dot{\lambda}^m) \\ -\mathbf{g}^\lambda(\dot{\mathbf{x}}^m, \dot{\lambda}^m) \end{array} \right\} \qquad (13.58)$$

This is the standard form of a Newton–Raphson algorithm applied to Eqs (13.55a,b). Evaluating each partial derivative term results in the following:

$$\frac{\partial \mathbf{g}^b}{\partial \dot{\mathbf{x}}} = \mathbf{A}^b \qquad (13.59a)$$

$$\frac{\partial \mathbf{g}^b}{\partial \dot{\lambda}} = -\mathbf{C}^b \mathbf{K} \qquad (13.59b)$$

$$\frac{\partial \mathbf{g}^\lambda}{\partial \dot{\mathbf{x}}} = \mathbf{L}^\sigma \mathbf{A}^\sigma \qquad (13.59c)$$

$$\frac{\partial \mathbf{g}^\lambda}{\partial \dot{\lambda}} = -\mathbf{L}^\sigma \mathbf{C}^\sigma \mathbf{K} + \mathbf{I} \qquad (13.59d)$$

Hence, the system equation (13.58) can now be rewritten explicitly as[43]

$$\begin{bmatrix} \mathbf{A}^b & -\mathbf{C}^b\mathbf{K} \\ \mathbf{L}^\sigma\mathbf{A}^\sigma & \mathbf{I}-\mathbf{L}^\sigma\mathbf{C}^\sigma\mathbf{K} \end{bmatrix} \left\{ \begin{array}{c} (\Delta\dot{\mathbf{x}}^m) \\ (\Delta\dot{\boldsymbol{\lambda}}^m) \end{array} \right\} = \left\{ \begin{array}{c} -(\mathbf{A}^b\dot{\mathbf{x}}^m - \dot{\mathbf{b}} - \mathbf{C}^b\mathbf{K}\dot{\boldsymbol{\lambda}}^m) \\ -(\mathbf{L}^\sigma\mathbf{A}^\sigma\dot{\mathbf{x}}^m - \mathbf{L}^\sigma\dot{\mathbf{b}}^\sigma - \mathbf{L}^\sigma\mathbf{C}^\sigma\mathbf{K}\dot{\boldsymbol{\lambda}}^m + \mathbf{I}\dot{\boldsymbol{\lambda}}^m) \end{array} \right\}$$

(13.60)

Further, in a global form this becomes simply

$$[\mathbf{A}^G]\{\Delta\dot{\mathbf{X}}^m\} = \{\mathbf{R}^m\}$$

(13.61)

where $[\mathbf{A}^G]$ is the system matrix, $\{\Delta\dot{\mathbf{X}}^m\}$ is a matrix of the generalized unknowns comprising of $\Delta\dot{\mathbf{x}}$ and $\Delta\dot{\boldsymbol{\lambda}}$. Additionally, $\{\mathbf{R}^m\}$ is the entire right-hand side vector for the mth iteration. The iteration scheme[55] for the solution of the unknowns makes use of the above equation in conjunction with the non-linear stress–strain relation ($\dot{\boldsymbol{\sigma}} = \mathbf{D}^{ep}\dot{\boldsymbol{\varepsilon}}$).

For certain problems, the system matrix $\mathbf{A}^G$ may not be recomputed for each iteration. This corresponds to the case of a modified Newton–Raphson scheme wherein the system matrix is only reassembled after a prescribed number of increments or iterations. However, for other more difficult non-linear problems, a full Newton–Raphson scheme may be adopted in which the system matrix is reassembled for every iteration.

It may be mentioned here that the variable stiffness method discussed in the previous section forms a special case of the more general Newton–Raphson schemes described above. In particular, the variable stiffness algorithm is recovered from Eq. (13.60) when the number of iterations is fixed at one and the equations are solved by first inverting $(\mathbf{I} - \mathbf{L}^\sigma\mathbf{C}^\sigma\mathbf{K})$ and then backsubstituting for $\Delta\dot{\boldsymbol{\lambda}}$.

13.5.5 An Observation on $\mathbf{L}^\sigma$ for Plane Strain[54,55]

It is noted that although there are three components of stress for an analysis in two dimensions in Eq. (13.41), the relation (13.44), which is employed in obtaining one of the governing equations, contains four components of stress including the non-zero stress in the out-of-plane direction for plane strain. The effect of this non-zero $\dot{\sigma}_3$ can be taken into account by properly modifying relation (13.5a). The following equation is at first written in terms of unreduced compliances b_{ij} as (see Chapter 4)

$$\dot{\varepsilon}_i = b_{ij}\dot{\sigma}_j^e \qquad (i,j = 1,2,3,6)$$

(13.62)

The condition of plane strain will now be satisfied if

$$\dot{\varepsilon}_3 = 0 = b_{3j}\dot{\sigma}_j^e = b_{3j}(\dot{\sigma}_j + \dot{\sigma}_j^0)$$

(13.63)

whence

$$\dot{\sigma}_3 = -\frac{1}{b_{33}}(b_{3\alpha}\dot{\sigma}_\alpha + b_{3j}\dot{\sigma}_j^0) \qquad (\alpha = 1,2,6; j = 1,2,3,6)$$

(13.64)

Equation (13.5a) can now be rewritten using Eqs (13.64) and (13.42a) as

$$\dot{\lambda} = L_\alpha \dot{\sigma}_\alpha + L_3 \dot{\sigma}_3 = \left(L_\alpha - \frac{b_{3\alpha}}{b_{33}} L_3 \right) \dot{\sigma}_\alpha + \beta_j K_j \dot{\lambda} \qquad (13.65a)$$

where

$$\beta_j = -\frac{b_{3j}}{b_{33}} L_3 \qquad (13.65b)$$

Hence,

$$\dot{\lambda} = \frac{1}{1 - \beta_j K_j} \left(L_\alpha - \frac{b_{3\alpha}}{b_{33}} L_3 \right) \dot{\sigma}_\alpha \qquad (\alpha = 1, 2, 6;\ j = 1, 2, 3, 6) \qquad (13.66)$$

which is the desired modified form of Eq. (13.5a).

13.6 INCREMENTAL COMPUTATIONS FOR VISCOPLASTICITY

The viscoplastic algorithm is essentially an initial strain approach and we can write the final system of equations for (13.47) in a form similar to Eq. (13.49), i.e.[38]

$$\mathbf{A}\,\Delta\mathbf{x} = \Delta\mathbf{b} + \mathbf{T}\,\Delta\boldsymbol{\varepsilon}^{\mathrm{vp}} \qquad (13.67)$$

where we have denoted the incremental quantities by Δ since in viscoplasticity the superior dots indicate the true time rate and the loading path has been divided into an arbitrary number of increments $\Delta\mathbf{b}$. Assuming that we have complete knowledge of the stresses $\boldsymbol{\sigma}^n$, displacements $\mathbf{u}^n$, viscoplastic strain $(\Delta\boldsymbol{\varepsilon}^{\mathrm{vp}})^n$ and $(\Delta\mathbf{b})^n$ at the nth state at time t, the simplest algorithm for obtaining the $(n+1)$ state, at time $t + \Delta t$, can be described as follows:[38]

1. Starting with the known values of the state variables at the nth state calculate the viscoplastic strain rate $\dot{\boldsymbol{\varepsilon}}^{\mathrm{vp}}$ from Eq. (13.28), i.e.

$$(\dot{\boldsymbol{\varepsilon}}^{\mathrm{vp}})^n = \mathbf{C}^{\mathrm{vp}} \boldsymbol{\sigma}^n \qquad (13.68)$$

2. Determine the change of the viscoplastic strain $\boldsymbol{\varepsilon}^{\mathrm{vp}}$ as

$$(\delta\boldsymbol{\varepsilon}^{\mathrm{vp}})^n \simeq (\dot{\boldsymbol{\varepsilon}}^{\mathrm{vp}})^n \Delta t \qquad (13.69)$$

3. Using the values of the known $(\Delta\mathbf{b})^{n+1}$ (i.e. time-dependent boundary loading) and $(\Delta\boldsymbol{\varepsilon}^{\mathrm{vp}})^{n+1} = (\Delta\boldsymbol{\varepsilon}^{\mathrm{vp}})^n + (\delta\boldsymbol{\varepsilon}^{\mathrm{vp}})^n$ compute the new right-hand side of Eq. (13.67), solve for the new $(\Delta\mathbf{x})^{n+1}$ and then calculate $(\Delta\mathbf{u})^{n+1}$, $(\Delta\boldsymbol{\varepsilon})^{n+1}$ and $(\Delta\boldsymbol{\sigma})^{n+1}$ from the equation $(\Delta\boldsymbol{\sigma})^{n+1} = \mathbf{D}^{\mathrm{e}}(\Delta\boldsymbol{\varepsilon} - \Delta\boldsymbol{\varepsilon}^{\mathrm{vp}})^{n+1}$.
4. Update the stresses and start a new time step $(n+2)$.

It should be noted that Eq. (13.67) need not be written in terms of $\Delta\mathbf{x}$, $\Delta\mathbf{b}$ and $\Delta\boldsymbol{\varepsilon}^{\mathrm{vp}}$. Since the theory allows for the path-dependent features that are present in elastoplasticity we could write Eq. (13.67) in terms of the current values of $\mathbf{x}$, $\mathbf{b}$ and $\boldsymbol{\varepsilon}^{\mathrm{vp}}$ at time t. If, however, we wish to solve elastoplasticity problems using a viscoplastic approach, then we are obliged to introduce incremental quantities

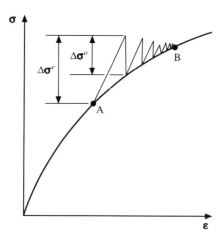

Figure 13.6 Diagrammatic representation of the iterative algorithm from A to B in the initial stress method for elastoplasticity.

into the equation. In such an algorithm the exact form of the viscoplastic flow parameters n (often $n = 1$) and γ in Eq. (13.27) is immaterial and time plays the role of a fictitious variable. For each increment of loading $\Delta\mathbf{b}$ an elastic solution is followed by a relaxation process (in time) back to a steady state equilibrium position. The time stepping is stopped when the stresses reach a point sufficiently close to the yield surface (i.e. when the steady state viscoplastic solution has been reached). Figures 13.6 and 13.7 illustrate the difference between the viscoplastic and initial stress methods for the analysis of elastoplastic solids.

The major problem associated with this approach is that the magnitude of the time step (Δt) that can be used is severely limited by stability and convergence requirements. These are discussed by Zienkiewicz and Cormeau.[14-16]

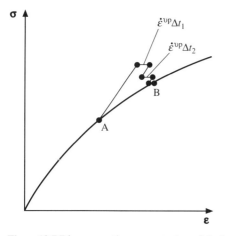

Figure 13.7 Diagrammatic representation of the iterative viscoplastic algorithm used in elastoplasticity.

13.7 APPLICATIONS TO OTHER RELATED SYSTEMS

There are many problems in engineering science where the non-linearities involved in the governing differential equation are similiar to those explored in the preceding sections. These include problems of non-Darcy flow through porous media, compressible flow of fluids, non-linear viscous flow, magnetic saturation problems, etc.,[65-70] where in addition to the boundary integrals a volume integral over the non-linear region may be introduced in the manner discussed in this chapter.

Boundary element methods applied to such problems make extensive use of the fact that the region over which non-linearities exist is usually quite small and therefore very efficient numerical solution techniques for non-linear problems can be developed using the BEM. For example, in a problem of finite deformation in solids, the large strain areas are usually confined to a small part of the solid which could be separated out as a zone and treated by a BEM large deformation algorithm.[71-73]

13.8 EXAMPLES

13.8.1 Three-dimensional Analysis of a Perforated Plate[48]

The plastic deformation of a perforated plate in tension is analysed under plane stress conditions. The volume distribution of initial stress is represented using either twenty-noded isoparametric volume cells or the point-based particular integral approach described in Chapter 7. Each representation is coupled with either the iterative or the variable stiffness solution process, leading once again to a total of four distinct algorithms. The material properties for the plate are:

$$E = 7000 \, \text{kg/mm}^2$$

$$\nu = 0.2$$

$$\sigma_0 = 24.3 \, \text{kg/mm}^2 \text{ (von Mises yield criterion)}$$

$$h = 224.0 \, \text{kg/mm}^2$$

The diameter of the circular hole, at the centre of the plate, is one-half of the width, and the thickness is one-fifth of the width. A quarter of the plate is discretized in two sub-regions, as shown in Fig. 13.8. The first region, containing the root of the plate, has 30 quadratic boundary elements. For the volume integral-based analysis the domain of this region is discretized using nine (twenty-node) isoparametric cells. In the particular integral analysis, particular integrals are defined at points corresponding to the cell nodes used in the volume integral analysis. The second region has 23 boundary elements. No volume discretization or definition of particular integrals is required in this region since it remains elastic throughout the analysis. Boundary conditions on both the front and back faces are assumed to be traction free.

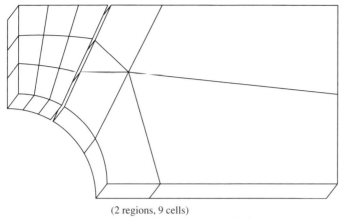

(2 regions, 9 cells)

Figure 13.8 Three-dimensional mesh of a perforated plate.

This problem was previously analysed experimentally by Theocaris and Marketos[74] and by Zienkiewicz[29] using the finite element method. The results obtained by the boundary element analysis are compared to these results in Figs 13.9 and 13.10. The stress–strain response at the root of the plate is shown in Fig. 13.9. The results obtained using the various BEM algorithms show good agreement with one another and with the variable stiffness FEM analysis. In Fig. 13.10, the stress distribution between the root and the free surface of the specimen is shown for a load of $2\sigma_m/\sigma_0 = 0.91$. Once again, excellent agreement is obtained among the four boundary element analyses. In order to evaluate the degree of convergence of the results, a mesh with 16 volume cells (or the particular integral equivalent) was studied. The results were virtually unchanged from those shown.

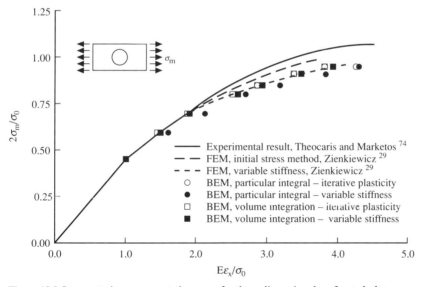

Figure 13.9 Stress–strain response at the root of a three-dimensional perforated plate.

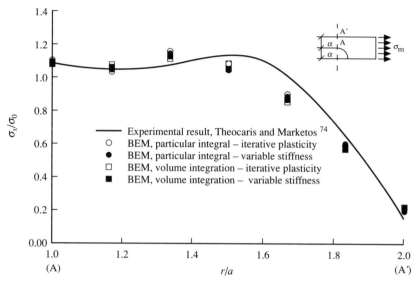

Figure 13.10 Stress distribution across a three-dimensional perforated plate near the collapse load (at $2\sigma_m/\sigma_0 = 0.91$).

The present analysis was carried out on the Cray-1 computer. The central processing unit (CPU) times for the four algorithms were:

Particular integral–iterative = 254 s
Particular integral–variable stiffness = 358 s
Volume integral–iterative = 272 s
Volume integral–variable stiffness = 361 s

13.8.2 Two-dimensional Analysis of a Perforated Plate[48]

The perforated plate of the previous problem is analysed using the particular integral-based, inelastic, two-dimensional, plane stress analysis. Three discretizations, shown in Fig. 13.11, are used to model the plate. Each mesh is divided into two sub-regions. The initial stress distribution in the inelastic region is defined using the particular integral representation. To define the particular integrals, 20, 41 and 65 nodes are used respectively in these discretizations.

A solution is obtained using the variable stiffness method, and the results for the finer particular integral representation are found to agree well with those of the particular integral-based variable stiffness method of the previous example. Interestingly the result for the two coarser representations also gave acceptable values to within 5 per cent.

Finally, the computational times of the above analysis were compared using the four algorithms. The second mesh in Fig. 13.11 was chosen for this purpose. Nine quadratic volume cells were constructed in the notch region for the volume integral base analysis utilizing the interior points shown in the figure. The CPU times on a Hewlett–Packard 9000 (series 520, 1MIP) computer

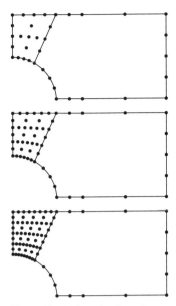

Figure 13.11 Discretization of a two-dimensional perforated plate.

were:

Particular integral–iterative $= 43.5\,\text{min}$
Particular integral–variable stiffness $= 49.1\,\text{min}$
Volume integral–iterative $= 59.1\,\text{min}$
Volume integral–variable stiffness $= 72.3\,\text{min}$

13.8.3 Residual Stress in a Steel Cylinder[55]

As the first application of quasi-static thermoplasticity, consider the residual stresses resulting from the sudden cooling of a steel cylinder. The cylinder is long and has a diameter of 2.5 in. It is initially at a temperature of 1250°F without any stresses and is subsequently cooled in an 80°F fluid. The rate of convection is governed by the film coefficient specified along its cylindrical surface. A film coefficient of $h = 12.24\ \text{in} \cdot \text{lb/s} \cdot \text{in}^2 \cdot {}^\circ\text{F}$ is selected to simulate a fairly rapid cooling process (see Chapter 9).

The material properties, corresponding to 1060 steel, are selected as

$$E = 30 \times 10^6\,\text{lb/in}^2$$

$$\nu = 0.3$$

$$\alpha = 6.0 \times 10^{-6}/{}^\circ\text{F}$$

$$k = 5.8\,\text{in} \cdot \text{lb/s} \cdot \text{in} \cdot {}^\circ\text{F}$$

$$\rho c_\epsilon = 283.0\,\text{in} \cdot \text{lb/in}^3 \cdot {}^\circ\text{F}$$

In addition, a thermally-sensitive von Mises constitutive model is employed with the yield stress at reference temperature, $\sigma_y = 48\,\text{k/in}^2$, and the melting temperature $T_{\text{melt}} = 1300°\text{F}$. No strain hardening is permitted to take place.

The axisymmetric BEM mesh utilized for this problem consists of six quadratic boundary elements and four quadratic volume cells. A time step size of 4.0 s. was found suitable for this problem, based upon the smallest significant element length as well as the diffusivity (see Chapter 9). This step size was further confirmed through convergence studies. In addition to the mesh described above, a more refined mesh with 16 volume cells was also constructed to verify the convergence of the results obtained from the present analysis. The Newton–Raphson algorithm was used.

The temperature response, as a function of time at the surface ($r = 1.25$) and at the centre ($r = 0.0$) of the cylinder at mid-height, is shown in Fig. 13.12. A rapid cooling takes place at the surface, as expected, whereas the centre cools more gradually. After about 80 s, the entire cylinder is close to room temperature and the analysis is terminated. The sudden cooling phenomenon gives rise to differential stress behaviour in the cylinder, as evident from Fig. 13.13. The axial stresses are plotted at the same locations (that is $r = 0.0$ and 1.25), showing the variation with time. The surface of the cylinder develops tensile stress initially, while the centre goes into compression. However, since the material is still very hot, yielding commences at a much lower value of stress and considerable plastic flow results. Eventually, after the first ten seconds, the thermal gradients diminish and the current yield stress value increases, reducing the amount of plastic flow. The

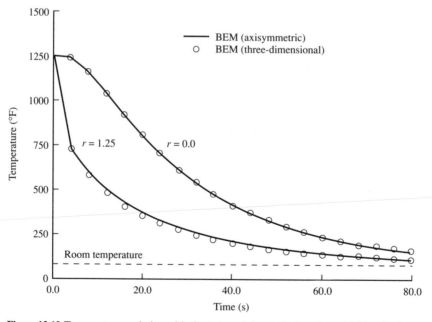

Figure 13.12 Temperature variation with time at $r = 0$ (centreline) and $r = 1.25$ (surface).

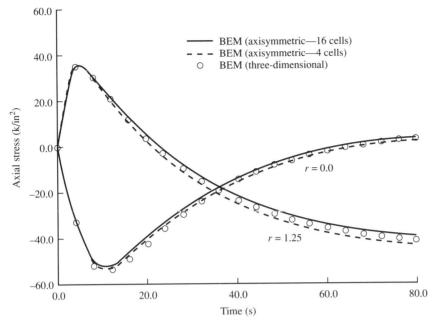

Figure 13.13 Axial stress variation with time at $r = 0$ and $r = 1.25$.

behaviour is altered as a result and elastic unloading begins to occur. Ultimately, after about 40.0 s, the surface goes into compression whereas the stress at the centre diminshes in value. At the stage where the analysis is terminated, the cylinder exhibits a compressive stress of 40.0 k/in^2 at the surface while the centre has a small amount of tensile stress.

13.8.4 Pressure Vessel Containing a Hot Fluid[46,47,55]

The next application studied is the thermoplastic analysis of an axisymmetric, steel pressure vessel subjected to internal pressure of a fluid at a high temperature. The combined effect of the mechanical and thermal loading produces a complex stress pattern within the wall of the vessel. Assuming the material obeys an elastoplastic stress–strain law, the internal pressure of the fluid is sufficiently increased to cause yielding, particularly around the welded connection between the shell and the neck. This static elastoplastic behaviour was studied by Zienkiewicz[29] using the FEM and by Henry and Banerjee[46,47] using an elastoplastic BEM formulation. In the present analysis,[55] the thermal effects (see Chapter 9) are superimposed on to the elastoplastic mechanical analysis to study the combined response and a Newton–Raphson solution procedure is used. The problem description, along with the dimensions of the vessel, is provided in Fig. 13.14. The pressure–temperature relation used in the analysis is represented graphically in Fig. 13.15 on page 365.

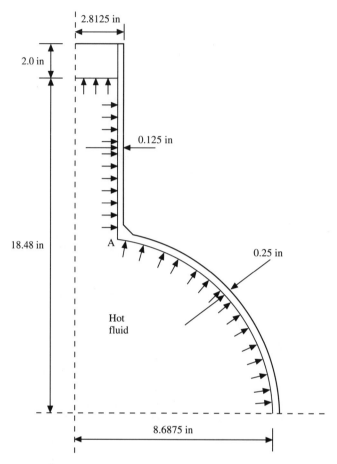

Figure 13.14 Pressure vessel containing hot fluid—problem description.

The vessel is made of steel and the following material properties apply:

$$E = 29.12 \times 10^6 \, \text{lb/in}^2$$

$$\nu = 0.3$$

$$\alpha = 6.0 \times 10^{-6} \, \text{in/in} \cdot {}^\circ\text{F}$$

$$\rho c_\epsilon = 283.0 \, \text{in} \cdot \text{lb/in}^3 \cdot {}^\circ\text{F}$$

$$k = 5.8 \, \text{in} \cdot \text{lb/s} \cdot \text{in} \cdot {}^\circ\text{F}$$

The thermally sensitive von Mises model is adopted with a yield stress at reference temperature of $\sigma_y = 40.54 \, \text{k/in}^2$ and no strain hardening. The melting temperature of steel is taken as $T_{\text{melt}} = 1400°\text{F}$. The temperature

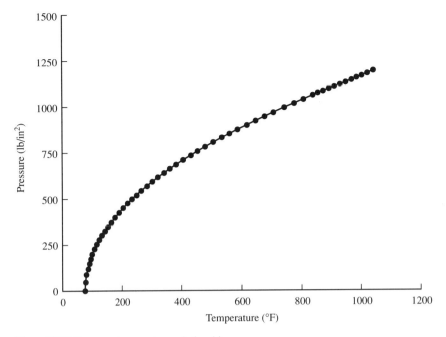

Figure 13.15 Pressure–temperature relationship.

of the fluid within the vessel is taken as 632°F and the corresponding internal pressure is 900 lb/in². The outside environment of the vessel is maintained at 70°F.

A six-region, axisymmetric BEM mesh, with 99 quadratic boundary elements and 12 volume cells, is employed to model the vessel, as depicted in Fig. 13.16. The volume cells are only placed in zones of potential yielding. Intuitively, the area around the welded connection would appear the likely zone for plasticity to occur. Hence, only region 4 is assumed to contain cells and all other regions are treated elastically. Of course, it must be verified at the completion of the analysis that, indeed, no plastic effects were present in these elastic regions.

First, an isothermal, steady state analysis is conducted in order to simulate an elastoplastic analysis and compare with the known solutions mentioned above. The results of the present BEM analysis for the vertical deflection of point A as a function of the applied pressure are plotted in Fig. 13.17 on page 367 which agree well with previous solutions given in Refs 29, 46 and 47.

Next, the non-isothermal conditions are assumed and the fluid is considered to attain the temperature and pressure states described earlier. The BEM results for the deflection of point A with pressure and temperature are shown in Fig. 13.17 along with the corresponding isothermal case for comparison. It becomes evident that the increase in fluid temperature reduces the value of the collapse pressure significantly.

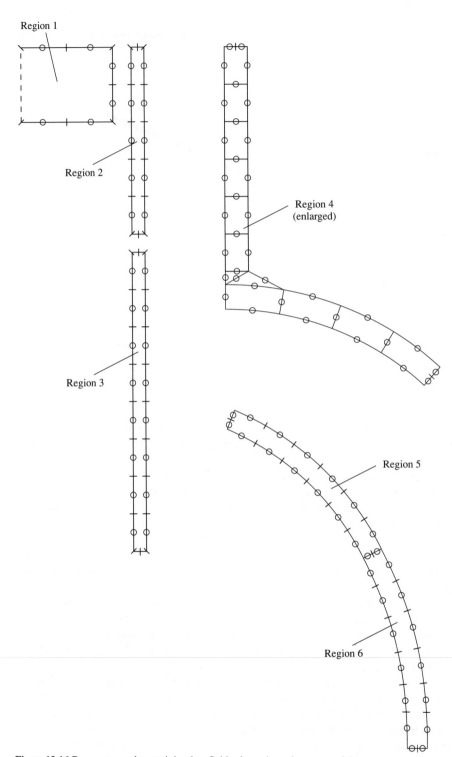

Figure 13.16 Pressure vessel containing hot fluid—boundary element model.

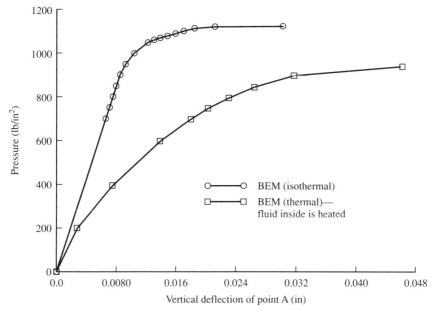

Figure 13.17 Vertical deflection of point A as a function of applied pressure and temperature.

13.9 CONCLUDING REMARKS

An attempt has been made in this chapter to demonstrate that the BEM is a useful problem-solving tool for non-linear problems in solid mechanics and, with imagination, the principles and methodologies outlined here could be applied to many other areas of engineering science. Some of these further extensions are discussed in Chapter 15.

REFERENCES

1. Drucker, D.C. (1951) A more fundamental approach to plastic stress–strain solutions, Proceedings of First U.S. National Congress in Applied Mechanics, pp. 487–491.
2. Fung, Y.C. (1965) *Foundations of Solid Mechanics*, Prentice-Hall, Englewood Cliffs, New Jersey.
3. Chen, W.F. (1975) *Limit Analysis and Soil Plasticity*, Elsevier, New York.
4. Banerjee, P.K. and Stipho, A.S. (1978) Associated and non associated constitutive relations for undrained behaviour of isotropic soil clay, *International Journal of Numerical Analysis Methods in Geomechanics*, Vol. 2, No. 1, pp. 35–56.
5. Prager, W. (1955) Theory of plasticity—a survey of recent achievement (James Clayton Lecture), *Proceedings of Institution of Mechanical Engineers*, Vol. 169, pp. 41–57.
6. Shields, H. (1959) On Prager's hardening rule, *ZAMP*, Vol. 9a, pp. 260–276.
7. Ziegler, H. (1959) A modification of Prager's hardening rule, *Quarterly Journal of Applied Mathematics*, Vol. 17, No. 1, pp. 55–65.
8. Mröz, Z. (1969) An attempt to describe the behaviour of metals under cyclic loads using a more general work hardening model, *Acta Mechanics*, Vol. 7, pp. 199–212.
9. Banerjee, P.K. and Yousif, N. (1986) A plasticity model for the mechanical behaviour of aniso-

tropically consolidated clays, *International Journal of Numerical Analysis Methods in Geomechanics*, Vol. 10, pp. 521–541.

10. Krieg, R.D. (1975), A practical two surface plasticity theory, *Journal of Applied Mechanics, ASME*, Vol. E42, pp. 641–646.

11. Mröz, Z. (1967) On the description of anisotropic work hardening, *Journal of Mechanical and Physical Solids*, Vol. 15, pp. 163–175.

12. Boley, B.A. and Weiner, J.H. (1960) *Theory of Thermal Stresses*, John Wiley, New York.

13. Perzyna, P. (1966) Fundamental problems in visco-plasticity, *Advances in Mechanics*, Vol. 9, pp. 243–277.

14. Zienkiewicz, O.C. and Cormeau, I.C. (1972) Visco-plasticity solution by finite element process, *Archives in Mechanics*, Vol. 24, pp. 5–6, 873–888.

15. Zienkiewicz, O.C. and Cormeau, I.C. (1974) Visco-plasticity, plasticity and creep in elastic solids—a unified numerical solution approach, *International Journal of Numerical Methods in Engineering*, Vol. 8, pp. 821–845.

16. Cormeau, I.C. (1975) Numerical stability in quasi-static elasto-visco-plasticity, *International Journal in Numerical Methods in Engineering*, Vol. 9, pp. 109–128.

17. Miller, A. (1976) An inelastic constitutive model for monotonic, cyclic and creep deformation: Part 1—equations development and analytical procedures, *Journal of Engineering Materials Technology, ASME*, Vol. 98, pp. 97–105.

18. Miller, A. (1976) An inelastic constitutive model for monotonic, cyclic and creep deformation: Part 2—applications to type 304 stainless steel, *Journal of Engineering Materials Technology, ASME*, Vol. 98, pp. 106–113.

19. Hart, E.W. (1976) Constitutive relations for nonelastic deformation of metals, *Journal of Engineeering Materials Technology, ASME*, Vol. 98, No. 3, pp. 193–202.

20. Lagneborg, R. (1972) A modified recovery-creep model and its evaluation, *Metal Science Journal*, Vol. 6, pp. 127–133.

21. Robinson, D.N. (1975) A candidate creep-recovery model of $2\frac{1}{2}$ Cr–I MO steel and its experimental implementation, Report No. ORNL-TM-5110, Oak Ridge National Laboratory.

22. Bodner, S.R. and Partom, Y. (1975) Constitutive equations for elastic visco-plastic strain hardening materials, *Journal of Applied Mechanics, ASME*, Vol. 42, No. 2, pp. 385–389.

23. Walker, K.P. (1981) Research and development program for nonlinear structural modeling with advanced time–temperature dependent constitutive relationships, NASA CR-165533.

24. Cassenti, B.N. and Thompson, R.L. (1983) Material response predictions for hot section gas turbine engine components, AIAA-83-2020, presented at the AIAA/SAE/ASME 19th Joint Propulsion Conference, Seattle, Washington, 27–29 June.

25. Lin, T.Y. (1967) Reciprocal theorem for displacements in inelastic bodies, *Journal of Composite Matter*, Vol. 1, pp. 144–151.

26. Lin, T.Y. (1969) *Theory of Inelastic Structures*, John Wiley, Chichester.

27. Reisner, H. (1931) Initial stresses and sources of initial stresses (in German), *ZAMM*, Vol. 11, pp. 1–8.

28. Zienkiewicz, O.C., Valliappan, S. and King, I.P. (1969) Elasto-plastic solution of engineering problems by initial stress, finite element approach, *International Journal of Numerical Methods in Engineering*, Vol. 1, pp. 75–100.

29. Zienkiewicz, O.C. (1977) *The Finite Element Method*, McGraw-Hill, London.

30. Kumar, V. and Mukherjee, S. (1977) A boundary-integral equation formulation for time-dependent inelastic deformation in metals, *International Journal of Mechanical Science*, Vol. 19, No. 12, pp. 713–724.

31. Mukherjee, S. and Kumar, V. (1978) Numerical analysis of time dependent inelastic deformation in metallic media using boundary integral equation method, *Journal of Applied Mechanics, ASME*, Vol. 45, No. 4, pp. 785–790.

32. Marjaria, M. and Mukherjee, S. (1979) Improved boundary integral equation method for time-dependent inelastic deformation in metals, *International Journal of Numerical Methods in Engineering*, Vol. 15, pp. 97–111.

33. Chaudonneret, M. (1977) Boundary integral equation method for visco-plasticity analysis (in French), *Journal de Mechanics Applique*, Vol. 1, No. 2, pp. 113–131.
34. Banerjee, P.K., Cathie, D.N. and Davies, T.G. (1979) Two and three-dimensional problems of elasto-plasticity, in *Development in Boundary Element Methods*, Eds P.K. Banerjee and R. Butterfield, Elsevier Applied Science Publishers, London.
35. Banerjee, P.K. and Davies, T.G. (1979) Analysis of some case histories of laterally loaded pile groups, in *Proceedings of International Conference on Numerical Methods in Offshore Piling*, Institution of Civil Engineers, London.
36. Banerjee, P.K. and Cathie, D.N. (1980) A direct formulation and numerical implementation of the boundary element method for two dimensional problems of elasto-plasticity, *International Journal of Mechanical Science*, Vol. 22, pp. 233–245.
37. Cathie, D.N. and Banerjee, P.K. (1980) Boundary element methods for axisymmetric plasticity, in *Innovative Numerical Methods for Applied Engineering Science*, Eds R.P. Shaw *et al.*, University of Virginia Press.
38. Banerjee, P.K. and Butterfield, R. (1981) *Boundary Element Methods in Engineering Science*, McGraw-Hill, London.
39. Raveendra, S.T. (1984) Advanced development of BEM for two and three dimensional nonlinear analysis, PhD Dissertation, State University of New York at Buffalo.
40. Banerjee, P.K., Wilson, R.B. and Miller, N. (1985) Development of a large BEM system for three-dimensional inelastic analysis, in *Advanced Topics in Boundary Element Analysis*, Eds T.A. Cruse, A.B. Pifko and H. Armen, AMD-Vol. 72, ASME, New York.
41. Banerjee, P.K. and Raveendra, S.T. (1986) Advanced boundary element of two- and three-dimensional problems of elastoplasticity, *International Journal of Numerical Methods in Engineering*, Vol. 23, pp. 985–1002.
42. Banerjee, P.K. and Raveendra, S.T. (1987) A new boundary element formulation for two-dimensional elastoplastic analysis, *Journal of Engineering Mechanics, ASCE*, Vol. 113, pp. 252–265.
43. Dargush, G.F. (1987) Boundary element methods for the analogous problems of thermomechanics and soil consolidation, PhD Dissertation, State University of New York at Buffalo.
44. Henry, D.P. (1987) Advanced development of the boundary element method for elastic and inelastic thermal stress analysis, PhD Dissertation, State University of New York at Buffalo.
45. Banerjee, P.K., Wilson, R.B. and Raveendra, S.T. (1987) Advanced applications of BEM to three-dimensional problems of monotonic and cyclic plasticity, *International Journal of Mechanical Science*, Vol. 29, No. 9, pp. 637–653.
46. Henry, D.P. and Banerjee, P.K. (1988) A thermoplastic BEM analysis of substructured axisymmetric bodies, *Journal of Engineering Mechanics, ASCE*, Vol. 113, No. 12, pp. 1880–1900.
47. Henry, D.P. and Banerjee, P.K. (1988) A variable stiffness type boundary element formulation for axisymmetric elastoplastic media, *International Journal of Numerical Methods in Engineering*, Vol. 26, pp. 1005–1027.
48. Henry, D.P. and Banerjee, P.K. (1988) A new BEM formulation for two- and three-dimensional elastoplasticity using particular integrals, *International Journal of Numerical Methods in Engineering*, Vol. 26, pp. 2079–2096.
49. Ahmad, S. (1986) Linear and nonlinear dynamic analysis by boundary element method, PhD Dissertation, State University of New York at Buffalo.
50. Ahmad, S. and Banerjee, P.K. (1988) Inelastic transient dynamic analysis of three dimensional problems by BEM, *International Journal of Numerical Methods in Engineering*, Vol. 29, pp. 371–390.
51. Banerjee, P.K., Henry, D.P. and Raveendra, S.T. (1989) Advanced inelastic analysis of solids by the boundary element method, *International Journal of Mechanical Science*, Vol. 31, No. 4, pp. 309–322.
52. Banerjee, P.K. and Henry, D.P. (1989) Advanced applications of BEM to inelastic analysis of solids, in *Developments in Boundary Element Methods*, Eds P.K. Banerjee and R.B. Wilson, Elsevier Applied Science Publishers, London.

53. Dargush, G.F. and Banerjee, P.K. (1991) Boundary element methods for three-dimensional thermoplasticity, *International Journal of Solids and Structures*, Vol. 28, No. 5, pp. 549–565.

54. Deb, A. (1991) Advanced development of the BEM for linear and nonlinear analyses of anisotropic solids, PhD Dissertation, State University of New York at Buffalo.

55. Chopra, M.B. (1992) Linear and nonlinear analyses of axisymmetric problems of thermomechanics and soil consolidation, PhD Dissertation, State University of New York at Buffalo.

56. Israil, A.S.M. (1991) Time-domain elastic and inelastic dynamic analysis of 2D solids by boundary element method, PhD Dissertation, State University of New York at Buffalo.

57. Israil, A.S.M. and Banerjee, P.K. (1992) Advanced development of BEM for two-dimensional dynamic elastoplasticity, *International Journal of Solids and Structures*, Vol. 29, No. 11, pp. 1433–1451.

58. Riccardella, P. (1973) An implementation of the boundary integral technique for plane problems of elasticity and elasto-plasticity, PhD Thesis, Carnegie Mellon University, Pittsburgh.

59. Mendelson, A. and Albers, L.U. (1975) Application of boundary integral equation method to elasto-plastic problems, in *Proceedings of ASME Conference on Boundary Integral Equation Methods*, Eds T.A. Cruse and F.J. Rizzo, Vol. AMD-II, ASME, New York.

60. Telles, J.C.F. (1983) The boundary element method applied to inelastic problems, in *Lecture Notes in Engineering*, Springer-Verlag, New York.

61. Chandra, A. and Mukherjee, S. (1984) A finite element analysis of metal-forming problems with an elastic-viscoplastic material model, *International Journal of Numerical Methods in Engineering*, Vol. 20, pp. 1613–1628.

62. Mukherjee, S (1982) *Boundary Elements in Creep Fracture*, Applied Science Publishers, London.

63. Comi, C. and Maier, G. (1992) Extremum convergence and stability properties of the finite increment problems in elastoplastic boundary element analysis, *International Journal of Solids and Structures*, Vol. 29, No. 2, pp. 249–270.

64. Maier, G. and Polizotto, C. (1987) A Galerkin approach to boundary element elastoplastic analysis, *Computer Methods in Applied Mechanics and Engineering*, Vol. 60, pp. 175–194.

65. Volker, R.E. (1909) Nonlinear flow in porous media by finite elements, *Proceedings of American Society C.E.*, Vol. 95(H76), pp. 2093–2114.

66. Ahmed, H. and Suneda, D.K. (1909) Nonlinear flow in porous media, *Proceedings of American Society C.E.*, Vol. 95(H76), pp. 1847–1859.

67. Winslow, A.M. (1907) Numerical solution of quasi-linear Poisson equation in non-uniform triangle mesh, *Journal of Computational Physics*, Vol. 1, pp. 149–172 .

68. Luu, T.S. and Coulmy, G. (1977) Method of calculating the compressible flow around an airfoil or a cascade up to the shockfree transonic range, *Computers and Fluids*, Vol. 5, pp. 201–275.

69. Karmaker, H.C. and Robertson, S.D.T. (1978) An integral equation formulation for electro-magnetic field analysis in electrical apparatus, presented at the IEEE PES Summer Meeting, Los Angeles, California, 16–21 July.

70. Banerjee, P.K. (1979) Nonlinear problems of potential flow, Chapter II in *Developments in Boundary Element Methods*, Vol. I, Eds P.K. Banerjee and R. Butterfield, Elsevier Applied Science Publishers, London.

71. Chandra, S. and Mukherjee, S. (1987) A boundary element analysis of metal extrusion processes, *ASME Journal of Applied Mechanics*, Vol. 54, pp. 336–340.

72. Chandra, S. and Saigal, S. (1991) A boundary element analysis of the axisymmetric extrusion process, *International Journal of Nonlinear Mechanics*, Vol. 26, pp. 1–13.

73. Okada, H., Rajiyah, H. and Atluri, S.N. (1990) A full tangent stiffness field–boundary–element formulation for geometric and material nonlinear problems of solid mechanics, *International Journal of Numerical Methods in Engineering*, Vol. 29, pp. 15–35.

74. Theocaris, P.S. and Marketos, E. (1964) Elasto-plastic analysis of perforated thin strips of a strain hardening material, *Journal of Mechanical and Physical Solids*, Vol. 12, pp. 377–390.

FOURTEEN

COMBINATION OF BOUNDARY ELEMENT METHODS WITH OTHER NUMERICAL METHODS

14.1 INTRODUCTION

So far in the book we have, we believe, demonstrated that BEMs are very efficient numerical tools for solving problems involving bulky two- and three-dimensional regions. On the other hand, finite element or finite difference methods are attractive in both finite regions and in regions where there are high degrees of either inhomogeneity or geometrical and material non-linearity. Therefore, for some problems it may be advantageous to use a combination of more than one method in what is often called a 'hybrid solution'.

In fact a very large group of problems in mechanics falls into the category where the use of either domain discretization (finite element or finite difference) or boundary discretization alone is inefficient. Problems such as:

- the interaction between a finite size structure and an extensive solid supporting material (e.g. soil–structure interaction problems) and
- problems in non-linear solid mechanics involving thin plates or shells embedded within fluid medium

have attracted the attention of many workers who have developed algorithms combining the domain discretization and integral methods of analysis.

The system matrices that arise in finite element methods are usually symmetric whereas the BEM algorithms described in earlier chapters lead to non-symmetric system matrices. Therefore if a small BEM system needs to be incorporated into a large finite element system, it is necessary for an efficient solution to modify the BEM formulation so that the matrices it generates are symmetric. The contrary case leads to no loss of efficiency since any large non-symmetric BEM system can accommodate a symmetric finite element one. In either case the two systems can be assembled together if care is taken to satisfy the interface compatibility conditions.

The BEM system equations can be made symmetric by using an energy minimization scheme to establish them rather than the collocation or point-matching scheme used in earlier chapters. One by-product of such an operation is that it enables a better estimate of errors to be made. However, for relatively small and medium size problems such solutions are rather more expensive than those using point matching and therefore as far as possible attempts should be made to achieve a non-symmetrical coupling in terms of the BEM boundary variables. For very large problems, of course, these symmetric formulations have distinct advantages.

14.2 BOUNDARY ELEMENT METHOD SOLUTIONS DERIVED BY AN ENERGY METHOD

14.2.1 Introduction

There is a considerable quantity of published literature[1-12] in which boundary integral statements are developed from energy considerations; e.g. the method of moments, the Galerkin methods and the Rayleigh–Ritz method. Although the use of any of these methods can lead to symmetric systems of matrices, the weighted residual approach[13-16] outlined below is probably preferable from an engineer's point of view since the physical process involved can be more readily understood. A reader should note that the weighted residual method outlined here is not the same as integration by parts used in the derivation of reciprocal identities in earlier chapters.

14.2.2 The General Theory of Weighted Residuals

To illustrate the weighted residual procedure we shall consider the determination of a function (u), which may be either scalar or a vector quantity within a region V bounded by S, defined by the general equation

$$L(u) = 0 \quad \text{in } V \tag{14.1}$$

subject to the boundary conditions

$$A(u) - f = 0 \quad \text{on } S_1$$

$$B(u) - g = 0 \quad \text{on } S_2 \tag{14.2}$$

where f and g are prescribed on $S = S_1 + S_2$. The operators L, A and B may be either differential or integral operators and are also either linear or non-linear in nature.

If u^0 is some approximation to u then Eqs (14.1) and (14.2) will not be satisfied exactly. Let us assume that errors involved are

$$L(u^0) = E_1$$

$$A(u^0) - f = E_2 \tag{14.3}$$

and $$B(u^0) - g = E_3$$

where E_1, E_2 and E_3 are residual error functions. To determine the approximate solution of u^0 some weighted integral of the errors, defined in Eqs (14.3), is set to zero, so that[13-17]

$$\int_V W^k L(u^0) \, dV + \int_{S_1} \overline{W}^k \lfloor A(u^0) - f \rfloor \, dS + \int_{S_2} \overline{\overline{W}}^k \lfloor (u^0) - g \rfloor \, dS = 0 \tag{14.4}$$

where W^k, $\overline{W}^k$ and $\overline{\overline{W}}^k$ are a set ($k = 1, 2, 3, \ldots, m$) of independent weighting functions.

Equation (14.4) is then the weighted residual statement for our problem. For a detailed account of this procedure see Zienkiewicz.[17]

14.2.3 The Indirect Boundary Element Method—A Special Class of Weighted Residual Method

Consider the potential flow problem of Chapter 3:

$$\frac{\partial^2 p}{\partial x_i \, \partial x_i} = 0 \quad \text{in } V \tag{14.5}$$

and $$p = f \quad \text{on } S_1 \quad \text{and} \quad \frac{\partial p}{\partial x_i} n_i = \frac{\partial p}{\partial n} = g \quad \text{on } S_2$$

The general weighted residual statement for Eq. (14.5) can be written as

$$\int_V W^k \left(\frac{\partial^2 p}{\partial x_i \, \partial x_i} \right) dV + \int_{S_1} \overline{W}^k (p - f) \, dS + \int_{S_2} \overline{\overline{W}}^k \left(\frac{\partial p}{\partial n} - g \right) dS = 0 \tag{14.6}$$

The volume integral can be transformed by noting that

$$W^k \frac{\partial^2 p}{\partial x_i \, \partial x_i} = \frac{\partial}{\partial x_i} \left(W^k \frac{\partial p}{\partial x_i} \right) - \frac{\partial W^k}{\partial x_i} \frac{\partial p}{\partial x_i} \tag{14.7}$$

Using the divergence theorem we can express Eq. (14.6) as[13-16]

$$-\int_V \frac{\partial W^k}{\partial x_i} \frac{\partial p}{\partial x_i} \, dV + \int_S W^k \frac{\partial p}{\partial n} \, dS + \int_{S_1} \overline{W}^k (p - f) \, dS + \int_{S_2} \overline{\overline{W}}^k \left(\frac{\partial p}{\partial n} - g \right) dS = 0 \tag{14.8}$$

in which we can once again use Eq. (14.7) with W^k and p interchanged and apply the divergence theorem to obtain

$$\int_V \frac{\partial^2 W^k}{\partial x_i \partial x_i} p \, dV - \int_S p \frac{\partial W^k}{\partial n} \, dS + \int_S W^k \frac{\partial p}{\partial n} \, dS$$

$$+ \int_{S_1} \overline{W}^k (p - f) \, dS + \int_{S_2} \overline{\overline{W}}^k \left(\frac{\partial p}{\partial n} - g \right) dS = 0 \qquad (14.9)$$

It should be noted that all integrations in these equations are carried out with respect to x_i and all variables are functions of x_i only.

We can describe the distribution of $p(x)$ in terms of a surface source distribution $\phi(\xi)$ via (see Chapter 3)

$$p = \int_S G\phi \, dS = N^r \phi^r \qquad (14.10)$$

where ϕ^r are the nodal values ϕ on the surface and the integration variable here is ξ_i. Similarly, $\partial p/\partial n$ is obtainable from, say,

$$\frac{\partial p}{\partial n} = \int_S F\phi \, dS = \frac{\partial N^r}{\partial n} \phi^r = M^r \phi^r \qquad (14.11)$$

If we choose the weighting functions W^k, $\overline{W}^k$ and $\overline{\overline{W}}^k$ such that[13-16]

$$\begin{aligned} W^k &= \tfrac{1}{2} N^k & x_i \text{ on } V \text{ and } S \\ \overline{W}^k &= -M^k & x_i \text{ on } S_1 \\ \overline{\overline{W}}^k &= N^k & x_i \text{ on } S_2 \end{aligned} \qquad (14.12)$$

we note immediately that the volume integral in Eq. (14.9) vanishes since W^k (that is N^k) satisfies the governing differential equation. Thus, substituting Eqs (14.10) to (14.12) in Eq. (14.9) we obtain[13-16]

$$K^{kr} \phi^r + F^k = 0 \qquad (14.13)$$

where

$$K^{kr} = \tfrac{1}{2} \left[\int_{S_2} (M^k N^r + N^k M^r) \, dS - \int_{S_1} (M^k N^r + N^k M^r) \, dS \right]$$

$$F^k = \int_{S_1} M^k f \, dS - \int_{S_2} N^k g \, dS$$

It is important to appreciate that there is a double integration process involved in generating the system of equations (14.13); the first one (the inner integral) is with respect to ξ [see Eqs (14.10) and (14.11)] and the second one (the outer integral) is with respect to x_i. The system matrix K in Eq. (14.13) is symmetric and can be combined with any symmetric finite element system matrix. This specific way of deriving a symmetric indirect formulation is due to Mustoe, Kelly and colleagues.[13-16]

Although such a symmetric IBEM algorithm appears to be a little more expensive computationally than the standard non-symmetric form it does lead to more accurate solutions, particularly near geometrical discontinuities.[15]

14.2.4 Symmetric Direct Boundary Element Formulation for Elasticity

DBEM formulations clearly do not follow the philosophy of trial function procedures described in the earlier section and therefore weighted residual procedures cannot be used to find a symmetric set of equations in this case.[16] However, for an elastic system (see Chapter 4), we can consider the total energy functional as

$$\Pi = \tfrac{1}{2} \int_V \sigma_{ij} \varepsilon_{ij} \, dV - \int_V u_i \psi_i \, dV - \int_{S_2} u_i g_i \, dS \qquad (14.14)$$

The volume integral now can be recast in terms of displacements by using the strain–displacement equations as

$$\tfrac{1}{2} \int_V \sigma_{ij} \varepsilon_{ij} \, dV = \tfrac{1}{4} \int_V (\sigma_{ij} u_{i,j} + \sigma_{ij} u_{j,i}) \, dV$$

$$= \tfrac{1}{4} \int_V [(\sigma_{ij} u_i)_{,j} + (\sigma_{ij} u_j)_{,i} - \sigma_{ij,j} u_i - \sigma_{ij,i} u_j] \, dV \qquad (14.15)$$

By applying the divergence theorem and the stress equilibrium equation to Eq. (14.15) we obtain

$$\tfrac{1}{2} \int_V \sigma_{ij} \varepsilon_{ij} \, dV = \frac{1}{2} \int_S u_i t_i \, dS - \frac{1}{2} \int_V u_i \psi_i \, dV \qquad (14.16)$$

Substituting Eq. (14.16) in Eq. (14.14) and assuming $\psi_i = 0$, for convenience, we get the total energy functional as[15]

$$\Pi = \tfrac{1}{2} \int_S u_i t_i \, dS - \int_S u_i g_i \, dS \qquad (14.17)$$

or, in matrix notation,

$$\Pi = \tfrac{1}{2} \int_S \mathbf{u}^T \mathbf{t} \, dS - \int_S \mathbf{u}^T \mathbf{g} \, dS \qquad (14.18)$$

By assuming a suitable variation of $\mathbf{u}$ and $\mathbf{t}$ in the form

$$\mathbf{u} = \mathbf{N} \mathbf{u}^n$$
$$\mathbf{t} = \mathbf{M} \mathbf{t}^n \qquad (14.19)$$

where $\mathbf{N}$ and $\mathbf{M}$ are shape functions, we can write Eq. (14.18) as

$$\Pi = \tfrac{1}{2} (\mathbf{u}^n)^T \left(\int_S \mathbf{N}^T \mathbf{M} \, dS \right) \mathbf{t}^n - (\mathbf{u}^n)^T \int_S \mathbf{N}^T \mathbf{g} \, dS \qquad (14.20)$$

It should be noted that the shape functions $\mathbf{N}$ and $\mathbf{M}$ need not, in general, be the same.

Now consider the direct boundary integral equation of Chapter 4:

$$c_{ij} u_i(\xi) = \int_S [t_i(x) G_{ij}(x, \xi) - F_{ij}(x, \xi) u_i(x)] \, dS \qquad (14.21)$$

which can be written, following the usual BEM procedure, as

$$\mathbf{A}\mathbf{u}^n = \mathbf{B}\mathbf{t}^n \qquad (14.22)$$

where the variations of $\mathbf{u}$ and $\mathbf{t}$ over the boundary elements are assumed to be identical to those in Eq. (14.19). If the surface has edges and corners it is necessary to eliminate the extraneous tractions at these nodes by the extrapolated values of traction components at other nodes of the elements meeting at these edges and corners to ensure that $\mathbf{B}$ becomes a square matrix. If we now rewrite Eq. (14.22) as

$$\mathbf{t}^n = \mathbf{B}^{-1}\mathbf{A}\mathbf{u}^n \qquad (14.23)$$

this equation can be used to eliminate $\mathbf{t}^n$ from Eq. (14.20). However, if this relationship is used the resulting formulation may not, due to numerical errors, satisfy equilibrium exactly.[15] This problem may be overcome by introducing an auxiliary equilibrium equation

$$\int_S \mathbf{t} \, dS = 0 \qquad (14.24)$$

the discretized form of which is

$$\left(\int_S \mathbf{M} \, dS \right) \mathbf{t}^n = \mathbf{Q}\mathbf{t}^n = 0 \qquad (14.25)$$

The nodal tractions $\mathbf{t}^n$ are found by linking together a modified form of Eq. (14.22), i.e.

$$\mathbf{A}\mathbf{u}^n = \mathbf{B}\mathbf{t}^n + \mathbf{Q}^T\lambda \qquad (14.26)$$

where (for two-dimensional problems) is a two-component vector of pseudo-Lagrange multipliers. Here the role of such multipliers is to introduce a controlled perturbation in Eq. (14.22) so that Eq. (14.25) can be satisfied exactly.

Equations (14.25) and (14.26) can now be written as

$$\begin{bmatrix} \mathbf{A} \\ \mathbf{0} \end{bmatrix} \{\mathbf{u}^n\} = \begin{bmatrix} \mathbf{B} & \mathbf{Q}^T \\ \mathbf{Q} & \mathbf{0} \end{bmatrix} \begin{Bmatrix} \mathbf{t}^n \\ \lambda \end{Bmatrix} \qquad (14.27)$$

from which the nodal tractions $\mathbf{t}^n$ can be determined by simple matrix algebra:

$$\mathbf{t}^n = \mathbf{E}\mathbf{u}^n \qquad (14.28)$$

If we substitute Eq. (14.28) in Eq. (14.20) the functional expression becomes[13–16]

$$\Pi = (\mathbf{u}^n)^T \mathbf{K}\mathbf{u}^n + (\mathbf{u}^n)^T \mathbf{F} \qquad (14.29)$$

where

$$\mathbf{K} = \tfrac{1}{2}\left(\int_S \mathbf{N}^T \mathbf{M} \, dS \right)\mathbf{E} \quad \text{and} \quad \mathbf{F} = -\int_{S_2} \mathbf{N}^T \mathbf{g} \, dS$$

The functional Π may now be minimized to obtain

$$\delta\Pi = \left(\frac{\partial\Pi}{\partial\mathbf{u}^n}\right)\delta\mathbf{u}^n = 0$$

and since this is valid for arbitrary variations $\delta\mathbf{u}^n$ we must have

$$\frac{\partial\Pi}{\partial\mathbf{u}^n} = 0 \tag{14.30}$$

which leads to the final system of equations

$$\mathbf{K}^0\mathbf{u}^n + \mathbf{F} = 0 \tag{14.31}$$

where

$$\mathbf{K}^0 = \tfrac{1}{2}\left\{\left(\int_S \mathbf{N}^T\mathbf{M}\,dS\right)\mathbf{E} + \left[\left(\int_S \mathbf{N}^T\mathbf{M}\,dS\right)\mathbf{E}\right]\right\}$$

Equation (14.31) is now a symmetric form of DBEM statement.[13-16]

It is interesting to note that an alternative symmetric DBEM formulation can be obtained by using a very simple method. The nodal forces $\mathbf{F}$ corresponding to a surface traction $\mathbf{t}$ on the boundary are given by

$$\mathbf{F} = \int_S \mathbf{N}^T\mathbf{t}\,dS \tag{14.32}$$

where $\mathbf{N}$ is the shape function matrix for displacements. Substituting the boundary shape functions (14.19) and the relationship between the nodal tractions and nodal displacements of Eq. (14.28) into Eq. (14.32) produces the force displacement relationship

$$\mathbf{F} = \left(\int_S \mathbf{N}^T\mathbf{M}\,dS\right)\mathbf{E}\mathbf{u}^n = \mathbf{K}'\mathbf{u}^n \tag{14.33}$$

In general the matrix $\mathbf{K}'$ is not symmetric and therefore the Maxwell–Betti reciprocal theorem will not be satisfied. In order to make the equations symmetric $\mathbf{K}'$ can be replaced by

$$\mathbf{K}^0 = \tfrac{1}{2}(\mathbf{K}' + \mathbf{K}'^T) \tag{14.34}$$

which leads to a formulation superficially identical to Eq. (14.31). However, Eq. (14.34) is an entirely intuitive modification of Eq. (14.33), its only justification being that it produces the same form of equation as that deduced from the energy method.[15] The main disadvantages of the above symmetric DBEM formulation are the elaborate housekeeping efforts at edges and corners that are required to make $\mathbf{B}$ matrix square and the cost of the inversion implied in Eq. (14.23). However, if the edges and corners are treated correctly the above method yields accurate solutions.

14.2.5 An Alternative Approach to Symmetric Boundary Element Method Algorithms

In Sec. 3.7 (Chapter 3) we derived an indirect formulation of the BEM for potential flow at any point, interior or exterior to the boundary S:

$$p^*(x) = \int_{S(\xi)} [G(x,\xi)\phi(\xi) - F(x,\xi)\mu(\xi)]\,dS(\xi) \tag{14.35}$$

where $\phi(\xi) = -[q(\xi) + \bar{q}(\xi)]$ and $\mu(\xi) = [p(\xi) - \bar{p}(\xi)]$ are two fictitious density functions. By differentiating the above integral representation we can determine the flux $q^*(x)$ at any point from (see Chapter 3)

$$q^*(x) = \int_{S(\xi)} [-F(\xi,x)\phi(\xi) + H(\xi,x)\mu(\xi)]\,dS(\xi) \tag{14.36}$$

Now consider the reciprocal relations between two independent states $[p(x), q(x)]$ and $[p^*(x), q^*(x)]$ existing in the same body. If sources inside the body are assumed to be zero for both states, we have

$$\int_{S(x)} [p^*(x)q(x) - q^*(x)p(x)]\,dS(x) = 0 \tag{14.37}$$

Substituting $p^*(x)$ and $q^*(x)$ from Eqs (14.35) and (14.36) into Eq. (14.37) we obtain

$$\int_{S(x)} \int_{S(\xi)} \{[G(x,\xi)\phi(\xi) - F(x,\xi)\mu(\xi)]q(x)$$

$$-[F(\xi,x)\phi(\xi) - H(\xi,x)\mu(\xi)]p(x)\}\,dS(x)\,dS(\xi) = 0 \tag{14.38}$$

By introducing appropriate variations of $\phi(\xi)$, $\mu(\xi)$, $p(x)$ and $q(x)$ using shape functions $\mathbf{N}(x)$ and $\mathbf{N}(\xi)$ over each boundary element ΔS we can discretize the above equation to the form

$$\left\{ \begin{matrix} \phi \\ \mu \end{matrix} \right\}^T \left[\begin{matrix} \Delta\mathbf{G} & -\Delta\mathbf{F} \\ -\Delta\mathbf{F}^T & -\Delta\mathbf{H} \end{matrix} \right] \left\{ \begin{matrix} \mathbf{q} \\ \mathbf{p} \end{matrix} \right\} = \mathbf{0} \tag{14.39}$$

where T stands for transpose,

$$\Delta\mathbf{G} = \int_{\Delta S(x)} \mathbf{N}(x) \int_{\Delta S(\xi)} G(x,\xi)\mathbf{N}^T(\xi)\,dS(\xi)\,dS(x)$$

while $\Delta\mathbf{F}$ and $\Delta\mathbf{H}$ are obtained from similar integrals over $\Delta S(x)$ and $\Delta S(\xi)$.

Since Eq. (14.39) should be valid for any arbitrary choices of the fictitious functions $\phi(\xi)$ and $\mu(\xi)$ we must have

$$\left[\begin{matrix} \Delta\mathbf{G} & -\Delta\mathbf{F} \\ -\Delta\mathbf{F}^T & -\Delta\mathbf{H} \end{matrix} \right] \left\{ \begin{matrix} \mathbf{q} \\ \mathbf{p} \end{matrix} \right\} = 0 \tag{14.40}$$

The above equation is clearly symmetric and therefore for very large problems the cost of the solution of equations reduces considerably. It is also of interest to note that while single integration over $\Delta S(x)$ or $\Delta S(\xi)$ of functions $F(x,\xi)$ and $H(x,\xi)$ are difficult, the double integration process used in the formation of the system matrices (14.40) makes them eminently suitable for numerical integration. Moreover, for the singular cases of the most difficult integrals involving the function $H(x, \xi)$, we can invoke the standard equipotential solution at boundary nodes (that is $\mathbf{q} = 0$, $\mathbf{p} = 1$) and obtain the value of the diagonal term of $\mathbf{\Delta H}$ indirectly.

The above formulation also allows for multiple unknown values of fluxes $\mathbf{q}$ at the edges and corners to be retained in the system equations. One may recall that in the collocation method used throughout the earlier chapters these somehow must be eliminated by extrapolation, by adding additional relations, etc. Thus, for a typical potential flow problem with n boundary nodes and m edges and corners, $\mathbf{\Delta G}$, $\mathbf{\Delta F}$, $\mathbf{\Delta F^T}$ and $\mathbf{\Delta H}$ will be $(m + n) \times (m + n)$, $(m + n) \times n$, $n \times (m + n)$ and $n \times n$ matrices respectively. It is possible, then, to select the required number of equations from Eq. (14.40) to form the final system of equations $\mathbf{Ax} = \mathbf{b}$, where $\mathbf{A}$ will be a symmetric matrix.

In recent years Maier and his colleagues[9-12] have discussed BEM solution procedures similar to the one discussed above for problems of elasticity and elasto-plasticity.

14.2.6 An Alternative Energy Approach to Symmetric IBEM

Consider the integral equation

$$K\phi = g \tag{14.41}$$

where K is an integral operator, ϕ an unknown function and g are the known boundary values. For self-adjoint K we can write a functional Π as

$$\Pi = \langle K\phi, \phi \rangle - 2\langle \phi, g \rangle \tag{14.42}$$

where $\langle \ \rangle$ denotes an inner product defined by

$$\langle u, v \rangle = \int_S uv \, \mathrm{d}S \tag{14.43}$$

If we adopt a distribution of ϕ such that

$$\phi = \mathbf{N^T} \boldsymbol{\phi}_n \tag{14.44}$$

where $\boldsymbol{\phi}_n$ are the nodal values of ϕ, we can rewrite Eq. (14.42) as

$$\Pi = \boldsymbol{\phi}_n^T \langle \mathbf{KN}, \mathbf{N^T} \rangle \boldsymbol{\phi}_n - 2\boldsymbol{\phi}_n^T \langle \mathbf{N}, g \rangle$$

By taking the variation of the functional with respect to $\boldsymbol{\phi}_n$ and equating $\partial \Pi / \partial \boldsymbol{\phi}_n$ to zero we get

$$\langle \mathbf{KN}, \mathbf{N^T} \rangle \boldsymbol{\phi}_n = \langle \mathbf{N}, g \rangle \tag{14.45}$$

Note that the integral form, say, of Eq. (10.45) written out explicitly is

$$\left[\int_{S(x)} \mathbf{N}(x) \int_{S(\xi)} K(x, \xi) \mathbf{N^T}(\xi) \, \mathrm{d}S(\xi) \, \mathrm{d}S(x) \right] \boldsymbol{\phi}_n = \int_{S(x)} \mathbf{N}(x) g(x) \, \mathrm{d}S(x) \tag{14.46}$$

which is a Galerkin formulation of the original boundary integral equation leading to a symmetric system matrix $\mathbf{A}$, a formulation that has been developed and used

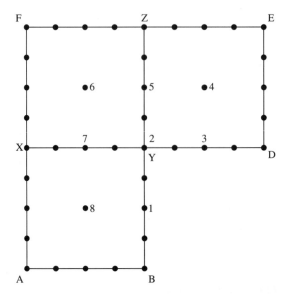

Figure 14.1 Discretization for multi-region symmetric BEM procedures.

by a number of workers, principally in electrical engineering.[1-8,18,19] A somewhat different approach for obtaining BEM system matrices by using the least square error on the boundary was described by Tottenham.[20]

14.3 EXAMPLES OF SOLVED PROBLEMS USING AN ENERGY APPROACH

14.3.1 Potential Flow in an L-Shaped Domain[16]

Figure 14.1 shows an L-shaped domain divided into three regions and analysed by the symmetric DBEM and IBEM procedures outlined in Secs 14.2.3 and 14.2.4 using quadratic elements. The problem is defined by the equation $\nabla^2 p = 0$ and boundary conditions along ABYD of $\partial p/\partial n = 0$, on DE of $p = 0$, along EZF of $\partial p/\partial n = 0$ and along AF of the potential $p = 1$. Table 14.1 shows that both of the symmetric BEM formulations gave results that were in very good agreement with an accurate solution obtained by Jaswan and Symm,[21] whereas a non-symmetric IBEM analysis would lead to much greater error near the geometric discontinuities. Table 14.2 shows the nodal values of the interface $\partial p/\partial n$ values along section XY and YZ. It can now be seen that conservation at the interfaces is being achieved in the nodal force sense rather than by point matching. It is also worth noting that the symmetric IBEM solution produces $\partial p/\partial n$ values for the two regions which are exactly equal and opposite along the interfaces, whereas the potential values in them are not exactly equal (Table 14.3).

Table 14.1 A comparison of symmetric IBEM and DBEM results for values of p

Node	Accurate solution	IBEM symmetric three-region	DBEM symmetric three-region
1	0.8640	0.8644	0.8587
2	0.6667	0.6733	0.6667
3	0.2972	0.2997	0.3023
4	0.2881	0.2883	0.2875
5	0.5680	0.5692	0.5706
6	0.8081	0.8039	0.8040
7	0.8514	0.8478	0.8488
8	0.9109	0.9063	0.9085
Maximum error (%)		0.85	1.6

Table 14.2 Interface $\partial p/\partial n$ values on the L-shaped domain by the symmetric BEM multi-region discretizations

	Symmetric DBEM		Symmetric IBEM	
Node	$\partial p/\partial n_1$	$\partial p/\partial n_2$	$\partial p/\partial n_1$	$\partial p/\partial n_2$
X	−0.0030	0.0043	−0.0052	0.0052
	−0.0111	0.0119		
	−0.0237	0.0170	−0.0180	0.0180
	−0.0280	0.0362		
	−0.0452	0.0555	−0.0296	0.0296
Y	−0.0990	0.0692	−0.1111	0.1111
	−0.1095	0.1480		
	−0.1198	0.1075	0.1616	0.1616
	−0.1115	0.1012		
	−0.0994	0.1077	−0.0984	0.0984
	−0.1045	0.1022	−0.0978	0.0978
Z	−0.1030	0.1015	−0.1088	0.1088

Table 14.3 Interface potential values given by the symmetric IBEM multi-region discretization

	Symmetric BEM	
Node	p_1	p_2
X	0.9752	0.9700
	0.8888	0.8850
	0.8104	0.8073
Y	0.7439	0.7456
	0.6247	0.6234
	0.5792	0.5795
	0.5588	0.5590
Z	0.5617	0.5562

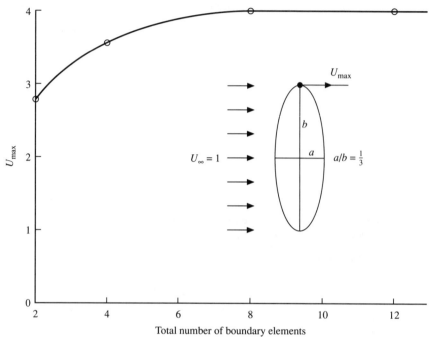

Figure 14.2 Potential flow past an elliptical obstruction.

14.3.2 A Cylinder of Elliptic Cross-Section in a Uniform Flow[19]

Figure 14.2 shows the convergence of a numerical solution of the problem using the Galerkin IBEM formulation outlined in Sec. 14.2.5 and a quadratic representation for the functions and geometry all due to Fried.[19] Hess and Smith (see Chapter 3) solved this problem very precisely using 180 nodes in a point collocation scheme. Fried's solution actually converged to three significant figures using only 16 nodes.

14.4 COMBINATION OF FEMs AND BEMs

14.4.1 Finite Element Formulations via Weighted Residuals

For all elasticity problems the weighted residual statement (14.4) becomes

$$\int_V W_i^k(\sigma_{ij,j+\psi_i})\,\mathrm{d}V + \int_{S_1} \overline{W}_i^k(u_i - f_i)\,\mathrm{d}S + \int_{S_2} \overline{\overline{W}}_i^k(t_i - g_i)\,\mathrm{d}S = 0 \qquad (14.47)$$

We can apply the divergence theorem again to the volume integral to obtain

$$\int_V W_i^k \sigma_{ij,j}\,\mathrm{d}V = -\int_V W_{i,j}^k \sigma_{ij}\,\mathrm{d}V + \int_{S_1} W_i^k t_i\,\mathrm{d}S \qquad (14.48)$$

Using Eq. (14.48) in Eq. (14.47) and assuming $\psi_i = 0$ for convenience, we obtain the 'weak formulation'

$$-\int_V W_{i,j}^k \sigma_{ij}\,\mathrm{d}V + \int_S W_i^k t_i\,\mathrm{d}S + \int_{S_1} \overline{W}_i^k(u_i - f_i)\,\mathrm{d}S + \int_{S_2} \overline{\overline{W}}_i^k(t_i - g_i)\,\mathrm{d}S = 0$$

$$(14.49)$$

If we utilize the symmetry of the tensor (σ_{ij}) and use the stress–strain and strain–displacement relations, we can express the volume integral in Eq. (14.49) as

$$\int_V W_{i,j}^k \sigma_{ij}\,\mathrm{d}V = \tfrac{1}{2}\int_V (W_{i,j}^k + W_{j,i}^k)\sigma_{ij}\,\mathrm{d}V$$

$$= \int_V \mu(W_{i,j}^k + W_{j,i}^k)\big[\alpha\delta_{ij}, u_{r,r} + \tfrac{1}{2}(u_{i,j} + u_{j,i})\big]\,\mathrm{d}V \quad (14.50)$$

where μ is the shear modulus and $\alpha = \nu/(1 - 2\nu)$.

By interpolating displacements in terms of the nodal values u_n^k for the kth node,

$$u_i = N_i^k u_n^k \tag{14.51}$$

we can rewrite Eq. (14.50) as[15]

$$\tfrac{1}{2}\int_V (W_{i,j}^k + W_{j,i}^k)\sigma_{ij}\,\mathrm{d}V = \left\{\int_V \mu(W_{i,j}^k + W_{j,i}^k)\big[\alpha\delta_{ij}N_{r,r}^k + \tfrac{1}{2}(N_{i,j}^k + N_{j,i}^k)\big]\,\mathrm{d}V\right\}u_n^k$$

$$(14.52)$$

It can be seen immediately that if W_i^k is selected so that $W_i^k = N_i^k$ Eq. (14.52) is symmetric with respect to the indices i and j. Therefore, if the weighting functions are chosen such that[15,17]

$$W_i^k = \begin{cases} -N_i^k & \text{in } V \\[2mm] 0 & \text{on } S_1 \end{cases}$$

$$\overline{W}_i^k = 0 \quad \text{on } S_1$$

$$\overline{\overline{W}}_i^k = N_i^k \quad \text{on } S_2$$

$$(14.53)$$

we can use Eq. (14.52) via Eq. (14.50) in Eq. (14.49) to obtain the symmetric set of linear equations

$$\left\{\int_V \mu(N_{i,j}^k + N_{j,i}^k)\big[\alpha\delta_{ij}N_{r,r}^k + \tfrac{1}{2}(N_{i,j}^k + N_{j,i}^k)\big]\,\mathrm{d}V\right\}u_n^k - \int_{S_2} N_i^k g_i\,\mathrm{d}S = 0 \quad (14.54a)$$

or simply

$$\mathbf{K}u_n + \mathbf{F} = 0 \tag{14.54b}$$

which is the displacement finite element formulation.[17]

14.4.2 Symmetric Coupling of Direct Boundary Element and Finite Element Methods

If the finite element region is designated as V_A and the BEM region V_B the system equation for the two regions following from the above can be written (in elasticity, for example) as[13-16,22-26]

$$\mathbf{K}_A \mathbf{u}_A + \mathbf{F}_A = 0 \tag{14.55a}$$

$$\mathbf{K}_B \mathbf{u}_B + \mathbf{F}_B = 0 \tag{14.55b}$$

where $\mathbf{u}_A$ and $\mathbf{F}_A$ are vectors of nodal displacements and forces within the domain V_A including the boundary nodes and $\mathbf{u}_B$ and $\mathbf{F}_B$ are the vectors of nodal displacements and forces on the boundary of the region V_B. Clearly, Eq. (14.55b) is the same form as any finite element system contribution from a new element. Therefore Eqs (14.55a,b) can be coupled by satisfying interface-displacement compatibility in the usual manner. Interface compatibility can only be ensured if the boundary shape function for displacements is identical to the variation of displacements in the adjoining finite elements. Equilibrium is satisfied in a nodal sense— simply that the sum of the nodal forces at every node has to be equal to the resultant external force at that node. In recent years Belytschko and co-workers[25,26] have considered other forms of variationally coupled FEMs and BEMs.

14.4.3 Symmetric Coupling of Indirect Boundary Element and Finite Element Methods

To apply the symmetric IBEM to the region V_B and the FEM to the region V_A, it is necessary to regard the common interface boundary as a part of the boundary S_1 where the potential p (in the potential flow problem, for example) is specified. The equivalent system equations for the region V_B given by Eq. (14.13) can be modified to

$$\begin{bmatrix} \mathbf{K} & \overline{\mathbf{K}} \\ \overline{\mathbf{K}}^T & 0 \end{bmatrix} \begin{Bmatrix} \phi \\ f \end{Bmatrix} = \begin{Bmatrix} \mathbf{F} \\ 0 \end{Bmatrix} \tag{14.56}$$

where $\mathbf{K}$, $\overline{\mathbf{K}}^T$ are defined by Eq. (14.13) in an obvious way. The second matrix equation in (14.56) simply states that the integral of ϕ over the boundary of V_B is zero (i.e. the auxiliary uniqueness condition discussed in Chapter 3).

Equation (14.56) can now be incorporated into a finite element system for V_A with nodeless variables ϕ and nodal variables f. Further details of this procedure are given by Kelly, Mustoe and Zienkiewicz.[13-16]

14.4.4 Examples

The machine component problem[16]   Figure 14.3 shows the geometry of a machine component that has been analysed by Kelly, Mustoe and Zienkiewicz[13-16] using (1) a finite element analysis with 95 eight-noded isoparametric elements, generating a total of 366 nodes of which 140 were boundary nodes, (2) a coupled DBEM

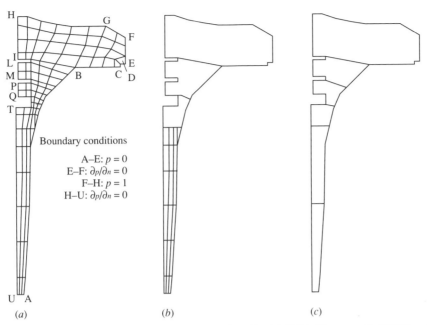

Boundary conditions

A–E: $p = 0$
E–F: $\partial p/\partial n = 0$
F–H: $p = 1$
H–U: $\partial p/\partial n = 0$

(a)   (b)   (c)

Figure 14.3 The machine component problem analysed by (a) FEM, (b) symmetric DBEM and FEM, (c) symmetric DBEM.

and FEM and (3) a symmetric DBEM. In the DBEM analysis the boundary was modelled using quadratic variation of the boundary potentials and the region subdivided into six sub-regions to facilitate the construction of a banded system matrix.

Figure 14.4 shows the distribution of the potential along the boundaries given by the three analyses. The finite element solutions were obtained in about one-third of the computing time required for the symmetric DBEM analysis. This is principally due to two reasons:

1. The symmetric analysis is about twice as expensive as the non-symmetric analysis for small problems.
2. For regions of this shape domain discretization schemes are shown to advantage because the bandwidth is small.

Waves incident upon a cylinder with a porous protecting wall[16]   The discretization for this problem is shown in Fig. 14.5 where the region between the cylinder and a finite region outside the porous protective wall was modelled by the FEM, beyond which symmetric DBEM modelling was used. The porous wall itself was simulated by a six-noded finite element to reproduce the flow condition through the wall; thus $\partial p/\partial n = K(p_2 - p_1)$ where p_1 and p_2 are the velocity potentials interior and exterior to the porous surface respectively. The wave forces on the cylinder and the porous wall are shown in Fig. 14.6 on page 387 for different values of the permeability ratio $(\beta/\beta_{\text{critical}})$ of the porous wall.

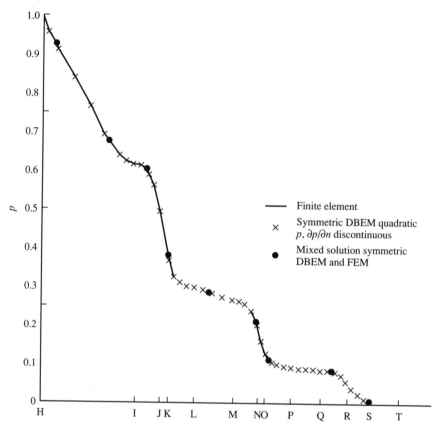

Figure 14.4 Machine component potential solution given by the symmetric DBEM (quadratic) procedure and the finite element method, along the boundary H to T.

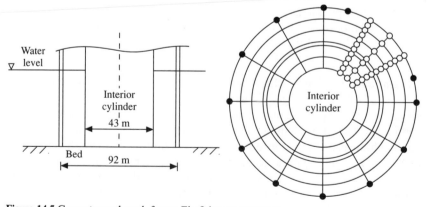

Figure 14.5 Geometry and mesh for an Ekofisk-type structure.

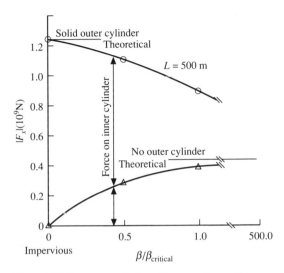

Figure 14.6 Forces on cylinder with protecting wall.

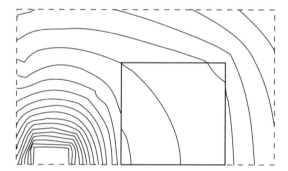

Figure 14.7 Equipotential lines near square conductor placed parallel to dielectric slab of relative permittivity 6.

A square conductor flanked by a dielectric slab[1]   Silvester and Hsieh described the coupling of the IBEM and FEM to model exterior field problems. Figure 14.7 shows the equipotential lines when a charged square metallic conductor is placed parallel to and near a rectangular dielectric slab of relative permeability 6. The region bounded by the dotted lines was represented by the FEM and the exterior region by a Galerkin formulation of the IBEM, outlined in Section 14.2.6.

14.5 FURTHER PRACTICAL EXAMPLES

14.5.1 The Problem of a Structural Framework Embedded in a Non-homogeneous Elastic Soil Mass[27-30]

In the mid-sixties the author became interested in the possibility that soil–structure interaction problems might be solved using a combination of a domain discretization method with the IBEM. Boundary element methods are not very attractive for modelling thin structural elements in bending, whereas, conversely, domain discretization methods are unattractive for massive three-dimensional solids. A combination of the two techniques would therefore be the most logical choice for problems of the following kind.

Figure 14.8 shows the solutions for a number of three-dimensional structures (pile groups with vertical or inclined piles) embedded in a three-dimensional solid

Text	Schematic diagram: Plan	Schematic diagram: Elevation	Lateral load for $\frac{1}{4}$ in deflection: Test	Lateral load for $\frac{1}{4}$ in deflection: Theory	Lateral deflection due to vertical load of 20 T/pile: Test	Lateral deflection due to vertical load of 20 T/pile: Theory
1	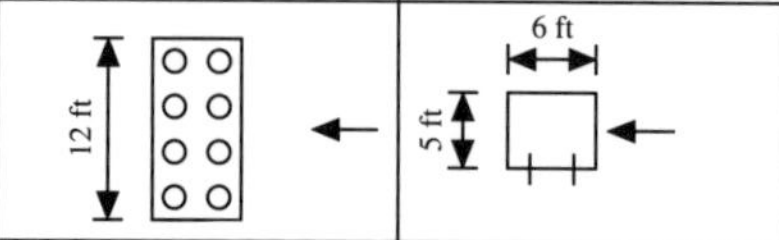		4.8	4.8	0.0	0.0
2	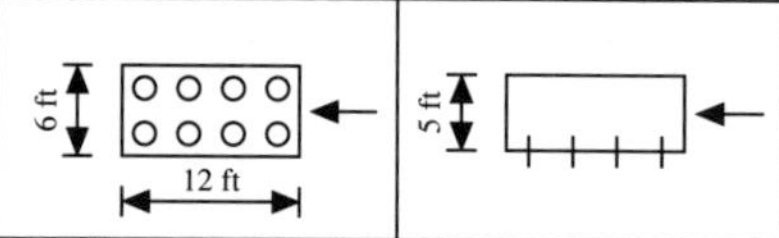		5.8	5.3	0.0	0.0
3	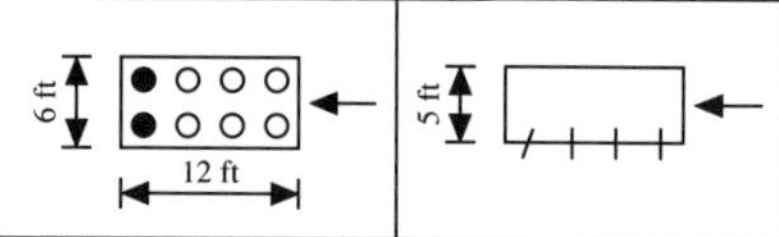		7.0	7.3	0.04	0.06
4	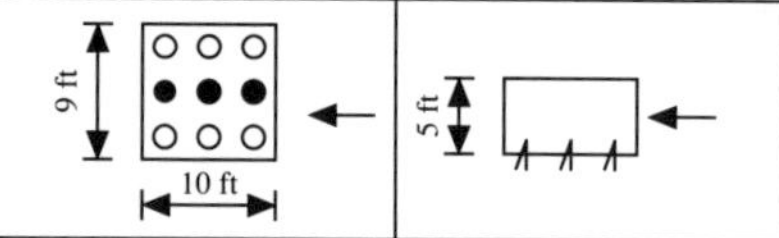		7.1	6.8	0.06	0.08
5	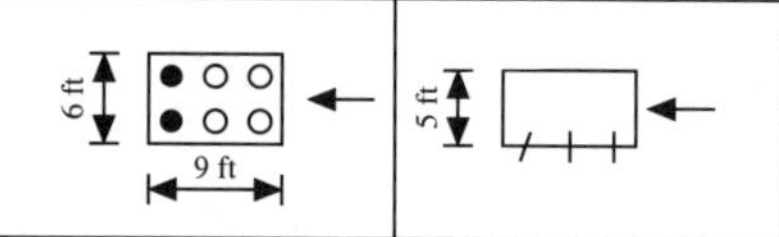		7.3	8.1	0.05	0.07
6	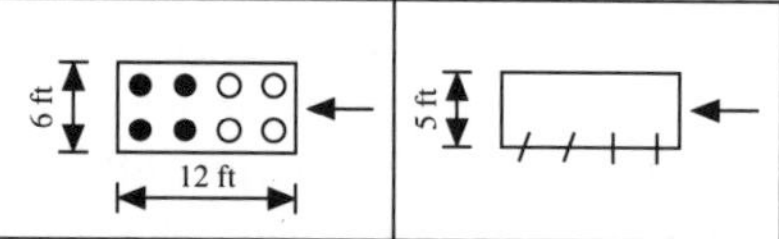		9.0	8.4	0.07	0.11
7	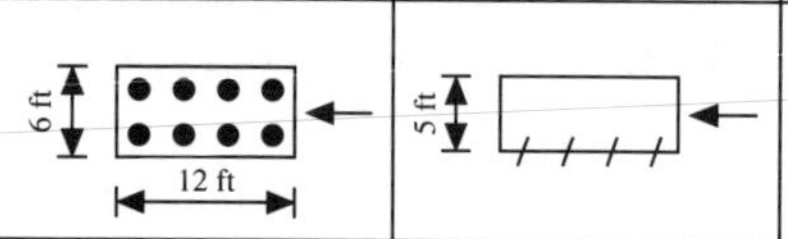		9.0	8.2	0.21	0.27
8	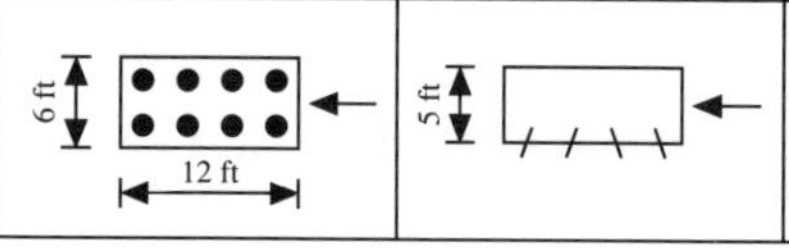		15.8	11.7	0.0	0.0

Figure 14.8 Comparisons with full-scale tests.

with a modulus of elasticity increasing linearly with the depth, compared with those obtained from a series of full-scale field tests. In these, the pile structure was modelled by a finite difference scheme and the extensive solid (soil) medium by the IBEM. An approximate point force solution[30] for the non-homogeneous solid was constructed so that it satisfied the ground-surface boundary conditions a priori and thus the only discretization needed was surface elements on the pile–soil interfaces.

14.5.2 Response of a T-shaped Bridge Pier Subject to SV Waves[24]

Figure 14.9 shows the deformation of an elastic T-shaped bridge pier embedded in a viscoelastic half-space subject to a time harmonic SV wave of oblique incidence. The model used for the analysis is depicted in Fig. 14.9a. The incident wave has a wavelength of L, and is polarized in the vertical plane obtained by rotating the $x_1 x_3$ plane by $30°$ around the x_3 axis. The incident angle is $20°$. The material constants for the bridge pier (denoted by subscripts f) and the ground (indicated by subscripts b) are as follows:

$$\hat{\lambda}_b(\omega) = \frac{2\nu_b}{1 - 2\nu_b} \hat{\mu}_b(\omega)$$

$$\lambda_f = \frac{2\nu_f}{1 - 2\nu_f} \mu_f$$

$$\frac{(1 + \nu_b)\hat{\mu}_b(\omega)}{(1 + \nu_f)\mu_f} = \frac{1}{100}(1 - 0.2i)$$

$$\frac{\rho_b}{\rho_f} = \frac{5}{6}$$

$$\nu_b = \frac{1}{3}$$

$$\nu_f = \frac{1}{6}$$

where λ, μ are Lame constants, ν the Poisson ratio, ρ the density and ω the frequency in radians.

Nishimura et al.[24] used the FEM for the structure and the BEM for the ground. A part of the free surface having a radius of approximately twice the incident wavelength is taken into account in the numerical analysis. The real and imaginary parts of the solution shown in Fig. 14.9b,c indicate the deformation at the moment when the peak (node) of the incident displacement wave reaches the origin of the coordinates. The sign of the incident field is chosen such that the real part of the displacement of the incident wave points into the ground at the origin.

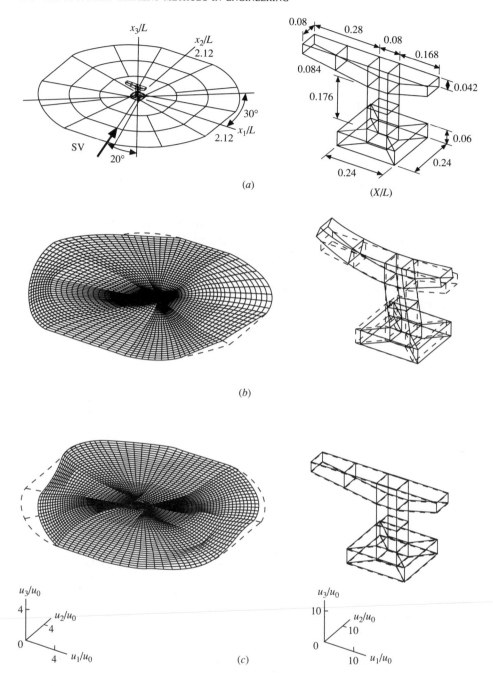

Figure 14.9 Deformations of an elastic T-shaped bridge pier–viscoelastic ground system subject to a time-harmonic SV wave of oblique incidence: (*a*) model used, (*b*) real part of solution, (*c*) imaginary part of solution.

14.6 CONCLUDING REMARKS

We have shown in this chapter how, by using the most advantageous features of different numerical methods, it is often possible to achieve the most satisfactory algorithm for solving many realistic engineering problems. Therefore the skilful analyst will utilize the relevant properties of each rather than slavishly extol the virtues of any single one of them. Additional examples of such combined analyses can be found in Refs 22, 23 and 31 to 34.

REFERENCES

1. Silvester, P.P. and Hsieh, M.S. (1971) Finite element solution of two-dimensional exterior field problem, *Proceedings of IEE*, Vol. 118, No. 12, pp. 1743–1747.
2. Silvester, P.P., Carpenter, G.J. and Wyatt, E.A. (1977) Exterior finite elements for two-dimensional field problems, with open boundaries, *Proceedings of IEE*, Vol. 124, No. 12, pp. 1267–1270.
3. Lean, M.H., Friedman, M. and Wexler, A. (1979) Advances in application of the boundary element method in electrical engineering, Chapter IX in *Developments in Boundary Element Methods*, Eds P.K. Banerjee and R. Butterfield, Volume 1, Elsevier Applied Science Publishers, London.
4. Silvester, P. and Hsieh, M.S. (1971) Projective solution of integral equations arising in electric and magnetic field problems, *Journal of Comparative Physics*, Vol. 8, pp. 73–82.
5. Hassan, M.A. and Silvester, P. (1977) Radiation and scattering by wire antenna. Structures near a rectangular plate reflector, *Proceedings of IEE*, Vol. 124, No. 5, pp. 429–535.
6. Gopinath, A. and Silvester, P. (1973) Calculation of inductance of finite length strips and its variation with frequency, *IEEE Transactions on Microwave Theory and Techniques*, Vol. MIT-21, No. 6, pp. 380–386 .
7. Jeng, G. and Wexler, A. (1977) Isoparametric, finite element variational solution of integral equations for three-dimensional fields, *International Journal of Numerical Methods in Engineering*, Vol. 11, pp. 455–471.
8. Jeng, G. and Wexler, A. (1978) Self-adjoint variational formulation of problems having a non-self adjoint operator, *IEEE Transactions on Microwave Theory and Techniques*, Vol. MIT-26, pp. 91–94.
9. Siritori, S., Maier, G., Novati, G. and Miccoil, S. (1992) A Galerkin symmetric BEM in elasticity: formulation and implementation, *International Journal of Numerical Methods in Engineering*, Vol. 35, pp. 255–282.
10. Comi, C. and Maier, G. (1992) Extremum, convergence and stability properties of the finite increment problems in elastoplastic boundary element analysis, *International Journal of Solids and Structures*, Vol. 29, No. 2, pp. 249–270.
11. Maier, G. and Polizotto, C. (1987) A Galerkin approach to boundary element elastoplastic analysis, *Computer Methods in Applied Mechanics and Engineering*, Vol. 60, pp. 175–194.
12. Maier, G., Diligenti, M. and Carini, A. (1991) A variational approach to boundary element elastodynamic analysis and extension to multi-domain problem, *Computer Methods in Applied Mechanics and Engineering*, Vol. 92, pp. 193–213.
13. Zienkiewicz, O.C., Kelly, D.W. and Bettess, P. (1977) The coupling of the finite element method and boundary solution procedure, *International Journal of Numerical Methods in Engineering*, Vol. 11, No. 12, pp. 355–375.
14. Kelly, D.W., Mustoe, G.G.W. and Zienkiewicz, O.G. (1978) On hierarchical order for trial functions based on the satisfaction of the governing equations, *Proceedings of International Symposium on Recent Advances in Boundary Element Methods*, Southampton University.
15. Mustoe, G.G.W. (1979) Coupling of boundary solution procedures and finite elements for continuum problems, PhD Thesis, University of Wales, University College, Swansea.

16. Kelly, D.W., Mustoe, G.G.W. and Zienkiewicz, O.G. (1979) Coupling of boundary element methods with other numerical methods, Chapter X in *Developments in Boundary Element Methods*, Eds P.K. Banerjee and R. Butterfield, Elsevier Applied Science Publishers, London.

17. Zienkiewicz, O.C. (1977) *The Finite Element Method*, McGraw-Hill, London.

18. Poggio, A.J. and Miller, E.K. (1973) Integral equation solutions of three-dimensional scattering problems, Chapter III in *Computers in Electromagnetics*, Ed. R. Mittra, Pergamon Press, Oxford.

19. Fried, I. (1968) Finite element analysis of problems formulated by an integral equation — application to potential flow, Report, University of Stuttgart, October.

20. Tottenham, H. (1979) The boundary element method for plates and shells, Chapter 8 in *Developments in Boundary Element Methods*, Eds P.K. Banerjee and R. Butterfield, Elsevier Applied Science Publishers, London.

21. Jaswan, M.A. and Symm, G.T. (1977) *Integral Equation Methods in Potential Theory and Elastostatics*, Academic Press, London.

22. Beer, G. (1983) Finite element boundary element and coupled analysis of unbounded problems in elastostatics, *International Journal of Numerical Methods in Engineering*, Vol. 19, pp. 567–580.

23. Beer, G. (1986) Implementation of combined boundary element and finite element analysis with applications in geomechanics, Chapter 7 in *Developments in Boundary Element Methods*, Vol. 4, Eds P.K. Banerjee and J.O. Watson, Elsevier Applied Science Publishers, London.

24. Nishimura, N., Kobayashi, S. and Kitahara, M. (1989) Advanced substructured analysis of elastodynamic wave propagation problems, Chapter 5 in *Industrial Applications of BEM*, Eds P.K. Banerjee and R.B. Wilson, Elsevier Applied Science Publishers, London.

25. Belytschko, T., Chang, H.S. and Lu, Y.Y. (1989) A variationally coupled finite element and boundary element method, *Computers and Structures*, Vol. 33, No. 1, pp. 17–20.

26. Belytschko, T. and Lu, Y.Y. (1991) Singular integration in a variationally coupled BE and FE method, *Journal of Engineering Mechanics, ASCE*, Vol. 117, No. 4, pp. 820–835.

27. Banerjee, P.K. and Butterfield, R. (1971) The problem of pile–cap group interaction, *Geotechnique*, Vol. 21, No. 3, pp. 135–142 .

28. Banerjee, P.K. and Driscoll, R.M.C. (1976) Three-dimensional analysis of raked pile groups, *Proceedings of Institute of Civil Engineers Research and Theory*, Vol. 61, pp. 653–671.

29. Banerjee, P.K. and Davies, T.G. (1979) Analysis of some reported case histories of laterally loaded pile groups, *International Conference on Numerical Methods in Offshore Piling*, Institute of Civil Engineers, London.

30. Banerjee, P.K. and Sen, R. (1987) Dynamic behavior of axially and laterally loaded piles and pile groups, Chapter 3 in *Developments in Soil Mechanics and Foundations Engineering*, Vol. 3, Eds P.K. Banerjee and R. Butterfield, Elsevier Applied Science, London.

31. Cruse, T.A., and Wilson, R.B. (1978) Advanced application of boundary integral equation methods, *Nuclear Engineering Design*, Vol. 46, pp. 223–234 .

32. Wilton, D.R. (1978) Acoustic radiation and scattering from elastic structures, *International Journal of Numerical Methods in Engineering*, Vol. 13, pp. 123–138.

33. Patel, J.S. (1978) Radiation and scattering from an arbitrary elastic structure using consistent fluid structure formulation, *Computers and Structures*, Vol. 9, pp. 287–291.

34. Olsen, K. and Hwang, L.S. (1971) Oscillations in a bay of arbitrary shape and variable depth, *Journal of Geophysical Research*, Vol. 76, No. 21, pp. 5048–5064.

FIFTEEN

VISCOUS FLOW AT LOW TO MEDIUM
REYNOLDS NUMBERS

15.1 INTRODUCTION

The great majority of fluid mechanics problems involve fluid regions that are
extensive and very often those that extend to infinity. Although the governing differential equations are usually highly non-linear, they can be reformulated in such
a way that the non-linear terms are only present over a localized part of the entire
region. Examples of such an algorithm have already been described in Chapter 13,
in which no interior discretization was needed within the dominant linear regions.
BEMs are, to this extent, unique among numerical methods since they can accommodate a linear region with infinitely distant boundaries without any discretization
whatsoever, a situation that occurs frequently in fluid mechanics. Following the
pioneering work of Hess and Smith (see Chapter 3) on potential flow, remarkable
progress has been made, particularly during the past fifteen years, in developing
the potential of the BEM in fluid mechanics. Most of this work is summarized
in the present chapter.

Integral equations governing flow of incompressible viscous fluids have been
available since the early portion of this century. Oseen,[1] in a most remarkable
monograph, develops exact integral representations, along with the corresponding infinite space fundamental solutions for both two- and three-dimensional
steady state and transient Navier–Stokes flows. Youngren and Acrivos[2] applied
numerical integration along with the method of collocation to these integral equations to examine steady Stokes flow past arbitrarily shaped objects. Naturally,
since Stokes flow was assumed, no volume integration was required.

The next major step was taken by Bush and Tanner,[3] who included the non-linear terms in the two-dimensional integral formulation as presented by Oseen for steady incompressible flow. An iterative non-linear scheme provided results for a number of interesting problems at low Reynolds number. Bush[4] extended this formulation in a significant way by incorporating fundamental solutions with linearized convective terms. However, the author reports that non-linear solutions could not be obtained beyond a Reynolds number of 1.0. More recently, Dargush and Banerjee[5–9] described the solution of the incompressible steady state Navier–Stokes flow by incorporating curvilinear boundary elements and by utilizing a Newton–Raphson algorithm for the solution of the set of non-linear equations. Similar formulation using linear elements was described by Tosaka and Kakuda.[10–11]

Several other researchers have also proposed numerical algorithms based upon integral equations for the solution of time-dependent incompressible viscous flow. Wu[12,13] began his investigations in the early seventies with a velocity–vorticity formulation, which separates the kinematics from the kinetics of the problem. Initially, integral equations were only written for the Poisson equation governing kinematics, while a finite difference scheme was employed for kinetics. Later,[13] more sophistication was added, and a diffusive fundamental solution was utilized in an integral representation for the kinetics. These procedures require the determination of boundary vorticity and, indirectly, the normal gradient of vorticity (i.e. second derivatives of velocity). The extension to three dimensions, however, is less attractive. In that case, the vorticity is no longer a scalar, and a total of six degrees of freedom result at each node.

A vorticity–stream function integral approach was developed by Onishi *et al.*[14] The kinetics are represented by a vorticity transport equation as is done by Wu. A Poisson equation relates the stream function and vorticity. This formulation is not generally applicable, since very seldom will boundary conditions be known in terms of the vorticity, stream function and their normal derivatives.

Tosaka and Onishi[15] provided a primitive variable formulation in terms of velocities, tractions and velocity gradients. The time derivatives in Navier–Stokes equations are replaced by a backward finite difference. The governing differential equations then become elliptic, and possess a well-known fundamental solution. In the ensuing integral representation, volume integration is required throughout the flow field, even for linear Stokes flow. Meanwhile, Piva *et al.*[16] and Piva and Morino[17] rederive the integral equations and fundamental solutions given by Oseen, but provide no numerics. Dargush and Banerjee[5–9] described the first comprehensive and exact BEM formulation for the time-dependent Navier–Stokes equation using a time-stepping convolution together with a Newton–Raphson algorithm for the non-linear solution.

15.2 GOVERNING EQUATIONS WITH STOKES APPROXIMATION

The governing equations for fluid flow have been known for more than a hundred years, and are presented in numerous textbooks.[18-23] Depending on various approximations these governing equations have been expressed in many different forms. However, the system can be thought of as containing only three basic relations, i.e. the three laws of conservation for physical systems: conservation of mass (continuity), conservation of momentum (Newton's second law) and conservation of energy (first law of thermodynamics):

Mass conservation:
$$\frac{D\rho}{Dt} + \rho\frac{\partial v_i}{\partial x_i} - \psi = 0 \qquad (15.1a)$$

Momentum conservation:
$$\rho\frac{Dv_i}{Dt} - \frac{\partial \sigma_{ij}}{\partial x_j} - f_i = 0 \qquad (15.1b)$$

Energy conservation:
$$\rho\frac{De}{Dt} + \frac{\partial q_i}{\partial x_i} - \sigma_{ij}\frac{\partial v_i}{\partial x_j} - \phi = 0 \qquad (15.1c)$$

where

ρ = mass density
v_i = local fluid velocity vector
σ_{ij} = stress tensor
e = internal energy
q_i = heat flux vector
ψ = mass source rate per unit volume
f_i = body force per unit volume
ϕ = heat source rate per unit volume
$\frac{D}{Dt}$ = material time derivative defined as $\frac{D}{Dt} = \frac{\partial}{\partial t} + v_i\frac{\partial}{\partial x_i}$

Next, constitutive relationships are introduced. In particular, consideration is limited to a homogeneous isotropic Newtonian fluid for which the shear stress is linearly proportional to the rate of deformation. For this kind of fluid the stress–strain relationship is specified by the equation

$$\sigma_{ij} = \mu\left(\frac{\partial v_i}{\partial x_j} + \frac{\partial v_j}{\partial x_i}\right) + \lambda\delta_{ij}\frac{\partial v_k}{\partial x_k} - \delta_{ij}p \qquad (15.2)$$

where μ and λ are coefficients of viscosity. This deformation law was first stated by Stokes in 1845.

It remains now to introduce the thermodynamic pressure p in Eq. (15.2). If the fluid is modelled as a perfect gas, the kinetic equation of state is simply

$$p = \rho R\theta \qquad (15.3)$$

where R is the gas constant and θ is the thermodynamic temperature. Fourier's law of heat conduction is also invoked, which for an isotropic medium becomes

$$q_i = -k\frac{\partial \theta}{\partial x_i} \qquad (15.4)$$

where k is the thermal conductivity. Finally, a relationship is needed for the internal energy e. From thermodynamics the enthalpy per unit mass, h, is defined by

$$h = e + \frac{p}{\rho} \tag{15.5}$$

where h is the enthalpy. In addition, if the specific heat at constant pressure, c_p, does not vary with temperature, then

$$h = c_p \theta \tag{15.6}$$

Substituting Eqs (15.2), (15.4), (15.5) and (15.6) into Eqs (15.1) leads to the following form of governing equations for the idealized thermoviscous fluid:

$$\frac{\mathrm{D}\rho}{\mathrm{D}t} + \rho \frac{\partial v_i}{\partial x_i} - \psi = 0 \tag{15.7a}$$

$$\rho \frac{\mathrm{D}v_i}{\mathrm{D}t} - (\lambda + \mu) \frac{\partial^2 v_j}{\partial x_i \partial x_j} - \mu \frac{\partial^2 v_i}{\partial x_j \partial x_j} + \frac{\partial p}{\partial x_i} - f_i = 0 \tag{15.7b}$$

$$\rho c_p \frac{\mathrm{D}\theta}{\mathrm{D}t} - k \frac{\partial^2 \theta}{\partial x_i \partial x_i} - \frac{\mathrm{D}p}{\mathrm{D}t} + \frac{p}{\rho}\psi - \Phi - \phi = 0 \tag{15.7c}$$

In Eq. (15.7c), Φ represents the viscous dissipation which is defined by

$$\Phi = \tau_{ij} \frac{\partial v_i}{\partial x_j} \tag{15.8}$$

and is responsible for heat generation in bearings and aerodynamic heating of spacecraft as they re-enter the earth's atmosphere. The viscous stresses are defined as

$$\tau_{ij} = \mu \left(\frac{\partial v_i}{\partial x_j} + \frac{\partial v_j}{\partial x_i} \right) + \lambda \delta_{ij} \frac{\partial v_k}{\partial x_k} \tag{15.9}$$

Summarizing, Eqs (15.7) along with (15.3) define a highly non-linear set of six equations in the six variables: velocity (v_i) (i has three values 1, 2, 3), pressure (p), temperature (θ) and density (ρ), of which five are assumed to be primary: v_i, p and θ. The remaining variable ρ is assumed known from the relation (15.3).

The first two equations of (15.7) are commonly referred to as the Navier–Stokes equations. In flows governed by these equations, one may find the velocity without using the energy equation. The solution of the Navier–Stokes equations is the central problem in fluid mechanics. Since several first-order convective terms are non-linear, the equations can only be solved exactly for the simplest of geometries and boundary conditions, even under steady conditions. Consequently, some approximations are necessary. For the fluid flow at low to medium Reynolds numbers, the fluid can be considered as incompressible and so its density is a constant; from the equation of continuity (15.7a), the divergence of velocity is zero.

The internal energy is directly proportional to the temperature, i.e.

$$e = c_v \theta \tag{15.10}$$

with c_v representing the specific heat at constant volume. As the result, the governing equations reduce to the following:

$$\frac{\partial v_i}{\partial x_i} = 0 \tag{15.11a}$$

$$\rho \frac{Dv_i}{Dt} - \mu \frac{\partial^2 v_i}{\partial x_j \partial x_j} + \frac{\partial p}{\partial x_i} - f_i = 0 \tag{15.11b}$$

$$\rho c_v \frac{D\theta}{Dt} - k \frac{\partial^2 \theta}{\partial x_i \partial x_i} - \Phi - \phi = 0 \tag{15.11c}$$

where Φ is the viscous dissipation

$$\Phi = \tau_{ij} \frac{\partial v_i}{\partial x_j} \tag{15.12}$$

and mass sources have been assumed absent. The viscous stresses are now defined as

$$\tau_{ij} = \mu \left(\frac{\partial v_i}{\partial x_j} + \frac{\partial v_j}{\partial x_i} \right) \tag{15.13}$$

By introducing a free stream velocity, U_i, and a velocity perturbation, u_i, the above governing equations can be rewritten as

$$\frac{\partial u_i}{\partial x_i} = 0 \tag{15.14a}$$

$$\rho \frac{\partial u_i}{\partial t} - \mu \frac{\partial^2 u_i}{\partial x_j \partial x_j} + \frac{\partial p}{\partial x_i} - \bar{f}_i = 0 \tag{15.14b}$$

$$\rho c_v \frac{\partial \theta}{\partial t} - k \frac{\partial^2 \theta}{\partial x_i \partial x_i} - \bar{\phi} = 0 \tag{15.14c}$$

with

$$\bar{f}_i = -\rho(U_j + u_j) \frac{\partial u_i}{\partial x_j} + f_i \tag{15.15}$$

$$\bar{\phi} = -\rho c_v (U_i + u_i) \frac{\partial \theta}{\partial x_i} + \Phi + \phi \tag{15.16}$$

In operator notation, the governing equations (15.14) are simply written as

$$L_{\alpha\beta} u_\beta + \bar{f}_\alpha = 0 \tag{15.17}$$

in which

$$u_\beta = \{v_i, p, \theta\}^\mathrm{T} \tag{15.18a}$$

$$\bar{f}_\beta = \{\bar{f}_i, 0, \bar{\phi}\}^\mathrm{T} \tag{15.18b}$$

For the low Reynolds numbers (slow flow) the convective terms are negligible, the governing equations for incompressible thermoviscous Stokes flow can then be obtained by eliminating the non-linear terms from Eqs (15.11):

$$\frac{\partial v_i}{\partial x_i} = 0 \tag{15.19a}$$

$$\rho \frac{\partial v_i}{\partial t} - \mu \frac{\partial^2 v_i}{\partial x_j \partial x_j} + \frac{\partial p}{\partial x_i} - \bar{f}_i = 0 \tag{15.19b}$$

$$\rho c_v \frac{\partial \theta}{\partial t} - k \frac{\partial^2 \theta}{\partial x_i \partial x_i} - \bar{\phi} = 0 \tag{15.19c}$$

in which

$$\bar{f}_i = -\rho v_j \frac{\partial v_i}{\partial x_j} + f_i \tag{15.20a}$$

$$\bar{\phi} = -\rho c_v v_i \frac{\partial \theta}{\partial x_i} + \Phi + \phi \tag{15.20b}$$

15.3 FUNDAMENTAL SOLUTIONS FOR STOKES FLOW

Consider, first, the effect of a unit point pulse force, i.e.

$$\bar{f}_i = \delta(x - \xi)\delta(t - \tau)e_i$$

The corresponding fundamental solution g_{ij}, g_{pj} must satisfy

$$\frac{\partial g_{ij}}{\partial x_j} = 0 \tag{15.21a}$$

$$\rho \frac{\partial g_{ij}}{\partial t} - \mu \frac{\partial^2 g_{ij}}{\partial x_k \partial x_k} + \frac{\partial g_{pj}}{\partial x_i} - \delta_{ij}\delta(x - \xi)\delta(t - \tau) = 0 \tag{15.21b}$$

The solutions of the above equations were given by Oseen:[1]

$$g_{ij}(x - \xi, t - \tau) = \delta_{ij} \frac{\partial^2 \phi(r, t')}{\partial x_k \partial x_k} - \frac{\partial^2 \phi(r, t')}{\partial x_i \partial x_j} \tag{15.22a}$$

$$g_{pj}(x - \xi, t - \tau) = \begin{cases} -\delta(t - \tau)\dfrac{\partial}{\partial x_j}\left(\dfrac{1}{4\pi r}\right) & \text{in three dimensions} \\[2ex] \delta(t - \tau)\dfrac{\partial}{\partial x_j}(\ln r) & \text{in two dimensions} \end{cases} \tag{15.22b}$$

where

$$
\phi(r,t) = \begin{cases} -\dfrac{1}{4\pi\rho r}\operatorname{erf}\dfrac{r}{2\sqrt{\nu t}} & \text{in three dimensions} \\[2ex] \dfrac{1}{2\pi\rho}\left[\ln r + \dfrac{1}{2}E_1\left(\dfrac{r^2}{4\nu t}\right)\right] & \text{in two dimensions} \end{cases}
$$

and $r^2 = y_i y_i$, $y_i = x_i - \xi_i$, $t' = t - \tau$ with $\nu = \mu/\rho$.

Next a unit point pulse heat source will be considered, i.e.

$$
\bar{\psi} = \delta(x - \xi)\delta(t - \tau)
$$

The corresponding fundamental solution $g_{\theta\theta}$ must satisfy the heat conduction equation

$$
\rho c_v \frac{\partial g_{\theta\theta}}{\partial t} - k\frac{\partial^2 g_{\theta\theta}}{\partial x_l \partial x_l} - \delta(x - \xi, t - \tau) = 0 \tag{15.23}
$$

the solution of which was given by Carslaw and Jaeger:[24]

$$
g_{\theta\theta}(x - \xi, t - \tau) = \begin{cases} \dfrac{1}{8\rho c_v(\pi\kappa t')^{3/2}}e^{-r^2/4\kappa t'} & \text{in three dimensions} \\[2ex] \dfrac{1}{4\pi k t'}e^{-r^2/4\kappa t'} & \text{in two dimensions} \end{cases} \tag{15.24}
$$

with $\kappa = k/\rho c_v$.

By integrating the above fundamental solutions from 0 to t we can obtain solutions for a step function in time for a three-dimensional problem:

$$
G_{ij}(x - \xi, t) = \delta_{ij}\frac{\partial^2\Phi}{\partial x_k \partial x_k} - \frac{\partial^2\Phi}{\partial x_i \partial x_j} \tag{15.25a}
$$

$$
G_{pj}(x - \xi, t) = \frac{\partial}{\partial x_j}\left(\frac{1}{4\pi r}\right) = -\frac{1}{4\pi r^2}\frac{y_j}{r} \tag{15.25b}
$$

$$
G_{\theta\theta}(x - \xi, t) = \frac{1}{4\pi k r}\operatorname{erfc}\frac{r}{2\sqrt{\kappa t}} \tag{15.25c}
$$

where

$$
\Phi(r,t) = -\frac{t}{4\pi\rho r}\left(1 - 4i^2\operatorname{erfc}\frac{r}{2\sqrt{\nu t}}\right)
$$

and for two-dimensional problems:

$$
G_{ij}(x - \xi, t) = \delta_{ij}\frac{\partial^2\Phi}{\partial x_k \partial x_k} - \frac{\partial^2\Phi}{\partial x_i \partial x_j} \tag{15.26a}
$$

$$
G_{pj}(x - \xi, t) = -\frac{\partial}{\partial x_j}\left(\frac{1}{2\pi}\ln r\right) = -\frac{1}{2\pi r}\frac{y_j}{r} \tag{15.26b}
$$

$$
G_{\theta\theta}(x - \xi, t) = \frac{1}{4\pi k}E_1\left(\frac{r^2}{4\kappa t}\right) \tag{15.26c}
$$

where

$$\Phi(r, t) = \frac{t}{4\pi\rho}\left[2\ln r + E_1\left(\frac{r^2}{4\nu t}\right) - E_2\left(\frac{r^2}{4\nu t}\right)\right]$$

$$y_i = x_i - \xi_i$$

$$r^2 = y_i y_i$$

$$\eta = \frac{r}{2\sqrt{\nu t}}$$

$$\nu = \frac{\mu}{\rho}$$

$$\text{erf}(z) = \frac{2}{\sqrt{\pi}}\int_0^z e^{-u^2}\, du$$

$$E_1(z) = \int_z^\infty \frac{e^{-u}}{u}\, du \quad \text{and} \quad E_2(z) = e^{-z} - zE_1(z)$$

When time $t = \infty$ the steady state solutions are recovered from the above equations. Thus for a three-dimensional problem,

$$G_{ij}(x - \xi) = \delta_{ij}\frac{\partial^2 \Phi^s}{\partial x_k \partial x_k} - \frac{\partial^2 \Phi^s}{\partial x_i \partial x_j} \tag{15.27a}$$

$$G_{pj}(x - \xi) = \frac{\partial}{\partial x_j}\left(\frac{1}{4\pi r}\right) \tag{15.27b}$$

$$G_{\theta\theta}(x - \xi) = \frac{1}{4\pi k r} \tag{15.27c}$$

where $\Phi^s(r) = r/(8\pi\mu)$ and for a two-dimensional problem,

$$G_{ij}(x - \xi) = \delta_{ij}\frac{\partial^2 \Phi^s}{\partial x_k \partial x_k} - \frac{\partial^2 \Phi^s}{\partial x_i \partial x_j} \tag{15.28a}$$

$$G_{pj}^s(x - \xi) = -\frac{\partial}{\partial x_j}\left(\frac{1}{2\pi}\ln r\right) \tag{15.28b}$$

$$G_{\theta\theta}^s(x - \xi) = -\frac{1}{2\pi k}\ln r \tag{15.28c}$$

where $\Phi^s(r) = -[1/(8\pi\mu)]r^2(\ln r - 1)$.

15.4 BOUNDARY INTEGRAL REPRESENTATIONS

The governing equations (15.14) multiplied by an arbitrary function $\tilde{g}_{\alpha\gamma}$ and integrated over time and space must remain equal to zero; i.e.

$$\langle \tilde{g}_{\alpha\gamma}, L_{\alpha\beta}u_{\beta} + \bar{f}_{\alpha}\rangle = \int_0^T \int_V \tilde{g}_{\alpha\gamma}(L_{\alpha\beta}u_{\beta} + \bar{f}_{\alpha})\,dV\,dt = 0 \qquad (15.29)$$

where the standard notation for the inner product of two functions has been introduced. Next, the divergence theorem can be applied repeatedly to applicable terms in Eq. (15.29) to transfer spatial as well as temporal derivatives from u_{β} to $\tilde{g}_{\alpha\gamma}$. Consequently, Eq. (15.29) becomes

$$\int_0^T \int_S \left\{ \tilde{g}_{ik}\left[\mu\left(\frac{\partial u_i}{\partial x_j} + \frac{\partial u_j}{\partial x_i}\right) - u_p\,n_i\right] - \left[\mu\left(\frac{\partial \tilde{g}_{ik}}{\partial x_j} + \frac{\partial \tilde{g}_{jk}}{\partial x_i}\right)n_j + \tilde{g}_{pk}n_i\right]u_i \right.$$
$$\left. + \tilde{g}_{\theta\theta}\left(k\frac{\partial u_\theta}{\partial x_i}n_i\right) - \left(k\frac{\partial \tilde{g}_{\theta\theta}}{\partial x_i\,\partial x_i}\right)u_\theta \right\}\,dS\,dt$$

$$+ \int_0^T \int_V (\tilde{g}_{ik}\bar{f}_i + \tilde{g}_{\theta\theta}\bar{f}_\theta)\,dV\,dt - \int_V (\tilde{g}_{ik}\rho u_i|_0^T + \tilde{g}_{\theta\theta}\rho c_v u_\theta|_0^T)\,dV$$

$$+ \int_0^T \int_V \left\{ \left[\rho\frac{\partial \tilde{g}_{ik}}{\partial t} + \mu\left(\frac{\partial^2 \tilde{g}_{ik}}{\partial x_j\,\partial x_j} + \frac{\partial^2 \tilde{g}_{jk}}{\partial x_i\,\partial x_i}\right) + \frac{\partial \tilde{g}_{pk}}{\partial x_i}\right]u_i \right.$$
$$\left. + \left(\frac{\partial \tilde{g}_{ik}}{\partial x_i}\right)u_p + \left(\rho c_v\frac{\partial \tilde{g}_{\theta\theta}}{\partial t} + k\frac{\partial^2 \tilde{g}_{\theta\theta}}{\partial x_i\,\partial x_i}\right)u_\theta \right\}\,dV\,dt = 0 \qquad (15.30)$$

where we have used u_p and u_θ for pressure p and temperature θ respectively.

In order to complete the derivation of the integral equation for the perturbed velocity and temperature at any point ξ, interior to S, at time $\tau < T$, the last volume integral in Eq. (15.30) must be reduced to $-u_\gamma(\xi, \tau)$. This is accomplished if

$$\langle \tilde{L}_{\beta\alpha}\tilde{g}_{\alpha k}, u_\beta\rangle = -u_k(\xi, \tau)$$

$$\langle \tilde{L}_{\beta\alpha}\tilde{g}_{\alpha\theta}, u_\beta\rangle = -u_\theta(\xi, \tau)$$

or after making use of the properties of the delta function

$$\tilde{L}_{\beta\alpha}\tilde{g}_{\alpha\gamma} + \delta_{\beta\gamma}\delta(x - \xi)\delta(t - \tau) = 0 \qquad (15.31)$$

where the differential operator $\tilde{L}_{\beta\alpha}$ has the definition

$$\tilde{L}_{\beta\alpha} = \begin{bmatrix} \delta_{ji}\rho\dfrac{\partial}{\partial t} + \mu\left(\delta_{ji}\dfrac{\partial^2}{\partial x_k\,\partial x_k} + \dfrac{\partial^2}{\partial x_j\,\partial x_i}\right) & \dfrac{\partial}{\partial x_j} & 0 \\[4ex] \dfrac{\partial}{\partial x_i} & 0 & 0 \\[4ex] 0 & 0 & \rho c_v\dfrac{\partial}{\partial t} + k\dfrac{\partial^2}{\partial x_i\,\partial x_i} \end{bmatrix}$$

Formally, $\tilde{L}_{\alpha\beta}$ is called the adjoint of the original incompressible thermo-viscous flow operator $L_{\alpha\beta}$ and $\tilde{g}_{\alpha\gamma}$, defined by Eq. (15.31), is the adjoint Green function. This function $\tilde{g}_{\alpha\gamma}$ can be obtained simply by transposing the fundamental solution $g_{\alpha\beta}$ presented in the previous section, i.e. (see also Chapters 8 and 9)

$$\tilde{g}_{\alpha\gamma}(x - \xi, t - \tau) = g_{\gamma\alpha}(\xi - x, \tau - t) = g_{\alpha\gamma}(\xi - x, \tau - t) \qquad (15.32)$$

with the final equivalence due to the symmetry of the operator in Eq. (15.31). Substitution of Eq. (15.32) into Eq. (15.30) produces the desired integral equation

$$c_{ij}(\xi)u_i(\xi, \tau) = \int_0^T \int_S \left[g_{ij}(\xi - x, \tau - t)t_i(x, t) - f_{ij}(\xi - x, \tau - t)u_i(x, t) \right] dS(x)\, dt$$

$$+ \int_0^T \int_V \left[g_{ij}(\xi - x, \tau - t)\bar{f}_i(x, t) \right] dV(x)\, dt$$

$$- \int_V \left[g_{ij}(\xi - x, \tau - t)\rho u_i(x, t)\big|_0^T \right] dV(x) \qquad (15.33)$$

and the corresponding heat equation becomes

$$c_{\theta\theta}(\xi)u_\theta(\xi, \tau) = \int_0^T \int_S [g_{\theta\theta}(\xi - x, \tau - t)t_\theta(x, t) - f_{\theta\theta}(\xi - x, \tau - t)u_\theta(x, t)]\, dS(x)\, dt$$

$$+ \int_0^T \int_V [g_{\theta\theta}(\xi - x, \tau - t)\bar{f}_\theta]\, dV(x)\, dt$$

$$- \int_V [g_{\theta\theta}(\xi - x, \tau - t)\rho c_v u_\theta(x, t)\big|_0^T]\, dV(x) \qquad (15.34)$$

where

$$t_i(x, t) = \mu \left[\frac{\partial u_i(x, t)}{\partial x_j} + \frac{\partial u_j(x, t)}{\partial x_i} \right] n_j(x) - u_p(x, t)\, n_i(x) \qquad (15.35a)$$

$$t_\theta(x, t) = k \frac{\partial u_\theta(x, t)}{\partial x_i} n_i(x) \qquad (15.35b)$$

$$f_{ij}(\xi - x, \tau - t) = \mu \left[\frac{\partial g_{ij}(\xi - x, \tau - t)}{\partial x_k} + \frac{\partial g_{jk}(\xi - x, \tau - t)}{\partial x_i} \right] n_k(x)$$

$$+ g_{pj}(\xi - x, \tau - t)n_i(x) \qquad (15.35c)$$

$$f_{\theta\theta}(\xi - x, \tau - t) = k \frac{\partial g_{\theta\theta}(\xi - x, \tau - t)}{\partial x_i} n_i(x) \qquad (15.35d)$$

As may be seen from Eqs (15.33) and (15.34), a portion of the non-linear effects are included in the body forces $\bar{f}_i$ and $\bar{f}_\theta$ in which the velocity and temperature gradients are involved. Consequently, this may produce difficulty in computation. Alternatively, the divergence theorem can be applied to the volume integrals of Eq. (15.33):[6–9]

$$\int_0^T \int_V (g_{ij}\bar{f}_i)\,\mathrm{d}V\,\mathrm{d}t = \int_0^T \int_V \left[g_{ij}\left(-\rho(U_k + u_k)\frac{\partial u_i}{\partial x_k} + f_i \right) \right] \mathrm{d}V\,\mathrm{d}t$$

$$= \int_0^T \int_S \left[-g_{ij}\rho(U_k + u_k)u_i n_k \right] \mathrm{d}S\,\mathrm{d}t$$

$$+ \int_0^T \int_V \left[\frac{\partial g_{ij}}{\partial x_k}\rho(U_k + u_k)\,u_i + g_{ij}f_i \right] \mathrm{d}V\,\mathrm{d}t \qquad (15.36a)$$

and similarly for Eq. (15.34):

$$\int_0^T \int_V (g_{\theta\theta}\bar{f}_\theta)\,\mathrm{d}V\,\mathrm{d}t = \int_0^T \int_V \left[g_{\theta\theta}\left(-\rho c_v(U_i + u_i)\frac{\partial u_\theta}{\partial x_i} + f_\theta \right) \right] \mathrm{d}V\,\mathrm{d}t$$

$$= \int_0^T \int_S \left[-g_{\theta\theta}\rho c_v(U_i + u_i)u_\theta n_i \right] \mathrm{d}S\,\mathrm{d}t$$

$$+ \int_0^T \int_V \left[\frac{\partial g_{\theta\theta}}{\partial x_i}\rho(U_i + u_i)u_\theta + g_{\theta\theta}f_\theta \right] \mathrm{d}V\,\mathrm{d}t \qquad (15.36b)$$

Then the required integral formulation can be rewritten as[6–9]

$$c_{ij}u_i = \int_0^T \int_S \left[g_{ij}t_i - f_{ij}u_i - g_{ij}\rho(U_k + u_k)\,n_k u_i \right] \mathrm{d}S\,\mathrm{d}t$$

$$+ \int_0^T \int_V \left[\frac{\partial g_{ij}}{\partial x_k}\rho(U_k + u_k)u_i + g_{ij}f_i \right] \mathrm{d}V\,\mathrm{d}t - \int_V (g_{ij}\rho u_i|_0^T)\,\mathrm{d}V \qquad (15.37a)$$

$$c_{\theta\theta}u_\theta = \int_0^T \int_S \left[g_{\theta\theta}t_\theta - f_{\theta\theta}u_\theta - g_{\theta\theta}\rho c_v(U_i + u_i)n_i u_\theta \right] \mathrm{d}S\,\mathrm{d}t$$

$$+ \int_0^T \int_V \left[\frac{\partial g_{\theta\theta}}{\partial x_i}\rho c_v(U_i + u_i)u_\theta + g_{\theta\theta}f_\theta \right] \mathrm{d}V\,\mathrm{d}t - \int_V (g_{\theta\theta}\rho c_v u_\theta|_0^T)\,\mathrm{d}V \qquad (15.37b)$$

Equations (15.37) can be rewritten in a compacted form as[6–9]

$$c_{\alpha\beta}u_\alpha = \int_S (g_{\alpha\beta} * t_\alpha - f_{\alpha\beta} * u_\alpha - g_{\alpha\beta} * t_\alpha^0)\,\mathrm{d}S + \int_V (d_{\alpha\beta k} * \sigma_{\alpha k}^0 + g_{\alpha\beta} * f_\alpha + g_{\alpha\beta}\rho u_\alpha^0)\,\mathrm{d}V$$

$$(15.38)$$

In the above equation, the Riemann convolution integral over time has been introduced, along with an initial condition volume integral involving u_α^o. In Eq. (15.38)

$$u_\alpha = \{u_i, \theta\}^\mathrm{T} \qquad t_\alpha = \{t_i, -q_n\}^\mathrm{T} \qquad f_\alpha = \{f_i, \phi\}^\mathrm{T}$$

and

$$c_{\alpha\beta} = \begin{bmatrix} c_{ij} & 0 \\ 0 & c_{\theta\theta} \end{bmatrix} \qquad g_{\alpha\beta} = \begin{bmatrix} g_{ij} & 0 \\ 0 & g_{\theta\theta} \end{bmatrix}$$

$$f_{\alpha\beta} = \begin{bmatrix} f_{ij} & 0 \\ 0 & f_{\theta\theta} \end{bmatrix} \qquad d_{\alpha\beta k} = \frac{\partial g_{\alpha\beta}}{\partial x_k}$$

$$\sigma_{\alpha k}^0 = [\rho(U_k + u_k)u_i, \ \rho c_v(U_k + u_k)\theta]^{\mathrm{T}}$$

$$t_\alpha^0 = \sigma_{\alpha k}^0 n_k$$

Here, $\sigma_{\alpha k}^0$ is labelled the generalized convective stress tensor,[6–9] while t_α^0 is the generalized convective traction. Both $\sigma_{\alpha k}^0$ and t_α^0 contain terms that are non-linear in the generalized velocities.

15.5 NUMERICAL IMPLEMENTATION

15.5.1 Introduction

In this section, boundary element technology is applied to unsteady incompressible viscous flow governed by the above equations. Recent developments in the fields of elastodynamics and thermoelasticity described in Chapters 9 and 10 are directly applicable for these transient problems, as are all of the numerical techniques discussed in Chapters 3, 4 and 5. The presentation below will concentrate on those aspects of the numerical implementation that differ from that detailed earlier. The implementation is limited to the descriptions for a two-dimensional case, although certainly the integral formulation expressed by Eq. (15.38) is equally valid in three dimensions.

15.5.2 Temporal Discretization

Consider the time integrals that appear on the right-hand side of Eq. (15.37a). Clearly, without any loss of precision, the time interval from zero to τ can be subdivided into N equal increments of duration $\Delta\tau$. Then, as in Chapters 8 to 10, the convolution time integrals can be written as a sum of N integrals. For example,

$$\int_0^\tau g_{ij}(\xi - x, \tau - t)t_j(x, t)\,dt = \sum_{n=1}^N \int_{(n-1)\Delta\tau}^{n\Delta\tau} g_{ij}(\xi - x, \tau - t)t_j(x, t)\,dt \qquad (15.39)$$

At this point, an approximation is introduced to facilitate the evaluation of these integrals. It is assumed that the field variables $t_j(x, t)$, $u_j(x, t)$, $t_j^0(x, t)$ and $\sigma_{kj}^0(x, t)$ are constant within each $\Delta\tau$ time increment, and consequently can be brought outside of the time integral. As a result,

$$\int_0^\tau g_{ij}(\xi - x, \tau - t)t_j(x, t) = \sum_{n=1}^N t_j^n(x) \int_{(n-1)\Delta\tau}^{n\Delta\tau} g_{ij}(\xi - x, \tau - t)\,dt \qquad (15.40)$$

with similar relations holding for the remaining time integrals of Eq. (15.37a). The quantity $t_j^n(x)$ denotes the traction at x for the nth time step. Only g_{ij} remains inside the integral. Since this function is known in explicit form from the fundamental solutions, the integration indicated in Eq. (15.40) can be performed analytically. Let[9]

$$G_{ij}^{N-n+1}(\xi - x) = \int_{(n-1)\Delta\tau}^{n\Delta\tau} g_{ij}(\xi - x, \tau - t)\,dt \qquad (15.41)$$

along with

$$F_{ij}^{N-n+1}(\xi - x) = \int_{(n-1)\Delta\tau}^{n\Delta\tau} f_{ij}(\xi - x, \tau - t)\,dt \qquad (15.42)$$

$$D_{ijk}^{N-n+1}(\xi - x) = \int_{(n-1)\Delta\tau}^{n\Delta\tau} d_{ijk}(\xi - x, \tau - t)\,dt \qquad (15.43)$$

The incorporation of these integrals into the integral equations of (15.37) produces[8,9]

$$
\begin{aligned}
c_{ij}(\xi)u_j(\xi, \tau) = \sum_{n=1}^N \Bigg\{ &\int_S \left[G_{ij}^{N-n+1}(\xi - x)t_j^n(x) - F_{ij}^{N-n+1}(\xi - x)u_j^n(x) \right. \\
&\left. - G_{ij}^{N-n+1}(\xi - x)t_j^{0\,n}(x) \right] dS(x) \\
&+ \int_V \left[D_{ijk}^{N-n+1}(\xi - x)\sigma_{jk}^{0\,n}(x) \right] dV(x) \Bigg\} \\
&+ \int_V \left[g_{ij}(\xi - x, \tau)\rho u_j(x, 0) \right] dV(x) \qquad (15.44)
\end{aligned}
$$

in which all of the time integrals have been replaced by summations. In fact, the summations indicated in Eq. (15.44), with the time increment superscript increasing for the field variables while decreasing for the kernels, represents the discretized version of a convolution integral (see Chapters 8, 9 and 10). As a result of this temporal discretization, Eq. (15.44) is no longer an exact integral representation for transient flow; however, the approximation can, in theory, be continually improved by taking smaller time increments or using higher-order variations in time, as shown in Chapter 10. The integrals in Eq. (15.37b) for the heat transfer can also be treated in an identical manner.

The kernel functions G^n, F^n, etc., in the above equations are none other than those described in Eqs (15.25) and (15.26) respectively for three- and two-dimensional problems with the following interpretation:

$$G_{\alpha\beta}^n(x - \xi) = G_{\alpha\beta}(x - \xi, n\Delta\tau) \qquad \text{for } n = 1$$

$$G_{\alpha\beta}^n(x - \xi) = G_{\alpha\beta}(x - \xi, n\Delta\tau) - G_{\alpha\beta}(x - \xi, (n-1)\Delta\tau) \quad \text{for } n > 1$$

For steady state $(t = \infty)$ Eq. (15.44) and the associated temperature equation arising from Eq. (15.37b) can be simplified to[6,7]

$$c_{ij}u_i = \int_S \left[G_{ij}t_i - F_{ij}u_i - G_{ij}\rho(U_k + u_k)n_k u_i \right] dS$$

$$+ \int_V \left[\frac{\partial G_{ij}}{\partial x_k} \rho(U_k + u_k)u_i \right] dV \tag{15.45a}$$

$$c_{\theta\theta}\theta = \int_S \left[G_{\theta\theta}q - F_{\theta\theta}\theta - G_{\theta\theta}\rho c_\varepsilon(U_k + u_k)n_k\theta \right] dS$$

$$+ \int_V \left[\frac{\partial G_{\theta\theta}}{\partial x_k} \rho c_\varepsilon(U_k + u_k)\theta \right] dV \tag{15.45b}$$

15.5.3 Spatial Discretization

The surface is subdivided into M three-noded line elements (for a two-dimensional problem), while the volume is composed of L six- and eight-noded cells (see Chapters 2 and 7). The field variables are represented in terms of nodal values and shape functions N_ω and M_ω. Then the discretized integral equation becomes

$$c_{ij}(\xi)u_j(\xi, \tau) = \sum_{n=1}^N \left[\sum_{m=1}^M \left(t_{j\omega}^n \int_{S_m} G_{ij}^{N-n+1} N_\omega \, dS \right. \right.$$

$$\left. - u_{j\omega}^n \int_{S_m} F_{ij}^{N-n+1} N_\omega \, dS - t_{j\omega}^{0n} \int_{S_m} G_{ij}^{N-n+1} N_\omega \, dS \right)$$

$$\left. + \sum_{l=1}^L \left(\sigma_{jk\omega}^{0n} \int_{V_l} D_{ijk}^{N-n+1} M_\omega \, dV \right) \right]$$

$$+ \sum_{l=1}^L \left[u_{j\omega}(0) \int_{V_l} \rho g_{ij}(\tau) M_\omega \, dV \right] \tag{15.46}$$

All of the integrands remaining in Eq. (15.46) are in terms of known functions. Consequently, the spatial integration required can now be performed numerically,[6-9,25-28] as discussed in Chapter 3.

Naturally, the multi-region concept is equally applicable to unsteady viscous flows. Each geometric modelling region (GMR) occupies a volume of space V_g, bounded by the surface S_g. Then, for any field point ξ within V_g, integration can be confined to V_g and S_g. Compatibility across GMR interfaces need only be enforced at the assembly stage. As a result, during integration the GMRs can be considered independently.[25]

For each of the unsteady kernels detailed in Sec. 15.3, the degree of singularity is not immediately evident; however, series expansions can be used to isolate the

Table 15.1 Kernel singularities (in two dimensions)

Kernel	Singularity order
G_{ij}^1	$\ln r$
G_{ij}^n for $n > 1$	Non-singular
F_{ij}^1	$1/r$
F_{ij}^n for $n > 1$	Non-singular
D_{ijk}^1	$1/r$
D_{ijk}^n for $n > 1$	Non-singular

behaviour as $\xi \to x$. The results of this exercise for the various unsteady kernels are shown in Table 15.1. Notice, in particular, that only the kernel functions for the initial time increment contain singularities. These kernels, which include the instant $t = \tau$, can be rewritten as the sum of steady and transient components. Then all of the singularities are contained in the steady state components, which are just the steady incompressible viscous flow kernels. The transient components $^{tr}G_{ij}^1$, $^{tr}F_{ij}^1$ and $^{tr}D_{ijk}^1$ are completely non-singular.

15.5.4 Numerical Solution

Now that the singularities in the unsteady kernel functions have been identified and associated with those of steady viscous flow, appropriate numerical integration algorithms can be defined. Considerable care must be exercised, since the final incompressible solution is very sensitive to errors introduced during integration. For an initial time increment $N = 1$, integration required is performed exactly as discussed for elasticity and potential flow,[9,25−28] except for the strongly singular surface integrals. In this latter case, the transient component $^{tr}F_{ij}^1$ is completely non-singular and easily integrable using standard Gaussian integration. The remainder of F_{ij}^1 is the steady component $^{ss}F_{ij}$, which can be calculated indirectly via the 'rigid body' method. Consequently, the total F_{ij}^1 singular integral is determined by summing the numerically integrated $^{tr}F_{ij}^1$ with $^{ss}F_{ij}$. A similar approach was taken in elastodynamics[29] (Chapter 10) and in thermoelasticity[30,31] (Chapter 9).

Beyond the first time step, however, additional integration is required. This integration involves the kernels G_{ij}^n, F_{ij}^n and D_{ijk}^n for $n > 1$. From Table 15.1, these are all non-singular. As a result, a much less sophisticated integration scheme is employed to obtain the required level of accuracy with fewer sub-segments and Gauss points. If the initial velocities are not uniform, then the non-singular initial condition integral of Eq. (15.44) must also be evaluated at each time step.

With kernel integration complete over both time and space, a set of non-linear algebraic equations can next be assembled by bringing the boundary and interior nodes together. For each time step N, a nodal collocation process yields

$$\sum_{n=1}^{N} \left(\mathbf{G}^{N-n+1} \mathbf{t}^n - \mathbf{F}^{N-n+1} \mathbf{u}^n - \mathbf{G}^{N-n+1} \mathbf{t}^{0n} + \mathbf{D}^{N-n+1} \sigma^{0n} \right) - \mathbf{\Gamma}^N \mathbf{u}^0 = \mathbf{0} \qquad (15.47)$$

where the last term is due to the initial conditions at $t = 0$.

All of the coefficient matrices in Eq. (15.47) contain independent blocks for each GMR in multi-region problems. However, for any well-posed problem, the boundary conditions and interface relations remove all but unknown components of $\mathbf{u}^N$ and $\mathbf{t}^N$. Furthermore, by solving Eq. (15.47) at each increment of time, all of the components of $\mathbf{u}^n, \mathbf{t}^n, \mathbf{t}^{0n}$ and σ^{0n} for $n < N$ are known from previous time steps. Then Eq. (15.47) can be rewritten at time $N\Delta\tau$ as

$$\mathbf{g}(\mathbf{x}) = \mathbf{A}\mathbf{x}^N - \mathbf{D}^1\sigma^{0N} + \mathbf{G}^1\mathbf{t}^{0N} - \mathbf{B}\mathbf{y}^N$$

$$-\sum_{n=1}^{N-1}(\mathbf{G}^{N-n+1}\mathbf{t}^n - \mathbf{F}^{N-n+1}\mathbf{u}^n - \mathbf{G}^{N-n+1}\mathbf{t}^{0n} + \mathbf{D}^{N-n+1}\sigma^{0n}) + \mathbf{\Gamma}^N\mathbf{u}^0 = \mathbf{0} \quad (15.48)$$

in which

$$\mathbf{x}^N = \text{nodal vector of unknowns with two } P \text{ components}$$

$$\mathbf{y}^N = \text{nodal vector of knowns with two } Q \text{ components}$$

while $\mathbf{A}$ and $\mathbf{B}$ are the associated coefficients obtained from $\mathbf{F}^1$ and $\mathbf{G}^1$. The $\mathbf{A}$ matrix now includes the compatibility relationships enforced on GMR interfaces. As a result, the GMR blocks in $\mathbf{A}$ are no longer independent; however $\mathbf{A}$ does remain block-banded.

The terms included in the summation of Eq. (15.48) represent the contribution of past events. This, along with the terms $\mathbf{B}\mathbf{y}^N$ and $\mathbf{\Gamma}^N\mathbf{u}^0$, can be simply evaluated once at each time step N with no need for iteration. Let

$$\mathbf{b}^N = -\mathbf{B}\mathbf{y}^N - \sum_{n=1}^{N-1}(\mathbf{G}^{N-n+1}\mathbf{t}^n - \mathbf{F}^{N-n+1}\mathbf{u}^n - \mathbf{G}^{N-n+1}\mathbf{t}^{0n} + \mathbf{D}^{N-n+1}\sigma^{0n}) + \mathbf{\Gamma}^N\mathbf{u}^0$$

$$(15.49)$$

Then Eq. (15.48) becomes the following non-linear set of algebraic equations:

$$\mathbf{g}(\mathbf{x}) = \mathbf{A}\mathbf{x}^N - \mathbf{D}^1\sigma^{0N} + \mathbf{G}^1\mathbf{t}^{0N} + \mathbf{b}^N = \mathbf{0} \quad (15.50)$$

From Eq. (15.49), for each step N, one new set of matrices $\mathbf{G}^N, \mathbf{F}^N, \mathbf{D}^N$ and $\mathbf{\Gamma}^N$ must be determined via integration and assembly. Integration, particularly the volume integration needed for $\mathbf{D}^N$ and $\mathbf{T}^N$, can be quite expensive.

As an alternative to the convolution approach defined above, a time-marching recurring initial condition algorithm can be employed. As discussed in Chapter 8, this has been utilized by a number of researchers for transient problems of heat conduction. For this latter approach, at time step N the entire contribution of past events is represented by an initial condition integral which utilizes $\mathbf{u}^{N-1}$ as the initial velocity. Thus, once again Eq. (15.50) is applicable, but now

$$\mathbf{b}^N = -\mathbf{B}\mathbf{y}^N + \mathbf{\Gamma}^1\mathbf{u}^{N-1} \quad (15.51)$$

Evidently, only the evaluation of $\mathbf{b}^N$ is different. The advantage of the recurring initial condition approach is that no integration is needed beyond the first time step. However, volume integration is required throughout the entire domain because of the presence of $\mathbf{u}^{N-1}$, even for linear problems in which volume integration would not normally be required.

In order to take full advantage of both methods, Dargush and Banerjee[8,9] utilize the convolution approach in linear regions and the recurring initial condition algorithm for the remaining non-linear GMRs which are filled with volume cells. Since $\mathbf{b}^{N}$ can be computed independently for each GMR, this new dual approach provides no particular difficulty.

An iterative algorithm, along the lines of those traditionally used for BEM elasto-plasticity, can be employed to solve the boundary value problem. This was the approach originally taken by Bush and Tanner. However, convergence is usually achieved only at low Reynolds number. More generally the interior equations must be brought into the system matrix, and a full or modified Newton–Raphson algorithm must be employed to obtain solutions at moderate Reynolds number. (Related 'variable stiffness' algorithms have also been introduced by Banerjee, Raveendra, and Henry for elastoplasticity.[32–34]) Symbolically, at any iteration k, from Eq. (15.50)

$$\left[\frac{\partial \mathbf{g}}{\partial \mathbf{x}}(\mathbf{x}^{\ell})\right]\left\{\Delta \mathbf{x}^{k}\right\} = -\left\{\mathbf{g}(\mathbf{x}^{k})\right\} \tag{15.52a}$$

and

$$\mathbf{x}^{k+1} = \mathbf{x}^{k} + \Delta \mathbf{x}^{k} \tag{15.52b}$$

With the full Newton–Raphson approach, $\ell = k$ and the system matrix must be formed and decomposed at each iteration (see also Chapter 13). Iteration continues until $\|\Delta \mathbf{x}\|/\|\mathbf{x}\|$ is reduced to a small quantity, where $\|\mathbf{x}\|$ represents the Euclidean norm of $\mathbf{x}$. For the modified Newton–Raphson algorithm, since the system matrix is not formed at every iteration, only backsubstitution is needed to determine $\Delta \mathbf{x}^{k}$.

Once the iterative process has converged, a number of additional boundary quantities of interest can be easily calculated. For example, lift and drag can be calculated by numerically integrating the known nodal traction and shape function products over the surface elements of interest. Low-order Gaussian quadrature is adequate for this integration, since all the functions are very well behaved.

Furthermore, at any boundary location, the pressure p, total stress σ_{ij} and strain rates $\partial u_j/\partial x_i$ can be determined by simultaneously solving the following relationships:[6]

$$\sigma_{ji}(\xi)n_j(\xi) = N_\omega(\zeta)t_{i\omega} \tag{15.53a}$$

$$\sigma_{ij}(\xi) - \mu\left[\frac{\partial u_i}{\partial x_j}(\xi) + \frac{\partial u_j}{\partial x_i}(\xi)\right] + p(\xi) = 0 \tag{15.53b}$$

$$\frac{\partial x_j}{\partial \zeta}\frac{\partial u_i}{\partial x_j}(\xi) = \frac{\partial N_\omega}{\partial \zeta}u_{i\omega} \tag{15.53c}$$

$$\frac{\sigma_{ii}(\xi)}{2} + p(\xi) = 0 \tag{15.53d}$$

It should be emphasized that, in two dimensions, Eqs (15.53) represent a set of nine independent equations which are written at the boundary point ξ and can be solved easily for p, σ_{ij} and $\partial u_j/\partial x_i$ at that point (see Chapter 17). Afterwards, boundary vorticity and dilatation can be obtained respectively from

$$\Omega = \frac{\partial u_2}{\partial x_1} - \frac{\partial u_1}{\partial x_2} \qquad (15.54a)$$

$$\Delta = \frac{\partial u_1}{\partial x_1} + \frac{\partial u_2}{\partial x_2} \qquad (15.54b)$$

Of course, for incompressible flow, the dilatation should be zero, but Eq. (15.54b) can be used as a check.

15.6 EXAMPLES OF APPLICATION

15.6.1 Transient Couette Flow[9]

Consider first the case of developing Couette flow between two plates, parallel to the xz plane, a distance h apart. Initially, both of the plates, as well as the fluid, are at rest. Then, beginning at time $t = 0$, the bottom plate is moved continuously with velocity V in the x direction. Due to the no-slip condition at the fluid–plate interface, Couette flow begins to develop as the vorticity diffuses. Eventually, when steady conditions prevail, the x component of the velocity assumes a linear profile.

The following exact solution to this unsteady problem is provided by Schlicting:[22]

$$v_x(y, t) = V \left\{ \sum_{n=0}^{\infty} \mathrm{erfc}(2n\eta_1 + \eta) - \sum_{n=0}^{\infty} \mathrm{erfc}[2(n+1)\eta_1 - \eta] \right\} \qquad (15.55a)$$

$$v_y(y, t) = 0 \qquad (15.55b)$$

where

$$\eta = \frac{y}{(4\mu t/\rho)^{1/2}}$$

$$\eta_1 = \frac{h}{(4\mu t/\rho)^{1/2}}$$

$$\mathrm{erfc}(z) = 1 - \mathrm{erf}(z) = 1 - \frac{2}{\pi^{1/2}} \int_0^z e^{-\gamma^2} d\gamma$$

All of the non-linear terms vanish, since both v_y and $\partial v_x/\partial x$ are zero.

The two-dimensional boundary element model, utilized for this problem, is displayed in Fig. 15.1. Four quadratic surface elements are employed, with one along each edge of the domain. A number of sampling points are included strictly to monitor response. Notice that the region of interest is arbitrarily truncated at

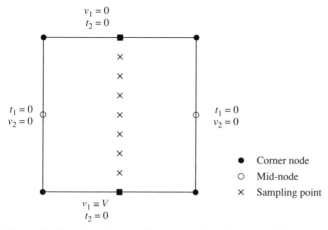

$v_1 = 0$
$t_2 = 0$

$t_1 = 0$
$v_2 = 0$

$t_1 = 0$
$v_2 = 0$

● Corner node
○ Mid-node
× Sampling point

$v_1 = V$
$t_2 = 0$

Figure 15.1 Transient Couette flow—boundary element model.

the planes $x = 0$ and $x = \ell$. All of the boundary conditions are also shown in Fig. 15.1. For the presentation of GP-BEST results, all quantities are normalized. Thus, $Y = y/h$, $T = ct/h^2$ and the horizontal velocity is v_x/V. Figure 15.2 provides the velocity profiles at four different times, using a time step $\Delta T = 0.025$ and the convolution approach. There is some error present at small times near the top plate, where the velocity is nearly zero. Results at $Y = 0.5$ versus time are shown in Fig. 15.3 for several values of the time step. Obviously, the correlation improves with a reduction in time step and $\Delta T = 0.025$ provides accurate velocities throughout the time history. However, even for a very large time step, the GP-BEST

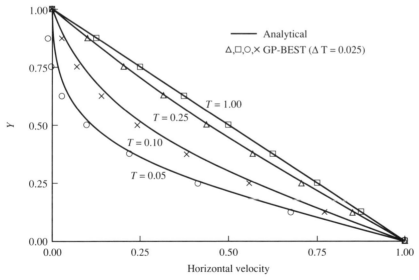

Figure 15.2 Transient Couette flow—velocity profile.

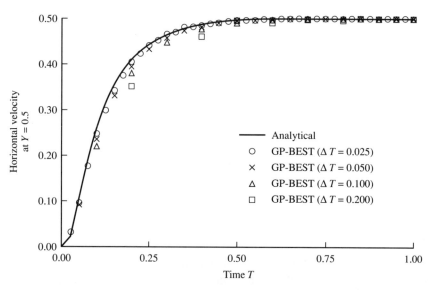

Figure 15.3 Transient Couette flow—time history via convolution.

solution shows no signs of instability. Error, evident in the initial portion, diminishes with time, and all values of ΔT produce the correct steady response. Further reduction of ΔT beyond 0.025 yields little benefit. Instead, mesh refinement in the y direction is needed, primarily to capture the short time behaviour. Figure 15.4 shows the GP-BEST results for a model with just two, equal length, elements along each vertical side. The correlation with the analytical solution is now excel-

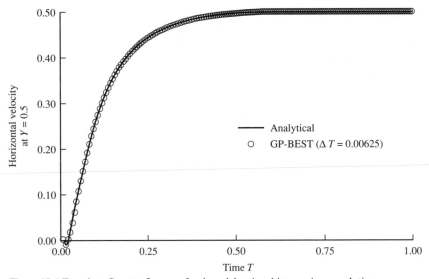

Figure 15.4 Transient Couette flow—refined model—time history via convolution.

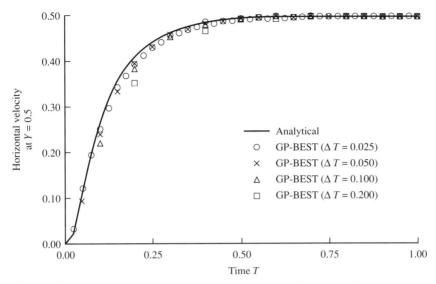

Figure 15.5 Transient Couette flow—time history via recurring initial condition.

lent. The time step selected for the refined model was based upon the general recommendation that $\Delta T \cong 0.05\ell_{min}^2/c$, where ℓ_{min} is the length of the smallest element.

The convolution approach, defined by Eqs (15.49) and (15.50), was used to obtain the results presented in Figs 15.2 to 15.4. Alternatively, the recurring initial condition algorithm can be invoked. In that case, complete volume discretization is required, even for this linear problem. For the model of Fig. 15.1, a single volume cell connecting the eight nodes is all that is required. The GP-BEST results for different values of ΔT are shown in Fig. 15.5. The solutions are good for the two smaller time step magnitudes; however, there is a slight degradation in accuracy from the convolution results.

15.6.2 Flow between Rotating Cylinders[9]

As the next example, the developing flow between rotating cylinders is analysed. The inner cylinder of radius r_i is stationary, while the outer concentric cylinder with radius r_o is given a tangential velocity V, beginning abruptly at time zero. The steady solution appears in Schlicting.[22] However, even for the transient case, the flow is purely circumferential. Thus, the governing Navier–Stokes equations reduce to

$$\mu\left(\frac{\partial^2 v_\theta}{\partial r^2} + \frac{1}{r}\frac{\partial v_\theta}{\partial r} - \frac{v_\theta}{r^2}\right) - \rho\frac{\partial v_\theta}{\partial t} = 0 \tag{15.56}$$

$$-\frac{\partial p}{\partial r} + \frac{v_\theta^2}{r} = 0 \tag{15.57}$$

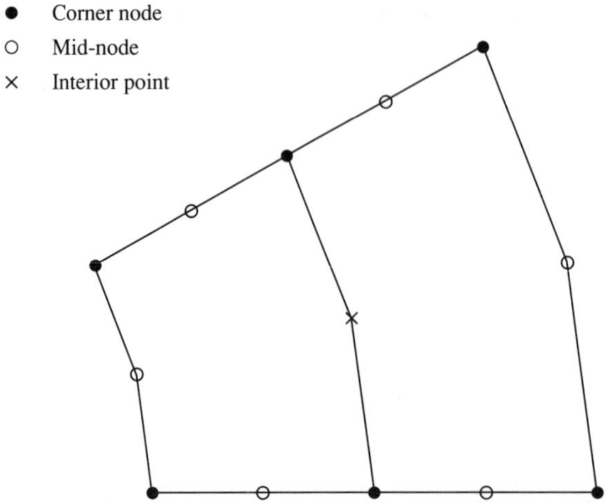

● Corner node
○ Mid-node
× Interior point

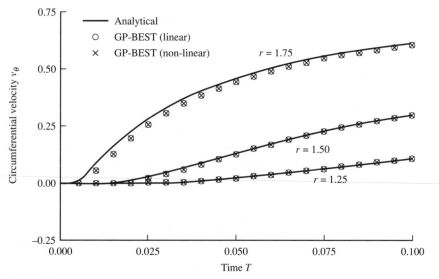

Figure 15.6 Flow between rotating cylinders—boundary element model.

in polar coordinates (r, θ, z). As discussed in Batchelor,[23] separation of variables can be used to obtain the analytical solution. Figure 15.6 depicts the boundary element model representing the region between the two cylinders. A 30° segment is isolated, with cyclic symmetry boundary conditions imposed along the edges $\theta = 0°$ and $\theta = 30°$. The inner radius is unity, while an outer radius of two is assumed. Unit values are also taken for the viscosity, density and V. The model consists of six quadratic elements and two quadratic cells. The cells, of course, are not needed for linear analysis utilizing the convolution approach.

GP-BEST analysis results are compared to the exact solution in Fig. 15.7

Figure 15.7 Flow between rotating cylinders—convolution.

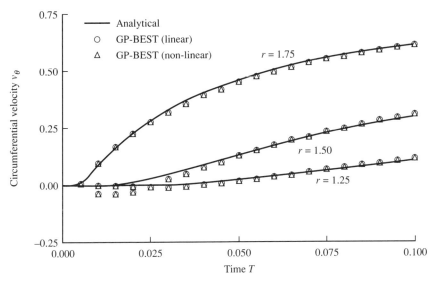

Figure 15.8 Flow between rotating cylinders—recurring initial condition.

for convolution and in Fig. 15.8 for the recurring initial condition algorithm. In both diagrams, results with and without the non-linear convective terms are plotted. The results are quite good throughout the time history with the convolution approach, while some noticeable error is present at early times for the recurring initial condition solutions. The linear and non-linear velocity profiles are nearly identical, as expected from the exact solution. However, unlike the previous example, the non-linear terms do not simply vanish from the integral equation written in Cartesian form. Instead, the non-linear surface and volume integrals must combine in the proper manner to produce the correct solution. Consequently, this problem provides a good test for the entire BEM formulation.

Relative run times are shown in Table 15.2 for the different analysis types. Obviously, the non-linear convolution approach is very expensive, since this involves volume integration at each time step. As a result, in the general implementation, convolution is only utilized in linear GMRs.

Table 15.2 Flow between rotating cylinders (run time comparisons)

Analysis type	Time-marching algorithm	Relative CPU time
Linear	Convolution	1.0
Non-linear	Convolution	25.8
Linear	Recurring initial condition	1.5
Non-linear	Recurring initial condition	1.8

15.6.3 Transient Driven Cavity Flow[6,9]

This example involves the initiation of flow in a square cavity.[35] An incompressible

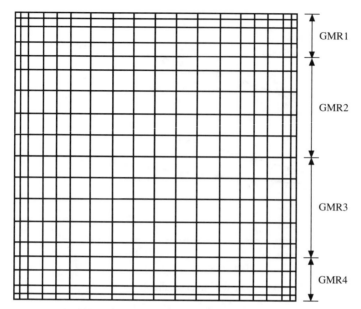

Figure 15.9 Driven capacity—boundary element model.

fluid of uniform density and viscosity is at rest within a unit square region. The velocities of the vertical sides and the bottom are fixed at zero throughout time. At time zero, the horizontal velocity of the top edge is suddenly raised to a value of 1000 and maintained at that level. A gradual transition of velocities is introduced near the top corners to provide continuity.

The four-region, 324-cell model shown in Fig. 15.9 is employed for the boundary element analysis. The resulting velocity vector plots at several times are shown in Fig. 15.10 for this case having a Reynolds number of 1000. The recurring condition algorithm was used. As in the previous two examples, the results lead directly to the steady solution after a sufficient number of time steps. The solution at $\tau = 12.5$ is within 1 per cent of the steady state boundary element results shown in Fig. 15.11, (on page 418), which in turn agree with Ghia *et al.*[36]

It should be noted that Tosaka and Kakuda[10] have run the transient driven cavity at $Re = 10\,000$. However, their results show signs of instability, even at relatively small times, and are compared to the steady solution of Ghia *et al.*[36] which also is not correct at this much higher Reynolds number. A valid solution in this *Re* range would necessitate the use of an extremely refined mesh, far beyond that employed by Tosaka and Kakuda[10] or Ghia *et al.*[36]

15.6.4 Converging Channel[6]

The two-dimensional incompressible flow through a converging channel also

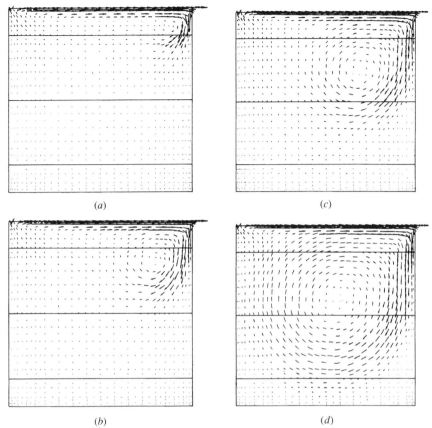

Figure 15.10 Driven capacity—velocity vectors at $Re = 1000$: (a) $t = 1.0$, (b) $t = 2.5$, (c) $t = 5.0$, (d) $t = 12.5$.

possesses a well-known analytical solution which is purely radial.[37] A comprehensive finite element study of this problem has been made by Gartling *et al.*[38]

The boundary element model is shown in Fig. 15.12. The mesh contains 96 cells and is divided into two regions. The boundary conditions were modelled using an exact specification of the boundary conditions appearing in the analytical solution (Fig. 15.12). Viscosity is unity, and tractions and density are incremented to reach higher Reynolds numbers. The Reynolds number for this problem is defined as $Re = \rho R_i V_2(R_i)/\nu$ where $V_2(R_i)$ is the maximum velocity in the region, which is -24.0 for the problem solved here.

Figure 15.13 illustrates the results for two Reynolds numbers, indicating good accuracy along the entire width of the channel. Not only are the velocities accurate, but the pressures and tractions are very accurate also.

It has been observed that finite element versions of this problem have several peculiarities which prevent the analytical solution from being reproduced. First of all, since velocities are often specified at the inlet and at the wall and centreline, an

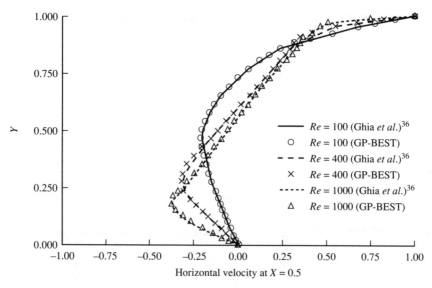

Figure 15.11 Driven capacity—centreline velocity profile.

ambiguous boundary condition specification results. Also, typically a parabolic 'fully developed' velocity profile is usually specified at the inlet. However, the non-linear solution has a flattened velocity distribution across the width of the channel (see Fig. 15.13). Hence, the analytical solution cannot be

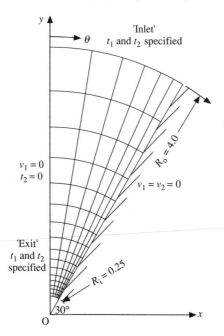

Figure 15.12 Converging channel—problem definition.

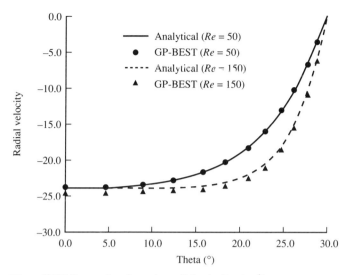

Figure 15.13 Converging channel—radial velocity at exit.

reproduced exactly if the 'fully developed' profile is specified at the inlet. Also, the finite element modellers of this problem usually leave out the traction distribution at the exit and specify zero tractions there. This also gives rise to non-radial flow.

15.6.5 Flow over a Cylinder

Next, an attempt of unconfined flow around an obstacle is considered. In particular, the oft-studied case of a unit diameter circular cylinder is examined. For example, Bush[4] obtained solutions up to $Re = 1$ with a boundary element approach. In the present analysis, higher Re flows are pursued and heat transfer is also included. The boundary element mesh is illustrated in Fig. 15.14. Notice that three distinct regions are evident. The smallest region, labelled GMR1, represents a thermoelastic thick-walled cylinder. Only the surface of the solid is discretized. The next region, GMR2, models a thermoviscous fluid in the vicinity of the cylinder. In GMR2 volume cells are required due to convective stresses.

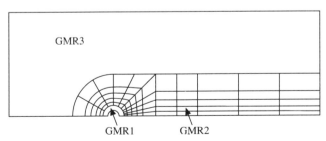

Figure 15.14 Flow over a cylinder—boundary element model.

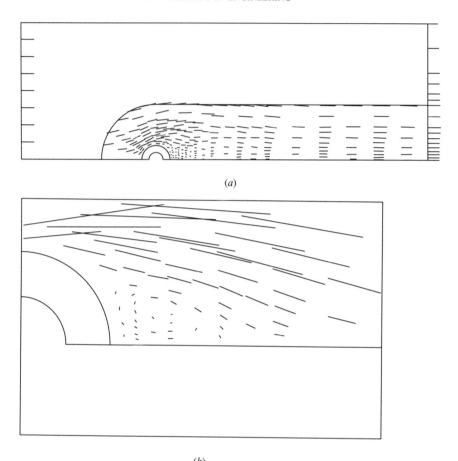

(a)

(b)

Figure 15.15 Flow over a cylinder $Re = 20$.

However, sufficiently remote from the cylinder, these convective stresses become negligible and once again a boundary-only region, in this case GMR3, is valid.

Steady state velocity vector plots are displayed in Figs 15.15 and 15.16 for $Re = 20$ and 40 respectively. The circulating zone, behind the cylinder, is clearly visible.

Additionally, the problem was extended to include thermal effects. The temperature of the fluid at inlet was specified as 100°C, while that at the inner surface of the hollow cylinder was maintained at 0°C. The effective heat transfer coefficient between the fluid and solid can then be obtained from the resulting temperature and flux at the outer surface of the cylinder. The distribution of the non-dimensional Nusselt number (Nu) around the circumference is plotted in Fig. 15.17 on page 422. These curves agree, at least, qualitatively with the experimental results of Eckert and Soehngen.[39] Of course, if the purpose of the analysis is to determine the temperature and stress in the solid, then there is really no need to complete the

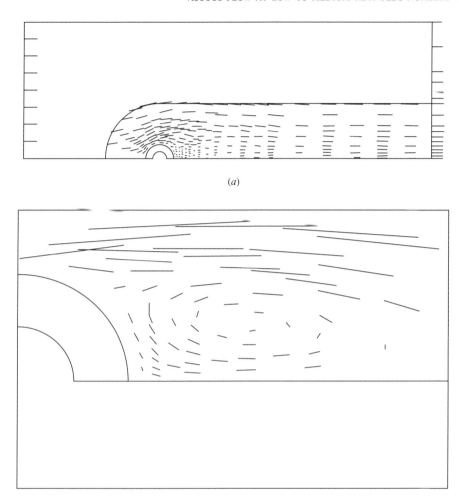

(a)

(b)

Figure 15.16 Flow over a cylinder $Re = 40$.

heat transfer coefficients. The desired solid temperatures and the stresses come directly out of the analysis.

When the Reynolds number is elevated above 40, convergence difficulties begin to appear, even with a full Newton–Raphson approach. At high Re, the convective terms start to dominate the viscous effects. The present formulation, which employs the Stokes flow fundamental solutions, is then no longer suitable.

15.7 CONCLUDING REMARKS

In this chapter, the boundary element method has been developed for transient

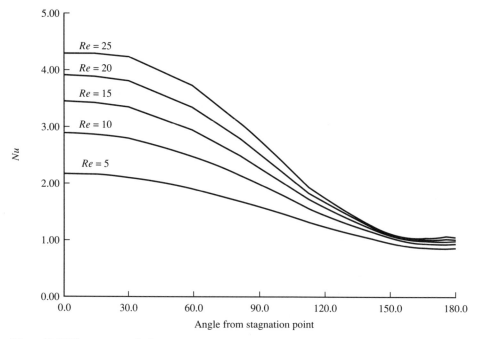

Figure 15.17 Flow over a cylinder—Nusselt number.

incompressible viscous flow. The integral formulation is written exclusively in terms of velocities and tractions. The evaluation of velocity gradients is not needed anywhere in the formulation. For Stokes flow, the non-linear terms vanish. In that case, a boundary-only solution results and no volume discretization is required. The current method is ideal for these problems. At higher velocities, the non-linear convective terms are important, but these are often confined to a small portion of the flow region. Even though volume cells, and subsequently volume integration, are needed in this non-linear portion, the method still provides for very cost-effective analyses.

The BEM has been described in a general form using, for instance, time marching which can be accomplished via a combination of convolution and the recurring initial condition approach. Iteration options for non-linear analyses include a full or modified Newton–Raphson approach, or, for mild linearities, direct iteration. The examples presented here indicate the level of precision that is obtained with the boundary element approach for transient incompressible flow. The stability of these solutions is equally impressive.

REFERENCES

1. Oseen, C.W. (1927) *Neuere Methoden und Ergebnisse in der Hydrodynamik*, Akad. Verlagsgellschaft, Leipzig.

2. Youngren, G.K. and Acrivos, A. (1975) Stokes flow past a particle of arbitrary shape: a numerical method of solution, *Journal of Fluid Mechanics*, Vol. 69, No. 2, pp. 377–403.
3. Bush, M.B. and Tanner, R.I. (1983) Numerical solution of viscous flow using integral equation methods, *International Journal of Numerical Methods in Fluids*, Vol. 3, pp. 71–92.
4. Bush, M.B. (1983) Modelling two-dimensional flow past arbitrary cylindrical bodies using boundary element formulations, *Applied Mathematical Modelling*, Vol. 7, pp. 386–394.
5. Dargush, G.F. and Banerjee, P.K. (1989) Development of an integrated BEM for hot fluid–structure interaction, *Transactions of the ASME*, Paper 89-GT-128 (1989); also appeared in *Journal of Engineering for Gas Turbine and Power, ASME*, Vol. 112, pp. 243–250 (1990).
6. Dargush, G.F. and Banerjee, P.K. (1991) A BEM for steady incompressible thermoviscous fluid flow, *International Journal of Numerical Methods in Engineering*, Vol. 31, pp. 1605–1626.
7. Dargush, G.F. and Banerjee, P.K. (1990) Advanced boundary element methods for steady incompressible thermoviscous flow, in *Developments in Boundary Element Methods*, Vol. 6, Eds P.K. Banerjee and L. Morino, Elsevier Applied Science Publishers, London.
8. Dargush, G.F. and Banerjee, P.K. (1990) A time-dependent incompressible viscous BEM for moderate Reynolds number, in *Developments in Boundary Element Methods*, Vol. 6, Eds P.K. Banerjee and L. Morino, Elsevier Applied Science Publishers, London.
9. Dargush, G.F. and Banerjee, P.K. (1991) A time dependent incompressible viscous BEM for moderate Reynolds numbers, *International Journal of Numerical Methods in Engineering*, Vol. 31, pp. 1627–1648.
10. Tosaka, N. and Kakuda, K. (1987) Numerical simulation of laminar and turbulent flows by using an integral equation, *Boundary Elements* IX, Eds A. Brebbia, W.L. Wendland and G. Kuhn, pp. 489–502, Springer-Verlag, Berlin.
11. Tosaka, N. (1987) Integral equation method for analysis of Newtonian and non-Newtonian flows, IUTAM Symposium on *Advanced Boundary Element Method*, San Antonio, Texas.
12. Wu, J.C. and Thompson, J.F. (1973) Numerical solutions of time-dependent incompressible Navier–Stokes equations using an integro-differential formulation, *Computers and Fluids*, Vol. 1, pp. 197–215.
13. Wu, J.C., Rizk, Y.M. and Sankar, H.L. (1984) Problems of time-dependent Navier–Stokes flow, Chapter 6, in *Developments in Boundary Element Methods*, Vol. 3, Eds P.K. Banerjee and S. Mukherjee, Elsevier Applied Science, London.
14. Onishi, K., Kuroki, T. and Tanaka, M. (1984) An application of boundary element method to incompressible laminar viscous flows, *Engineering Analysis*, Vol. 1, No. 3, pp. 122–127.
15. Tosaka, N. and Onishi, K. (1986) Boundary integral equation formulations for unsteady incompressible viscous fluid flow by time-differencing, *Engineering Analysis*, Vol. 3, No. 2, pp. 101–104.
16. Piva, R., Graziani, G. and Morino, L. (1987) Boundary integral equation method for unsteady viscous and inviscid flows, IUTAM Symposium on *Advanced Boundary Element Method*, San Antonio, Texas.
17. Piva, R. and Morino, L. (1987) Vector Green's function method for unsteady Navier–Stokes equations, *Meccanica*, Vol. 22, pp. 76–85.
18. Lamb, H. (1932) *Hydrodynamics*, 6th edn, Dover, New York.
19. White, F.M. (1974) *Viscous Fluid Flow*, McGraw-Hill, New York.
20. Yih, C.S. (1969) *Fluid Mechanics*, McGraw-Hill, New York.
21. Ladyzhenskaya, O.A. (1969) *The Mathematical Theory of Viscous Incompressible Flow*, Gordon and Breach Science Publishers, New York.
22. Schlicting, H. (1955) *Boundary Layer Theory*, McGraw-Hill, New York.
23. Batchelor, G.K. (1967) *An Introduction to Fluid Dynamics*, Cambridge University Press, Cambridge.
24. Carslaw, H.S. and Jaeger, J.C. (1959) *Conduction of Heat in Solids*, 2nd edn, Oxford University Press.
25. Banerjee, P.K., Wilson, R.B. and Miller, N. (1988) Advanced elastic and inelastic three-dimensional analysis of gas turbine engine structures by BEM, *International Journal of Numerical Methods in Engineering*, Vol. 26, pp. 393–411.
26. Mustoe, G.G.W. (1984) Advanced integration schemes over boundary elements and volume cells

for two- and three-dimensional nonlinear analysis, in *Developments in Boundary Elements Methods*, Vol. III, Eds P.K. Banerjee and S. Mukherjee, Elsevier Applied Science Publishers, London.

27. Dargush, G.F. and Banerjee, P.K. (1989) Advanced development of the boundary element method for steady-state heat conduction, *International Journal of Numerical Methods in Engineering*, Vol. 28, pp. 2123–2142.

28. Honkala, K.A. (1992) Boundary element methods for two-dimensional coupled thermoviscous flow, PhD Dissertation, State University of New York at Buffalo.

29. Ahmad, S. and Banerjee, P.K. (1988) Transient elastodynamic analysis of three-dimensional problems by BEM, *International Journal of Numerical Methods in Engineering*, Vol. 26, No. 8, pp. 1560–1580.

30. Dargush, G.F. and Banerjee, P.K. (1989) Development of a boundary element method for time-dependent planar thermoelasticity, *International Journal of Solids and Structures*, Vol. 25, pp. 999–1021.

31. Dargush, G.F. and Banerjee, P.K. (1990) Boundary element method in three-dimensional thermo-elasticity, *International Journal of Solids and Structures*, Vol. 26, pp. 199–216.

32. Banerjee, P.K., Henry, D.P. and Raveendra, S.T. (1989) Advanced inelastic analysis of solids by BEM, *International Journal of Mechanical Science*, Vol. 31, No. 4, pp. 309–322.

33. Banerjee, P.K. and Raveendra, S.T. (1987) A new boundary element formulation for two-dimensional elastoplastic analysis, *Journal of Engineering Mechanics, ASCE*, Vol. 113, No. 2, pp. 252–265.

34. Henry, D.P. and Banerjee, P.K. (1988) A variable stiffness type boundary element formulation for axisymmetric elastoplastic media, *International Journal for Numerical Methods in Engineering*, Vol. 25, pp. 1005–1027.

35. Burggraf, O. (1966) Analytical and numerical studies of the structure of steady separated flows, *Journal of Fluid Mechanics*, Vol. 24, Part 1, pp. 113–151.

36. Ghia, U., Ghia, K.N. and Shin, C.T. (1982) High-*Re* solutions for incompressible flow using the Navier–Stokes equations and a multigrid method, *Journal of Comparative Physics*, Vol. 48, pp. 387–411.

37. Millsaps, K. and Pohlhausen, K. (1953) Thermal distributions in Jeffery–Hamel flows between nonparallel plane walls, *Journal of the Aeronautical Sciences*, March, pp. 187–196.

38. Gartling, D.K., Nickell, R.E. and Tanner, R.E. (1977) A finite element convergence study for accelerating flow problems, *International Journal of Numerical Methods in Engineering*, Vol. 11, pp. 1155–1174.

39. Eckert, E.R.G. and Soehngen, E. (1952) Distribution of heat-transfer coefficients around circular cylinders in crossflow at Reynolds numbers from 20 to 500, *Transactions of the ASME*, Vol. 74, pp. 343–347.

CONVECTIVE FLUID FLOW AND HEAT TRANSFER

16.1 INTRODUCTION

At high fluid velocities, the convective terms in the Navier–Stokes equations tend to dominate. As a result, boundary element formulations, outlined in Chapter 15, employing Stokes kernels begin to be unsuccessful at high Reynolds numbers, since these fundamental solutions model the effects of viscosity but not convection. For these types of flow more of the physics of the problem must be brought into the linear operator and an appropriate Green function must be developed. This concept was clearly understood by Oseen[1,2] in the early portion of the twentieth century. In his (1927) monograph, Oseen[2] developed exact integral expressions for Navier–Stokes equations using a convective steady state fundamental solution. Unfortunately, since this was well before the advent of the computer, he was unable to do much with his formulations beyond some approximate solutions at very low Reynolds numbers. In the present chapter, the work of Oseen and the ideas underlying it are resurrected to form the basis for an attractive boundary element method for high-speed flows.

16.2 GOVERNING EQUATIONS WITH OSEEN APPROXIMATION

As a starting point, reference values for each of the primary variables except the temperature are introduced in an effort to produce a linearized differential operator. Thus let

$$v_i = U_i + u_i \tag{16.1a}$$

$$p = p_0 + \tilde{p} \tag{16.1b}$$

$$\rho = \rho_0 + \tilde{\rho} \tag{16.1c}$$

in which U_i, p_0 and ρ_0 are constant reference values and u_i, $\tilde{p}$ and $\tilde{\rho}$ are the perturbations. Substituting Eqs. (16.1) into Eqs (15.7) yields, after some manipulation,

$$\frac{D_0\tilde{\rho}}{Dt} + \rho_0\frac{\partial u_i}{\partial x_i} = -u_i\frac{\partial \tilde{\rho}}{\partial x_i} - \tilde{\rho}\frac{\partial u_i}{\partial x_i} + \psi \tag{16.2a}$$

$$\rho_0\frac{D_0 u_i}{Dt} - (\lambda + \mu)\frac{\partial^2 u_j}{\partial x_i \partial x_j} - \mu\frac{\partial^2 u_i}{\partial x_j \partial x_j} + \frac{\partial \tilde{p}}{\partial x_i} = -\rho u_j\frac{\partial u_i}{\partial x_j} - \tilde{\rho}\frac{D_0 u_i}{Dt} + f_i \tag{16.2b}$$

$$\rho_0 c_v\frac{D_0\theta}{Dt} - k\frac{\partial^2\theta}{\partial x_i \partial x_i} - \frac{D_0\tilde{p}}{Dt} = -\rho c_v u_i\frac{\partial\theta}{\partial x_i} - \tilde{\rho}c_v\frac{D_0\theta}{Dt} + u_i\frac{\partial\tilde{p}}{\partial x_i}$$

$$-\frac{p}{\rho}\psi + \Phi + \phi \tag{16.2c}$$

where now a new linear differential operator is introduced:

$$\frac{D_0}{Dt} = \frac{\partial}{\partial t} + U_i\frac{\partial}{\partial x_i} \tag{16.3}$$

Now, in Eqs (16.2), the entire left-hand side involves a linear differential operator with constant coefficients. Terms on the right-hand side of Eqs (16.2) are nonlinear, and can for the present be considered as pseudo body forces and sources of unknown magnitude. Then, the linearized governing equations become

$$\frac{D_0\tilde{\rho}}{Dt} + \rho_0\frac{\partial u_i}{\partial x_i} = \bar{\psi} \tag{16.4a}$$

$$\rho_0\frac{D_0 u_i}{Dt} - (\lambda + \mu)\frac{\partial^2 u_j}{\partial x_i \partial x_j} - \mu\frac{\partial^2 u_i}{\partial x_j \partial x_j} + \frac{\partial \tilde{p}}{\partial x_i} = \bar{f}_i \tag{16.4b}$$

$$\rho_0 c_v\frac{D_0\theta}{Dt} - k\frac{\partial^2\theta}{\partial x_i \partial x_i} - \frac{D_0\tilde{p}}{Dt} = \bar{\phi} \tag{16.4c}$$

in which

$$\bar{\psi} = -u_i\frac{\partial\tilde{\rho}}{\partial x_i} - \tilde{\rho}\frac{\partial u_i}{\partial x_i} + \psi \tag{16.5a}$$

$$\bar{f}_i = -\rho u_j\frac{\partial u_i}{\partial x_j} - \tilde{\rho}\frac{D_0 u_i}{Dt} + f_i \tag{16.5b}$$

$$\bar{\phi} = -\rho c_v u_i\frac{\partial\theta}{\partial x_i} - \tilde{\rho}c_v\frac{D_0\theta}{Dt} + u_i\frac{\partial\tilde{p}}{\partial x_i} - \frac{p}{\rho}\psi + \Phi + \phi \tag{16.5c}$$

Under some circumstances the system equations (16.4) can be simplified further for several problems as follows:

- Convective heat transfer
- Convective potential flow
- Convective incompressible thermoviscous flow
- Convective compressible thermoviscous flow

The governing equations of the above problems will be presented in the following sub-sections separately. The fundamental solutions and boundary integral representations will also be presented in the subsequent sections.

16.3 CONVECTIVE HEAT TRANSFER

16.3.1 Governing Equation

If the flow field is known or can be approximated, the fluid temperature can be determined directly from the scalar convective diffusion equation

$$\rho c_\epsilon \frac{D_0 \theta}{Dt} - k \frac{\partial^2 \theta}{\partial x_i \partial x_i} - \bar{\phi} = 0 \tag{16.6}$$

where $\bar{\phi}$ is a non-linear term defined by

$$\bar{\phi} = -\rho c_\epsilon u_i \frac{\partial \theta}{\partial x_i} + \phi$$

The linear part of the differential equation (16.6), i.e. the first two terms on the left-hand side of Eq. (16.6), represents heat transfer as a medium moving with a velocity U_i.

16.3.2 Fundamental Solutions

For an instantaneous point source of heat defined by

$$\bar{\phi} = \delta(x - \xi)\delta(t - \tau)$$

in a medium moving with velocity U_i, the fundamental solutions can be expressed as

$$g(x - \xi, t - \tau) = \begin{cases} \dfrac{1}{8\rho c_\epsilon (\pi \kappa t')^{3/2}} e^{-r_u^2/4\kappa t'} & \text{in three dimensions} \\[3mm] \dfrac{1}{4\pi k t'} e^{-r_u^2/4\kappa t'} & \text{in two dimensions} \end{cases} \tag{16.7}$$

in which

$$r_u^2 = (y_i - U_i t')(y_i - U_i t') \quad \text{and} \quad \kappa = k/\rho c_\epsilon$$

The above solution, which is taken from Carslaw and Jaeger,[3] can be further manipulated for a maintained heat source

$$\bar{\phi} = \delta(x - \xi)H(t)$$

in a moving medium, to produce the following temperature distribution:[4,5,6]

$$G(x - \xi, t)$$

$$= \int_0^t g(x - \xi, t - \tau) \, d\tau$$

$$= \begin{cases} \dfrac{1}{8\pi kr} e^{U_k y_k/2\kappa} \left(e^{Ur/2\kappa} \operatorname{erfc} \dfrac{r + Ut}{2\sqrt{\kappa t}} + e^{-Ur/2\kappa} \operatorname{erfc} \dfrac{r - Ut}{2\sqrt{\kappa t}} \right) & \text{in three dimensions} \\[3mm] \dfrac{1}{4\pi k} e^{U_k y_k/2\kappa} \psi(r, t) & \text{in two dimensions} \end{cases}$$

$$(16.8)$$

where $U^2 = U_i U_i$ and

$$\psi(r, t) = \int_0^t \frac{e^{-r^2/4\kappa t' - U^2 t'/4\kappa}}{t'} \, dt' = K_0\left(\frac{Ur}{2\kappa}\right) + K_0\left(\ln\frac{Ut}{r}, \frac{Ur}{2\kappa}\right) \qquad (16.9)$$

$K_0(\alpha)$ is the modified Bessel function of the second kind of order zero and $K_0(\alpha, \beta)$ is so-called incomplete MacDonald function defined by[4]

$$K_\nu(\alpha, \beta) = \pi \frac{\beta^\nu}{A_\nu} \int_1^{\cosh \alpha} e^{-\beta x}(x^2 - 1)^{\nu - 1/2} \, dx$$

$$= \pi \frac{\beta^\nu}{A_\nu} \int_0^\alpha e^{-\beta \cosh x} \sinh^{2\nu} x \, dx$$

$$(16.10)$$

where the coefficient

$$A_\nu = 2^\nu \Gamma(\nu + \tfrac{1}{2}) \Gamma(\tfrac{1}{2})$$

The steady state response can be obtained from Eqs (16.8) if one lets $t \to \infty$, when these simplify to

$$G^s(x - \xi) = G(x - \xi, \infty) = \begin{cases} \dfrac{1}{4\pi kr} e^{(U_k y_k - Ur)/2\kappa} & \text{in three dimensions} \\[3mm] \dfrac{1}{2\pi k} e^{U_k y_k/2\kappa} K_0\left(\dfrac{Ur}{2\kappa}\right) & \text{in two dimensions} \end{cases}$$

$$(16.11)$$

It is of interest to compare Eqs (16.11) with a stationary counterpart, with $U = 0$. The stationary response is symmetric about the source point; however, the convective response is magnified ahead of the source point, but greatly reduced behind it. This latter phenomenon is just the Doppler effect applied to moving heat sources.

The two-dimensional transient convective diffusion kernel can also be formed by expressing the incomplete MacDonald function in a series. Thus,

$$
G(x - \xi, t) =
\begin{cases}
\dfrac{1}{4\pi k} e^{U_k y_k / 2\kappa} \displaystyle\sum_{n=0}^{\infty} \dfrac{(-U^2 t/4\kappa)^n}{n!} E_{n+1}\left(\dfrac{r^2}{4\kappa t}\right) & \text{for small } t \\[2em]
\dfrac{1}{2\pi k} e^{U_k y_k / 2\kappa} K_0\left(\dfrac{Ur}{2\kappa}\right) - \dfrac{1}{4\pi k} e^{U_k y_k / 2\kappa} & \text{for large } t \\[1em]
\times \displaystyle\sum_{n=0}^{\infty} \dfrac{(-r^2/4\kappa t)^n}{n!} E_{n+1}\left(\dfrac{U^2 t}{4\kappa}\right)
\end{cases}
$$

$$(16.12)$$

where E_{n+1} is the exponential integral of order $n + 1$. These Green functions have $\ln r$ type singularity for a two-dimensional problem and $1/r$ type singularity for a three-dimensional problem.

16.3.3 Boundary Integral Representations

We now multiply both sides of Eq. (16.6) by $\tilde{g}$ and integrate by parts twice over the volume V. This will generate an equation expressing θ in terms of the boundary information on S and derivatives of $\tilde{g}$. Thus we have

$$
\int_0^T \int_S \left[-\tilde{g}\left(-k\frac{\partial \theta}{\partial x_i} n_i + \rho c_\epsilon U_i \theta n_i \right) + \left(-k\frac{\partial \tilde{g}}{\partial x_i} n_i \right)\theta \right] dS\,dt
$$

$$
+ \int_0^T \int_V [\tilde{g}\bar{\phi}]\,dV\,dt - \rho c_\epsilon \int_V [\tilde{g}\theta|_0^T]\,dV
$$

$$
+ \int_0^T \int_V \left(\rho c_\epsilon \frac{D_0 \tilde{g}}{Dt} + k\frac{\partial^2 \tilde{g}}{\partial x_i \partial x_i} \right)\theta\,dV\,dt = 0 \qquad (16.13)
$$

with n_i defined as the unit normal to the surface S at x. To complete the derivation of the integral equation for any point ξ interior to S at time τ, the last volume integral appearing in Eq. (16.13) must be reduced to $\theta(\xi, \tau)$. This is accomplished if

$$
\rho c_\epsilon \frac{D_0 \tilde{g}}{Dt} + k\frac{\partial^2 \tilde{g}}{\partial x_i \partial x_i} + \delta(x - \xi)\delta(t - \tau) = 0 \qquad (16.14)
$$

The Green function $\tilde{g}$ defined by Eq. (16.14) is once again the adjoint of the original Green function. This adjoint Green function can be obtained simply by suitably transposing the fundamental solution presented above.

$$
\tilde{g}(x - \xi, t - \tau) = g(\xi - x, \tau - t)
$$

Substituting Eq. (16.14) into Eq. (16.13) produces the desired integral equation

$$c(\xi)\theta(\xi,\tau) = \int_0^T \int_S [f(\xi - x, \tau - t)\theta(x,t) - g(\xi - x, \tau - t)q_n(x,t)] \, \mathrm{d}S(x) \, \mathrm{d}t$$
$$+ \int_0^T \int_V [g(\xi - x, \tau - t)\bar{\phi}(x,t)] \, \mathrm{d}V(x) \, \mathrm{d}t + \int_V [g(\xi - x, \tau)\rho c_\epsilon \theta(x,0)] \, \mathrm{d}V(x)$$

$$(16.15)$$

in which

$$q_n(x,t) = -k\frac{\partial\theta(x,t)}{\partial x_i}n_i(x) + \rho c_\epsilon U_i \theta(x,t)n_i(x)$$

$$f(\xi - x, \tau - t) = -k\frac{\partial g(\xi - x, \tau - t)}{\partial x_i}n_i(x)$$

and $c(\xi)$ is constant whose value once again depends on the location of ξ. The last integral represents the effect of the initial temperature distribution throughout the region of interest.

For simplicity, Eq. (16.15) can be rewritten, as in Chapter 8, using Riemann convolution as

$$c\theta = \int_S (f * \theta - g * q_n) \, \mathrm{d}S + \int_V (g * \bar{\phi} + g\rho c_\epsilon \theta^0) \, \mathrm{d}V \qquad (16.16)$$

in which θ^0 presents the initial condition.

As may be recalled from Eq. (16.4c), a portion of the convective effects are included in the heat source $\bar{\phi}$. Assuming for the moment that this is the only non-zero component of $\bar{\phi}$, then the volume integral in (16.16) can be rewritten as

$$\int_V (g * \bar{\phi}) \, \mathrm{d}V = -\int_V \left(g * \rho c_\epsilon u_i \frac{\partial\theta}{\partial x_i} \right) \mathrm{d}V \qquad (16.17)$$

Applying the divergence theorem to the right-hand side of Eq. (16.17) produces[5,6]

$$\int_V \left(g * \rho c_\epsilon u_i \frac{\partial\theta}{\partial x_i} \right) \mathrm{d}V = \int_S (g * \rho c_\epsilon u_i \theta n_i) \, \mathrm{d}S - \int_V \left(\frac{\partial g}{\partial x_i} * \rho c_\epsilon u_i \theta \right) \mathrm{d}V$$

Thus, finally Eq. (16.16) becomes[7,8]

$$c\theta = \int_S (f * \theta - g * q_n') \, \mathrm{d}S + \int_V (d_i * q_i^0 + g\rho c_\epsilon \theta^0) \, \mathrm{d}V \qquad (16.18)$$

where

$$q_n' = q_n + \rho c_\epsilon u_i \theta n_i = -k\frac{\partial\theta}{\partial n} + \rho c_\epsilon v_n \theta$$

$$d_i = \frac{\partial g}{\partial x_i}$$

$$q_i^0 = \rho c_\epsilon u_i \theta$$

16.3.4 Numerical Implementation

After the usual temporal and spatial discretization, the boundary integral equation becomes (see Chapters 8, 9, 10 and 15)

$$
\begin{aligned}
c\theta = \sum_{n=1}^{N} & \left[\sum_{m=1}^{M} \left(\theta_\omega^n \int_{S_m} F^{N-n+1} N_\omega \, dS - q_{n\omega}^n \int_{S_m} G^{N+n-1} N_\omega \, dS \right) \right. \\
& \left. + \sum_{l=1}^{L} \left(q_{i\omega}^{0n} \int_{V_l} D_i^{N-n+1} M_\omega \, dV \right) \right] + \sum_{l=1}^{L} \left(\rho c_\epsilon \theta_\omega^0 \int_{V_l} g^N M_\omega \, dV \right)
\end{aligned}
$$

(16.19)

where once again the time integration has been performed analytically and

$$
G^n = G(\xi - x, n\Delta\tau)
$$
$$
F^n = F(\xi - x, n\Delta\tau)
$$

can be obtained from Eq. (16.8) with appropriate change of arguments, as explained in Chapter 15.

As noted earlier, the behaviour of the convective fundamental solutions G and F can be quite different from their stationary counterparts. Consequently, new integration algorithms must be devised to perform the required quadrature in an accurate and efficient manner. The following three categories are distinguished when integrating an element m:

I. The source point does not lie on element m.
II. The source point lies on element m, but only the weakly singular G kernel is involved.
III. The source point lies on element m, and the strongly singular F kernel is involved.

In general, for non-singular integration of both the G and F kernels of category I, sub-segmentation is employed along with standard Gaussian quadrature formulae. The sub-segmentation pattern and Gaussian order are established on the basis of the non-dimensional parameters α and β, where

$$
\alpha = \frac{Ur}{2\kappa} \qquad \beta = \frac{U_k y_k}{2\kappa}
$$

For integration of elements far upstream of the source point, only a minimal number of Gauss points are required. At the other extreme are elements just downstream of the source. In that case, graded sub-segmentation is utilized along with high-order formulae.

Several alternatives are possible for evaluation of category II integrals. The first involves subsegmentation and high-order Gauss quadrature. This approach is quite inefficient, although, for large problems, integration cost is not significant. A second alternative combines analytical and numerical integration. In this case, a small flat sub-segment tangential to the singular element is identified and integrated analytically, the remaining portion is then handled numerically.

Thus the length of this subsegment S^* can be determined from

$$S^* \leq \frac{2\kappa}{U}\alpha^*$$

and the desired integral is expressible as

$$\int_{S^*} e^{-\beta} K_0(\alpha) N_\omega(\zeta)\, dS(x) = \frac{2\kappa}{U} \int_0^{\alpha^*} e^{-k\alpha} K_0(\alpha) N_\omega(\zeta(\alpha))\, d\alpha \qquad (16.20)$$

where $k = U_k y_k / Ur$ is constant over the flat segment. The integral in Eq. (16.20) can be then evaluated analytically as shown in Shi and Banerjee.[7,8]

After the evaluation of integrals of types I and II, the ensuing nodal collocation process produces the usual global set of equations

$$[F]\{\theta\} - [G]\{q_n\} = \{0\} \qquad (16.21)$$

The diagonal terms of F have not been completely determined due to the strongly singular nature of F. These diagonal contributions can not be calculated by imposing a uniform 'equipotential' field on the body directly because the 'equipotential' and zero flux condition does not exist any more. Instead, Eq. (16.21) can be written as

$$[F]\{1\} - [G]\{\rho c_\epsilon U_n\} = 0 \qquad (16.22)$$

where $U_n = U_i n_i$.

Using Eq. (16.22), the desired diagonal terms, F^d, can be obtained from the summation of the off-diagonal terms of $[F]$ and the vector $\{[G]\{\rho c_\epsilon U_n\}\}$. This can be expressed as[7,8]

$$F^d = -\sum_{\substack{p=1 \\ p \neq d}}^{P} F + \sum_{q=1}^{Q} \rho c_\epsilon G U_n \qquad (16.23)$$

The indirect calculation for diagonals of F kernels can also be extended to calculate the diagonals of G kernels which are also difficult to obtain by numerical integration for the high convective velocity U_i. This is possible, provided θ and the corresponding q_n are taken to be known quantities, as evaluated from known simple temperature fields which satisfy the governing equation for an arbitrary body in the convective heat transfer problem. The required number of these temperature fields clearly depends on the numbers of quantities to be evaluated as the unknowns in Eq. (16.22). In addition to the uniform temperature field used earlier, the temperature field for any arbitrary body can be taken as

$$\theta = x_2$$
$$\theta = e^{U_k x_k / \kappa}$$

where the existence of x_1 direction free-stream velocity is assumed. The corresponding flux fields on the boundary which could maintain these temperature fields could be computed easily. It should be noted here that only one set of these solutions is needed when ξ is on the mid-node of an element or ξ on an extreme node of an element which is smooth at ξ, and both sets of solutions are needed when ξ is on an extreme node of an element which is not smooth at ξ. In the latter case, the solution of a system of simultaneous equations is required.

16.4 CONVECTIVE POTENTIAL FLOW

16.4.1 Introduction

Compressible potential flow is one of the most important fields of aerodynamics. For sufficiently large Reynolds numbers the important viscous effects are often confined to an infinitesimal thin boundary layer adjacent to the surface of a body and its wake. Outside the wake and the vortical region near the boundary the flow is essentially irrotational. This fact was first observed by Prandtl in 1904.

16.4.2 Governing Equations

The linearized governing differential equation of convective potential flow for iso-tropic, homogeneous space can be written as

$$\frac{1}{c_0^2}\frac{\mathrm{D}_0^2\phi'}{\mathrm{D}t^2} - \frac{\partial^2\phi'}{\partial x_i\,\partial x_i} = \bar{\psi} \qquad (16.24)$$

where ϕ' = velocity perturbation potential defined as $u_i = \partial\phi'/\partial x_i$
u_i = velocity perturbation
U_i = reference velocity
c_0 = speed of sound
γ = ratio of specific heat at constant pressure to that at constant volume
$\bar{\psi}$ = pseudo mass source rate per unit mass which is defined by

$$\bar{\psi} = -\frac{\gamma-1}{c_0^2}\left(\frac{\partial\phi'}{\partial t} + U_i u_i + \frac{1}{2}u^2\right)\frac{\partial u_j}{\partial x_j} - \frac{2}{c_0^2}u_i\frac{\partial u_i}{\partial t} - \frac{2}{c_0^2}\left(U_i u_j + \frac{1}{2}u_i u_j\right)\frac{\partial u_i}{\partial x_j} \quad (16.25)$$

where $u^2 = u_i u_i$.

The equation governing the compressible potential flow has different charac-ters in different flow regimes. For a transient problem, the governing equation is hyperbolic for all Mach numbers and solutions can be obtained using a time-marching procedure. The situation is very different when a steady flow is assumed, the equation is then elliptic when the flow is subsonic and hyperbolic when the flow is supersonic. One of the most important and distinctive features of supersonic flow is the fact that shock waves occur in the flow field.

16.4.3 Fundamental Solutions

Equation (16.24) is a well-known convective scaler wave equation. However, the fundamental solution in the convective form does not appear to exist in the literature although some discussions can be found in Goldstein[9] and Morse and Feshbach.[10]

Three-dimensional flow   The fundamental solutions of Eq. (16.24) due to an instantaneous source $\delta(x - \xi)\delta(t - \tau)$ in a medium moving with a velocity U_i can be obtained by using Laplace and triple Fourier transform as[9,10]

$$g(x - \xi, t - \tau) = \begin{cases} \dfrac{c_0}{4\pi} \dfrac{\delta(t' - r_0/c_0)}{\sqrt{(U_k y_k)^2 + c_0^2 \beta^2 r^2}} & M_\infty < 1 \\[4mm] \dfrac{c_0}{4\pi} \dfrac{H(U_k y_k - c_0 \beta r)}{\sqrt{(U_k y_k)^2 - c_0^2 \beta^2 r^2}} \left[\delta\left(\dfrac{t' - r_0}{c_0}\right) + \delta\left(\dfrac{t' - r_1}{c_0}\right) \right] & M_\infty > 1 \end{cases}$$

(16.26)

where r_0 and r_1 are two values of r given by

$$r_{0,1} = \frac{U_k y_k \pm \sqrt{(U_k y_k)^2 - c_0^2 \beta^2 r^2}}{c_0 \beta^2} \qquad M_\infty > 1$$

where M_∞ is the free-stream Mach number and $\beta^2 = |1 - M_\infty^2|$. The solution for the unit step source can be obtained from Eqs (16.26) by integrating over τ, thus:[7,11]

$$\begin{aligned} G(x - \xi, t) &= \int_0^t g(x - \xi, t - \tau) \, d\tau \\[2mm] &= \begin{cases} \dfrac{c_0}{4\pi} \dfrac{H(t - r_0/c_0)}{\sqrt{(U_k y_k)^2 + c_0^2 \beta^2 r^2}} & M_\infty < 1 \\[4mm] \dfrac{c_0}{4\pi} \dfrac{H(U_k y_k - c_0 \beta r)}{\sqrt{(U_k y_k)^2 - c_0^2 \beta^2 r^2}} \left[H\left(t - \dfrac{r_0}{c_0}\right) + H\left(t - \dfrac{r_1}{c_0}\right) \right] & M_\infty < 1 \end{cases} \end{aligned}$$

(16.27)

The steady state response can be derived from the above by letting $t \to \infty$, when Eqs (16.27) simplify to:[7,11]

$$\begin{aligned} G^s(x - \xi) &= G(x - \xi, \infty) \\[2mm] &= \begin{cases} \dfrac{c_0}{4\pi} \dfrac{1}{\sqrt{(U_k y_k)^2 + c_0^2 \beta^2 r^2}} & M_\infty < 1 \\[4mm] \dfrac{c_0}{2\pi} \dfrac{H(U_k y_k - c_0 \beta r)}{\sqrt{(U_k y_k)^2 - c_0^2 \beta^2 r^2}} & M_\infty > 1 \end{cases} \end{aligned}$$

(16.28)

The nature of the above solutions changes substantially depending on whether M_∞ is greater or less than 1. For the supersonic case, the surface bounding the region reached by a disturbance starting from a given point is called the Mach surface or characteristic surface which is defined by the Heaviside function in Eqs (16.28). The properties of supersonic flow described above give it a character that is quite different from that of the subsonic flow. If a subsonic flow meets any obstacle, the presence of this obstacle affects the flow in a space, both upstream and downstream, and the effects of the obstacle is zero only asymptotically at an infinite distance from it. A supersonic flow, however, is incident 'blindly' on an obstacle; the effect of the latter extends only downstream, and in all the upstream the flow does not see the obstacle. Finally it is of interest to note that a $1/r$ singularity appears in the above equations.

Two-dimensional flow   Similar to the three-dimensional case, the response for an instantaneous point source is considered first:

$$g(x - \xi, t - \tau) = \frac{1}{2\pi} \frac{H(t' - r_u/c_0)}{\sqrt{t'^2 - r_u^2/c_0^2}} \tag{16.29}$$

where

$$r_u^2 = (y_i - U_i t')(y_i - U_i t')$$

The convective response for the unit step source is[7,11]

$$G(x - \xi, t) = \frac{1}{2\pi} \int_0^t \frac{H(t' - r_u/c_0)}{\sqrt{t'^2 - r_u^2/c_0^2}} \, d\tau$$

$$= \begin{cases} \dfrac{H(t - r_0/c_0)}{2\pi\beta} \ln \dfrac{c_0^2\beta\sqrt{t^2 - r_u^2/c_0^2} + c_0^2\beta^2 t + U_k y_k}{\sqrt{(U_k y_k)^2 + c_0^2\beta^2 r^2}} & M_\infty < 1 \\[20pt] \dfrac{H(U_k y_k - c_0\beta r)}{2\pi\beta} \left\{ H\!\left(t - \dfrac{r_0}{c_0}\right) \left[\sin^{-1} \dfrac{c_0^2\beta^2 t - U_k y_k}{\sqrt{(U_k y_k)^2 - c_0^2\beta^2 r^2}} + \dfrac{\pi}{2}\right] \right. \\[20pt] \left. -H\!\left(t - \dfrac{r_1}{c_0}\right) \left[\sin^{-1} \dfrac{c_0^2\beta^2 t - U_k y_k}{\sqrt{(U_k y_k)^2 - c_0^2\beta^2 r^2}} - \dfrac{\pi}{2}\right] \right\} & M_\infty > 1 \end{cases} \tag{16.30}$$

in which the variables r_0 and r_1 were defined earlier. The steady state response can be obtained from Eqs. (16.30) by taking limit $t \to \infty$, when the above simplifies to[7,11]

$$G^s(x - \xi) = \begin{cases} -\dfrac{1}{2\pi\beta} \ln \dfrac{\sqrt{(U_k y_k)^2 + c_0^2\beta^2 r^2}}{c_0\beta} & M_\infty < 1 \\[15pt] \dfrac{1}{2\beta} H(U_k y_k - c_0\beta r) & M_\infty > 1 \end{cases} \tag{16.31}$$

In the case of steady two-dimensional flow, the characteristic surfaces will now be replaced by characteristic lines (on simply characteristics) in the plane of the flow. Through any point in this plane passes two characteristics, which intersect the stream line through this point at Mach angle α. The downstream branches of the characteristics bound the region of the flow where perturbations starting from the point can take effect.

These functions have an ln r singularity for the subsonic flow and are non-singular for supersonic flow.

16.4.4 Boundary Integral Representations

The desired integral representation for convective compressible potential flow can be derived directly from the governing differential equation by integration by part and substituting the adjoint fundamental solution in the usual manner:

$$c(\xi)\phi(\xi,\tau) = \int_0^T \int_S \left[g(\xi - x, \tau - t)u'_n(x,t) - f(\xi - x, \tau - t)\phi(x,t) \right] dS(x) \, dt$$

$$+ \int_0^T \int_V \left[g(\xi - x, \tau - t)\bar{\psi}(x,t) \right] dV(x) \, dt$$

$$- \frac{1}{c_0^2} \int_V \left[\frac{D_0 g(\xi - x, \tau)}{Dt} \phi(x,0) - g(\xi - x, \tau)\frac{D_0 \phi(x,0)}{Dt} \right] dV(x) \quad (16.32)$$

where

$$u'_n(x,t) = u_n(x,t) - \frac{U_n}{c_0^2} \frac{D_0 \phi(x,t)}{Dt}$$

$$f(\xi - x, \tau - t) = \frac{\partial g(\xi - x, \tau - t)}{\partial n} - \frac{U_n}{c_0^2} \frac{D_0 g(\xi - x, \tau - t)}{Dt}$$

The boundary integral equation (16.32) can be rewritten in our usual more compact notation

$$c\phi = \int_S \left[g * u'_n - f * \phi \right] dS + \int_V (g * \bar{\psi}) \, dV \quad (16.33)$$

The $*$ in Eq. (16.33) once again symbolizes a Riemann convolution integral. If body source is absent and the small perturbation approximation is introduced, the volume integral in Eq. (16.33) no longer remains and the above representation then simplifies to

$$c\phi = \int_S (g * u'_n - f * \phi) \, dS \quad (16.34)$$

while for steady conditions this reduces to[7,11]

$$c\phi = \int_S (G^s u'_n - F^s \phi) \, dS \quad (16.35)$$

where

$$u'_n = u_n - \frac{U_n U_i}{c_0^2} u_i$$

$$F^s = \frac{\partial G^s}{\partial n} - \frac{U_n U_i}{c_0^2} \frac{\partial G^s}{\partial x_i}$$

However, it is generally not convenient to apply boundary values u'_n in solving the physical problems. We shall discuss this below.

16.4.5 Numerical Implementation[7,11]

In this section, the discussions will be restricted to two-dimensional steady subsonic and supersonic flow problems. The standard methodology employed for spatial discretization can be employed and as a result the boundary integral equation for steady state convective compressible potential flow can now be written as

$$c(\xi)\phi(\xi) =$$

$$\sum_{m=1}^{M} \left[\int_{S_m} G(x(\zeta) - \xi) N_\omega(\zeta) u'_{n\omega} \, dS(x(\zeta)) - \int_{S_m} F(x(\zeta) - \xi) N_\omega(\zeta) \phi_\omega \, dS(x(\zeta)) \right]$$

$$(16.36)$$

The integrands remaining in Eq. (16.36) are known in explicit form from the fundamental solutions,

$$G(x - \xi) = G^s(\xi - x) \qquad (16.37a)$$

$$F(x - \xi) = \frac{\partial G^s(\xi - x)}{\partial x_i} n_i(x) - \frac{U_n U_i}{c_0^2} \frac{\partial G^s(\xi - x)}{\partial x_i} \qquad (16.37b)$$

The positioning of the nodal variables outside the integrals is a key step then the integrands of Eq. (16.36) contain only known functions, which can be evaluated numerically. A method is applied here to transform u'_n into boundary values u_n and ϕ. The shape function can be used to transform these values:

$$u'_n(x) - u_n(x) = -\frac{U_n U_i}{c_0^2} u_i(x) \qquad (16.38a)$$

$$\frac{\partial N_\omega(\zeta)}{\partial \zeta} \phi_\omega = \frac{\partial x_i}{\partial \zeta} u_i(x) \qquad (16.38b)$$

Solving the above equations, one obtains[7,11]

$$u_n(x) = u_i(x) n_i(x) = -\frac{c_0^2}{U_n^2} [u'_n(x) - u_n(x)] - \frac{1}{U_n} [U_2 n_1(x) - U_1 n_2(x)] J^{-1} \frac{\partial N_\omega(\zeta)}{\partial \zeta} \phi_\omega$$

$$(16.38c)$$

in which

$$n_1(x) = J^{-1} \frac{\partial x_2}{\partial \zeta}$$

$$n_2(x) = -J^{-1} \frac{\partial x_1}{\partial \zeta}$$

and J is the determinant of the Jacobian matrix. Rearranging Eq. (16.38c), one gets

$$u'_n(x) = \left(1 - \frac{U_n^2}{c_0^2}\right) u_n(x) - \frac{U_n}{c_0^2} [U_2 n_1(x) - U_1 n_2(x)] J^{-1} \frac{\partial N_\omega(\zeta)}{\partial \zeta} \phi_\omega \qquad (16.39)$$

Substituting Eq. (16.39) into Eq. (16.36) produces the required discretized form

$$c\phi = \sum_{m=1}^{M} \left\{ u_{n\omega} \int_{S_m} \left(1 - \frac{U_n^2}{c_0^2}\right) G N_\omega \, dS \right.$$

$$\left. - \phi_\omega \int_{S_m} \left[F N_\omega + \frac{U_n}{c_0^2} (U_2 n_1 - U_1 n_2) J^{-1} \frac{\partial N_\omega}{\partial \zeta} G \right] dS \right\} \qquad (16.40)$$

The schemes utilized for integration over boundary elements as exemplified in Eq. (16.40) differ due to the distinctive nature of kernel functions G and F for subsonic and supersonic cases. Considerable care must be exercised during integration. This is particularly true for supersonic flow, in which the F kernel contains a delta function, for the integration algorithm must be much more sophisticated than those developed for subsonic flow. For supersonic flow the G kernel is non-singular; therefore much less sophisticated integration schemes can be employed to obtain the required level of accuracy with relatively few sub-segments and Gauss points.

For the integration of the F kernel (which is a delta function) a numerical integration is no longer possible and analytical integration must be carried out. In order to explain this scheme, the problem is simplified to only one involving x_1 direction free-stream U_1. Thus, the G kernel can be written as

$$G(x - \xi) = \frac{1}{2\beta} [H(\beta y_2 - y_1) - H(\beta y_2 + y_1)] \qquad (16.41)$$

From Eq. (16.37b), the F kernel can be obtained as

$$F(x - \xi) = \tfrac{1}{2} [(n_2 + \beta n_1)\delta(y_1 - \beta y_2) - (n_2 - \beta n_1)\delta(y_1 + \beta y_2)] \qquad (16.42)$$

In the above equation the arguments of delta function $y_1 - \beta y_2$ and $y_1 + \beta y_2$ represent the two characteristic lines of Mach cone. Thus, for any element m, the required integral can be written as

$$\int_{S_m} F(x - \xi) N_\omega \, dS = \tfrac{1}{2} \int_{S_m} (n_2 + \beta n_1) N_\omega \delta(y_1 - \beta y_2) \, dS$$

$$- \tfrac{1}{2} \int_{S_m} (n_2 - \beta n_1) N_\omega \delta(y_1 + \beta y_2) \, dS \qquad (16.43)$$

Only the first integral in above equation will be examined. To assist in this endeavour, the following two distinct cases can be identified:

1. The characteristic line does not cross the element m.
2. The characteristic line crosses the element m.

In the first case, the integral is zero according to the definition of the delta function. Turning next to case 2, the characteristic line and element m may have one or two intersections (for quadratic element) which can be located by solving the following equation:

$$y_1 - \beta y_2 = 0$$

In the local coordinate system, the above equation becomes

$$a\zeta^2 + b\zeta + c = 0 \tag{16.44}$$

By imposing the shape function we can express the coordinates y_i as

$$y_i = x_i(\zeta) - \xi_i = N_\omega(\zeta)x_{i\omega} - \xi_i \tag{16.45}$$

The coefficients of Eq. (16.44) can then be defined as

$$a = 2z_1 - 4z_2 + 2z_3 \qquad b = -3z_1 + 4z_2 - z_3 \qquad c = z_2 - \eta$$

where $z_\omega = x_{1\omega} - \beta x_{2\omega}$ and $\eta = \xi_1 - \beta\xi_2$.

Of the two roots of equation (16.44) ζ_1 and ζ_2, only the solutions $0 \le \zeta_i \le 1$ are relevant. Once the intersections are found, the desired integral is obtained as

$$\frac{1}{2}\int_{S_m}(n_2 + \beta n_1)N_\omega\delta(y_1 - \beta y_2)\,\mathrm{d}S = \frac{1}{2a}\sum_{i=1}^{2}[n_2(\zeta_i) + \beta n_1(\zeta_i)]N_\omega(\zeta_i)J(\zeta_i) \qquad 0 \le \zeta_i \le 1 \tag{16.46}$$

where J is known as the Jacobian (determinant) of the transformation. Following the same procedure, the second integral in Eq. (16.43) can be determined easily.

For subsonic flow the numerical methodology is almost identical to those discussed for the convective heat transfer. Moreover, for both subsonic and supersonic flows the strongly singular diagonal term of the F matrix can be obtained by the standard 'equipotential' method. It is then a simple operation to rearrange the final system matrix into known and unknown components and set up the final system of simultaneous equations for the subsonic case:

$$[A]\{x\} - [B]\{y\} = \{0\} \tag{16.47}$$

which can be solved for $\{x\}$ and thus all ϕ and u_n components are now known on S.

Because of the hyperbolic nature of the governing equation, the supersonic problem is more like an initial value problem rather than a boundary value problem. Consequently, a marching procedure must be performed in space. Initial data ϕ and $\partial\phi/\partial n$ are prescribed along the line $x = 0$ (see Fig. 16.1) and the solution is advanced in the x direction subject to wall boundary conditions and an appropriate condition at the upper boundary y_{max}. By using this procedure, the quantities on the boundary can be determined sequentially using the prescribed initial data surface (inlet surface), where both ϕ and $\partial\phi/\partial n$ must be specified.

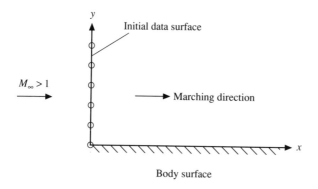

Figure 16.1 Coordinate system for a marching problem.

On the other hand, the values u_n on the outlet surface remain undetermined, because their G coefficients are all zero, which means that the points at the outlet can only receive influences from other points within the upstream Mach cone. The boundary conditions at exit are not required but are determined from the remaining boundary solutions.

16.5 CONVECTIVE INCOMPRESSIBLE THERMOVISCOUS FLOW

16.5.1 Governing Equations

The differential equations, governing the behaviour of an incompressible thermoviscous fluid in the presence of a free-stream velocity U_i, can be written as

$$\frac{\partial u_i}{\partial x_i} = 0 \qquad (16.48a)$$

$$\rho \frac{\mathrm{D}_0 u_i}{\mathrm{D}t} - \mu \frac{\partial^2 u_i}{\partial x_j\, \partial x_j} + \frac{\partial p}{\partial x_i} - \bar{f}_i = 0 \qquad (16.48b)$$

$$\rho c_v \frac{\mathrm{D}_0 \theta}{\mathrm{D}t} - k \frac{\partial^2 \theta}{\partial x_i\, \partial x_i} - \bar{\psi} = 0 \qquad (16.48c)$$

where u_i, once again, represents the velocity perturbation. In Eqs (16.48), the effective body forces and sources are defined as

$$\bar{f}_i = -\rho u_j \frac{\partial u_i}{\partial x_j} + f_i \qquad (16.49a)$$

$$\bar{\psi} = -\rho c_v u_i \frac{\partial \theta}{\partial x_i} + \psi \qquad (16.49b)$$

These equations are of course identical to those presented in Chapter 15, except that now the convective terms $\rho U_j \partial u_i/\partial x_j$ and $\rho c_v U_i \partial \theta/\partial x_i$ are included in the linear differential operator. Fundamental solutions based upon Eqs (16.48) will contain the character of the flow field at high velocities.

16.5.2 Fundamental Solutions

Three-dimensional flow   Consider, first, the effect of a unit pulse point force in a moving medium ($U_i \neq 0$); then the convective viscous fundamental solution g_{ij}, g_{pj} can be obtained from those of Stokes flow, and in which a new system of coordinates is introduced by Galilean transformation $y_i \sim y_i - U_i t'$. Thus we get

$$g_{pj}(x - \xi, t - \tau) = -\frac{\partial}{\partial x_j}\left[\frac{\delta(t - \tau)}{4\pi r}\right] \tag{16.50a}$$

$$g_{ij}(x - \xi, t - \tau) = \delta_{ij}\frac{\partial^2 \phi(r_u, t')}{\partial x_k \partial x_k} - \frac{\partial^2 \phi(r_u, t')}{\partial x_i \partial x_j} \tag{16.50b}$$

where

$$r_u^2 = (y_i - U_i t')(y_i - U_i t')$$

and the function ϕ was defined in Chapter 15.

The effect of a unit step force in the i direction acting at the point ξ in a moving medium can be obtained from Eqs (16.50) by integrating over τ:

$$G_{pj}(x - \xi, t) = \int_0^t g_{pj}(x - \xi, t - \tau)\,\mathrm{d}\tau = -\frac{\partial}{\partial x_j}\left(\frac{1}{4\pi r}\right) \tag{16.51a}$$

$$G_{ij}(x - \xi, t) = \int_0^t g_{ij}(x - \xi, t - \tau)\,\mathrm{d}\tau = \delta_{ij}\frac{\partial^2 \Phi^U}{\partial x_k \partial x_k} - \frac{\partial^2 \Phi^U}{\partial x_i \partial x_j} \tag{16.51b}$$

where

$$\Phi^U(r, t) = \int_0^t \phi(r_u, t - \tau)\,\mathrm{d}\tau = -\frac{1}{4\pi\rho}\int_0^t \frac{1}{r_u}\,\mathrm{erf}\frac{r_u}{2\sqrt{\nu(t - \tau)}}\,\mathrm{d}\tau$$

The above equation can be written explicitly as[7,12,13]

$$G_{ij}(x - \xi, t) = \frac{1}{8\pi\mu r}\delta_{ij}\, e^{U_k y_k/2\nu}\left[e^{-Ur/2\nu}\mathrm{erfc}\frac{r - Ut}{2\sqrt{\nu t}} + e^{Ur/2\nu}\mathrm{erfc}\frac{r + Ut}{2\sqrt{\nu t}}\right] + \frac{\partial \psi_i}{\partial x_j} \tag{16.52a}$$

$$G_{pj}(x - \xi, t) = -\frac{\partial}{\partial x_j}\left(\frac{1}{4\pi r}\right) \tag{16.52b}$$

where

$$\psi_i = \frac{1}{8\pi\rho Ur}\left[\frac{U_i r - U y_i}{Ur - U_k y_k}\left(1 - e^{-(Ur - U_k y_k)/2\nu}\mathrm{erfc}\frac{r - Ut}{2\sqrt{\nu t}} - \frac{r - Ut}{r_u}\mathrm{erf}\frac{r_u}{2\sqrt{\nu t}}\right)\right.$$

$$\left. + \frac{U_i r + U y_i}{Ur + U_k y_k}\left(1 - e^{(Ur + U_k y_k)/2\nu}\mathrm{erfc}\frac{r + Ut}{2\sqrt{\nu t}} - \frac{r + Ut}{r_u}\mathrm{erf}\frac{r_u}{2\sqrt{\nu t}}\right)\right]$$

As t approaches infinity, the steady state fundamental solutions reduce to (first derived by Oseen[2])

$$G_{ij}^s(x - \xi) = G(x - \xi, \infty) = \delta_{ij} \frac{\partial^2 \Phi}{\partial x_k \partial x_k} - \frac{\partial^2 \Phi}{\partial x_i \partial x_j} \tag{16.53a}$$

$$G_{pj}^s(x - \xi) = G_{pj}(x - \xi, \infty) = -\frac{\partial}{\partial x_j}\left(\frac{1}{4\pi r}\right) \tag{16.53b}$$

where

$$\Phi(x - \xi) = \frac{1}{4\pi \rho U}\left[\ln \frac{Ur - U_k y_k}{2\nu} + E_1\left(\frac{Ur - U_k y_k}{2\nu}\right)\right]$$

Two-dimensional flow   The fundamental solutions for an impulse body force can be obtained from those of Stokes flow, i.e.

$$g_{pj}^U(x - \xi, t - \tau) = \frac{\partial}{\partial x_j}\left[\frac{\delta(t - \tau)}{2\pi}\ln r\right] \tag{16.54a}$$

$$g_{ij}^U(x - \xi, t - \tau) = \delta_{ij}\frac{\partial^2 \phi(r_u, t')}{\partial x_k \partial x_k} - \frac{\partial^2 \phi(r_u, t')}{\partial x_i \partial x_j} \tag{16.54b}$$

in which the function ϕ was defined in Chapter 15.

Next, consider the effect of unit step forces acting in the j direction at the points (ξ_1, ξ_2, x_3) where the range on x_3 is from $-\infty$ to $+\infty$. Thus, the total applied force is a line load and the desired solution can be obtained by integrating the three-dimensional Green function along the entire x_3 direction. The response, at any point $(x_1, x_2, 0)$, is given by[7,12,13]

$$G_{ij}(x - \xi, t) = \frac{1}{4\pi\mu}\frac{U_i U_j}{U^2}e^{U_k y_k / 2\nu}\psi_\nu(r, t) + \left(\delta_{ij}\frac{U_k}{U}\frac{\partial \phi}{\partial x_k} - \frac{U_i}{U}\frac{\partial \phi}{\partial x_j} - \frac{U_j}{U}\frac{\partial \phi}{\partial x_i}\right) \tag{16.55a}$$

$$G_{pj}(x - \xi, t) = \frac{\partial}{\partial x_j}\left(\frac{1}{2\pi}\ln r\right) \tag{16.55b}$$

where

$$\psi_\nu(r, t) = \int_0^t \frac{e^{-r^2/4\nu t' - U^2 t'/4\nu}}{t'}\, dt' = K_0\left(\frac{Ur}{2\nu}\right) + K_0\left(\ln\frac{Ut}{r}, \frac{Ur}{2\nu}\right)$$

$$\phi(x - \xi, t) = \frac{1}{2\pi \rho U}\left[\ln\frac{r}{r_u} + \frac{1}{2}e^{U_k y_k / 2\nu}\psi_\nu(r, t) - \frac{1}{2}E_1\left(\frac{r_u^2}{4\nu t}\right)\right]$$

$$r_u^2 = (y_i - U_i t)(y_i - U_i t)$$

$K_0(\alpha)$ is the modified Bessel function of the second kind of order zero and $K_0(\alpha, \beta)$ is the so-called incomplete MacDonald function defined earlier.

At steady state the above fundamental solution somewhat simplifies to[7,12,13]

$$G_{ij}^s(x - \xi) = G_{ij}(x - \xi, \infty)$$
$$= \frac{1}{2\pi\mu} \frac{U_i U_j}{U^2} e^{U_k y_k / 2\nu} K_0\left(\frac{Ur}{2\nu}\right) + \left(\delta_{ij} \frac{U_k}{U} \frac{\partial \phi^s}{\partial x_k} - \frac{U_i}{U} \frac{\partial \phi^s}{\partial x_j} - \frac{U_j}{U} \frac{\partial \phi^s}{\partial x_i}\right) \quad (16.56a)$$

$$G_{pj}^s(x - \xi) = G_{pj}(x - \xi, \infty) = \frac{\partial}{\partial x_j}\left(\frac{1}{2\pi} \ln r\right) \quad (16.56b)$$

where

$$\phi^s(x - \xi) = \phi(x - \xi, \infty) = \frac{1}{2\pi\rho U}\left[\ln r + e^{U_k y_k / 2\nu} K_0\left(\frac{Ur}{2\nu}\right)\right]$$

16.5.3 Boundary Integral Representation

The boundary integral representation can be derived in the usual manner and written as

$$c_{ij}u_i = \int_S (g_{ij} * t_i - f_{ij} * u_i + g_{ij} * t_i^0) \, dS$$
$$+ \int_V (d_{ijk} * \sigma_{ik} + g_{ij} * f_i + g_{ij}\rho u_i^0) \, dV \quad (16.57)$$

where

$$t_i(x, t) = \mu\left[\frac{\partial u_i(x, t)}{\partial x_j} + \frac{\partial u_j(x, t)}{\partial x_i}\right] n_j(x) - p(x, t)n_i(x)$$

$$\sigma_{ik}(x, t) = \rho u_i(x, t)u_k(x, t)$$

$$t_i^0(x, t) = \sigma_{ik}(x, t)n_k(x)$$

$$f_{ij}(\xi - x, \tau - t) = \mu\left[\frac{\partial g_{ij}(\xi - x, \tau - t)}{\partial x_k} + \frac{\partial g_{jk}(\xi - x, \tau - t)}{\partial x_i}\right] n_k(x)$$

$$+ g_{pj}(\xi - x, \tau - t)n_i(x) + \rho u_n g_{ij}(\xi - x, \tau - t)$$

$$d_{ijk}(\xi - x, \tau - t) = \frac{\partial g_{ij}(\xi - x, \tau - t)}{\partial x_k}$$

and $u_i^0(x) = u_i(x, 0)$ is the initial condition.

16.5.4 Numerical Implementation

After temporal and spatial discretization, the above boundary integral representation for a two-dimensional problem can be written as[6,13,14,15]

$$c_{ij}u_i^N =$$

$$\sum_{n=1}^{N}\left[\sum_{m=1}^{M}\left(t_{i\omega}^n\int_{S_m}G_{ij}^{N-n+1}N_\omega\,\mathrm{d}S-u_{i\omega}^n\int_{S_m}F_{ij}^{N-n+1}N_\omega\,\mathrm{d}S-t_{i\omega}^{0n}\int_{S_m}G_{ij}^{N-n+1}N_\omega\,\mathrm{d}S\right)\right.$$

$$\left.+\sum_{l=1}^{L}\left(\sigma_{ik\omega}^{0n}\int_{V_l}D_{ijk}^{N-n+1}M_\omega\,\mathrm{d}V\right)\right]+\sum_{l=1}^{L}\left(\rho u_{i\omega}^0\int_{V_l}g_{ij}^N M_\omega\,\mathrm{d}V\right)\tag{16.58}$$

where

$$G_{ij}^n=G_{ij}(\xi-x,n\Delta\tau)$$

with similar expressions holding for the remaining terms.

For steady conditions Eq. (16.58) reduces to[8,13,14,15]

$$c_{ij}u_i=\sum_{m=1}^{M}\left(t_{i\omega}\int_{S_m}G_{ij}N_\omega\,\mathrm{d}S-u_{i\omega}\int_{S_m}F_{ij}N_\omega\,\mathrm{d}S-t_{\alpha\omega}^0\int_{S_m}G_{ij}N_\omega\,\mathrm{d}S\right)$$

$$+\sum_{l=1}^{L}\left(\sigma_{ik\omega}^0\int_{V_l}D_{ijk}M_\omega\,\mathrm{d}V\right)\tag{16.59}$$

where M and L are the total number of surface elements and volume cells, respectively.

The integration schemes for convective thermoviscous incompressible flow are quite similar to those presented earlier for convective heat transfer, because of the similarity of their convective fundamental solutions. One notices that the singular part of the convective kernel G_{ij} is exactly the same as that of convective heat transfer which is $e^{-\beta}K_0(\alpha)$, but here, instead of κ, the constant ν is introduced. Thus in this case α and β become

$$\alpha=\frac{Ur}{2\nu}\qquad\beta=\frac{U_k y_k}{2\nu}$$

Similar to convective heat transfer, the indirect method can also be employed to evaluate the diagonal blocks of G_{ij}. One set of equilibrated known velocity solutions is

$$u_1=C_1\left(x_1+\frac{1}{2}\frac{U_1}{\nu}x_2^2\right)+C_2 x_2$$

$$u_2=-C_1 x_2\tag{16.60}$$

$$u_p=C_3$$

where C_i ($i=1,2,3$) are arbitrary constants and the existence of only the x_1 direction free-stream velocity is assumed. Using the above equations we get the traction solutions:

$$t_1=2\mu C_1 n_1+\mu\left(C_1\frac{U_1}{\nu}x_2+C_2\right)n_2-C_3 n_1$$

$$t_2=\mu\left(C_1\frac{U_1}{\nu}x_2+C_2\right)n_1-2\mu C_1 n_2-C_3 n_2\tag{16.61}$$

The strongly singular integrals involving F_{ij} cannot be evaluated numerically very easily with sufficient precision. Fortunately, the standard indirect 'rigid body' method, is applicable, and leads to the accurate determination of the singular block of the F matrix.

Once the spatial discretization and numerical integration are completed, a system of non-linear algebraic equations can be developed to permit the solution of the boundary value problem as outlined in Chapters 4 and 15.

16.6 CONVECTIVE COMPRESSIBLE THERMOVISCOUS FLOW

An alternative formulation for the mass conservation equation (16.4a) can be developed noting that the speed of sound $c = \sqrt{dp/d\rho}$ depends on the relationship between pressure and density. Thus the final form of governing equations for compressible thermoviscous flow can now be rewritten as

$$\frac{1}{c^2}\frac{D_0^2\tilde{p}}{Dt^2} - \frac{\partial^2\tilde{p}}{\partial x_i\partial x_i} - \bar{\Omega} = 0 \qquad (16.62a)$$

$$\rho_0\frac{D_0 u_i}{Dt} - (\lambda + \mu)\frac{\partial^2 u_j}{\partial x_i\partial x_j} - \mu\frac{\partial^2 u_i}{\partial x_j\partial x_j} + \frac{\partial\tilde{p}}{\partial x_i} - \bar{f}_i = 0 \qquad (16.62b)$$

$$\rho_0 c_p\frac{D_0\theta}{Dt} - k\frac{\partial^2\theta}{\partial x_i\partial x_i} - \frac{D_0\tilde{p}}{Dt} - \bar{\phi} = 0 \qquad (16.62c)$$

where

$$\bar{\Omega} = \left[\frac{D_0}{Dt} - (\lambda + 2\mu)\frac{\partial^2}{\partial x_i\partial x_i}\right]\bar{\psi} - \frac{\partial\bar{f}_i}{\partial x_i} + \frac{\eta}{c^2}\frac{\partial^2}{\partial x_i\partial x_i}\left(\frac{D_0\tilde{p}}{Dt}\right) \qquad (16.63)$$

and $\eta = \lambda + 2\mu/\rho_0$. Note that, the third term which is the viscous effect in the pressure equation is included in the body source $\bar{\Omega}$ since its contribution can be assumed to be small (the coefficient η/c^2 is small). The first two equations of (16.62) are one-way coupled, and the first one is independent of the others while the mass and momentum balance operators are coupled by the inclusion of both velocity (u_i) and pressure ($\tilde{p}$). The fundamental solutions and integral formulation based upon Eqs (16.62) can be found in Dargush, Banerjee and Shi.[7,13]

16.7 EXAMPLES OF APPLICATION

16.7.1 Convective Heat Transfer

The Burger problem[7,8] The classic uniaxial linear Burger problem provides an excellent test of the convective heat transfer formulations. The heat flows in the x direction with uniform velocity U. Meanwhile, the temperature is specified as

T_0 at the inlet and zero at the outlet. The length of the flow field is L. The analytical solution for the above boundary condition

$$T(0) = T_0$$
$$T(L) = 0$$

is given by Schlicting[16] as

$$T = T_0 \left\{ \frac{1 - \exp[Pe(x/L - 1)]}{1 - \exp(-Pe)} \right\}$$

where $Pe = UL/\kappa$ is Peclet number.

The boundary element model employs 18 quadratic surface elements encompassing the rectangular domain. The elements are graded, providing a very fine discretization near the exit, where T varies substantially for large Pe. Results are shown in Fig. 16.2. Excellent agreement with the analytical solution is obtained for this boundary-only analysis, even for the highly convective case of $Pe = 1000$.

16.7.2 Convective Compressible Potential Flow

Flow over a wedge[7,11,13]   An example of flow produced by a two-dimensional wedge (half angle $\theta = 15°$) moving at a Mach number of $M_\infty = 2.0$ is considered. The BEM model with half-symmetry uses 224 quadratic elements on the boundary. At a large distance from the wedge, the unperturbed flow will generally be assumed to be uniform and directed along the x_1-axis (the horizontal axis), i.e.

$$v_i = U_i \qquad \text{or} \qquad u_i = 0$$
$$\phi' = 0$$

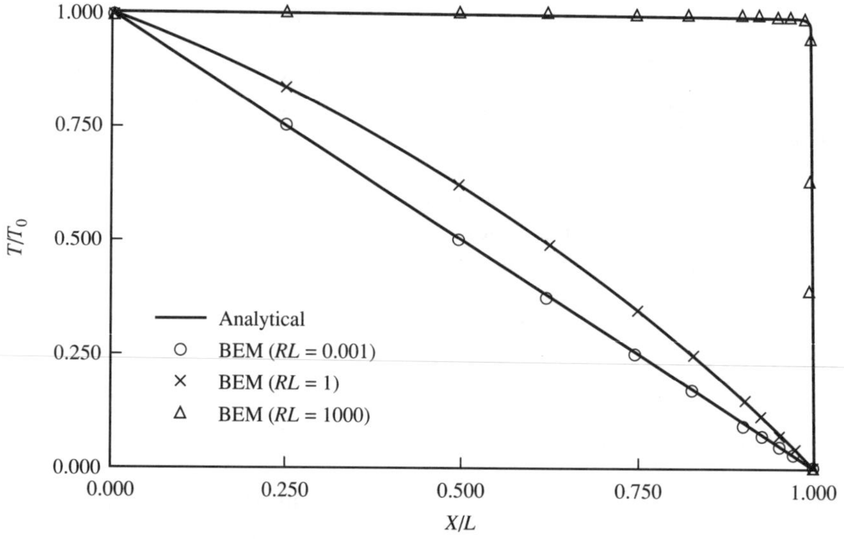

Figure 16.2 Thermal Burger problem—convective fundamental solutions.

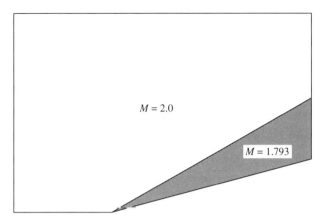

Figure 16.3 Supersonic wedge flow—Mach contours.

On the other hand, the velocity components normal to the surface of wedge are equal to zero, i.e.

$$v_n = 0 \qquad \text{or} \qquad u_n = -U_n$$

For the supersonic case ($M_\infty = 2.0$), the flow is conical. The boundary conditions are the surface tangency requirement at the wedge surface and free-stream conditions outside the shock front. One must include enough space in the computational domain so that the shock wave can form naturally and not be affected by the boundary conditions at other boundaries of the computational domain. The results of Mach contours are shown in Fig. 16.3. The 'shock front' is inclined at the Mach angle $\alpha = \sin^{-1}(1/M_\infty)$ to the horizontal axis. It is interesting to note that the velocity of flow behind the 'shock front' decreases. Figure 16.4 shows the discontinuity of the pressure wave solution.

Flow through a channel[7,11,13]   In this example, the supersonic flow in a channel with compression and expansion ramps is solved. The mesh and boundary condition for the case $M_\infty = 1.3$ are given in Fig. 16.5.

Steady state potential and local Mach number contours for the supersonic case obtained using the marching procedure in space are shown in Figs 16.6a,b and 16.7a,b on pages 449 and 450 respectively. All clearly show the generation of an oblique shock wave at the compression ramp, its reflection off the top wall of the channel and its interaction with the expansion shock produced by the downstream ramp. The velocity vectors calculated in this region of flow are shown in Fig. 16.8 on page 451 and clearly show the effects described above. Figure 16.6a,b shows plots of the potential and pressure coefficient distribution along both upper and lower surfaces respectively.

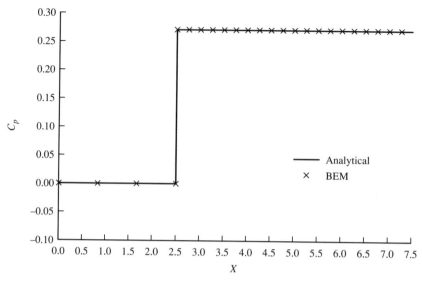

Figure 16.4 Pressure coefficient distribution.

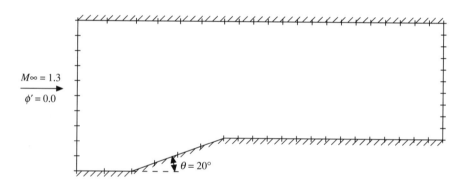

Figure 16.5 Mesh of channel flow—62 elements.

16.7.3 Convective Viscous Flow

The Burger problem[13–15]   The classic uniaxial linear Burger problem provides an excellent test of the convective incompressible viscous flow formulations. The fluid flows in the x_1 direction with uniform velocity U. Meanwhile, the x_2 component of velocity is specified as U_0 at the inlet and zero at the outlet. The length of the flow

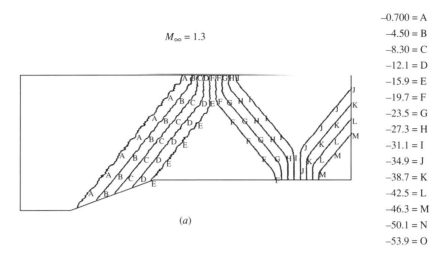

$M_\infty = 1.3$

$-0.700 = A$
$-4.50 = B$
$-8.30 = C$
$-12.1 = D$
$-15.9 = E$
$-19.7 = F$
$-23.5 = G$
$-27.3 = H$
$-31.1 = I$
$-34.9 = J$
$-38.7 = K$
$-42.5 = L$
$-46.3 = M$
$-50.1 = N$
$-53.9 = O$

(a)

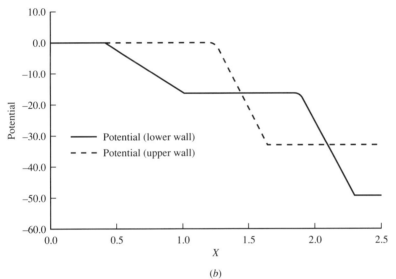

(b)

Figure 16.6 Channel flow: (a) potential contours, (b) potential distributions.

field is L. The analytical solution for the above boundary condition

$$u_2(0) = U_0$$
$$u_2(L) = 0$$

is given by Schlicting[16] as

$$u_2 = U_0\left\{\frac{1 - \exp[Re(x/L - 1)]}{1 - \exp(-Re)}\right\}$$

where $Re = UL/\nu$ is the free-stream Reynolds number.

The boundary element model employs 18 quadratic surface elements which

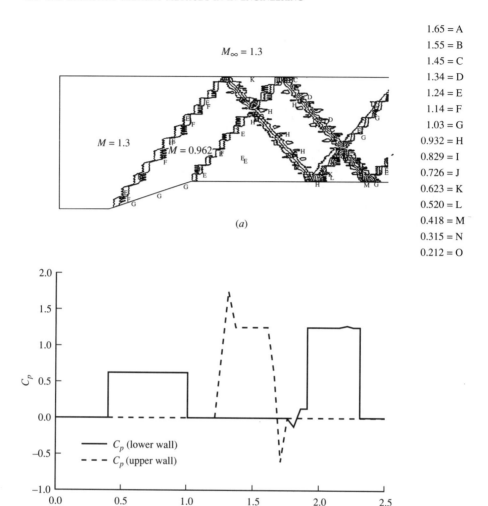

Figure 16.7 Supersonic channel flow: (*a*) Mach contours, (*b*) pressure coefficient distributions.

encompass the rectangular domain. The elements are graded, providing a very fine discretization near the exit, where u_2 varies substantially for large Re. Results are shown in Fig. 16.9. Excellent correlation with the analytical solution is obtained for this boundary-only analysis, even for the highly convective case of $Re = 1000$.

It should be noted that traditionally finite difference and finite element methods have a difficult time dealing with the convective terms present in this problem. Generally, upwinding techniques must be introduced to produce stable, accurate solutions. On the other hand, with the present convective boundary element

$M_\infty = 1.3$

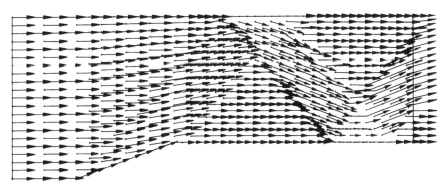

Figure 16.8 Supersonic channel flow—velocity vectors.

approach the kernel functions contain an analytical form of upwinding. As a result, very precise results can be obtained.

Flow around objects[5,13,15]   As final examples, the incompressible viscous flow around a circular cylinder and airfoils are considered. Historically, the cylinder problem has received considerable attention, probably because, despite the geometric simplicity, most of the important characteristics of general flows are present. Oseen[2] and Bairstow et al.[17] were among the earliest investigators to consider application of the linearized equation of motion.

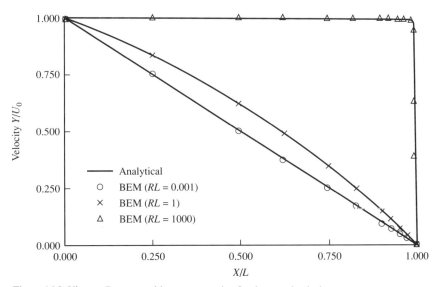

Figure 16.9 Viscous Burger problem—convective fundamental solutions.

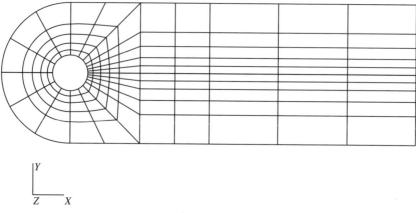

Figure 16.10 Flow over a cylinder—full mesh.

Here the analysis is conducted as an exterior problem utilizing the boundary element mesh displayed in Fig. 16.10. At the surface of the cylinder, the non-slip condition is imposed, while at infinity the perturbation velocity u_i is specified as zero. Three types of analyses were carried out:

- the Oseen flow with a boundary only mesh
- full nonlinear Navier-Stokes solution with boundary mesh and volume cells using Oseen kernels, and
- a nonlinear analysis using Oseen kernels and a single layer of volume cells around the cylinder.

Steady state velocity vector plots from the full non-linear Newton–Raphson solutions at $Re = 40$ and 100 are shown in Figs 16.11 and 16.12 respectively. For this problem, the free-stream Reynolds number is defined by $Re = UD/\nu$

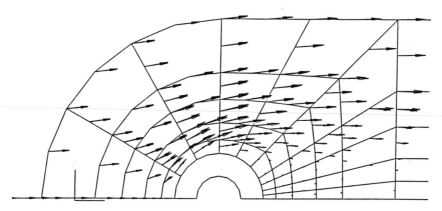

Figure 16.11 flow over a cylinder—non-linear solution versus linear Oseen solution.

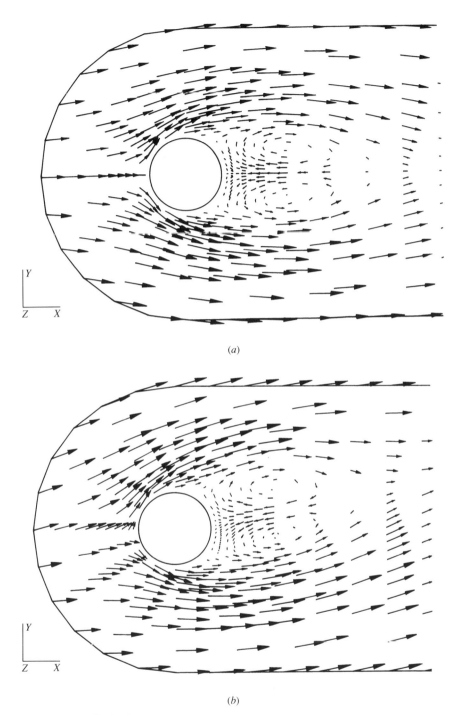

(a)

(b)

Figure 16.12 Full cylinder: (a) angle of attack = 0, (b) angle of attack = 10.

with the diameter D of the cylinder. The computations displayed show that behind the cylinder a pair of vortices are formed, symmetrical about the centreline. These are very small at first, but increase in size and become elongated downstream as the Reynolds number increase. An important aspect of real flow can also be described at least qualitatively with the aid of the Oseen solutions. Included in Fig. 16.11 are the solutions of Oseen flow which is obtained with a boundary discretization only. Notice that the two results deviate substantially only in a small region around the cylinder. This suggests that the Oseen solution is a suitable starting point for a full non-linear analysis in which the non-linear volume integrals do not contribute much at all!

Besides the general character of the flow, one is often interested in the drag induced on the body. From the boundary element analysis, drag can be determined by simply integrating the t_1 component of the traction over the surface of the cylinder. Low-order Gaussian quadrature is applicable since the traction varies at most quadratically along each element. Figure 16.13 shows the drag on the cylinder plotted against Re; obviously the full Navier–Stokes solution gives the correct drag coefficients but the Oseen solution gives the correct drag only at low Re numbers. However, a solution with only one row of volume cells gives quite acceptable results.

Figure 16.14 shows the Oseen solution for flow through a cascade of airfoils. Note that the physics of high-speed flow is captured by a boundary-only analysis. The velocity distributions at four vertical sections along the flow reproduces the expected behaviour of high-speed flow.

16.8 CONCLUDING REMARKS

The principal focus of the present chapter is the development of the basic ingredients of boundary element methods for the convective thermoviscous fluid flow. This was necessary because only rudimentary formulations were available in the published literature. Starting from the development of governing equations based upon the Oseen approximation, in Sec. 16.2, for coupled compressible thermoviscous flow, these equations are simplified for heat transfer, compressible potential flow, incompressible thermoviscous flow and compressible thermoviscous flow.

Sections 16.3 to 16.6 describe fundamental solutions and BEM formulations. Beginning with the derivation of infinite space fundamental solutions, both two-dimensional and three-dimensional, steady-state and transient flows are considered. With these new fundamental solutions and boundary element formulations it may now be possible to solve many practical problems. It was shown that for supersonic problems, the kernels changed from the elliptic to the hyperbolic form; consequently, a 'marching procedure' in space must be applied. As the result, the shock wave, the most important phenomena for supersonic flow, has been successfully captured without using any internal subdivision.

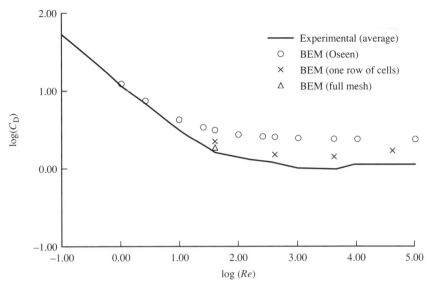

Figure 16.13 Flow over a cylinder—drag versus Reynolds number.

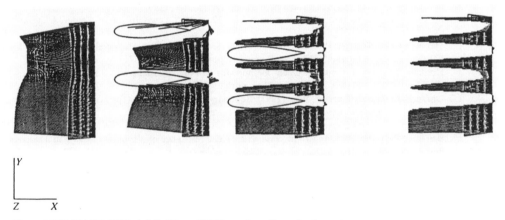

Figure 16.14 NACA-0018 airfoils ($Re = 10\,000$; angle = 0)—velocity vectors.

Although the steady state Oseen kernels are not new, they are rederived in a more general form and applied in BEM. These new kernels embody more of the physics of high Reynolds number flows and as a result much of the character of the problem can be captured with a linear boundary-only analysis. In the numerical example for flow around a cylinder, the circulation behind the cylinder is shown by the boundary-only solution. With some added volume integrals which takes care of non-linearities a complete solution of the Navier–Stokes equation for high-speed flows is now feasible. Although some research workers,[8,18–30,31] particularly in the area of aero-acoustics, have utilized some of these convective solutions in the past, this chapter develops them in a way that would make them applicable to a general engineering problem.

REFERENCES

1. Oseen, C.W. (1911) Über die Stokes'sche Formel und über eine verwandte Aufgabe in der Hydro-dynamik II, *Ark. F. Mat., Astr. Och Fysik*, Vol. 7.
2. Oseen, C.W. (1927) *Neuere Methoden und Ergebnisse in der Hydrodynamik*, Akad. Verlagsge-sellschaft, Leipzig, Germany.
3. Carslaw, H.S. and Jaeger, J.C. (1947) *Conduction of Heat in Solids*, Clarendon Press, Oxford.
4. Agrest, M.M. and Maksimov, M.S. (1971) *Theory of Incomplete Cylindrical Functions and Their Applications*, Springer-Verlag, New York.
5. Dargush, G.F. and Banerjee, P.K. (1991) A boundary element method for steady incompressible thermoviscous flow, *International Journal of Numerical Methods in Engineering*, Vol. 31, pp. 1605–1626.
6. Dargush, G.F. and Banerjee, P.K. (1991) A time dependent viscous BEM for moderate Reynolds numbers, *International Journal of Numerical Methods in Engineering*, Vol. 31, pp. 1627–1648.
7. Shi, Y. (1992) Fundamental solutions and boundary element formulations for convective fluid flow, PhD Dissertation, State University of New York at Buffalo.
8. Shi, Y. and Banerjee, P.K. (1993) Boundary element methods for convective heat transfer, *Computer Methods in Applied Mechanics and Engineering* (to appear).
9. Goldstein, M.E. (1976) *Aeroacoustics*, McGraw-Hill, New York.
10. Morse, P.M and Feshbach, H. (1978) *Methods of Theoretical Physics*, McGraw-Hill, New York.
11. Shi, Y. and Banerjee, P.K. (1993) Boundary element methods for convective potential flow. *Computer Methods in Applied Mechanics*, Vol. 105, pp. 261–284.
12. Dargush, G.F., Banerjee, P.K. and Shi, Y. (1992) Development of an integrated BEM approach to fluid-structure interaction, Final Report NAG-712, NASA Lewis Research Center, Cleveland, Ohio.
13. Dargush, G.F., Banerjee, P.K. and Shi, Y. (1990) Development of an integrated BEM approach for hot fluid structure interaction, NASA Annual Report, Grant NAG3-712.
14. Honkala, K. (1992) Boundary element methods for two-dimensional coupled thermoviscous flow, PhD Dissertation, State University of New York at Buffalo.
15. Dargush, G.F. and Banerjee, P.K. (1993) Boundary element methods for nonlinear Oseen flow, *Computer Methods in Applied Mechanic and Engineering* (to appear).
16. Schlicting, H. (1955) *Boundary Layer Theory*, McGraw-Hill, New York.
17. Bairstow, L., Cave, B.M. and Lang, E.D. (1923) The resistance of a cylinder moving in a viscous fluid, *Philosophical Transactions of Royal Society (London) A*, Vol. 223, pp. 383–432.
18. Bush, M.B. (1983) Modelling two-dimensional flow past arbitrary cylindrical bodies using bound-ary element formulations, *Applied Mathematical Modelling*, Vol. 7, pp. 386–394.
19. Fang, Z. and Paraschivoiu, I. (1991). Boundary linear integral method for compressible potential flows, *International Journal of Numerical Methods in Fluids*. Vol. 13, pp. 699–711.
20. Hunt, B. and Hewitt, B.L. (1986) The indirect boundary integral formulation for elliptic, hyperbolic and non-linear fluid flows, in *Developments in Boundary Element Methods*, Vol. 4, Eds P.K. Banerjee and J.O. Watson, Elsevier Applied Science, London.
21. Kandil, S.A. and Hu, H. (1988) Unsteady transonic airfoil computation using the integral solution of full-potential equation, Chapter 10 in *Boundary Elements*, Vol. 2, Ed. C.A. Brebbia, Springer-Verlag.
22. Morino, L. (1973) Unsteady compressible potential flow around lifting bodies general theory, *AIAA*, paper No. 73-196, Vol. 13, 526 pp.
23. Morino, L., Bharadvaj, B.K., Freedman, M.I. and Tseng, K. (1987) BEM for wave equation with boundary in arbitrary motion and applications to compressible potential aerodynamics of airplanes and helicopters, in *Advanced Boundary Elements Methods*, Ed. T.A. Cruse, Springer-Verlag, Berlin.
24. Tanaka, Y., Honma, T. and Kaji, I. (1986) On mixed boundary element solutions of convection-diffusion problems in three dimensions, *Applied Mathematical Modelling*, Vol. 10, pp. 170–175.
25. Tseng, K. and Morino, L. (1982) Nonlinear Green's function for unsteady transonic flow, in *Transonic Aerodynamics*, Ed. D. Nixon, pp. 565–603. AIAA, New York.

26. Yang, Z. (1987) Integral equation method for compressible potential flows, in *Proceedings of the 1st Japan–China Symposium on Boundary Element Methods*, Eds M. Tanaka and Q.H. Du, Pergamon Press.

27. Ffowcs Williams, J.E. and Hawkings, D.L. (1969) Sound generated by turbulence and surfaces in arbitrary motion, *Philosophical Transactions of Royal Society*, Vol. A264, pp. 321–342.

28. Farassat, F. and Brentner, S. (1988) The uses and abuses of the acoustic analogy in helicopter rotor noise prediction, *AHS Journal*, Vol. 33, No. 1, pp. 29–36.

29. Farassat, F. (1981) Linear acoustic formulas for calculation of rotating blade noise, *AIAA Journal*, Vol. 19, No. 8, pp. 1122–1130.

30. Farassat, F. (1977) Discontinuities in aerodynamics and aeroacoustics: the concept and applications of generalized derivatives, *Journal of Sound and Vibration*, Vol. 55, No. 2, pp. 165–193.

31. Morino, L. and Tseng, K. (1990) A general theory of unsteady compressible potential flows with applications to aeroplanes and helicopters. Chapter 6 in *Boundary Element Methods in Nonlinear Fluid Dynamics*, Ed. P.K. Banerjee and L. Morino, Elsevier Applied Science, London, pp. 183–245.

SEVENTEEN

COMPUTER IMPLEMENTATION OF BOUNDARY ELEMENT METHODS

17.1 INTRODUCTION

No matter how powerful or elegant a numerical method may be, its full potential can only be realized when it has been programmed efficiently. In this respect, the BEM probably requires greater effort from the programmer and less from the prospective user than do most finite element methods. The first generation of BEM programs were not very efficient since they were mainly developed by workers as exercises in the process of investigating the method. However, the situation has altered considerably in recent years with the emergence of several large general purpose computer codes. One of the pioneering efforts towards this end was the implementation of Lachat and Watson[1,2] who laid the foundation for most subsequent developments. Currently, several advanced computer codes, such as BEST-3D,[3,4] GP-BEST,[5] BEASY,[6] EZBEA,[7] BETSY[8,9] and DBETSY,[10,11] are available for BEM analyses. Nevertheless, more development is still necessary for these programs to approach the level of widespread acceptability enjoyed by the major finite element systems such as NASTRAN, MARC, ANSYS, ABAQUS, ADINA, NISA, etc.

In this chapter, the basic steps involved in developing a simple BEM program are outlined. In addition, a sample computer program called DBEM (see Appendix B), is discussed in order to highlight some salient features of developing a BEM code. Having studied the fundamental principles as well as the simple program presented here, readers with established programming skills may proceed to develop their own systems.

In its current form, DBEM permits the solution of problems of static elasticity and steady state heat conduction for two-dimensional and axisymmetric geometries. It requires the user to specify a boundary element model, made up of surface elements and associated nodes assuming a quadratic variation of geometry and functional variables. Appropriate material properties are required to be input and a set of boundary and loading conditions must be imposed for a well-posed boundary value problem. In addition, DBEM is able to solve problems with certain body force loading such as gravitational (i.e. self-weight) and centrifugal loads.

Throughout the course of this chapter, references are made to the appropriate locations (i.e. sub-routines) in DBEM that perform the task under consideration. This side-by-side study of a BEM program provides a better understanding of the theoretical and numerical aspects.

17.2 STRUCTURE OF A BOUNDARY ELEMENT PROGRAM

The following logical sequence of steps is basic to all BEM analyses:

1. Generation of input data defining the geometry of the surface elements.
2. Integration of kernel-shape function products to generate the system matrices.
3. Assembly of equations for each sub-region.
4. Solution of the system of equations to generate the unknown boundary data.
5. Backsubstitution of boundary data into the integrals at interior points to obtain interior information.

The task of developing a BEM program is facilitated by adopting a modular approach based on these primary subdivisions.

In particular, for the program DBEM, the different tasks are performed by separate modules controlled by the main program (i.e. MAIN) in the file DBEM. These primary modules (or sub-routines) are listed below along with their functions:

Name	Function
INPUT	Input data specification
INTGRA	Integration of kernel-shape function products
ASMBLR	Assembly, solution and results output
INTSTR	Interior displacement and stress evaluation

This program is capable of analysing problems of a relatively small nature with a limit imposed on the maximum number of nodes and elements. These limits are listed below:

Maximum number of nodes = 200
Maximum number of elements = 100 (quadratic)
Maximum number of surfaces = 10
Maximum number of interior points = 100

In addition, only single-region bodies with one material type may be studied. The

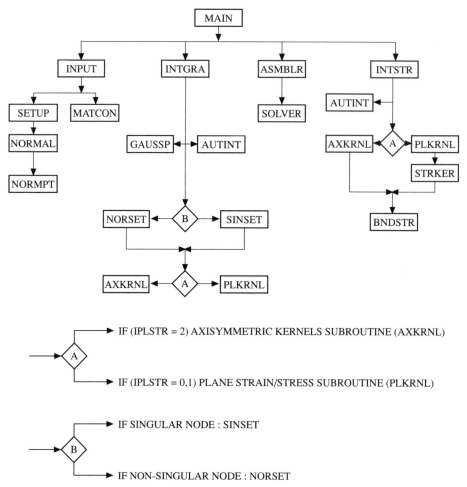

Figure 17.1 Flow chart for program DBEM.

prescription of the boundary and loading conditions must be done with respect to a global coordinate system. However, having mastered the features of this simple computer code and closely studied the fundamental theoretical background of the previous chapters, readers will be in a better position to develop their own BEM systems with enhanced capacity and more general features. The complete structure of DBEM is effectively illustrated through the use of a tree diagram (see Fig. 17.1) showing the various sub-routines and their hierarchical locations.

17.3 SPECIFICATION AND GENERATION OF INPUT DATA

In the specification of the geometry, the boundaries may be represented by line

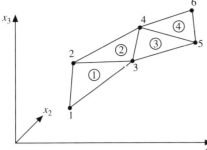

Figure 17.2 Elements and nodal connectivities.

elements in two dimensions and surface elements in three dimensions, defined by their nodal coordinates and some specified variation of the surface geometry. It is also necessary to have a global numbering system for elements and nodes so that a directory for each element and its connectivity (through the nodes) with the adjoining elements may be specified. The node numbers for each element should be specified in either a clockwise or an anti-clockwise manner when viewed in the direction of the outward normal. Thus a flat, triangular element is defined by three nodes as follows (Fig. 17.2):

Element number	Node numbers
2	2 4 3
3	3 4 5
etc.	

This entry shows that the element numbers 2 and 3 are connected via common nodes 3 and 4. Therefore, when the integrations have been performed on these elements, coefficients associated with the values of the function at these nodes are summed during the formation of the system matrices. This means that only one set of unknowns is being allowed for each node (i.e. there is a continuous distribution of functions through the nodes). Discontinuity of either the geometry or the functions can be allowed for by assigning different numbers to coincident nodes, which must be organized internally, leading to a significant housekeeping effort.

A global boundary condition matrix needs to be defined for each surface node of the system so that a directory of surface boundary conditions can be constructed. A similar matrix is also necessary to identify the interface nodes between two regions. A good computer program will of course have a comprehensive input data checking facility so that only a correctly specified data set is passed through for execution.

The data input specification for DBEM is now described. These data are read in the sub-routine INPUT and echoed to the user for visual verification at the start of the result file. The sub-routine INPUT is now given:

DBEM - INPUT DATA SPECIFICATION

1.0 Title Specification:

NLINE	Specifies the number of lines in the title; Maximum of 8 lines.
TITLE	Title Description; each line of title must contain a maximum of 80 characters

2.0 Case (problem type) Specification:

NDEG, IPLSTR, IPROB

NDEG = 1	Steady state heat conduction
NDEG = 2	Static elasticity
IPLSTR = 0	Plane strain analysis
IPLSTR = 1	Plane stress analysis
IPLSTR = 2	Axisymmetric analysis
IPROB = 0	No body force loading
IPROB = 1	Gravitational body force specified
IPROB = 2	Centrifugal body force specified

3.0 Material Properties Specification:

3.1(a) For steady state heat conduction problem (NDEG = 1)

COND   Conductivity

3.1(b) For elastostatic problem (NDEG = 2) with no body force loading (IPROB = 0)

EMOD, PR   Elastic Modulus, Poisson's Ratio

3.1(c) For elastostatic problem (NDEG = 2) with body force loading (IPROB ≠ 0)

EMOD, PR, RHO   Elastic Modulus, Poisson's Ratio, Mass Density

3.2(a) If Gravitational body force loading is specified (IPROB = 1)

ACCN   Acceleration due to gravity

3.2(b) If Centrifugal body force loading is specified (IPROB = 2)

OMEGA   Rotational Speed (radians/sec)

4.0 Geometry Specification:

4.1 Total nodes and elements

NTNODE, NTELEM

NTNODE = Total number of nodes in the entire boundary

NTELEM = Total number of elements in the entire boundary

4.2 Total surfaces

NSEG   Number of surfaces making up the boundary

NSEG = 1 for singly connected

NSEG > 1 for multiply connected

4.3 Loop over all surfaces (ISEG from 1 to NSEG)

4.3.1 **NNODE, NELEM**

NNODE = Number of nodes on the current surface (ISEG)

NELEM = Number of elements on the current surface (ISEG)

4.3.2 Read Coordinates
INOD, CORD(INOD,1), CORD(INOD,2)
INOD = Node Number
CORD (INOD,1) = X (or R) coordinate of node INOD
CORD (INOD,2) = Y (or Z) coordinate of node INOD
Note 1. : Element Connectivity is internally generated;
nodes must be input in sequential manner for each element
(from 1 to NELEM)
Note 2. : No elements should be specified on the axis of symmetry
for Axisymmetry

5.0 Boundary Conditions Specification:
  NBCC   Total number of boundary conditions
  5.1 For heat conduction (NDEG = 1):
    NEL, NB, IT3, VAL3
  5.2 For elasticity (NDEG = 2):
    NEL, NB, IT1, VAL1, IT2, VAL2
    NEL = Element Number
    NB = Node number within element (1,2 or 3)
    IT1 = Type of boundary condition in X or R direction
    IT2 = Type of boundary condition in Y or Z direction
    IT1/IT2 = 1 Traction specified
    IT1/IT2 = 2 Displacement specified
    IT3 = 1 Flux specified
    IT3 = 2 Temperature specified
    VAL1, VAL2, VAL3 = Values of boundary conditions

    Note: Default boundary conditions are traction free and insulated

6.0 Interior point data:
  NIP Total number of interior points where
    displacements and stresses are desired
  If (NIP ≠ 0): for each point K from 1 to NIP
  XIP(K), YIP(K)
  XIP(K) = X (or R) coordinate of node K
  YIP(K) = Y (or Z) coordinate of node K

End of Input Data

The sub-routine INPUT calls upon a number of other sub-routines to perform certain tasks. These are listed below along with their functions:

Name	Function
INPUT	Input data specification
MATCON	Computation of material constants
SETUP	Set up sizes of arrays; maximum element length; normals at the nodes

AUTCD   Automatic generation of coordinates
NORMAL   Compute normal at a node
NORMPT   Compute direction cosines at a node

17.4 INTEGRATION OF THE KERNEL-SHAPE FUNCTION PRODUCTS

17.4.1 Introduction

In order to form the system matrices it is necessary to evaluate the integrals in the discretized boundary integral equations (here body forces are assumed to be zero for convenience); i.e. for the elastostatic problems from Chapter 4:

$$c_{ij}(\xi_0^p)u_i(\xi_0^p) = \sum_{q=1}^{N}\left(\int_{\Delta S_q}\left\{G_{ij}[x(\eta),\xi_0^p]\sum_{k=1}^{n}N^k(\eta)J(\eta)\,\mathrm{d}S(\eta)\right\}t_i^k\right.$$

$$\left. - \int_{\Delta S_q}\left\{F_{ij}[x(\eta),\xi_0^p]\sum_{k=1}^{n}N^k(\eta)J(\eta)\,\mathrm{d}S(\eta)\right\}u_i^k\right) \tag{17.1}$$

where N is the total number of boundary elements and n is the total number of nodes on the qth boundary element.

It is implied in Eq. (17.1) that the Cartesian coordinates x_i of an arbitrary point on the qth element are defined in terms of the nodal Cartesian coordinates x_i^k ($k = 1, 2, \ldots, n$ for example) and the shape functions $N^k(\eta)$:

$$x_i(\eta) = N^k(\eta)x_i^k \tag{17.2}$$

where the shape functions $N^k(\eta)$ can be linear (flat elements), quadratic, cubic, quartic, etc. (see Chapter 2). By differentiating Eq. (17.2) with respect to η (the local axis directed along the line boundary element) we obtain a vector defining the tangent to the element at the point η:

$$s_i(\eta) = \frac{\mathrm{d}x_i}{\mathrm{d}\eta} = \frac{\mathrm{d}N^k(\eta)}{\mathrm{d}\eta}x_i^k \tag{17.3}$$

The Jacobian $J(\eta)$ is equal to $\sqrt{(\mathrm{d}x_i/\mathrm{d}\eta)(\mathrm{d}x_i/\mathrm{d}\eta)}$ and the normals are given by either $(s_2, -s_1)$ or $(-s_2, s_1)$ depending on which side of the element the elastic body lies.

For a surface boundary element in a three-dimensional analysis we can similarly construct a local axes system η_j ($j = 1, 2$), and differentiating Eq. (17.2) with respect to η_j vectors tangent to the coordinate lines are obtained as

$$s_{ij}(\eta) = \frac{\mathrm{d}x_i}{\mathrm{d}\eta_j} = \frac{\mathrm{d}N^k(\eta)}{\mathrm{d}\eta_j}x_i^k \tag{17.4}$$

A vector normal to the element at η can be obtained by taking the cross-product of the two vectors defined by Eq. (17.4), and the Jacobian is the modulus of this cross-product. Having thus defined the Jacobian $J(\eta)$ and the normal $n_i(\eta)$ we can now attempt to integrate the kernel-shape function products in Eq. (17.1).

In DBEM, this task is controlled by the sub-routine INTGRA. The other sub-routines called by INTGRA and their respective functions are tabulated below:

Name	Function
INTGRA	Integration control
AUTINT	Set up the order of integration (NGP, NSUB)
	NGP = number of Gauss points;
	NSUB = number of subsegments
GAUSSP	Set up Gauss point locations and weights
NORSET	Set up sub-segmentation for normal integration
SINSET	Set up sub-segmentation for singular integration
AXKRNL	Integration of boundary displacement kernels
	for the axisymmetric case
PLKRNL	Integration of boundary displacement kernels
	for the two-dimensional case

17.4.2 Evaluation of the Non-singular Integrals

When the field point ξ_0^p in Eq. (17.1) is not on the node k the kernel, shape function and Jacobian product remains bounded and therefore can be integrated numerically by ordinary Gaussian integration formulae with weight function unity:

$$\int_{-1}^{1} f(\xi)\,d\xi = \sum_{i=1}^{m} A_i f(\xi) \tag{17.5a}$$

$$\int_{-1}^{1} f(\xi_1, \xi_2)\,d\xi_1\,d\xi_2 = \sum_{i=1}^{m_1} \sum_{j=1}^{m_2} A_i^{m_1} B_j^{m_2} f(\xi_1, \xi_2) \tag{17.5b}$$

for the line and surface area integrals respectively. Explicit details of such integration formulae can be found in Chapter 2. It is clearly necessary to map our boundary elements to unit intervals, so that the formulae (17.5a,b) can be used to evaluate the kernel-shape function and Jacobian products.

Since the evaluation of the non-singular integrals occupies a significant amount of the computational time it is necessary to optimize this whenever possible. This may be accomplished (e.g. in Lachat and Watson[1,2,12]) by specifying a maximum upper bound to the error in the numerical integrations. In simple terms this implies that the order of the integrations should be varied depending on the ratio of the distance between the 'loaded' boundary element and the field point to a characteristic dimension of the 'loaded' element and the strength of the singularity. A comprehensive discussion of approximate formulae can be found in Lachat and Watson[1,2,12] and Mustoe.[13] Watson[12] discusses additionally an integration scheme over infinite boundary elements.

Details of the numerical implementation of the non-singular integration scheme can be further understood by studying the sub-routines AUTINT and GAUSSP in DBEM. The actual integration process is performed in the following sub-routines : AXKRNL (for axisymmetry) and PLKRNL (for plane stress/ strain), after setting up the order of the integration and the number of sub-segments in sub-routine NORSET.

17.4.3 Evaluation of the Singular Integrals

When the field point ξ_0^p coincides with the node k the integration scheme outlined above cannot be used. For a two-dimensional analysis the integral involving G_{ij} can be split into a singular part and a non-singular part, i.e.

$$\int_{\Delta S}[G_{ij}N^k(\eta)J(\eta)]\,\mathrm{d}S\ t_i^k = A\delta_{ij}\int_{\Delta S}[\ln rN^k(\eta)J(\eta)]\,\mathrm{d}S\ t_i^k + \int_{\Delta S}[G_{ij}^0N^k(\eta)J(\eta)]\,\mathrm{d}S\ t_i^k$$

(17.6)

where A is a constant and the second integral containing G_{ij}^0 is non-singular and can therefore be evaluated as before. The first integral, however, has a logarithmic singularity and needs some special attention.

This particular singular integral in Eq. (17.6) can be calculated using a quadrature formula due to Stroud and Secrest[14] (see Chapter 2):

$$\int_0^1 \log\left(\frac{1}{\xi}\right)f(\xi)\,\mathrm{d}\xi = \sum_{i=1}^m w_if(\xi_i)$$

(17.7)

which integrates Eq. (17.7) exactly when $f(\xi)$ is any polynomial whose order is not greater than $(2m-1)$. To use this formula, each boundary element is divided into several sub-elements so that the singular point is always at the origin O of Eq. (17.7). Unfortunately, singular integrals involving the function F_{ij} cannot be evaluated in this manner. However, by noting that the shape function is unity at the singular node (say at node 1), we can write

$$\int_{\Delta S}(F_{ij}N^1J)\,\mathrm{d}S\ u_j^1 = \int_{\Delta S}[F_{ij}(1-N^1)J]\,\mathrm{d}S\ u_j^1 - \int_{\Delta S}(F_{ij}J)\,\mathrm{d}S\ u_j^1$$

(17.8)

The integral can then be treated using the following method outlined in Banerjee and Butterfield[15] (page 379).

The first integral, which is non-singular since $(1-N^1)$ approaches zero at the singular point and the integrand remains reasonably well behaved, can be evaluated numerically but the second integral must be calculated analytically. For flat boundary elements, these integrals are easy to evaluate. For quadratic and cubic boundary elements (i.e. curved boundary elements), however, it is necessary to split this second integral into a part involving a flat tangent plane through the singular point which can be calculated analytically and a part involving the curvature of the boundary element which can be evaluated numerically. The discontinuity term c_{ij} is then added to these coefficients to give the dominant diagonal block of coefficients. It is of course possible to use the above-mentioned procedure for integrals involving G_{ij} as well as for integrals involving field points close to the element under integration.

However, a more elegant method of evaluating the contributions from the above integral and the free term c_{ij} in Eq. (17.1) is to consider the rigid-body displacement modes discussed in earlier chapters and also in Refs 1, 2, 12 and 15. Watson[12] has given a particularly lucid account of this idea applied to the calculation of the singular contributions over infinite boundary elements.

For the three-dimensional case, integrals involving F_{ij} can be evaluated by any of the methods outlined above, i.e. via either Eq. (17.8) or utilizing rigid body displacements. As far as integrals involving G_{ij} are concerned, when the singular point is on the corner of a triangular (degenerate quadrilateral) element (see Fig. 17.3a) the integrations can be carried out with respect to the coordinates $\bar{\eta}_1$ and $\bar{\eta}_2$. Note that, since the lines $\bar{\eta}_1 = +1$ and $\bar{\eta}_2 = -1$ converge to the field point, the product $(G_{ij}NJ)$ tends to a finite limit as the field point is approached and therefore the formula (17.5b) may be used. When the field point is on a corner node of a rectangle or a mid-side node (see Fig. 17.3b), however, the element must be subdivided into triangular sub-elements and integration over each sub-element is carried out with respect to the axes shown in Fig. 17.3a. Further discussions on these integration schemes can be found in Refs 1, 12, 13 and 15, as well as Chapter 3.

The singular integrations, in the DBEM, for the two-dimensional case only are performed in the sub-routines AXKRNL (for axisymmetry) and PLKRNL (for plane stress/strain) based upon the variable ISING. A non-zero value of ISING

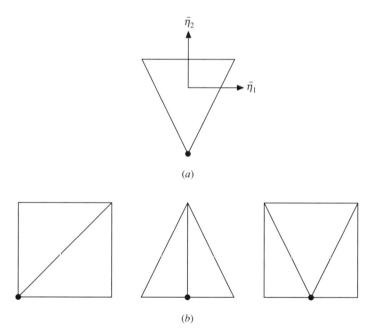

(a)

(b)

Figure 17.3 Sub-elements for integration over surface elements.

indicates that the node under consideration during the integration process is singular in nature. Next, the order of this node on the element is identified (1, 2 or 3) and the appropriate Gauss points, weights and sub-elements are set up in the sub-routine SINSET. In addition, the singular diagonal contributions via the rigid body displacement technique are calculated in the sub-routine INTGRA at the completion of the assembly of the GMAT and FMAT kernels. The 'inflation mode' computations for the jump terms in the axisymmetric case are also performed in INTGRA.

17.5 SYSTEM MATRIX ASSEMBLY

The first step in the assembly process is the reduction of the rectangular matrix of F_{ij} integrals to a square matrix. This matrix becomes the prototype of the final system matrix. The columns of the matrix are transformed or replaced, as required by the boundary conditions, as the assembly process proceeds.

Since the boundary nodes are shared by a number of elements it is necessary to compile an element connectivity matrix which contains a directory of these nodal connections. Thus after the integration over individual elements the coefficients relating to the common nodes can be summed and put in the appropriate location in the system matrices for each region. The assembly process for potential flow analysis was examined in detail in Chapter 3. The procedure for elastostatics is similar except that the scalar coefficients at each node are replaced by 2×2 blocks of coefficients for the two-dimensional problem. Thus, after the collocation process is completed, Eq. (17.1) can be expressed, in matrix notation, as:

$$[G]\{t\} - [F]\{u\} = \{0\} \tag{17.9}$$

where

$[G] =$ unassembled matrix of G coefficients
$[F] =$ assembled matrix of F coefficients, with c_{ij} terms absorbed into the diagonals
$\{t\} =$ global nodal traction vector
$\{u\} =$ global nodal displacement vector

The singular contributions to the $[F]$ matrix, i.e. the diagonal blocks, are determined using the 'rigid body' technique or the 'inflation mode' technique. As explained in Chapters 3 and 4, Eq. (17.9) is scaled such that the coefficients of $[G]$ and $[F]$ matrices are of the same orders of magnitude.

The final assembly of the system equations is carried out in three steps:

1. Transformation of columns of the matrices to appropriate local coordinate systems and incorporation of boundary conditions.
2. Incorporation of compatibility and equilibrium conditions on interfaces between regions.
3. Application of specified displacements and tractions (and/or temperature or flux).

If local boundary conditions are prescribed, the corresponding displacements and tractions are suitably transformed into the local system on a nodal basis. Thus,

$$t_i^g = a_{ij}t_j^l \tag{17.10a}$$

$$u_i^g = a_{ij}u_j^l \tag{17.10b}$$

where t_i^g and u_i^g are the global components of tractions and displacements respectively, t_i^l and u_i^l are the corresponding local components at that node and a_{ij} is the transformation tensor. Upon rearranging the matrices for coefficients corresponding to known and unknown variables, the system equation (17.9) can be expressed as

$$[\mathbf{A}]\{\mathbf{X}\} = \{\mathbf{b}\} \tag{17.11}$$

A particular feature of the assembly process deserves special comment. In multi-region problems the system matrix is not full. Rather, as shown in Chapter 3, it can be considered to be consisting of several sub-matrices, some of which are fully populated while others are completely zero. Only the non-zero portions of the system equations should be used during the solution. In addition, columns are re-ordered such that the variables from two adjoining regions are as close to each other as possible to improve the numerical conditioning during the solution process.

The task of system assembly is performed in the sub-routine ASMBLR in the program DBEM. At the completion of the integration (sub-routine INTGRA) the coefficients are stored in the arrays GMAT and FMAT. Subsequently, in ASMBLR, FMAT is transformed into the system matrix by applying the boundary conditions, rearranging the coefficients for the known and unknown variables and reordering for better conditioning of the matrix. The known right-hand side vector {b}, stored as BVEC in DBEM, is formed out of GMAT. The solution scheme is discussed in the next section.

17.6 SOLUTION OF THE SYSTEM EQUATIONS

Unless some special approach is adopted in the discretization process, a BEM analysis generates a non-symmetric fully populated matrix for a single region and a non-symmetric block-banded system matrix for multiple regions. For small and medium size problems, the time required for the solution of such a system of equations does not exceed that required for the formation of the system matrices. For very large problems, however, the equation solution time dominates the computing costs.

The numerical formulation of an n-zone problem will be represented by a system of equations, conveniently expressed in block form (see Chapter 3) where each block row of the matrix corresponds to an individual zone. This matrix is more densely populated than that normally solved by iterative methods. Furthermore, although it is diagonally strong it is not necessarily diagonally dominant and iterative convergence can not be assured. Such a system of equations is therefore best solved by a direct reduction.

For a given number of equations, the smaller the bandwidth, the less the amount of computational effort required whether the equations are block-banded or not. The minimum block bandwidth of the coefficient matrix in Eq. (17.12) below corresponds to an ordering of the zones wherein the maximum difference between any two adjacent zones is minimized. This is again analogous to the scheme used in finite element analyses for element ordering. In the matrix of Eq. (17.12) the maximum difference between adjacent zones is q.

$$
\begin{bmatrix}
A_1 & U_{11} & U_{12} & . & . & . & U_{1q} & 0 & . & 0 \\
L_{11} & A_2 & U_{21} & U_{22} & . & . & . & U_{2q} & . & . \\
L_{12} & L_{21} & A_3 & U_{31} & U_{32} & . & . & . & U_{3q} & 0 \\
. & L_{22} & L_{31} & . & . & . & . & . & . & . \\
. & . & L_{32} & . & . & . & . & . & . & . \\
. & . & . & . & . & . & . & . & . & . \\
L_{1q} & . & . & . & . & . & . & . & . & . \\
0 & L_{2q} & . & . & . & . & . & . & . & . \\
. & . & L_{3q} & . & . & . & . & . & . & . \\
0 & . & 0 & . & . & . & . & . & . & A_n
\end{bmatrix}
\{Q\} =
\begin{Bmatrix}
d_1 \\
d_2 \\
d_3 \\
. \\
. \\
. \\
. \\
. \\
. \\
d_n
\end{Bmatrix}
\tag{17.12}
$$

Several efficient routines for solving the block-banded system of equations (17.12) have been described in Refs 1 to 11. These are based upon Gaussian elimination analogous to established routines for solving banded systems by triangular decomposition, with the principal exception that block (matrix) multiplication and division operations are performed instead of single-entry multiplication and division. The following features are important in a BEM solver:

1. Advantage is taken of that block form of the coefficient matrix, thereby avoiding wasted storage space which would result from the use of a routine designed for single-entry banded matrices.
2. Housekeeping is easier when the identification of the position of each entry within a block is retained, instead of being converted to identify the position relative to the complete matrix.
3. Since only a few blocks are operated on at any stage, considerable saving of core storage is achieved by reserving the remaining blocks in backing store.

The single-region program DBEM of course utilizes a simple Gaussian elimination scheme for the solution process (sub-routine SOLVER).

17.7 CALCULATIONS AT INTERIOR POINTS

Once the boundary solution is obtained, the displacement and stress may be obtained at selected interior points by expressing the interior integral identities in the form of Eq. (17.1), which are then integrated numerically and the known boundary solution is substituted. The cost of evaluating such integrals for interior points is often high. For example, the cost of calculating the system equation at a boundary node using Eq. (17.1) is almost identical to that at an interior point.

Using the direct BEM it is sometimes more efficient to divide a homogencous region into a number of sub-regions so that the initial solution of the system equations provides the displacement and tractions at a number of interface nodes which, if carefully chosen, are then useful interior results. Lachat and Watson[1] used this idea to advantage and described solutions at that time (1976) for a number of three-dimensional problems of considerable geometrical complexity.

Often in a problem the stresses at a surface nodes are all that are required. For such surface stress computations, one can use the procedure described below.

Assuming that we are interested in finding the stress and strain at a point A which lies on a boundary element and has intrinsic coordinates η_i (see Fig. 17.4), the following relationships may be written:

$$t_i = \sigma_{ij} n_j \qquad (17.13a)$$

$$\sigma_{ij} = C_{ijkl} \left(\frac{u_{k,l} + u_{l,k}}{2} \right) \qquad (17.13b)$$

$$\frac{\partial u_i}{\partial \eta_k} = u_{i,j} \frac{\partial x_j}{\partial \eta_k} \qquad (17.13c)$$

$$\frac{\partial u_i}{\partial \eta_\ell} = \sum_{k=1}^{n} \frac{\partial N^k}{\partial \eta_\ell} u_i^k \qquad (17.13d)$$

where C_{ijkl} is the elastic constitutive tensor, x_{j,η_k} are the directional derivatives, n is the total number of nodes in the element, N^k is the shape function, u_i^k is the nodal value of u_i.

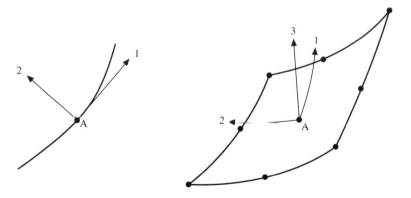

Figure 17.4 Local axes through a surface point A.

17.7.1 Two-dimensional Problems

Using an intrinsic coordinate η, Eqs (17.13a,b,c) can be combined to form a matrix equation:

$$
\begin{bmatrix}
n_1 & 0 & n_2 & 0 & 0 & 0 & 0 \\
0 & n_2 & n_1 & 0 & 0 & 0 & 0 \\
1 & 0 & 0 & -\lambda - 2\mu & 0 & 0 & -\lambda \\
0 & 1 & 0 & -\lambda & 0 & 0 & -\lambda - 2\mu \\
0 & 0 & 1 & 0 & -\mu & -\mu & 0 \\
0 & 0 & 0 & -n_2 & 0 & n_1 & 0 \\
0 & 0 & 0 & 0 & -n_2 & 0 & n_1
\end{bmatrix}
\begin{Bmatrix}
\sigma_{11} \\
\sigma_{22} \\
\sigma_{12} \\
u_{1,1} \\
u_{2,1} \\
u_{1,1} \\
u_{2,2}
\end{Bmatrix}
=
\begin{Bmatrix}
t_1 \\
t_2 \\
0 \\
0 \\
0 \\
u_{1,\eta} \\
u_{2,\eta}
\end{Bmatrix}
\quad (17.14)
$$

where n_1 and n_2 are the unit normals on the boundary at the point A, i.e.

$$n_1 = x_{2,\eta} \quad \text{and} \quad n_2 = -x_{1,\eta}$$

The stress and strain at point A can be obtained by inverting the matrix of Eq. (17.14) and then multiplying the inverted matrix by the right-hand side vector. For this purpose, the right-hand side vector is obtained by using Eq. (17.13d) along with Eq. (17.2).

17.7.2 Three-dimensional Problems

For this case, the intrinsic coordinates become (η_1, η_2), as shown in Fig. 17.4. The relationships given by Eqs (17.13) still hold true and, once again, Eqs (17.13a,b,c) may be combined to form the matrix equation:

$$[S]\{p\} = \{q\} \quad (17.15)$$

where $[S]$ is a 15×15 matrix containing unit normals, a 3×3 unit matrix and material constants; $\{p\}$ is the unknown vector of σ_{ij} and $\partial u_i / \partial \xi_j$; and $\{q\}$ is a vector containing the tractions t_i and local coordinates of the displacements at point A. Similar to the two-dimensional case, the stress and strain at A can be obtained by inverting $[S]$ and multiplying it by the right-hand side vector derived from Eqs (17.2) and (17.13d).

For potential flow problems, the flux on the surface cam also be determined without integration by using the following relationships for a two-dimensional analysis:

$$\sum_{k=1}^{n} N^k(\eta) q^k = n_i q_i^k \quad (17.16a)$$

$$\sum_{k=1}^{n} \frac{\partial N^k}{\partial \eta} \theta^k = X_{i,\eta} q_i^k \quad (17.16b)$$

where q_i^k are the nodal values of flux and θ^k is the nodal value of θ.

In the program DBEM, the interior computations are controlled by the sub-routine INTSTR. This routine, in turn, calls upon the following sub-routines to perform certain specific tasks:

Name	Function
INTSTR	Interior computations control
AUTINT	Set up the order of integration (NGP, NSUB)
AXKRNL	Integration of interior displacement and stress kernels for the axisymmetric case
PLKRNL	Integration of interior displacement kernels for the two-dimensional case
STRKER	Integration of interior stress kernels for the two-dimensional case
BNDSTR	Surface stress computations

17.8 SAMPLE DATA AND OUTPUT

This section presents sample input data with solutions to some simple problems using the program DBEM. Having studied these illustrative examples, the reader should have a better understanding of the input data formats and may then go on to solve more complex problems.

17.8.1 A Simple Tension Problem

This example consists of a unit cube under plane strain conditions subjected to uniform tension in the x direction. The boundary element discretization is shown in Fig. 17.5a, indicating the use of eight nodes and four elements. The applied boundary conditions are also shown symbolically. The material properties for this problem are selected as: elastic modulus $(E) = 100.0$ lb/in^2 and Poisson's ratio $(\nu) = 0.0$; a traction load (t_x) of 100.0 lb/in^2 is applied as shown. Several points are placed in the interior to monitor the response. The input data set for this problem is given below:

Input Data for Simple Tension

```
2
PROBLEM # 1
SIMPLE TENSION TEST
2 0 0
100.0 0.0
8 4
1
8 4
1 0.0 0.0
2 0.5 0.0
```

```
3 1.0 0.0
4 1.0 0.5
5 1.0 1.0
6 0.5 1.0
7 0.0 1.0
8 0.0 0.5
6
2 1 1 100. 1 0.
2 2 1 100. 1 0.
2 3 1 100. 1 0.
4 1 2 0. 1 0.
4 2 2 0. 2 0.
4 3 2 0. 1 0.
5
0.50 0.25
0.25 0.50
0.50 0.50
0.75 0.50
0.50 0.75
```

The results obtained from the DBEM for this example are presented below for certain selected boundary nodes and interior points. The corresponding analytical solutions are cited alongside the numerical results.

Element	Node	DBEM	Analytical
(1) u_x:			
1	2	0.49968	0.5000
2	4	0.99929	1.0000
(2) t_x:			
4	7	−99.971	−100.000
4	8	−100.010	−100.000

For the interior point with coordinates (0.25, 0.50), the displacement u_x is found to be 0.2497 and the stress (σ_{xx}) is 99.98 lb/in^2 while the exact solutions are 0.25 and 100.0 lb/in^2 respectively.

17.8.2 Uniform Radial Expansion

In this example, a hollow cylinder of inner radius 1.0 in and a thickness of 1.0 in is subjected to uniform internal pressure. Assuming plane strain conditions in the axial direction, only purely radial expansion is permitted. Thus, the axisymmetric

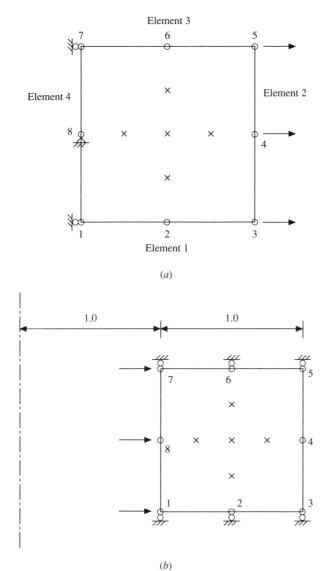

Figure 17.5 BEM model for (*a*) simple tension, (*b*) radial expansion.

analysis option in the DBEM is selected and the BEM discretization pattern is displayed in Fig. 17.5*b*. The material properties are chosen as: $E = 100 \text{ lb/in}^2$ and $\nu = 0.3$, and a pressure of 100 lb/in^2 is applied on the inner surface (element 4). Once again, interior points are placed at certain locations within the cylinder. The input data set for this problem now follows:

Input Data for Uniform Radial Expansion

```
2
PROBLEM # 2
RADIAL EXPANSION TEST
2 2 0
100.0 0.3
8 4
1
8 4
1 1.0 0.0
2 1.5 0.0
3 2.0 0.0
4 2.0 0.5
5 2.0 1.0
6 1.5 1.0
7 1.0 1.0
8 1.0 0.5
9
1 1 1 0. 2 0.
1 2 1 0. 2 0.
1 3 1 0. 2 0.
3 1 1 0. 2 0.
3 2 1 0. 2 0.
3 3 1 0. 2 0.
4 1 1 100. 1 0.
4 2 1 100. 1 0.
4 3 1 100. 1 0.
5
1.50 0.25
1.25 0.50
1.50 0.50
1.75 0.50
1.50 0.75
```

The numerical solutions at certain nodes from the DBEM are listed below along with the exact solutions:

Element	Node	DBEM	Analytical
(1) u_r:			
1	1	1.9026	1.9066
1	2	1.4142	1.4151
1	3	1.2131	1.2133

(2)$\sigma_{\theta\theta}$:

2	3	66.65	66.66
4	8	166.46	166.66

The solution at an interior point (1.50, 0.25) from the DBEM yields: $u_r = 1.415$; $\sigma_{rr} = -26.47$; $\sigma_{\theta\theta} = 92.14$, in comparison to the exact solutions: $u_r = 1.4155$; $\sigma_{rr} = -25.93$; $\sigma_{\theta\theta} = 92.59$ respectively.

17.8.3 One-dimensional Heat Conduction

In order to test the facility for solving steady state heat transfer analyses in the DBEM, a problem of one-dimensional heat conduction is studied. The unit cube under plane strain utilized in the simple tension problem (see Fig. 17.5a) is used once again for this case. The only material property required is conductivity which is assumed to be 1.0. The temperature of the lower edge (element 1) is maintained at 0°F while that of the upper edge (element 3) is at 100°F. A number of interior points are also placed within the cube similar to the previous problems. The sample input data set for this case is shown below (note that the geometry definition is identical to that of the problem in Sec. 17.8.1, and is not repeated):

Input Data for One-Dimensional Heat Conduction

2
PROBLEM # 3
ONE-DIMENSIONAL HEAT CONDUCTION
1 0 0
1.0
......
...... (geometry definition not repeated)
......
6
1 1 2 0.0
1 2 2 0.0
1 3 2 0.0
3 1 2 100.0
3 2 2 100.0
3 3 2 100.0
......
...... (interior points locations not repeated)
......

The results obtained from the DBEM compare very well with the expected exact solutions to this problem. For instance, the temperature at nodes 4 and 8 is obtained as 50.00°F and those for the interior points at (0.50, 0.25) and (0.50, 0.75) as 24.99°F and 75.01°F respectively.

17.8.4 Radial Heat Conduction

The final example is an axisymmetric problem of purely radial heat conduction. The hollow cylinder from the problem in Sec. 17.8.2 is now subjected to a thermal gradient between the inner and outer surfaces. Selecting, once again, the conductivity as 1.0, the inner surface (element 4) is now maintained at a temperature of $100°$F while the outer surface (element 2) is at $0°$F. The input data set for this problem is shown below, excluding the geometry definition which is identical to that of the problem in Sec. 17.8.2 (see Fig. 17.5b):

Input Data for Radial Heat Conduction

```
2
PROBLEM # 4
RADIAL HEAT CONDUCTION
1 2 0
1.0
......
...... (geometry definition not repeated)
......
6
2 1 2 0.0
2 2 2 0.0
2 3 2 0.0
4 1 2 100.0
4 2 2 100.0
4 3 2 100.0
......
...... (interior points locations not repeated)
......
```

The analytical solution for the temperature at $r = 1.5$ (nodes 2 and 6) is computed to be $41.504°$F while the result obtained from the DBEM is $41.506°$F. Similarly, at an interior point with radius $r = 1.25$, the exact solution for temperature is $67.807°$F while DBEM yields the value of $67.87°$F.

REFERENCES

1. Lachat, J.C. and Watson, J.O. (1976), Effective numerical treatment of boundary integral equations: a formulation for three-dimensional elastostatics, *International Journal of Numerical Methods in Engineering*, Vol. 10, pp. 991–1005.
2. Lachat, J.C. and Watson, J.O. (1975), Progress in the use of boundary integral equations, illustrated by examples, *Computer Methods in Applied Mechanics and Engineering*, Vol. 10, pp. 273–289.
3. Banerjee, P.K. and Wilson, R.B. (1991) *BEST-3D User's Manual*, NASA Contract Report 185298.
4. Wilson, R.B., Miller, N. and Banerjee, P.K. (1989) Advanced applications of BEM to gas turbine

engine structures, Chapter 1 in *Developments in Boundary Element Methods*, Vol. V, Eds P.K. Banerjee and R.B. Wilson, Elsevier Applied Science Publishers, London.

5. GP-BEST (1991) *GP-BEST User's Manual*, Boundary Element Software Technology Corporation, PO Box 310, Getzville, New York 14068, USA.

6. BEASY (1989), *BEASY User's Manual*, Computational Mechanics Ltd, Southampton, England.

7. EZBEA (1987), *EZBEA User's Documentation*, Ver. 2.2, Caterpillar, Inc., Peoria, Illinois, USA.

8. Kuhn, G., Lobel, G. and Sichert, W. (1987), Erstellung des Programms BETTI zur Berechnung der stationaren und instationaren Warmeleitung mittels BEM, Final report, Forschungsbericht des Forschungskuratoriums Maschinenbau (FKM), Frankfurt, Germany.

9. Kuhn, G. and Mohrmann, W. (1983), Boundary element methods in elastostatics: theory and applications, *Applied Mathematical Modelling*, Vol. 7, pp. 97–105.

10. Mohrmann, W. (1984) *DBETSY—die Boundary Element Methode in der industriellen Berechnung*, VDI - Ber., No. 537, pp. 627–650.

11. Butenschon, H.J., Mohrmann, W. and Bauer, W. (1989), Advanced stress analysis by a commercial BEM code, Chapter 8 in *Developments in Boundary Element Methods*, Vol. V, Eds P.K. Banerjee and R.B. Wilson, Elsevier Applied Science Publishers, London.

12. Watson, J.O. (1979), Advanced implementation of boundary element method in two and three-dimensional elastostatics, Chapter 3 in *Developments in Boundary Element Methods*, Vol. I, Eds P.K. Banerjee and R. Butterfield, Elsevier Applied Science Publishers, London.

13. Mustoe, G.G. (1984) Advanced integration schemes over boundary elements and volume cells for two and three-dimensional nonlinear analysis', in *Developments in Boundary Element Methods*, Vol. IV, Eds, P.K. Banerjee and S. Mukherjee, Elsevier Applied Science, London.

14. Stroud, A.H. and Secrest, D. (1966), *Gaussian Quadrature Formulae*, Prentice-Hall, Englewood Cliffs, New Jersey.

15. Banerjee, P.K. and Butterfield, R. (1981) *Boundary Element Methods in Engineering Science*, McGraw-Hill, London.

A

DERIVATION OF FUNDAMENTAL SOLUTIONS

A.1 INTRODUCTION

At the heart of the boundary element method is the fundamental solution. Each problem has its own fundamental solution, which, typically, describes the response at a point and time to a source acting at another point and time. The success of the BEM lies in the ability to determine the appropriate fundamental solution for the given problem.

In addition to requiring a fundamental solution, the BEM analyst may desire to have more than one fundamental solution available. This would allow the analyst to choose the one with superior computational benefits. For example, in the case of buoyancy-driven flow, a fundamental solution without buoyancy is available (i.e. the stokeslet), and buoyancy effects therefore needs to be treated through a volume integration. However, if the buoyancy is incorporated, or coupled, into the fundamental solution, two benefits emerge: (1) an interior mesh is no longer required for Stokes flow and (2) the non-linear convergence behaviour would be substantially improved.[1] Hence, a more complete fundamental solution of the differential operator for a given problem may be desirable.

The majority of the boundary element formulations have relied heavily on the fundamental solutions that have already been published. The fundamental solutions to Laplace's equation, equations of elasticity, etc., have been exploited by boundary element analysts and are described in Chapters 3 and 4 respectively. However, when solutions to new problems and equations are sought, the lack of the appropriate fundamental solution will discourage the most enthusiastic

BEM developer. What is desperately needed is a systematic way of deriving new fundamental solutions.

This appendix[1] will cover the relevant aspects necessary to understand how some fundamental solutions can be derived in a systematic manner. Rigorous theory is not employed; the presentation is more practical than theoretical. First, some background concepts from linear algebra are discussed. Then, Hörmander's method[2] for combining PDEs is presented in terms of linear algebra concepts. Finally, simpler fundamental solutions are combined in two ways to yield more complex solutions—by convolution and by the 'difference trick'. Examples are used throughout to illustrate the concepts.

A.2 ELEMENTS OF LINEAR ALGEBRA

Since Hörmander's method is heavily rooted in linear algebra, some pertinent linear algebra and matrix concepts are presented here. Elementary texts[3,4] can be consulted for more detail. Although simple algebraic matrices are used here, the extension to partial differential operators is made in the following section.

The determinant of a 1×1 matrix $[A] = [a_{11}]$ is merely $\det[A] = a_{11}$. Now consider the 2×2 matrix

$$[A] = \begin{bmatrix} a_{11} & a_{12} \\ a_{21} & a_{22} \end{bmatrix} = \begin{bmatrix} 5 & -4 \\ 6 & 2 \end{bmatrix} \tag{A.1}$$

Let $[B] = [b_{ij}]$ be the reduced matrix obtained by deleting row i and column j from matrix $[A]$. Then the minor belonging to member a_{ij} is $m_{ij} = \det[b_{ij}]$. The minor matrix $[M]$ is composed of all the minors of $[A]$. For the above example, $m_{11} = \det[2] = 2$, $m_{12} = \det[6] = 6$, etc., so that

$$[M] = [m_{ij}] = \begin{bmatrix} 2 & 6 \\ -4 & 5 \end{bmatrix} \tag{A.2}$$

If a new matrix $[C]$ is constructed, whose terms are given by $c_{ij} = (-1)^{(i+j)} m_{ij}$, then the matrix is called a 'cofactor matrix', or a 'matrix of cofactors'. The terms c_{ij} are the cofactors corresponding to the terms a_{ij}. For example, $c_{11} = (-1)^{(1+1)}(2) = 2$, and

$$[C] = [c_{ij}] = \begin{bmatrix} 2 & -6 \\ 4 & 5 \end{bmatrix} \tag{A.3}$$

The adjoint of $[A]$ is the transpose of the matrix of cofactors, denoted by $\mathrm{adj}[A] = [C]^{\mathrm{T}}$:

$$\mathrm{adj}[A] = [c_{ij}]^{\mathrm{T}} = \begin{bmatrix} 2 & 4 \\ -6 & 5 \end{bmatrix} \tag{A.4}$$

The determinant of an $n \times n$ matrix, when $n \geq 2$, can be found by multiplying a row of $[A]$ by its corresponding row in $[C]$, or a column of $[A]$ by its corresponding column in $[C]$. For instance, using the first row, $\det[A] = a_{11}c_{11} + a_{12}c_{12} =$

$(5)(2) + (-4)(-6) = 34$. Equivalently, using the first column, $\det[A] = a_{11}c_{11} + a_{21}c_{21} = (5)(2) + (6)(4) = 34$.

The inverse of a matrix $[A]$ is denoted by $[A]^{-1}$. The relationship between these is

$$[A]^{-1}[A] = [I] \tag{A.5}$$

where $[I]$ is the identity matrix of order $n \times n$. Although this relationship is valid, it does not allow for effective calculation of $[A]^{-1}$, which is crucial in solving a system of equations.

An important theorem of linear algebra states that

$$[A][\mathrm{adj}[A]] = \det[A][I] \tag{A.6}$$

and therefore

$$[A]^{-1} = \frac{\mathrm{adj}[A]}{\det[A]} \tag{A.7}$$

It is important to understand how to apply this in solving a system of equations. If a solution of the system of equations given by $[A]\{X\} = \{E\}\psi$ is sought, where $[A]$ and $\{E\}$ are known and ψ is merely a scalar multiplier, then the solution is given by

$$\{X\} = [A]^{-1}\{E\}\psi = \left[\frac{\mathrm{adj}[A]}{\det[A]}\right]\{E\}\psi = [\mathrm{adj}[A]]\{E\}\phi \tag{A.8}$$

where ϕ is just a scalar multiplier in the equation $\det[A]\phi = \psi$.

For example, consider the sample matrix used above. A solution is sought for $[A]\{X\} = \{E\}\psi$ given by

$$\begin{bmatrix} 5 & -4 \\ 6 & 2 \end{bmatrix} \{X\} = \begin{Bmatrix} 1 \\ 0 \end{Bmatrix} 17 \tag{A.9}$$

First, the adjoint matrix is determined, and then the determinant:

$$\mathrm{adj}[A] = \begin{bmatrix} 2 & 4 \\ -6 & 5 \end{bmatrix} \qquad \text{and} \qquad \det[A] = 34 \tag{A.10}$$

Next, the scalar ϕ is determined:

$$\det[A]\phi = \psi \quad \rightarrow \quad (34)\phi = 17 \quad \rightarrow \quad \phi = 0.5 \tag{A.11}$$

Finally, the solution of $\{X\}$ is given by

$$\{X\} = \mathrm{adj}[A]\{E\}\phi \quad \rightarrow \quad \{X\} = \begin{bmatrix} 2 & 4 \\ -6 & 5 \end{bmatrix} \begin{Bmatrix} 1 \\ 0 \end{Bmatrix} (0.5) = \begin{Bmatrix} 1 \\ -3 \end{Bmatrix} \tag{A.12}$$

Although this may seem like a laborious way to solve a system of equations (and it is), its benefit will be demonstrated in the next section. The main point to be derived from this section is that the solution of a system of equations can be effected through the manipulation of the adjoint matrix and the determinant.

A.3 HÖRMANDER'S METHOD

In the derivation of integral equations, we are confronted with a linear operator equation that defines the appropriate fundamental solution. For example, consider

$$\tilde{L}_{ij}\tilde{u}_{jk} + \delta_{ik}\delta(x - \xi)\delta(t - \tau) = 0 \qquad (A.13)$$

Although this appears straightforward, in reality it is quite complex. $\tilde{L}_{ij}$ is really an $n \times n$ operator made up of several partial differential operators. The goal is to find the tensor $\tilde{u}_{jk}$, which is the fundamental solution for the given problem.

Hörmander described a method for doing this, which is now known as 'Hörmander's Method'. Actually, Hörmander's method does not provide $\tilde{u}_{jk}$ directly, but merely allows for the important first step: it allows several partial differential operators to be combined into a scalar differential operator. Often, it is easier to work with a complex scalar equation than several coupled differential equations. The second step in finding $\tilde{u}_{jk}$ involves the solution of the scalar differential equation, which is taken up in the next section.

Hörmander's method will be presented by illustrating a specific situation—that of steady Stokes flow without heat transfer. Extension to more complex situations is straightforward. The linear operator equation for Stokes flow is given by (see Chapter 15)

$$\tilde{L}_{ij}\tilde{u}_{jk} + \delta_{ik}\delta(x - \xi) = 0 \qquad (A.14)$$

where $i, j, k = 1, 2$. In matrix form, the equation is

$$[\tilde{L}][\tilde{u}] = -[I]\delta(x - \xi) \qquad (A.15)$$

The objective is to find $[\tilde{u}]$. The linear operator is

$$\tilde{L}_{ij} = [\tilde{L}] = \begin{bmatrix} \mu\nabla^2 & 0 & -\dfrac{\partial}{\partial x_1} \\[2mm] 0 & \mu\nabla^2 & -\dfrac{\partial}{\partial x_2} \\[2mm] \dfrac{\partial}{\partial x_1} & \dfrac{\partial}{\partial x_2} & 0 \end{bmatrix} \qquad (A.16)$$

Following the procedure laid down in Sec. A.2, the first step is to determine the matrix of cofactors $[C]$, whose terms are given by $c_{ij} = (-1)^{(i+j)}m_{ij}$, where the terms m_{ij} are the minors corresponding to terms in $\tilde{L}_{ij}$. The cofactor matrix for

Eq. (A.16) is

$$
C_{ij} = [C] =
\begin{bmatrix}
\dfrac{\partial^2}{\partial x_2^2} & -\dfrac{\partial^2}{\partial x_1 \partial x_2} & -\mu\nabla^2 \dfrac{\partial}{\partial x_1} \\[2ex]
-\dfrac{\partial^2}{\partial x_1 \partial x_2} & \dfrac{\partial^2}{\partial x_1^2} & -\mu\nabla^2 \dfrac{\partial}{\partial x_2} \\[2ex]
\mu\nabla^2 \dfrac{\partial}{\partial x_1} & \mu\nabla^2 \dfrac{\partial}{\partial x_2} & \mu\nabla^4
\end{bmatrix}
\tag{A.17}
$$

The determinant of $[\tilde{L}]$ can be determined by multiplying the terms in the first column of $[L]$ by the corresponding terms in the first column of $[C]$. Hence,

$$
\det[\tilde{L}] = (\mu\nabla^2)\left(\frac{\partial^2}{\partial x_2^2}\right) + (0)\left(-\frac{\partial^2}{\partial x_1 \partial x_2}\right) + \left(\frac{\partial}{\partial x_1}\right)\left(\mu\nabla^2 \frac{\partial}{\partial x_1}\right) = \mu\nabla^4 \tag{A.18}
$$

The adjoint matrix is merely the transpose of $[C]$:

$$
\mathrm{adj}[\tilde{L}] = [C]^{\mathrm{T}} =
\begin{bmatrix}
\dfrac{\partial^2}{\partial x_2^2} & -\dfrac{\partial^2}{\partial x_1 \partial x_2} & \mu\nabla^2 \dfrac{\partial}{\partial x_1} \\[2ex]
-\dfrac{\partial^2}{\partial x_1 \partial x_2} & \dfrac{\partial^2}{\partial x_1^2} & \mu\nabla^2 \dfrac{\partial}{\partial x_2} \\[2ex]
-\mu\nabla^2 \dfrac{\partial}{\partial x_1} & -\mu\nabla^2 \dfrac{\partial}{\partial x_2} & \mu^2\nabla^4
\end{bmatrix}
\tag{A.19}
$$

At this point, all the necessary matrices have been determined. Now, the theorem of linear algebra (Sec. A.2) will be applied to find the fundamental solution. We see the matrix $[\tilde{u}]$ that is the solution to Eq. (A.15). In this case, the term $-\delta(x - \xi)$ is merely a scalar multiplier of the identity matrix. Thus, it fulfills the role of ψ. Using the results of Sec. A.2, the fundamental solution is given by

$$
[\tilde{u}] = [\mathrm{adj}[\tilde{L}]][I]\phi = [\mathrm{adj}[\tilde{L}]]\phi \tag{A.20}
$$

where

$$
\det[\tilde{L}]\phi = -\delta(x - \xi) \tag{A.21}
$$

For the particular problem of Stokes flow,

$$
\mu\nabla^4\phi = -\delta(x - \xi) \tag{A.22}
$$

Equation (A.22) is the significant result of applying Hörmander's method. It says that if a solution to the scalar equation (A.22) can be found, then the fundamental solution $[\tilde{u}]$ can be determined from Eq. (A.20). In other words, the problem of solving Eq. (A.15), which is a matrix (or tensor) equation, has been transformed to the problem of solving Eq. (A.22), which is a scalar equation. Hörmander's method does not assert that a fundamental solution can be found.

It merely provides a method of simplifying the problem to one that may allow solution.

At this point, it is logical to finish the solution of Eq. (A.22), since it is commonly known as the biharmonic equation:

$$\phi = \frac{-1}{8\pi\mu} r^2 \ln (r) \tag{A.23}$$

where $r^2 = y_i y_i$, and $y_i = x_i - \xi_i$. The differentiation of ϕ can be carried out in Eqs (A.19) and (A.20), so that all the terms in the stokeslet fundamental solution are known.

Although it is best to have the fundamental solution ϕ in explicit form, it is not always necessary. Recall that the reason for using Hörmander's method in the first place is to determine $\tilde{u}_{ij}$. Equation (A.20) indicates that $\tilde{u}_{ij}$ is related, through terms in $\tilde{L}_{ij}$ to ϕ. If the linear operator $\tilde{L}_{ij}$ is composed exclusively of partial derivatives, then $\tilde{u}_{ij}$ can be found from partial derivatives of ϕ. In this case, having ϕ in explicit form is not necessary. This may be useful when expressions for $\partial\phi/\partial x_i$ or $\partial\phi/\partial t$ are available, but integrating them to get ϕ is a monumental task.

A.4 FUNDAMENTAL SOLUTIONS FOR SCALAR EQUATIONS

A.4.1 Introduction

The previous section used Hörmander's method to combine the coupled partial differential operator into a single scalar differential operator. The next step involves solving for the scalar fundamental solution. This latter step is the subject of the remaining sections of this appendix.

Let the differential operator of the scalar equation be denoted by $\mathcal{L}$. In the equation $\mu\nabla^4\phi = -\delta(x - \xi)$, the linear operator is $\mathcal{L} = \mu\nabla^4$. Often, $\mathcal{L}$ is composed of several factors. For instance, $\mathcal{L} = \mu\nabla^4 = \mu\nabla^2\nabla^2$. Another example is $\mathcal{L} = \nabla^2(\mu\nabla^2 - \rho\,\partial/\partial t)$. In each of these examples, the linear operator contains two simpler factors. This raises the question: if the fundamental solutions to the factors in $\mathcal{L}$ are known, can they be combined in some way to yield the fundamental solution for the more complex $\mathcal{L}$? The answer to this question is 'yes'. This will be covered in the sections on 'convolution' and the 'difference trick'.

A.4.2 Convolution

Consider a linear differential operator equation, along with its known fundamental solution ϕ:

$$\mathcal{L}\phi = -\delta(x - \xi) \tag{A.24}$$

An important use of convolution theory is in solving the general problem

$$\mathcal{L}\psi = f \tag{A.25}$$

where f is a function of space and time. The solution ψ to Eq. (A.25) is given by

$$\psi = f * \phi \tag{A.26}$$

which is 'the convolution of f and ϕ'. Texts such as Morse and Feshbach[5] and Kanwal[6] contain much detail about convolution theory.

Convolution theory will now be applied to the biharmonic equation

$$\nabla^4 \phi = -\delta(x - \xi) \tag{A.27}$$

Consider first, the simpler Poisson equation, along with its fundamental solution:

$$\nabla^2 \phi_p = -\delta(x - \xi) \tag{A.28a}$$

$$\phi_p = \frac{-1}{2\pi} \ln(r) \tag{A.28b}$$

where $r^2 = y_i y_i$ and $y_i = x_i - \xi_i$.

The following steps can be verified:

$$\nabla^2(\nabla^2 \phi) = -\delta(x - \xi) = \nabla^2 \phi_p \tag{A.29}$$

$$\nabla^2(\nabla^2 \phi - \phi_p) = 0 \tag{A.30}$$

$$\nabla^2 \phi = \phi_p \tag{A.31}$$

By convolution theory, the solution to Eq. (A.31) is

$$\phi = \phi_p * \phi_p \tag{A.32a}$$

$$\phi = \left[\frac{-1}{2\pi} \ln(r)\right] * \left[\frac{-1}{2\pi} \ln(r)\right] \tag{A.32b}$$

Thus, solution of the biharmonic operator is reduced to the self-convolution of the Poisson fundamental solution.

In practice, convolution theory is difficult to apply. Direct integration of Eq. (A.32b) can be difficult, even for the simplest problems. Fourier transforms and Laplace transforms could be used to carry out the necessary integration.

A.4.3 The Difference Trick

Ortner[7] discussed a method of combining two simpler fundamental solutions in a way that yielded a more complex fundamental solution. He called the method 'the difference trick'. Although it is a simple concept, it can be the stepping stone from simpler to more complex fundamental solutions.

To illustrate, consider the problem

$$\mathcal{L}\phi = (\nabla^2 - a_1)(\nabla^2 - a_2)\phi = -\delta \tag{A.33}$$

for which the fundamental solution ϕ is sought. The terms a_1 and a_2 are constants. Two factors in Eq. (A.33) are clearly seen. To illustrate, we will suppose that the fundamental solution ϕ is not known, while the fundamental solutions to the components in Eq. (A.33) are known:

$$(\nabla^2 - a_1)\phi_1 = -\delta \qquad \text{(A.34}a)$$

$$(\nabla^2 - a_2)\phi_2 = -\delta \qquad \text{(A.34}b)$$

Combining Eqs (A.33) and (A.34) leads to

$$\phi_1 = (\nabla^2 - a_2)\phi \qquad \text{(A.35}a)$$

$$\phi_2 = (\nabla^2 - a_1)\phi \qquad \text{(A.35}b)$$

while the difference of Eqs (A.35a) and (A.35b) gives

$$\phi_1 - \phi_2 = (a_1 - a_2)\phi \qquad \text{(A.36}a)$$

or

$$\phi = \frac{\phi_1 - \phi_2}{a_1 - a_2} = \frac{\phi_1}{a_1 - a_2} + \frac{\phi_2}{a_2 - a_1} \qquad \text{(A.36}b)$$

Since ϕ_1 and ϕ_2 are both known fundamental solutions, the solution ϕ is determined through this simple difference formula. However, it should be noted that there are two limitations in this example:

1. The two differential operators in Eq. (A.33) were identical.
2. The constants cannot be equal, that is $a_1 \neq a_2$ [see Eq. (A.36b)].

The first limitation can be relaxed. It is not a requirement to have the same operators in both equations (A.33). However, it is desirable to have the difference result in ϕ by itself, as in Eq. (A.36a). Because it may not be possible to do so, some auxiliary manipulation may be required.

Consider, for instance, the scalar equation that results when Hörmander's method is applied to the linear operator equation for transient Stokes flow of Chapter 15:

$$\mathcal{L}\phi = \nabla^2 \left(\mu\nabla^2 - \rho\frac{\partial}{\partial t} \right)\phi = -\delta(x - \xi)\delta(t - \tau) \qquad \text{(A.37)}$$

The component fundamental solutions are known to be

$$\nabla^2 \phi_1 = -\delta(x - \xi)\delta(t - \tau) \qquad \text{(A 38}a)$$

$$\phi_1 = \frac{-1}{2\pi} \ln (r)\delta(t - \tau) \qquad \text{(A.38}b)$$

$$\phi_1 = \left(\mu\nabla^2 - \rho\frac{\partial}{\partial t} \right)\phi \qquad \text{(A.38}c)$$

and

$$\left(\mu\nabla^2 - \rho\frac{\partial}{\partial t}\right)\phi_2 = -\delta(x - \xi)\delta(t - \tau) \tag{A.39a}$$

$$\phi_2 = \frac{1}{4\pi\mu(t - \tau)}e^{-r^2/[4\nu(t-\tau)]}H(t - \tau) \tag{A.39b}$$

$$\phi_2 = \nabla^2\phi \tag{A.39c}$$

with $\nu = \mu/\rho$. By taking the difference of Eqs (A.38c) and (A.39c), the differential equation for ϕ is

$$\mu\phi_2 - \phi_1 = \rho\frac{\partial\phi}{\partial t} \tag{A.40}$$

This time, ϕ is not just a simple difference between ϕ_1 and ϕ_2, but there is an additional time derivative. The integration is carried out in the following manner:

$$\phi = \frac{1}{\rho}\int[\mu\phi_2(t - \tau) - \phi_1(t - \tau)]\,dt \tag{A.41}$$

$$\phi = \int\frac{1}{4\pi\rho(t - \tau)}e^{-r^2/[4\nu(t-\tau)]}H(t - \tau)\,dt - \int\frac{-1}{2\pi\rho}\ln(r)\delta(t - \tau)\,dt \tag{A.42}$$

$$\phi = \left\{\frac{1}{2\pi\rho}\ln(r) + \frac{1}{4\pi\rho}E_1\left[\frac{r^2}{4\nu(t - \tau)}\right]\right\}H(t - \tau) \tag{A.43}$$

To illustrate how the difference trick is extended to three factors, consider the scalar equation

$$\mathcal{L}\phi = (\nabla^2 - a_1)(\nabla^2 - a_2)(\nabla^2 - a_3)\phi = -\delta \tag{A.44}$$

The component fundamental solutions are assumed to be known:

$$(\nabla^2 - a_1)\phi_1 = -\delta \tag{A.45a}$$

$$(\nabla^2 - a_2)\phi_2 = -\delta \tag{A.45b}$$

$$(\nabla^2 - a_3)\phi_3 = -\delta \tag{A.45c}$$

Thus,

$$\phi_1 = (\nabla^2 - a_2)(\nabla^2 - a_3)\phi \tag{A.46a}$$

$$\phi_2 = (\nabla^2 - a_1)(\nabla^2 - a_3)\phi \tag{A.46b}$$

$$\phi_3 = (\nabla^2 - a_1)(\nabla^2 - a_2)\phi \tag{A.46c}$$

By rearranging Eqs (A.45) to (A.46), the fundamental solution ϕ can be derived:

$$\phi = \frac{\phi_1}{(a_1 - a_2)(a_1 - a_3)} + \frac{\phi_2}{(a_2 - a_1)(a_2 - a_3)} + \frac{\phi_3}{(a_3 - a_1)(a_3 - a_2)} \qquad (A.47)$$

Comparing Eqs (A.36b) and (A.47), the recurrence relation can be discerned, which is useful to extend the technique to equations with more factors.

Although the difference trick has only been illustrated with three examples, the extension to more complex operators and more factors in $\mathcal{L}$ is straightforward. Even with its limitations, this method is a key step to building more complex families of fundamental solutions from simpler, known fundamental solutions.

A.5 CONCLUDING REMARKS

Several methods have been detailed in an effort to determine some fundamental solutions. The presentation has been more practical than what has appeared previously in the literature. The first step involves finding a single scalar differential equation that captures the essence of the system of coupled partial differential equations (Hörmander's method). The second step involves solving the scalar equation for the scalar fundamental solution. Once the scalar fundamental solution is determined, the tensor fundamental solution can be simply derived. Some scalar fundamental solutions can be found in the literature; some can be derived through convolution, while others can be determined from the difference trick. Methods such as these are crucial to extend the BEM to other coupled engineering problems that do not process known fundamental solutions.

REFERENCES

1. Honkala, K. (1992) Boundary element methods for two-dimensional, coupled, thermoviscous flow, PhD Dissertation, State University of New York at Buffalo.
2. Hörmander, L. (1964) *Linear Partial Differential Operators*, Springer-Verlag, Berlin.
3. Isaak, S. and Manougian, M.N. (1976) *Basic Concepts of Linear Algebra*, W.W. Norton and Company, New York.
4. Strang, G. (1980) *Linear Algebra and Its Applications*, 2nd edn, Academic Press, London.
5. Morse, P.M. and Feshbach, H. (1953) *Methods of Theoretical Physics*, McGraw-Hill, New York.
6. Kanwal, R.P. (1983) *Generalized Functions: Theory and Technique*, Academic Press, New York.
7. Ortner, N. (1987) Methods of construction of fundamental solutions of decomposable linear differential operators, in *Boundary Elements IX*, Vol. 1, Eds C.A. Brebbia, W.L. Wendland and G. Kuhn, pp. 79–97, Springer-Verlag, Berlin.

OBTAINING ACCOMPANYING SOFTWARE

A $3\frac{1}{4}$ inch disk, in UNIX TAR format, containing the programs listed in Chapter 17 is available free of charge, by sending this cut-out page (photocopies not acceptable) to the McGraw-Hill offices listed below.

The Marketing Department
McGraw-Hill Book Company Europe
Shoppenhangers Road
Maidenhead
Berkshire
SL6 1UQ
England, UK

or

Ms Sue Bell
McGraw-Hill Inc./43rd Floor
1221 Avenue of the Americas
New York
NY 10020
USA

Please include full details of your name and postal address when returning this page to McGraw-Hill.

Although every effort has been made to ensure the reliability and accuracy of the disk, the Publisher cannot guarantee that the disk, when used for its intended or any other purpose, is free from error or defect.

INDEX

493